COMMON CORE EDITION

COURSE
4

CORE-PLUS
MATHEMATICS
Preparation for Calculus

Christian R. Hirsch • James T. Fey • Eric W. Hart
Harold L. Schoen • Ann E. Watkins

with

Beth E. Ritsema • Rebecca K. Walker • Brin A. Keller
Robin Marcus • Arthur F. Coxford

McGraw Hill Education

Cover (t)John Slater/Digital Vision/Getty Images, (tc)Pixtal/AGE Fotostock, (bc)Nick Rowe/Getty Images, (b)OJO Images/Getty Images

mheonline.com

This material is based upon work supported, in part, by the National Science Foundation under grant no. ESI 0137718. Opinions expressed are those of the authors and not necessarily those of the Foundation.

Send all inquiries to:
McGraw-Hill Education
STEM Learning Solutions Center
8787 Orion Place
Columbus, OH 43240

ISBN: 978-0-07-665790-2
MHID: 0-07-665790-6

Core-Plus Mathematics
Preparation for Calculus
Course 4 Student Edition

Printed in the United States of America.

1 2 3 4 5 6 7 8 9 DOW 18 17 16 15 14

McGraw-Hill is committed to providing instructional materials in Science, Technology, Engineering, and Mathematics (STEM) that give all students a solid foundation, one that prepares them for college and careers in the 21st century.

Core-Plus Mathematics Development Team

Senior Curriculum Developers

Christian R. Hirsch (Director)
Western Michigan University

James T. Fey (Emeritus)
University of Maryland

Eric W. Hart
Maharishi University of Management

Harold L. Schoen (Emeritus)
University of Iowa

Ann E. Watkins
California State University, Northridge

Contributing Curriculum Developers

Beth E. Ritsema
Western Michigan University

Rebecca K. Walker
Grand Valley State University

Brin A. Keller
Michigan State University

Robin Marcus
University of Maryland

Arthur F. Coxford (deceased)
University of Michigan

Principal Evaluator

Steven W. Ziebarth
Western Michigan University

Advisory Board

Diane Briars (formerly)
Pittsburgh Public Schools

Jeremy Kilpatrick
University of Georgia

Robert E. Megginson
University of Michigan

Kenneth Ruthven
University of Cambridge

David A. Smith
Duke University

Mathematical Consultants

Deborah Hughes-Hallett
University of Arizona

Stephen B. Maurer
Swarthmore College

William McCallum
University of Arizona

Doris Schattschneider
Moravian College

Richard Scheaffer
University of Florida

Evaluation Consultant

Norman L. Webb
University of Wisconsin-Madison

Collaborating Teachers

Mary Jo Messenger
Howard County Public Schools, Maryland

Valerie Mills
Oakland County Schools, Michigan

Jacqueline Stewart
Okemos, Michigan

Technical and Production Coordinator

James Laser
Western Michigan University

Support Staff

Angela Reiter
Hope Smith
Matthew Tuley
Teresa Ziebarth
Western Michigan University

Graduate Assistants

Allison BrckaLorenz
Christopher Hlas
University of Iowa

Michael Conklin
University of Maryland

AJ Edson
Nicole L. Fonger
Rob Kipka
Diane Moore
Western Michigan University

Undergraduate Assistants

Cassie Durgin
University of Maryland

Rachael Lund
Jessica Tucker
Western Michigan University

Core-Plus Mathematics, CCSS Edition
Field-Test Sites

The CCSS Edition of *Core-Plus Mathematics* builds on the strengths of the 1st and 2nd editions, which were shaped by multi-year field tests in 49 high schools in Alaska, California, Colorado, Georgia, Idaho, Iowa, Kentucky, Michigan, Missouri, Ohio, South Carolina, Texas, and Wisconsin. Each text is the product of a three-year cycle of research and development, pilot testing and refinement, and field testing and further refinement. Special thanks are extended to the following teachers and their students who participated in the most recent testing and evaluation of Course 4: *Preparation for Calculus.*

Hickman High School
Columbia, Missouri

Sandra Baker
Michelle Johnson
Rachelle Kunce
Ryan Pingrey

Holland Christian High School
Holland, Michigan

Brian Lemmen
Mike Verkaik

Malcolm Price Lab School
Cedar Falls, Iowa

Megan Balong
James Maltas

Riverside University High School
Milwaukee, Wisconsin

Cheryl Brenner

Rock Bridge High School
Columbia, Missouri

Patricia Avery

Sauk Prairie High School
Prairie du Sac, Wisconsin

Mary Walz

Washington High School
Milwaukee, Wisconsin

Anthony Amoroso

Development and evaluation of the student text materials, teacher support materials, assessments, and computer software for *Core-Plus Mathematics* was funded through a series of grants from the National Science Foundation to the Core-Plus Mathematics Project (CPMP). We express our appreciation to NSF and, in particular, to our program officer John Bradley for his long-term trust, support, and input.

We are also grateful to Texas Instruments and, in particular, to Dave Santucci and Cara Kugler for collaborating with us by providing classroom sets of graphing calculators to field-test schools.

As seen on page iii, CPMP has been a collaborative effort that has drawn on the talents and energies of teams of mathematics educators at several institutions. This diversity of experiences and ideas has been a particular strength of the project. Special thanks is owed to the exceptionally capable support staff at these institutions, particularly to Angela Reiter, Hope Smith, Matthew Tuley, and Teresa Ziebarth at Western Michigan University.

We are grateful to our Advisory Board, Diane Briars (formerly Pittsburgh Public Schools), Jeremy Kilpatrick (University of Georgia), Robert E. Megginson (University of Michigan), Kenneth Ruthven (University of Cambridge), and David A. Smith (Duke University) for their ongoing guidance and advice. We also acknowledge and thank Norman L. Webb (University of Wisconsin-Madison) for his advice on the design and conduct of our field-test evaluations.

Special thanks are owed to the following mathematicians: Deborah Hughes-Hallett (University of Arizona), Stephen B. Maurer (Swarthmore College), William McCallum (University of Arizona), Doris Schattschneider (Moravian College), and to statistician Richard Scheaffer (University of Florida) who reviewed and commented on units as they were being developed, tested, and refined.

Our gratitude is expressed to the teachers and students in the evaluation sites listed on page iv. Their experiences using the revised *Core-Plus Mathematics* units provided constructive feedback and suggested improvements that were immensely helpful.

Finally, we want to acknowledge Catherine Donaldson, Angela Wimberly, Justin Moyer, Michael Kaple, Karen Corliss, and their colleagues at McGraw-Hill Education who contributed to the design, editing, and publication of this program.

The first three courses in *Core-Plus Mathematics* provide a significant common core of broadly useful mathematics aligned with the Common Core State Standards for Mathematics (CCSS) and intended for all students. They were developed to prepare students for success in college, in careers, and in daily life in contemporary society.

Course 4: *Preparation for Calculus* continues the preparation of STEM-oriented (science, technology, engineering, and mathematics) students for success in college mathematics, especially calculus.

A separate alternative fourth-year capstone course, *Transition to College Mathematics and Statistics*, is intended for students planning to major in college programs that do *not* require calculus.

Core-Plus Mathematics is a problem-based, inquiry-oriented program that builds upon the theme of mathematics as reasoning and sense-making. Through investigations of real-life contexts, students develop a rich understanding of important mathematics that makes sense to them and which, in turn, enables them to make sense out of new situations and problems.

Course 4 of the *Core-Plus Mathematics* program shares many of the mathematical and instructional features of Courses 1–3.

- **Integrated Content** Course 4 continues to advance students' understanding of mathematics along interwoven strands of algebra and functions, statistics and probability, geometry and trigonometry, and discrete mathematics. The primary focus is on advanced topics in the algebra/functions and geometry/trigonometry strands. These strands are unified by fundamental themes, by common topics, by mathematical practices, and by mathematical habits of mind.

- **Mathematical Modeling** The problem-based curriculum emphasizes mathematical modeling including the processes of data collection, representation, interpretation, prediction, and simulation. The modeling perspective permits students to experience mathematics as a means of making sense of quantitative data and problems involving the dynamics of change that is foundational to the study of calculus.

- **Access and Challenge** The curriculum is designed to make mathematics accessible to more STEM-oriented students while at the same time challenging the most able students. Differences in student performance and interest can be accommodated by the depth and level of abstraction to which core topics are pursued, by the nature and degree of difficulty of applications, and by providing opportunities for student choice of homework tasks and projects.

- **Technology** Numeric, graphic, and symbolic manipulation capabilities such as those found in *CPMP-Tools*® and on many graphing calculators are assumed and appropriately used throughout the curriculum. *CPMP-Tools* is a suite of software tools that provide powerful aids to learning mathematics and solving mathematical problems. (See page xvii for further details.) This use of technology permits the curriculum and instruction to emphasize multiple linked representations (verbal, numerical, graphical, and symbolic) and to focus on goals in which mathematical thinking and problem solving are central.

- **Active Learning** Instructional materials promote active learning and teaching centered around collaborative investigations of problem situations followed by teacher-led whole-class summarizing activities that lead to analysis, abstraction, and further application of underlying mathematical ideas and principles. Students are actively engaged in exploring, conjecturing, verifying, generalizing, applying, proving, evaluating, and communicating mathematical ideas.

- **Multi-dimensional Assessment** Comprehensive assessment of student understanding and progress through both curriculum-embedded formative assessment opportunities and summative assessment tasks support instruction and enable monitoring and evaluation of each student's performance in terms of mathematical processes, content, and dispositions.

- **Flexibility** Course 4 is intended for students planning to pursue programs in the mathematical, physical, and biological sciences and engineering, or other programs requiring calculus. For students intending to pursue programs in the social, management, and some of the health sciences, or other programs that do not require calculus, the *Transition to College Mathematics and Statistics* course is recommended. Information about this alternative course can be found in the *Implementing Core-Plus Mathematics* guide.

Integrated Mathematics

Core-Plus Mathematics is an international-like curriculum. It replaces the traditional Algebra-Geometry-Advanced Algebra/Trigonometry-Precalculus sequence of high school mathematics courses with a sequence of courses that features a coherent and connected development of important mathematics drawn from four strands. In Course 4, the primary focus is on algebra and functions, vector and three-dimensional geometry, and trigonometry.

The *Algebra and Functions* strand develops student ability to recognize, represent, and solve problems involving relations among quantitative variables. Central to the development is the use of functions as mathematical models. The key algebraic models in the curriculum are linear, exponential, power, polynomial, logarithmic, rational, and circular functions. Modeling with systems of equations, both linear and nonlinear, is developed. Students further their fluency in *symbolic manipulation*—rewriting expressions in equivalent forms to reveal important information about the corresponding function and to solve equations and inequalities. They also extend their skills in *symbolic reasoning*—making inferences about symbolic relations and connections between symbolic representations and graphical, numerical, and contextual representations.

The primary goal of the *Geometry and Trigonometry* strand is to develop visualization skills and reasoning and the ability to build, interpret, and apply mathematical models involving shape and motion. In Course 4, concepts and methods of geometry, algebra, and trigonometry become increasingly intertwined in the development of models for describing and analyzing motion in two dimensions and surfaces in three dimensions.

The primary role of the *Statistics and Probability* strand is to develop student ability to analyze data intelligently, to recognize and measure variation, and to understand the patterns that underlie probabilistic situations. Graphical methods of data analysis and experience with the collection and interpretation of real data are featured. Special attention is given to the use of logarithms for linearizing bivariate data and fitting models.

The *Discrete Mathematics* strand develops student ability to solve problems using recursion, systematic counting methods (combinatorics), and prove statements using either the Principle of Mathematical Induction or the Least Number Principle.

These strands are developed within focused and coherent units connected by fundamental ideas such as mathematical modeling, functions, visualization, geometric transformations and symmetry, and data analysis. The strands also are connected across units by CCSS's mathematical practices and by mathematical habits of mind. These important mathematical practices include disposition toward, and proficiency in:

- making sense of problems and persevering in solving them
- reasoning both quantitatively and algebraically
- constructing sound arguments and critiquing the reasoning of others
- using mathematics to model problems in everyday life, society, and in careers
- selecting and using appropriate tools, especially technological tools (graphing calculator, spreadsheet, computer algebra system, statistical packages, and dynamic geometry software)
- communicating precisely and expressing calculations with an appropriate precision
- searching for and making use of patterns or structure in mathematical situations
- identifying structure in repeated calculations, algebraic manipulation, and reasoning patterns

Additionally, mathematical habits of mind such as visual thinking, recursive thinking, searching for and explaining patterns, making and checking conjectures, reasoning with multiple representations, inventing mathematics, and providing sound arguments are integral to each strand.

Important mathematical ideas are frequently revisited through this attention to connections within and across strands, enabling students to develop a robust and connected understanding of, and proficiency with, mathematics and its mindful practices.

Active Learning and Teaching

The manner in which students encounter mathematical ideas can contribute significantly to the quality of their learning and the depth of their understanding. *Core-Plus Mathematics* units are designed around multi-day lessons centered on big ideas. Each lesson includes 2–5 mathematical investigations that engage students in a four-phase cycle of classroom activities, described in the following paragraph—*Launch, Explore, Share and Summarize*, and *Check Your Understanding*. This cycle is designed to engage students in investigating and making sense of problem situations, in constructing important mathematical concepts and methods, in generalizing and proving mathematical relationships, and in communicating, both orally and in writing, their thinking and the results of their efforts. Most classroom activities are designed to be completed by students working collaboratively in groups of two to four students.

The Launch phase of a lesson promotes a teacher-led class discussion of a problem situation and of related questions to think about, setting the context for the student work to follow and providing important information about students' prior knowledge. In the second or Explore phase, students investigate more focused problems and questions related to the launch situation. This investigative work is followed by a teacher-led class discussion in which students summarize mathematical ideas developed in their groups, providing an opportunity to construct a shared understanding of important concepts, methods, and justifications. Finally, students are given tasks to complete on their own, to check their understanding of the concepts and methods of the lesson.

Each lesson also includes homework tasks to engage students in applying, connecting, reflecting on mathematical practices, extending, and reviewing their mathematical understanding. These *On Your Own* tasks are central to the learning goals of each lesson and are intended primarily for individual work outside of class. Selection of tasks should be based on student performance and the availability of time and technology access. Students can exercise some choice of tasks to pursue, and at times they should be given the opportunity to pose their own problems and questions to investigate.

Formative and Summative Assessment

Assessing what students know and are able to do is an integral part of *Core-Plus Mathematics*. There are opportunities for formative assessment in each phase of the instructional cycle. Initially, as students pursue the investigations that comprise the curriculum, the teacher is able to informally assess student understanding of mathematical processes and content and their disposition toward mathematics. At the end of each investigation, a class discussion to Summarize the Mathematics provides an opportunity for the teacher to assess levels of understanding that various groups of students and individuals have reached as they share, explain, and discuss their findings. Finally, the Check Your Understanding tasks and the tasks in the On Your Own sets provide further opportunities for formative assessment of the level of understanding of each individual student. Quizzes, in-class tests, take-home assessment tasks, and extended projects are included in the teacher resource materials for summative assessments.

Also included in the Course 4 teacher resource materials for each unit are two Preparing for Undergraduate Mathematics Placement (PUMP) sets of multiple-choice items providing practice in skills and reasoning strategies commonly assessed on college mathematics placement tests.

UNIT **1** Families of Functions

Families of Functions extends student understanding of linear, exponential, quadratic, power, and circular functions to model data patterns whose graphs are transformations of basic patterns; and develops understanding of operations on functions useful in representing and reasoning about quantitative relationships.

Topics include linear, exponential, quadratic, power, and trigonometric functions; data modeling; translation, reflection, and stretching of graphs; and addition, subtraction, multiplication, division, and composition of functions.

UNIT **2** Vectors and Motion

Vectors and Motion develops student understanding of two-dimensional vectors and their use in modeling linear, circular, and other nonlinear motion.

Topics include concept of vector as a mathematical object used to model situations defined by magnitude and direction; equality of vectors, scalar multiples, opposite vectors, sum and difference vectors, dot product of two vectors, position vectors and coordinates; and parametric equations for motion along a line and for motion of projectiles and objects in circular and elliptical orbits.

UNIT **3** Algebraic Functions and Equations

Algebraic Functions and Equations reviews and extends student understanding of properties of polynomial and rational functions and skills in manipulating algebraic expressions and solving polynomial and rational equations, and develops student understanding of complex number representations and operations.

Topics include polynomials, polynomial division, factor and remainder theorems, operations on complex numbers, representation of complex numbers as vectors, solution of polynomial equations, rational function graphs and asymptotes, and solution of rational equations and equations involving radical expressions.

UNIT **4** Trigonometric Functions and Equations

Trigonometric Functions and Equations extends student understanding of, and ability to reason with, trigonometric functions to prove or disprove potential trigonometric identities and to solve trigonometric equations; develops student ability to geometrically represent complex numbers and their operations and to find powers and roots of complex numbers expressed in trigonometric form.

Topics include fundamental trigonometric identities, sum and difference identities, double-angle identities; periodic solutions of trigonometric equations; definitions of secant, cosecant, and cotangent functions; absolute value and trigonometric form of complex numbers, De Moivre's Theorem, and roots of complex numbers.

UNIT **5** Exponential Functions, Logarithms, and Data Modeling

Exponential Functions, Logarithms, and Data Modeling extends student understanding of exponential and logarithmic functions to the case of natural exponential and logarithmic functions, solution of exponential growth and decay problems, and use of logarithms for linearization and modeling of data patterns.

Topics include exponential functions with rules in the form $f(x) = Ae^{kx}$, natural logarithm function, linearizing bivariate data and fitting models using log and log-log transformations.

UNIT **6** Surfaces and Cross Sections

Surfaces and Cross Sections extends student ability to visualize and represent three-dimensional shapes using contours, cross sections, and reliefs, and to visualize and represent surfaces and conic sections defined by algebraic equations.

Topics include using contours to represent three-dimensional surfaces and developing contour maps from data; sketching surfaces from sets of cross sections; conics as planar sections of right circular cones and as loci of points in a plane; three-dimensional rectangular coordinate system; sketching surfaces using traces, intercepts and cross sections derived from algebraically-defined surfaces; and surfaces of revolution and cylindrical surfaces.

UNIT 7 Concepts of Calculus

Concepts of Calculus develops student understanding of fundamental calculus ideas through explorations in a variety of applied problem contexts and their representations in function tables and graphs.

Topics include instantaneous rates of change, linear approximation, area under a curve, and applications to problems in physics, business, and other disciplines.

UNIT 8 Counting Methods and Induction

Counting Methods and Induction extends student ability to count systematically and solve enumeration problems in a variety of real-world and mathematical settings, and develops understanding of, and ability to carry out, proofs by mathematical induction and by use of the Least Number Principle.

Topics include systematic listing, counting trees, the Multiplication Principle of Counting, the Addition Principle of Counting, combinations, permutations, selections with repetition; the Binomial Theorem, Pascal's triangle, combinatorial reasoning; the General Multiplication Rule for Probability; proof by mathematical induction; and arguments using proof by contradiction and the Least Number Principle.

TABLE OF CONTENTS

UNIT **3** Algebraic Functions and Equations

UNIT **4** Trigonometric Functions and Equations

UNIT 7 Concepts of Calculus

UNIT 8 Counting Methods and Induction

TABLE OF CONTENTS

Have you ever wondered ...

- How climate scientists are able to predict potential effects of global warming such as rise in sea levels and pollution in the atmosphere?

- How operators of ships account for forces like wind and water current when navigating a planned course?

- How the rules for polynomial, exponential, rational, and circular functions can be customized to fit data and graph patterns that look familiar but cannot be matched by the familiar rules?

- Why "contour maps" are so useful in hiking?

- How calculators and computer software actually determine best-fitting models for nonlinear data patterns?

- How many different Web site passwords or ATM PINs are possible, and how long it would take a hacker to try them all?

- What your chances are of winning a particular state lottery?

- How maps and models of surfaces of distant planets like Mars are constructed?

- What calculus is all about and how it goes beyond the algebra and geometry that you have studied in previous courses?

- How coordinates can be used to represent mathematical ideas in three dimensions or even in four dimensions?

The mathematics you will learn in *Core-Plus Mathematics* Course 4: *Preparation for Calculus* will help you answer questions like these.

Because real-world situations and problems like those above often involve data, shape, quantity, change, or chance, you will study concepts and methods from several interwoven strands of mathematics. In particular, you will develop an understanding of broadly useful ideas—especially from algebra and functions, geometry and trigonometry, but also from statistics and probability and discrete mathematics. In the process, you will also see and use many connections among these strands.

In this course, you will learn important mathematics as you investigate and solve interesting problems. You will develop the ability to reason and communicate about mathematics as you are actively engaged in understanding and applying mathematics. You will often be learning mathematics in the same way that many people work in their jobs—by working in teams and using technology to solve problems.

In the 21st century, anyone who faces the challenge of learning mathematics or using mathematics to solve problems can draw on the resources of powerful information technology tools. Calculators and computers can help with calculations, drawing, and data analysis in mathematical explorations and solving mathematical problems.

Graphing calculators and computer software tools will be useful in your work on many of the investigations in *Core-Plus Mathematics*. Set as one of your goals to learn how to make strategic decisions about the choice and use of technological tools in solving particular problems.

The curriculum materials include computer software called *CPMP-Tools* that will be of great help in learning and applying the mathematical topics in Course 4. You can access the software at www.wmich.edu/cpmp/CPMP-Tools/.

The software toolkit includes four families of programs:

- Algebra—The software for work on algebra problems includes a spreadsheet and a computer algebra system (CAS) that produces tables and graphs of functions, manipulates algebraic expressions, and solves equations and inequalities.

- Geometry—The software for work on geometry problems includes an interactive drawing program for constructing, measuring, and manipulating geometric figures, including vectors, in a coordinate or coordinate-free environment.

- Statistics—The software for work on data analysis and probability problems provides tools for graphic display and analysis of data, simulation of probabilistic situations, and mathematical modeling of quantitative relationships.

- *Discrete Mathematics*—The software provides tools for creating and analyzing discrete mathematical models.

In addition to these general purpose tools, *CPMP-Tools* includes files of most data sets essential for work on problems in *Core-Plus Mathematics* Course 4. When the opportunity to use computer tools in an investigation seems appropriate, select the *CPMP-Tools* menu corresponding to the content of, and your planned approach to, the problem. Then select the submenu items corresponding to the appropriate mathematical operations and required data set(s).

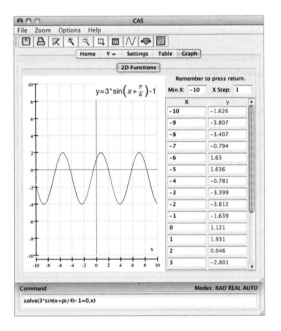

As in Courses 1–3, you will continue to learn a lot of useful mathematics and it is going to make sense to you. You will also deepen your understanding of fundamental ideas and methods that support future coursework in calculus and in other advanced mathematics courses. You are going to strengthen your skills in working collaboratively on problems and communicating with others as well. You are also going to strengthen your skills in using technological tools strategically and effectively. Finally, you will continue to have plenty of opportunities to be creative and inventive. Enjoy!

Families of Functions

In earlier courses of *Core-Plus Mathematics*, you developed a toolkit of mathematical functions for describing and reasoning about variables in a wide variety of problem situations. For example, the sine and cosine functions from trigonometry can be used to model the motion of passenger capsules on amusement park rides like the Millennium Wheel in London's Jubilee Gardens.

In the lessons of this unit, you will extend your skill in use of basic function families to model more complex relationships among variables.

The key ideas will be developed in four lessons.

LESSONS

1 Function Models Revisited

Review properties of situations in which linear, exponential, power, and quadratic functions are useful models of data patterns and relationships between quantitative variables. Review table, graph, and symbolic rule patterns of those basic function families.

2 Customizing Models by Translation and Reflection

Develop skill in modifying rules for functions to produce models for data patterns whose graphs are related to those of familiar functions by vertical and horizontal translations and reflections.

3 Customizing Models by Stretching and Compressing

Develop skill in modifying rules for functions to produce models for data patterns whose graphs are related to those of familiar functions by vertical and horizontal compressing and stretching.

4 Combining Functions

Use arithmetic operations and composition to build function models for more complex relationships between variables.

Function Models Revisited

One of the most important and controversial problems in Earth and space science today is measuring, understanding, and predicting global warming. There is deep concern that the average annual surface temperature on Earth has been increasing over the past century. This change will have important consequences for industry, agriculture, and personal lifestyles.

The graph that follows shows the pattern of change in average world temperature over the past 162 years. While the average global temperature has increased by less than a degree, this is still a large amount relative to historical data. This recent temperature increase is four to five times faster than any other climate change in the past millennium.

Annual Deviation from the 1961–1990 Average Global Temperature

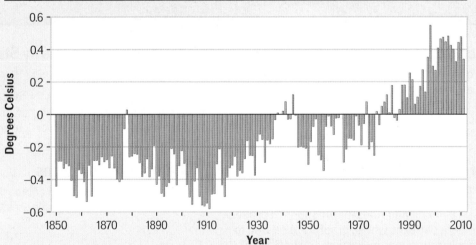

Source: Met Office Hadley Centre, UK. "Contains public sector information licensed under the Open Government Licence v1.0".

Many scientists believe that the most likely variable contributing to the increase in world temperature is greenhouse gases that reduce radiation of energy from Earth's surface into space. The graph below gives data on change in atmospheric greenhouse gases over more than 1,000 years.

Atmospheric Concentrations of Carbon Dioxide, 1000–2011

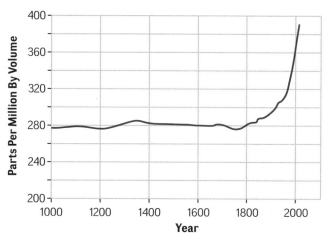

Source: www.earth-policy.org/datacenter/xls/book_fpep_ch8_3.xlsx

THINK ABOUT THIS SITUATION

The challenge for atmospheric scientists is deciding how current trends in greenhouse gas amounts and world temperature change should be projected into the future. Different projections imply different corrective actions.

a In what ways do you imagine that future global warming could change Earth's atmosphere and your own life?

b Examine the Annual Deviation from the 1961–1990 Average Global Temperature graph on the previous page. What do you notice? What questions do you have about the information displayed?

c Based on the data given in that graph, what strategy for projecting change in global temperature would make most sense to you?

d Examine the Atmospheric Concentrations of Carbon Dioxide graph above. What do you notice? What questions do you have about the displayed information?

e Based on the data given in the graph above, what strategy for projecting change in atmospheric carbon dioxide makes most sense to you?

In the first three courses of *Core-Plus Mathematics*, you investigated several important **families of functions** that are useful in describing and predicting patterns of change. In the investigations of this lesson, you will review key properties of the most important function families. Then in subsequent lessons, you will learn ways to modify those basic functions to model more complex situations.

Modeling Atmospheric Change

As different scientists have studied the historical records of temperature and carbon dioxide data, they have proposed different scenarios for the future of global warming. Each is based on certain assumptions about the best models for patterns of change. You can get an overview of the issues by visiting: www.ncdc.noaa.gov/cmb-faq/globalwarming.html.

As you work on the problems of this investigation, look for answers to this question:

> *What problem conditions and data patterns suggest use of linear, power, inverse variation, and exponential functions in modeling different aspects of atmospheric change?*

Predicting Change in Temperature and Greenhouse Gases Data giving Earth's surface temperature are collected from several sources: land-based weather stations, weather balloons sent up regularly by those stations, ships and fixed buoys in the ocean, and orbiting satellites. These data are combined to estimate Earth's annual average temperature, currently about 58°F.

1 The rate at which the average Earth temperature is changing is controversial. Many scientists believe it is rising, but estimates vary from an increase of about 0.05°F to 0.15°F per decade.

a. Write rules that predict the annual average temperature *x* decades from now for three different rate-of-increase estimates: 0.05°F, 0.10°F, and 0.15°F per decade. Draw sketches of the various models on the same coordinate system, indicating clearly the rules corresponding to each graph.

b. Use the low (0.05°F per decade) and high (0.15°F per decade) rate-of-change rules to write calculations, equations, or inequalities whose solutions would answer the following questions. Then answer each question using methods that seem appropriate—estimation using tables and graphs or exact solution using algebraic reasoning or a computer algebra system (CAS).

   i. What will the average Earth temperature be 50 years from now?

   ii. What will the average Earth temperature be 65 years from now?

   iii. When will the average Earth temperature reach 60°F?

   iv. How long will the average Earth temperature remain below 59°F?

2 Atmospheric carbon dioxide (CO_2) is believed to be a primary factor in global warming. Levels of CO_2 are increasing because human activities, such as burning fossil fuels, send more into the atmosphere than natural biological processes remove. Estimates in 2012 suggested that Earth's atmosphere contained about 393 parts per million (ppm) of CO_2 with another 2 ppm added each year. (**Source:** co2now.org)

    a. Based on findings of the 2012 study, what function would estimate atmospheric CO_2 at any time x years after 2012?

    b. Using your model from Part a:

        i. what level of atmospheric CO_2 can be expected in the year 2025?

        ii. in what year can we expect atmospheric CO_2 to reach 400 ppm?

    c. Suppose that when the atmospheric CO_2 reaches 400 ppm, a way is found to reduce CO_2 emissions from human activity and increase biological processes that extract CO_2 from the atmosphere.

        i. What linear rate of change would be necessary to bring atmospheric CO_2 back to the 2012 level in 20 years?

        ii. What function could be used to estimate atmospheric CO_2 at any time x years after corrective action began?

3 Data suggest that the rate of increase in atmospheric CO_2 has not been constant. Suppose that the recent increase of 2 ppm is expressed as a percent and that future increases (from the 2012 level of 393 ppm) occur at the same percent rate.

    a. What is the current percent rate of increase?

    b. What function should be used to estimate atmospheric carbon dioxide x years in the future if we assume growth from 393 ppm at the constant percent rate found in Part a?

    c. Compare estimates of the increase in atmospheric CO_2 for years 2019 and 2029 under assumptions of the two different models—increase at a constant rate of 2 ppm per year versus increase at the percent rate calculated in Part a.

    d. Based on your model comparisons in Part c, explain why you think you might trust predictions of one model over the other. Explain why neither model should be trusted for very long-term predictions.

Carbon Dating of Past Events It has been estimated that in the 10,000 years since the end of the last ice age, the annual average temperature of Earth has increased by about 9°F and atmospheric carbon dioxide has increased by at least 50%. Scientists arrived at such estimates by analyzing material that has been trapped deep in very old glaciers and on the floors of lakes and oceans for thousands of years.

One of the interesting problems in such work is estimating the age of deposits that are uncovered by core drilling. A common technique is called *carbon dating*. Carbon occurs in all living matter in several forms. The most common forms (carbon-12 and carbon-13) are stable; a third form, carbon-14, is radioactive and decays at a rate of 1.2% per century.

By measuring the proportion of carbon-14 in a scientific sample and comparing that figure to the proportion in living matter, it is possible to estimate the time when the matter in the sample was last alive. Despite the very small amounts of carbon-14 involved (less than 0.000000001% of total carbon in living matter), modern instruments can make the required measurements.

4. Suppose that drilling into a lake bottom produces a piece of wood which, according to its mass, would have contained 5 nanograms (5 billionths of a gram) of carbon-14 when the wood was alive. Use the fact that this radioactive carbon decays continuously at a rate of about 1.2% per century to analyze the sample.

   a. How much of that carbon-14 would be expected to remain:

      i. 1 century later?

      ii. 2 centuries later?

      iii. x centuries later?

   b. Estimate the half-life of carbon-14.

   c. Estimate the time when the wood was last alive if the sample contained:

      i. only 3 nanograms of carbon-14.

      ii. only 1 nanogram of carbon-14.

Glaciers, Polar Ice Caps, and Global Warming Glacier formations hold clues to the past. One of the ominous and spectacular predictions about global warming is that the melting of polar ice caps and expansion of ocean water will cause sea levels to rise and flood cities along all ocean shores. According to the Environmental Protection Agency, the most likely scenario is a 34-cm rise in sea levels by the year 2100. Such a change would flood large parts of low-lying countries like the Netherlands and Bangladesh and areas such as coastal Florida.

Estimates of such a rise in the sea level depend on measurements of glacier volumes and ocean surface areas. Earth is approximately a sphere with these properties:

- Oceans cover approximately 70% of Earth's surface.

- The Greenland and Antarctic ice sheets cover nearly 6 million square miles.

- Those ice sheets contain nearly 7 million cubic miles of ice.

- The water in those ice sheets is only 2% of all water on the planet.

Alaska's Exit Glacier has receded several hundred meters from this post planted in 1978.

5 In making estimates of the size of Earth (and other spherical planets as well), it is useful to have formulas showing the circumference, surface area, and volume of a sphere as functions of the diameter or radius. Sometimes it is useful to modify those relationships to show the radius or diameter required to give specified circumference, surface area, or volume.

a. Which of the following functions gives *circumference* of a sphere, which gives *surface area*, and which gives *volume* in terms of the radius r? Be prepared to explain the clues that help in matching each function to the corresponding measurement.

   i. $f(r) = 4\pi r^2$

   ii. $g(r) = \frac{4}{3}\pi r^3$

   iii. $h(r) = 2\pi r$

b. What patterns would you expect in graphs of spherical circumference, surface area, and volume as functions of radius?

c. Earth is not a perfect sphere, but nearly so, with average radius of about 4,000 miles.

    i. What is the approximate surface area of Earth's oceans? (Oceans cover about 70% of Earth's surface.)

    ii. What volume of water would be required to raise the level of those oceans by 3 feet? Assume that raising the level would not change the surface area of the ocean significantly.

d. What rise in ocean levels would be caused by the total melting of the Greenland and Antarctic ice caps? Again, assume that the surface area of the oceans would not change significantly. The two ice caps contain 7 million cubic miles of ice. The volume of water from melting ice is approximately 92% of the volume of the ice.

e. Earth is the fifth largest planet in the solar system. The largest planet, Jupiter, has a radius roughly 11 times the radius of Earth. The radius of Mars is roughly half that of Earth.

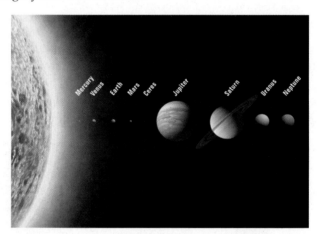

    i. How do the circumference, surface area, and volume of Jupiter compare to the corresponding measures of Earth?

    ii. How do the circumference, surface area, and volume of Mars compare to the corresponding measures of Earth?

Gravitation On and Near Earth's Surface The silent force of gravity influences almost every aspect of life on and near the surface of Earth. The gravitational force that holds all of us anchored to Earth's surface diminishes as one moves up into the atmosphere.

6 In general, the gravitational force of attraction between any two masses is directly proportional to the product of the masses and inversely proportional to the square of the distance between their centers.

a. Describe the pattern of change in gravitational force as:

    i. the distance between two planetary bodies increases.

    ii. one or both of the bodies increase in mass.

b. Which of the following formulas matches the given information about the force between masses m_1 and m_2 with centers located at a distance d apart?

$$F = k(m_1 m_2 - d^2) \qquad F = k\left(\frac{m_1 m_2}{d^2}\right) \qquad F = -k\left(\frac{m_1 m_2}{d^2}\right)$$

c. Use algebraic reasoning to determine how the gravitational force of attraction between two bodies will change:

    i. if the mass of one body increases by a factor of c.

    ii. if the distance between the masses increases by a factor c.

SUMMARIZE THE MATHEMATICS

In this investigation, you modeled aspects and consequences of atmospheric change and gravitational force with various types of functions.

a Which contexts considered in this investigation were examples of:

- linear functions?
- exponential functions?
- direct variation functions?
- inverse variation functions?

b What problem conditions or data patterns suggested each type of function in Part a as probably the most appropriate model for the relationship between variables?

Be prepared to share your ideas and reasoning with the class.

 CHECK YOUR UNDERSTANDING

Coyotes are mammals similar to wolves and dogs that we think of as living in wild habitats of the western United States. However, they are very adaptable carnivores and now appear in urban areas as far east as Washington, D.C., and New York.

    There is limited data on the actual numbers of coyotes now living in east coast states, but suppose there are now about 1,500 such animals living in Delaware, Maryland, Virginia, and Washington, D.C.

a. Assume that the coyote population in these areas increases by 15% per year. What function will predict the population n years from now?

b. Assume instead that the population increases by 250 coyotes per year. What function will predict the population n years from now?

c. Since coyotes are predators that feed on other wild animals like fox, geese, raccoons, and deer, it might be reasonable to assume that the population of those prey species would be inversely proportional to the population of coyotes. Under that assumption, which of the following functions would provide a model for the relationship between the population of deer d and the population of coyotes c in city and suburban areas around Washington, D.C., over the next few years? Explain.

$$d(c) = 10{,}000 - c \qquad d(c) = \frac{10{,}000}{c} \qquad d(c) = 10{,}000c$$

It's All in the Family

Solving the problems of Investigation 1 required use of a variety of different functions to model patterns of change in variables that are central to Earth and space science. In each case, the algebraic rules for the functions involved specific numerical constants to match specific problem conditions.

For example, the linear models proposed to predict change in average global temperature included $T(n) = 0.05n + 58$ and $T(n) = 0.15n + 58$, with n representing number of decades in the future. The first function assumes the low-end rate of increase—0.05°F per decade—and the second function assumes the high-end rate of increase—0.15°F per decade. Both functions assumed a value of 58°F for average global temperature.

The two models for predicting global temperature are linear functions. All members of that family of functions can be represented with rules in the form $y = ax + b$. The values of a and b that define any specific member in the family are called **parameters**. As you know from prior studies, the specific members of other function families—like exponential, polynomial, and rational functions— are also defined by specifying values of parameters.

As you complete this investigation, look for answers to this question:

How do the parameters in the basic function families provide tools for matching function models to specific problem conditions, data patterns, and graphs?

To explore how the parameters in the basic function families affect graphs and tables of values, it is helpful to use software like the *CPMP-Tools* computer algebra system (CAS) that accepts function definitions like $f(x) = ax^2 + bx + c$, allows you to move "sliders" controlling the values of the parameters, and quickly produces corresponding graphs and tables.

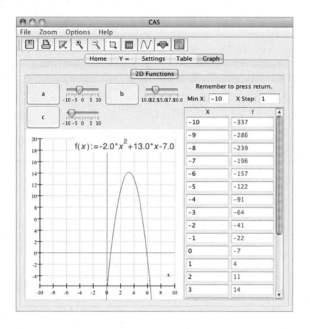

1 On an extended copy of the following functions toolkit table, for each function family indicated:

   a. Describe the domains and ranges of typical functions in the family.

   b. Produce graphs of specific examples.

   i. Describe how the patterns in graphs of $(x, f(x))$ are related to parameters in the rules.

   ii. Note key points of the graphs that would help you make quick sketches of the graphs.

   iii. Record sketches of typical functions in each family.

   c. Describe the patterns of change to be expected in tables of $(x, f(x))$ values and the relationship of those patterns to parameters in the rules.

   d. Describe any maximum or minimum points of the graphs.

   e. Describe any symmetries of the graphs.

   f. Describe any horizontal and/or vertical asymptotes of the graphs.

   g. How would you use the words "increasing" and/or "decreasing" to describe the patterns of change exhibited in your sample graphs and tables of values?

Toolkit of Basic Function Families

Function Family Features

I. Linear Functions

$f(x) = ax + b$

domain:

range:

maximum/minimum value(s) (if any):

symmetries (if any):

asymptotes (if any):

II. Exponential Functions

$f(x) = a(b^x), a \neq 0, b > 1$ or $0 < b < 1$

domain:

range:

maximum/minimum value(s) (if any):

symmetries (if any):

asymptotes (if any):

III. Power Functions

$f(x) = ax^n, a \neq 0$ and n a positive integer

domain:

range:

maximum/minimum value(s) (if any):

symmetries (if any):

asymptotes (if any):

IV. Inverse Variation Functions

$f(x) = \dfrac{a}{x^n}, a \neq 0$ and n a positive integer

domain:

range:

maximum/minimum value(s) (if any):

symmetries (if any):

asymptotes (if any):

V. Quadratic Functions

$f(x) = ax^2 + bx + c, a \neq 0$

domain:

range:

maximum/minimum value(s) (if any):

symmetries (if any):

asymptotes (if any):

VI. Circular Functions

$s(x) = a \sin x, a \neq 0$

domain:

range:

maximum/minimum value(s) (if any):

symmetries (if any):

asymptotes (if any):

VII. Common Logarithmic Functions

$f(x) = a \log_{10} x, a \neq 0$

domain:

range:

maximum/minimum value(s) (if any):

symmetries (if any):

asymptotes (if any):

$c(x) = a \cos x, a \neq 0$

domain:

range:

maximum/minimum value(s) (if any):

symmetries (if any):

asymptotes (if any):

2 Look back at your completed functions toolkit table. Assuming the domain of each function is the set of whole numbers, where possible write a recursive formula for the function:

a. using the words *NOW* and *NEXT*. Specify the starting value.

b. using the f_n and f_{n-1} notation. Specify f_0.

SUMMARIZE THE MATHEMATICS

In this investigation, you analyzed a variety of functions that can occur when parameters are changed in the rules for the families of linear, exponential, quadratic, power, inverse variation functions, circular functions, and logarithmic functions. What are the most striking similarities and differences that appear when comparing rules, graphs, and tables of the following pairs of function families?

a Linear and exponential functions

b Linear and quadratic functions

c Exponential and inverse variation functions

d Exponential and power functions

e Sine and cosine functions and all other function families

f Logarithmic and exponential functions with the same base

Be prepared to share your comparisons and thinking with the class.

 CHECK YOUR UNDERSTANDING

Identify families of functions that should be considered as models for relationships between variables in the following cases. Explain reasons for each model choice.

a. Variables x and y are related as shown by data in this table.

x	−3	−2	−1	0	1	2	3
y	11.1	8.2	5.3	2.4	−0.5	−3.4	−6.3

b. The graph of $y = f(x)$ is a curve that is symmetric about the vertical line $x = 2$, with a maximum point at (2, 5).

c. The relationship between y and x satisfies the rule $NEXT = 0.8NOW$ (with $y = 20$ when $x = 0$).

d. The graph of $y = f(x)$ is a curve that has asymptotes $y = 0$ and $x = 0$.

APPLICATIONS

These tasks provide opportunities for you to use and strengthen your understanding of the ideas you have learned in the lesson.

Public health officials in many countries are worrying about the possibility that diseases like SARS and avian influenza might jump from mammals and birds to humans and then spread in deadly fashion around the world. Suppose that data in the following table give numbers of cases in an outbreak of bird flu. Use the data to answer the questions in Tasks 1 and 2. To put the number of bird flu cases in perspective, the world population is over 7 billion.

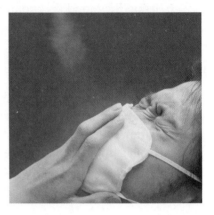

Time t (in weeks)	0	1	2	3	4	5	6
Cases $C(t)$ (in 1,000s)	2.5	3.1	3.9	4.9	6.1	7.6	9.5

1 Explore consequences of using a linear function to model the bird flu data pattern.

   a. What is the best-fit linear model for the change in number of flu cases over time?

   b. What numbers of cases are predicted for times 12, 18, and 24 weeks from the start of the outbreak of this influenza epidemic?

   c. How long will it take for 1,000,000 people to contract the disease?

   d. What do the coefficient of the independent variable and the constant term in your linear model say about the predicted spread of the disease?

   e. What reasons might you have for questioning the validity of the linear model for predicting the rate of spread of the epidemic?

2 Explore consequences of using an exponential function to model the bird flu data pattern.

   a. What is the best-fit exponential model for change in number of flu cases over time?

   b. What numbers of cases are predicted for times 12, 18, and 24 weeks from the start of the outbreak of this influenza epidemic?

   c. How long will it take for 1,000,000 people to contract the disease?

   d. What do the parameters in your exponential model say about the predicted spread of the disease?

   e. What reasons might you have for questioning the validity of the exponential model for predicting the rate of spread of the epidemic?

Many businesses use vans and small trucks for deliveries and for support of service and construction work. Owning such vehicles requires accounting calculations that involve variables like *operating costs*, *depreciation*, and *inflation*.

Use what you know about functions, equations, and inequalities to answer the questions in Tasks 3–5.

3 Suppose that a company estimated that the annual operating cost of its business delivery van would be $5,000 for insurance plus $0.65 per mile for gas, oil, and maintenance.

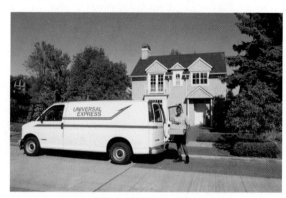

    a. What function shows how total annual operating cost depends on number of miles driven x?

    b. Write and solve equations or inequalities that match these questions about the van operating cost.

        i. How many miles can the van be driven if the annual operating cost is to be less than $15,000?

        ii. If the company records show a total operating cost of $20,000 for one year, how many miles must have been driven with the van?

    c. If inflation increases the operating cost of the van at a rate of about 5% per year and the current annual operating cost is $12,500, what function could be used to predict the annual operating cost at a time n years later? What assumption(s) are you making?

4 When a company buys a new van for $40,000, the resale value of the vehicle can be expected to decrease at a rate of 20% per year.

    a. What function can be used to predict the resale value of the van n years after its purchase?

    b. How long can the company keep the van and still expect to resell it for at least $15,000?

5 Accounting rules allow companies to recognize expense from wear and tear on equipment like a delivery van by reducing its value in business records each year. This accounting expense, called *depreciation*, reduces the company's taxable income and saves the company money.

    a. If a company buys a $40,000 delivery van and plans to fully depreciate it by a "straight line" method over 8 years, what is the annual depreciation allowance?

    b. What function gives the depreciated value of the van on the company's books at any time n years after its purchase?

Cylinders are one of the most common shapes for liquid storage tanks. The volume of a cylinder is a function of its radius r and height h given by the formula $V = \pi r^2 h$. The surface area of a cylinder (including top and bottom) is a function of the same dimensions with the formula $A = 2\pi rh + 2\pi r^2$. Use these relationships to help answer the questions in Tasks 6–9.

6 Suppose that a home oil tank has a radius of 2 feet and a height of 5 feet.

a. Find the volume of that tank:

  i. in cubic feet.

  ii. in gallons. (There are about 7.5 gallons to a cubic foot.)

b. Find the surface area of the tank.

c. Suppose that a large cylindrical oil storage tank has radius 20 feet and height 50 feet.

  i. Will its volume be 10 times, 100 times, or 1,000 times greater than the home oil tank with radius 2 feet and height 5 feet?

  ii. Will its surface area be 10 times, 100 times, or 1,000 times as much as the home oil tank with radius 2 feet and height 5 feet?

  iii. Show how algebraic reasoning can provide answers to parts i and ii without actually calculating the volume or surface area of the larger tank.

7 The functions $V_r = (\pi r^2)(5)$ and $V_h = \pi(2^2)h$ show how volume of a cylinder depends on radius when height is fixed at 5 feet and on height when radius is fixed at 2 feet.

a. If the design for a cylinder specifies a radius of 2 feet and a height of 5 feet, which change in the design will produce the greatest change in volume: (1) increasing the radius by 1 foot; or (2) increasing the height by 1 foot?

b. How do the algebraic rules for $V_r = (\pi r^2)(5)$ and $V_h = \pi(2^2)h$ help to explain your answer to the comparison question in Part a?

8 The surface area of Earth's oceans is about 361 million square kilometers, and they contain about 1.347 billion cubic kilometers of water. Use these facts to answer the following questions about the effects of glaciers on ocean sea levels.

a. In the greatest ice age, sea levels were about 135 meters lower than they are today. About how much more water could have been contained in the glaciers of that ice age than exists now? (Assume that lowering the depth of the ocean would not appreciably change its surface area. Recall that 1 meter is 0.001 kilometer.)

b. What is the average depth of the oceans?

9 Suppose that you need to design a cylindrical tank with fixed volume of 10,000 cubic feet.

a. Write an equation showing the relationship among radius r, height h, and this fixed volume.

b. Solve the equation in Part a for r to show how the required radius depends on the choice of height. Then use this equation to find the required radius if the height is 20 feet.

c. Solve the equation in Part a for h to show how the required height depends on the choice of radius. Then use this equation to find the required height if the radius is 15 feet.

d. Describe in words the proportionality relationships between r and h that are expressed by the equations in Parts b and c by completing sentences like these:

   i. For a cylindrical tank with volume 10,000 cubic feet, the radius is _____ proportional to _____ with constant of proportionality _____ .

   ii. For a cylindrical tank with volume 10,000 cubic feet, the height is _____ proportional to _____ with constant of proportionality _____ .

10 There are many important situations in which variables are related by a function. Following are six graphs and descriptions of several such situations. Match the descriptions in Parts a–f to Graphs I–VI that seem to fit them best. Then for each situation:

- explain why the graph makes sense as a model of the relationship between variables.

- describe the function family (if any) that would probably provide a good model for the relationship and say as much as you can about the values of the parameters in the model.

Graphs:

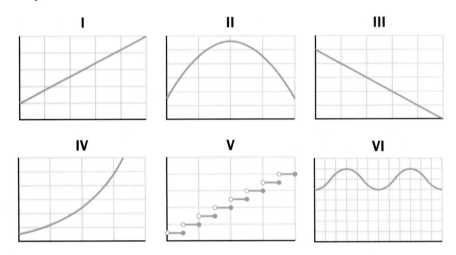

I II III

IV V VI

Situations:

a. When a football team's punter kicks the ball, the ball's height changes as time passes from kick to catch. What pattern seems likely to relate time and height?

b. The senior class officers at Lincoln High School decided to order and sell souvenir baseball caps with the school insignia and name on them. One supplier said it would charge $150 to create the design and then an additional $6 for each cap made. How would the total cost of the order be related to the number of caps in the order?

c. The number of hours between sunrise and sunset changes as days pass in each year. What pattern seems likely to relate day of the year and hours of sunlight?

d. In planning a bus trip to Florida for spring break, a travel agent worked on the assumption that each bus would hold at most 40 students. How would the number of buses be related to the number of student customers?

e. When the Riverside High School sophomore class officers decided to order and sell T-shirts with the names of everyone in their class on the shirts, they checked with a sample of students to see how many would buy a T-shirt at various proposed prices. How would sales be related to price charged?

f. The population of the world has been increasing for as long as records have been available. What pattern of population growth has occurred over that time?

11 Graph V in Applications Task 10 is an example of a **piecewise-defined function**. The rule for the function is different for different intervals of the domain.

a. Assume tick marks on the x-axis represent 20 units and tick marks on the y-axis represent 10 units. Complete this rule for the function $f(x)$ whose graph is Graph V.

$$f(x) = \begin{cases} 5 & \text{for } 0 < x \le 10 \\ \vdots \end{cases}$$

b. Write the absolute value function $v(x) = |x|$ as a piecewise linear function.

12 For which function(s) $f(x)$ in Applications Task 10 is it the case that if r is any value in the range, there is exactly one x in the domain such that $f(x) = r$? Such functions are called **one-to-one functions**.

a. What test could you apply to the graph of a function to determine if it is a one-to-one function? Explain.

b. Explain why every one-to-one function $f(x)$ has an inverse.

c. Explain why if a function $f(x)$ has an inverse, then $f(x)$ is one-to-one.

CONNECTIONS

These tasks will help you to build links between mathematical topics you have studied in the lesson and to connect those topics with other mathematics that you know.

13 The diameter of a circle or sphere is double the radius. Use this fact to write formulas for these sphere measurements in terms of diameter d instead of radius r.

a. Circumference   b. Surface area   c. Volume

14 If two right prisms are similar with scale factor k, what can you say about the relationship of:

a. the perimeters of the bases of the two prisms?

b. the heights of the two prisms?

c. the surface areas of the two prisms?

d. the volumes of the two prisms?

15 Suppose the function $C(t) = 393(1.005^t)$ predicts the level (in parts per million) of atmospheric carbon dioxide at a time t years in the future. Then finding the time when atmospheric CO_2 is expected to reach 450 ppm requires solving the equation $393(1.005^t) = 450$. Give reasons that justify each step in the following solution of that equation using logarithms.

$$393(1.005^t) = 450$$
$$(1.005^t) = 1.145 \qquad (1)$$
$$\log 1.005^t = \log 1.145 \qquad (2)$$
$$t \log 1.005 = \log 1.145 \qquad (3)$$
$$t = \frac{\log 1.145}{\log 1.005} \qquad (4)$$
$$t \approx 27.1 \qquad (5)$$

16 Graphs of *periodic functions*, like variations of the sine and cosine functions, with domain the set of real numbers have translation symmetry. Examine the graph below showing ocean depth at a retaining wall as a function of time in hours.

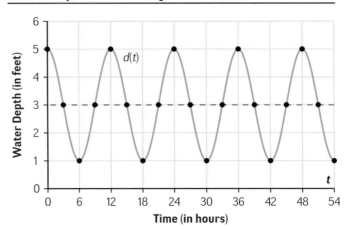

Ocean Depth at Retaining Wall

a. Assuming the pattern of the graph continues to the left and to the right, what is the smallest magnitude of a translation that maps the curve onto itself?

b. What other types of symmetry are present in this graph?

c. The dashed red line on the graph is called the *midline* of the graph. Why does this description make sense?

d. How is the amplitude of this function related to the midline?

e. What is the midline of the graphs of $s(x) = \sin x$ and $c(x) = \cos x$? How is the midline related to the amplitude of each of these functions?

17 Any function $f(n)$ with domain the set of whole numbers is called a **sequence**.

a. Any **arithmetic sequence** can be described recursively by a formula of the form $f(n) = f(n - 1) + d$.

　i. What rule shows the value of $f(n)$ for any n directly, when d and $f(0)$ are given?

　ii. To which family of functions do arithmetic sequences belong?

b. Any **geometric sequence** can be described recursively by a formula of the form $f(n) = r \cdot f(n - 1)$.

　i. What rule shows the value of $f(n)$ for any n directly, when r and $f(0)$ are given?

　ii. To which family of functions do geometric sequences belong?

18 Suppose that a game involves tossing two fair coins until both turn up "heads."

a. What is the probability that you will toss the coins:

　i. one time?　　　　　　　**ii.** two times?

　ii. three times?　　　　　　**iv.** n times?

b. If $p(n)$ gives the probability that the first occurrence of "two heads" will be on toss n, to which family of functions does $p(n)$ belong?

REFLECTIONS

These tasks provide opportunities for you to re-examine your thinking about ideas in the lesson and your use of mathematical practices and habits of mind that are useful throughout mathematics.

19 What clues do you find most helpful in deciding on the family of functions likely to provide a good model for the relationship between variables in any particular situation?

20 News stories that involve consideration of change in some variable over time or the relation between two or more variables often use phrases like "growing exponentially," "periodic," or "directly or inversely related." What do you think people generally mean when they use each of those descriptive terms? How do the common usages relate to the technical mathematical usage?

21 What is the difference between a variable and a parameter in algebraic expressions like $a + bx$ or $a(b^x)$?

22 Some function families have the property that function values are increasing or decreasing for the entire domain. For other function families, function values increase on some intervals and decrease on others. Sort the function families in the chart on pages 11 and 12 by these criteria.

EXTENSIONS

These tasks provide opportunities for you to explore further or more deeply the mathematics you are learning.

23 Lighter-than-air transportation in zeppelins, dirigibles, and blimps has been in use for over 100 years. During the 1920s and '30s, hydrogen-filled dirigibles carried paying passengers across the Atlantic Ocean. Blimps are now used primarily as air-borne advertising signs and occasionally as platforms for televising sporting events.

Suppose that the shape of a blimp could be approximated as a cylinder with hemispherical caps on each end, so that a side-view profile consists of a rectangle with congruent semicircles on either end.

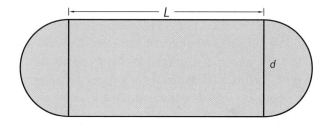

a. What formula gives the total volume V of such a shape as a function of length L and diameter d as shown above?

b. What formula gives the total surface area A of such a shape as a function of L and d?

24 The atmosphere of Earth is divided into several regions. For example, the troposphere extends from Earth's surface to an altitude of about 7 miles and the stratosphere extends from an altitude of 7 miles to about 30 miles. Recall that the radius of Earth is about 4,000 miles.

   a. What is the approximate volume of Earth plus its troposphere?

   b. What is the approximate volume of the troposphere alone?

   c. What is the approximate volume of the stratosphere alone?

   d. The entire atmosphere of Earth extends to an altitude of about 400 miles above Earth's surface. If a satellite in circular Earth orbit were to be at the top of the atmosphere, what would be the length of its orbit?

   e. How much larger is the area of the surface formed by the outer boundary of Earth's atmosphere than the surface of Earth at sea level?

25 Exponential models for population growth have graphs that show continually rising population values, with the slope of the graph getting steeper and steeper as time passes. In many situations, there are limits to growth that cause the rate of population increase to eventually slow. For example, if a colony of ants starts growing in a space with limited food and water, the number of ants at any time w weeks later might be modeled by a function like this:

$$P(w) = \frac{10{,}000}{1 + 49(0.85^w)}$$

   a. What number of ants does the model give for the start of the experiment?

   b. What is the shape of the graph of $P(w)$ and what does that shape say about the pattern of increase in the ant population?

   c. What seems to be the upper limit of the ant population and how can that limit be found without use of any calculations or graphs, simply analyzing the rule algebraically?

26 Coulomb's law in physics states that the electrical force F (in newtons) between two charged objects is directly proportional to the product of the charges q_1 and q_2 (which can be positive or negative, and are measured in coulombs) and inversely proportional to the square of the distance between the charges d (in meters). The constant of proportionality is 9×10^9. As a general rule, opposite charges attract and like charges repel.

   a. What formula shows how F depends on q_1, q_2, and d?

   b. Suppose that a point charge of 3.0 coulombs and another of 2.0 coulombs are separated by a distance of 0.5 meters. What is the magnitude of the force between the charges? Is the force an attracting or a repelling force?

c. Suppose that a point charge of −3.0 coulombs and another of 2.0 coulombs are separated by a distance of 0.25 meters. What is the magnitude of the force between the charges? Is the force an attracting or a repelling force?

d. Suppose that a point charge of 0.003 coulombs and another of −0.0008 coulombs attract each other by a force of 600 newtons. How far apart are the centers of those charges?

27 Piecewise-defined functions (see Applications Task 11) have a number of important applications. For example, income tax owed may be written as a piecewise linear function of income. The 2012 income tax for a single tax filer is determined using the following table.

Taxable Income	Income Tax
$0–$8,700	10% of income
$8,700–$35,350	15% of income over $8,700, plus $870.00
$35,350–$85,650	25% of income over $35,350, plus $4,867.50

a. Last year, James earned $8,700 of taxable income. How much should he pay in income tax?

b. What would his income tax be if he earned $20,000 of taxable income?

c. Write a piecewise linear function $T(x)$ for income tax as a function of taxable income.

d. Sketch a graph of your tax function. What features does your graph have?

REVIEW

These tasks provide opportunities for you to review previously learned mathematics and to refine your skills in using that mathematics.

28 Examine the following graphs on xy-coordinate axes.

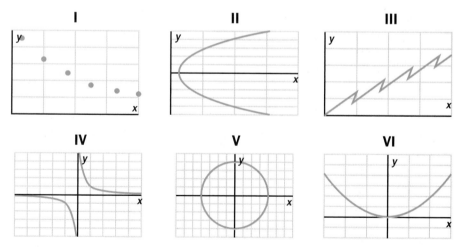

a. Which graphs show relationships between variables in which y is a function of x?

b. For those that are functions, which have inverses?

29 The graph of $f(x)$ below shows Mikayla's height in relation to the ground while she took a ride on a roller coaster.

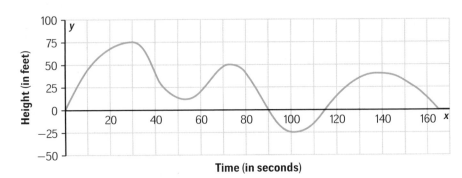

Time (in seconds)

a. What is the value of $f(20)$? What does it tell you about Mikayla's ride?

b. For what x values is $f(x) = 50$? What do these values tell you about Mikayla's ride?

c. What are the zeroes of $f(x)$?

d. What is the minimum value of $f(x)$ and at what point(s) during the ride does it occur?

e. What is the maximum value of $f(x)$ and at what point(s) during the ride does it occur?

30 The population of an experimental fruit fly colony is given by the function $p(t) = 30(2^t)$, with t representing time in days since the start of the experiment. How long will it take for the population to reach 1,200 fruit flies?

31 Rewrite each expression in standard polynomial form.

a. $(2x + 5)^2 + 10$

b. $-3x(4x^2 - 6) - 10x + 7$

c. $19x - x^2 + x(3x - 8)$

d. $(2 - 3x^2)(2 + 3x^2)$

32 Write an equation for the line satisfying each set of conditions.

a. Contains the points $(-1, -3)$ and $(-7, 12)$

b. Contains the points $(5, 0)$ and $(0, 7)$

c. Is parallel to the line with equation $y = \frac{3}{2}x + 2$ and contains the point $(6, 0)$

33 Evaluate each absolute value expression.

a. $|-3| + 2$

b. $|-3 + 2|$

c. $-5|-7|$

d. $-|8|$

e. $24 - |35|$

34 Miguel is riding on a Ferris wheel that has a 10-m radius and turns in a counterclockwise direction. Miguel is currently in the "3 o'clock" position. Consider Miguel's position in relation to the vertical line through the center of the wheel.

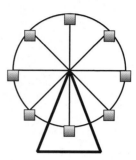

a. Determine his directed distance from the vertical line after he has rotated 90° counterclockwise.

b. Determine his directed distance from the vertical line after he has rotated 225° counterclockwise.

c. For what rotations (between 0° and 360°) will Miguel be 5 meters to the left or right of the vertical line through the center of the wheel?

d. Which graph below shows Miguel's directed distance from the vertical line through the center of the wheel as he travels through one revolution?

e. Write a function rule for the graph you chose in Part d.

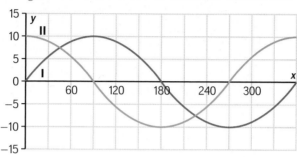

35 For each pair of figures below, describe the transformation that will map Figure I onto Figure II. Then write a coordinate rule $(x, y) \rightarrow (?, ?)$ for that transformation.

a.

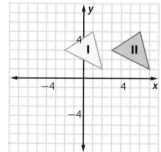

b.

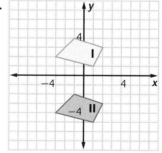

c.

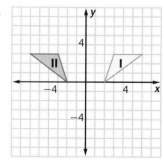

d.

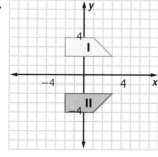

Customizing Models by Translation and Reflection

When it is important to chill a bottle of water or juice quickly, you can put it in the freezer compartment of a refrigerator or in an ice-filled cooler. You might expect the drink bottle to cool steadily from room temperature to the temperature of the freezer or the cooler. But it turns out that the temperature drops in a pattern like that shown in the following graph.

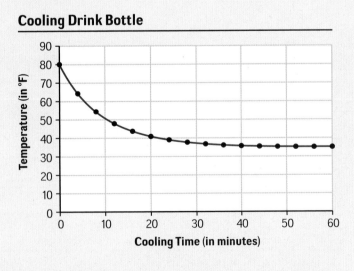

Cooling Drink Bottle

(x-axis: Cooling Time (in minutes), 0 to 60)
(y-axis: Temperature (in °F), 0 to 90)

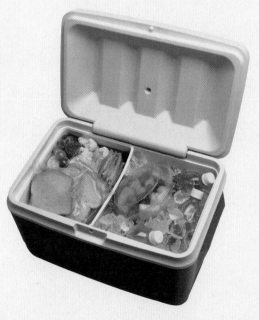

In this lesson, you will begin study of strategies for using familiar functions and transformations of their graphs to develop models for patterns that are variations of the basic linear, quadratic, exponential, circular, and absolute value functions. Trial-and-error testing of options and statistical regression methods are often effective strategies for finding function models of data patterns. But it is also helpful to know some general principles for modifying and combining the rules of basic function families to build new models for more complex situations.

INVESTIGATION

Vertical Translation

The graph that shows cooling of a drink bottle looks a lot like that of an exponential decay function whose graph has been shifted up about 35 units.

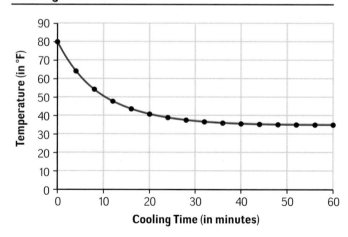

Cooling Drink Bottle

As you work on the problems of this investigation, look for answers to these questions:

What are the connections between rules for functions whose graphs are related by vertical translation?

How can functions with graphs obtained by vertical translation expand the supply of models for important patterns of variation?

To get ideas about how to find models for patterns that are vertical shifts of familiar graphs, it helps to look at some simple cases.

1 Shown below is the graph of an important special function called the *absolute value function*.

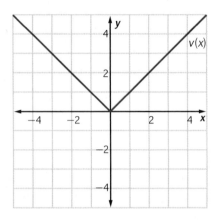

The rule for the **absolute value function** $v(x)$ is expressed symbolically as $v(x) = |x|$. Based on the pattern of values in the graph, how would you describe the rule for calculating $|x|$ for any value of x?

2 The next diagram shows the graph of $v(x)$ along with two variations.

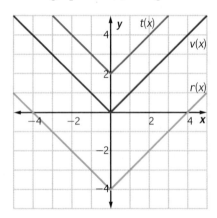

a. Complete a copy of the following table of values for $v(x)$, $r(x)$, and $t(x)$ for a sample of values of the independent variable x.

x	−4	−3	−2	−1	0	1	2	3	4
r(x)		−1							
v(x)	4	3	2	1	0	1	2	3	4
t(x)		5							

b. How are the values of $r(x)$ and $t(x)$ related to those of $v(x)$?

c. What rules seem to match the patterns in the graphs of $r(x)$ and $t(x)$?

d. What coordinate rules for geometric transformations $(x, y) \rightarrow (?, ?)$ map the graph of $v(x)$ onto:

    i. the graph of $r(x)$?   **ii.** the graph of $t(x)$?

3 In each part below, modify the given rule for $f(x)$ to produce related functions $g(x)$ and $h(x)$.

- The graph of $g(x)$ is congruent to, but translated 5 units upward from, the graph of $f(x)$.

- The graph of $h(x)$ is congruent to, but translated 3 units downward from, the graph of $f(x)$.

Then check your ideas using technology. Be prepared to explain *why* your rules for $g(x)$ and $h(x)$ do what is requested in each case.

a. $f(x) = x^2 - 4x$ **b.** $f(x) = 2(1.5^x)$ **c.** $f(x) = \dfrac{3}{x}$

4 **Properties of Vertical Translations** The questions of this problem ask you to compare properties of functions whose graphs are related by vertical translation. Consider the following graph of a function $f(x)$.

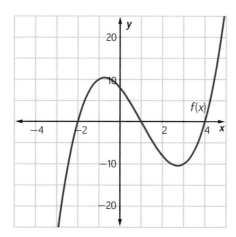

On a copy of the diagram:

a. sketch graphs of $g(x) = f(x) + 10$ and $h(x) = f(x) - 5$.

b. locate the local maximum and minimum points of each function and label those points on the graphs with their approximate coordinates.

c. locate the zeroes of each function and label corresponding points on the graphs with their approximate coordinates.

d. locate the y-intercept of each function and label those points on the graphs with their approximate coordinates.

e. Combine results of your work on Parts a–d and your experience in work on Problems 1 and 2 to formulate conjectures that complete these sentences.

   i. When $g(x) = f(x) + k$, the local maximum and minimum points of $g(x)$ … .

   ii. When $g(x) = f(x) + k$, the zeroes of $g(x)$ … .

   iii. When $g(x) = f(x) + k$, the y-intercept of $g(x)$ … .

Be prepared to explain *why* your conjectures are reasonable generalizations of the particular results in Parts a–d. That is, *why* is it reasonable to believe that each property will hold for any functions $f(x)$ and $g(x)$ whose graphs are related by vertical translation?

5 Consider again the cooling drink bottle pattern shown in the graph below.

Cooling Drink Bottle

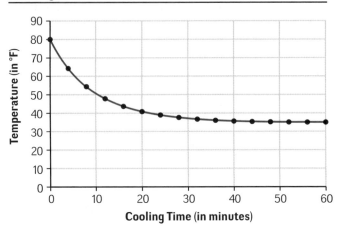

a. The following table gives a sample of (*cooling time, temperature*) values that show how temperature of the drink bottle decreases with the passage of time.

Cooling Time *t* (in minutes)	0	10	20	30	40	50	60
Temperature *b(t)* (in °F)	80	51	41	37	35.5	35.2	35.1

Use a curve-fitting tool to find an exponential model for the data pattern. Why is there a relatively poor match between the data and the regression model?

b. Devon and Carissa were still convinced that an exponential function would provide a useful model for the cooling pattern. It looked to them as if such a model would fit if the graph could be translated downward by about 35°. What evidence in the table and graph supports that reasoning?

c. Use a curve-fitting tool to find the "best-fitting" exponential model for the data pattern relating (*cooling time, temperature − 35°*).

d. Use the result of Part b and what you have observed in your work on Problems 1–3 to develop and test a model for the original (*cooling time, temperature*) data pattern. Then explain why your new model is a better match for the shape of the cooling graph than the regression equation found in Part a.

In this investigation, you explored connections between rules of functions whose graphs are related by vertical translations.

a If each point on the graph of $h(x)$ is k units above a corresponding point on the graph of $g(x)$, how will the rules for $h(x)$ and $g(x)$ be related? If "k units above" is replaced with "k units below," how does that affect your answer?

b If $g(x) = f(x) + k$ for all x, how are the locations of the maximum and minimum points, the zeroes, and the y-intercepts of two functions $f(x)$ and $g(x)$ related?

c How can you find models for data patterns that appear to be vertical translations of graphs for familiar functions?

Be prepared to share your ideas and reasoning with the class.

✓ CHECK YOUR UNDERSTANDING

Study the graph of a function $f(x)$ shown below.

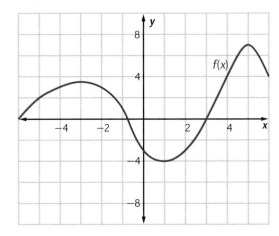

a. On a copy of the graph above, sketch and label graphs of $g(x) = f(x) + 3$ and $h(x) = f(x) - 6$.

b. Estimate and label the points corresponding to these function values:

   i. $g(2)$ **ii.** $g(-3)$

   iii. $h(4)$ **iv.** $h(-1)$

c. Estimate coordinates of local maximum and minimum points of all three functions.

d. Estimate coordinates of y-intercepts of all three functions.

e. Estimate the zeroes of all three functions.

Reflection Across the *x*-Axis

When you take a chilled drink bottle out of a refrigerator or cooler, it immediately begins warming toward room temperature. Once again you might expect that warming to occur at a steady rate, but the temperature will actually increase as shown in the following graph.

Warming Drink Bottle

Because it is reasonable to assume that warming and cooling should occur in similar patterns, it makes sense to believe that some variation of an exponential decay function would provide a model for the data pattern shown in the graph.

As you work on the problems of this investigation, look for answers to this question:

> *How can functions with graphs obtained by reflection across the x-axis*
> *expand the supply of models for important patterns of variation?*

1 Discuss with classmates how you could combine reflection and translation of an exponential decay function graph so that it would fit the pattern of (*warming time, temperature*) data for the warming drink bottle.

2 Once again, it helps to study some relatively simple cases to develop ideas that can be applied in more complex situations. On a copy of the graphs of the functions on the following page:

a. sketch a graph of the reflection across the *x*-axis.

b. give the rule for the function whose graph is the reflection image of the given function.

c. write a coordinate rule $(x, y) \rightarrow (?, ?)$ that maps each graph onto its reflection image across the *x*-axis.

Graph I

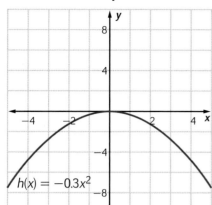

$h(x) = -0.3x^2$

Graph II

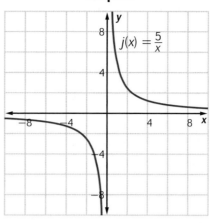

$j(x) = \frac{5}{x}$

Graph III

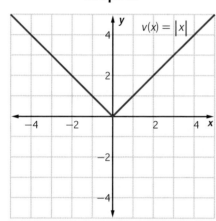

$v(x) = |x|$

Graph IV

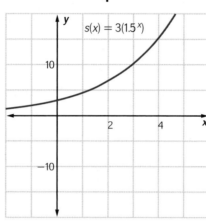

$s(x) = 3(1.5^x)$

Graph V

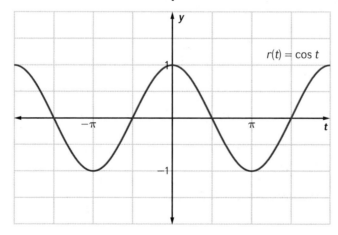

$r(t) = \cos t$

3 Use results of your work in Investigation 1 to sketch graphs for the following variations of the functions in Problem 2. Then check your ideas using technology.

a. $p(x) = -0.3x^2 + 5$

b. $w(x) = -\frac{5}{x} + 2$

c. $q(x) = -|x| - 3$

d. $u(x) = -3(1.5^x) + 12$

e. $k(t) = \cos t - 1$

4 **Properties of Reflection Across the *x*-axis** Suppose you are told that functions $f(x)$ and $g(x)$ have graphs that are images of each other by reflection across the *x*-axis.

  a. What relationship would you expect between the local maximum and minimum points of the two functions?

  b. What relationship would you expect between the zeroes of the two functions?

  c. What relationship would you expect between the *y*-intercepts of the two functions?

  d. Illustrate your answers to Parts a, b, and c by sketching graphs of $f(x) = x^4 - 3x^3 - x^2 + 3x$ and the function $g(x)$ whose graph is the image of $f(x)$ reflected across the *x*-axis. (A window of $-2 \le x \le 4$ and $-10 \le y \le 10$ will show key features of both graphs.)

  e. What is the rule for $g(x)$?

  Be prepared to explain why it is reasonable to believe that each of your conjectures in Parts a–c will hold for any functions $f(x)$ and $g(x)$ whose graphs are related by reflection across the *x*-axis.

5 Now consider again the function that models the pattern of change in temperature of a drink bottle after it has been taken from an icy cooler.

Warming Drink Bottle

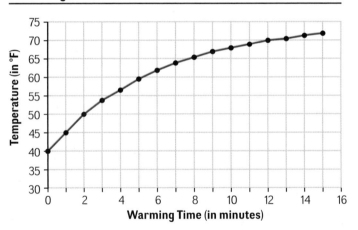

  a. The following table shows a representative sample of data points.

Warming Time *t* (in minutes)	0	3	6	9	12	15
Temperature *d*(*t*) (in °F)	40	54	62	67	70	72

  The temperature of the drink seems to be headed toward a room temperature of about 75°F. Use an exponential curve-fitting tool to find the rule for a function that models the (*warming time, 75° − temperature*) pattern.

  b. Now identify a combination of reflections and vertical translations that will transform the graph of your exponential function in Part a to a shape and location that matches the original drink bottle warming graph. Use those ideas to write a rule for the function $d(t)$ that gives the temperature of the drink *t* minutes after it has been taken from the cooler.

SUMMARIZE THE MATHEMATICS

In this investigation, you explored the connections between rules of functions whose graphs were reflections of each other across the *x*-axis. You then used those ideas to develop models for variations on familiar data patterns.

a Suppose $f(x)$ and $g(x)$ are functions whose graphs are reflection images of each other across the *x*-axis. How will the rules of those functions be related?

b How are the locations of maximum and minimum points, zeroes, and *y*-intercepts of two functions related if their graphs are reflection images of each other across the *x*-axis?

c What clues would you look for to see when reflections and vertical translations could be combined to produce models of data patterns that are similar to the basic function families that you know?

Be prepared to share your thinking with the class.

 CHECK YOUR UNDERSTANDING

For each of these functions, without use of technology:

- sketch a graph of the function over the specified domain interval.
- sketch the function whose graph is the reflection image across the *x*-axis of the given function. Give its rule.

a. $p(x) = -|x|$ graphed on the interval $[-3, 3]$

b. $h(x) = x^2 - 1$ graphed on the interval $[-3, 3]$

c. $d(x) = -1.5x + 2$ graphed on the interval $[-4, 4]$

d. $j(x) = \dfrac{4}{x^2}$ graphed on the interval $[-5, 5]$

Horizontal Translation

Slingshots have been used as hunting weapons for thousands of years, but a new version is now available as a toy for propelling water balloons.

Suppose that the following graph shows the trajectories of a series of water balloon shots.

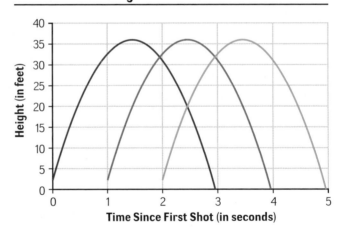

The three graphs look identical in shape but translated horizontally from each other.

As you work on the problems of this investigation, look for an answer to this question:

How are the rules of functions $f(x)$ and $g(x)$ related if their graphs are related by horizontal translation?

Once again, it helps to analyze some simple examples of functions with graphs related by horizontal translation to get some ideas about the patterns relating rules of any such pairs of functions.

1. Consider the absolute value function $f(x) = |x|$ and two variations $g(x) = |x + 3|$ and $h(x) = |x - 3|$.

   a. Graphs of $f(x)$, $g(x)$, and $h(x)$ are shown below. Before completing Parts b–d, make a conjecture about the match of function rules and graphs.

   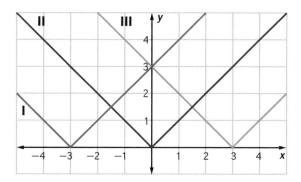

   b. Complete a copy of this table that compares values for the three functions.

x	−4	−3	−2	−1	0	1	2	3	4
$f(x) = |x|$									
$g(x) = |x + 3|$									
$h(x) = |x - 3|$									

   c. Match the functions $f(x)$, $g(x)$, and $h(x)$ to the graphs in the diagram in Part a. Then give rules in the form $(x, y) \rightarrow (?, ?)$ for geometric transformations that map the graph of $f(x)$ onto the graphs of $g(x)$ and $h(x)$.

   d. If k is some positive number, how do you think the graphs of the following functions will be related to the graph of $y = |x|$? Test your ideas using technology and see if you can use your answers to Parts b and c to explain why things happen as they do.

   i. $y = |x + k|$

   ii. $y = |x - k|$

2. Now consider variations on the basic quadratic $q(x) = x^2$. Match these functions with their graphs in the diagram at the right. Then test your ideas to see if you can explain why things happen as they do.

   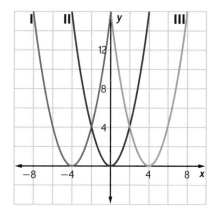

   a. $f(x) = x^2$

   b. $g(x) = (x - 4)^2$

   c. $h(x) = (x + 4)^2$

3 Now look back at the graphs of water balloon shots at the beginning of this investigation. Suppose the height of the first shot is a function of time in flight with rule $h(t) = -16t^2 + 48t$. The second water balloon was shot 1 second after the first and the third balloon was shot 2 seconds after the first.

   a. The function $j(t)$ gives height over time of the second water balloon shot. What is the rule for $j(t)$?

   b. The function $k(t)$ gives height over time of the third water balloon shot. What is the rule for $k(t)$?

4 Next consider further variations on the basic quadratic $q(x) = x^2$. Match these functions with their graphs in the diagram at the right. Then test your ideas to see if you can explain why things happen as they do.

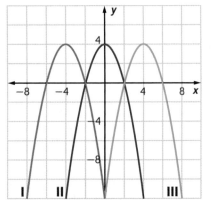

   a. $f(x) = -x^2 + 4$

   b. $g(x) = -(x - 4)^2 + 4$

   c. $h(x) = -(x + 4)^2 + 4$

5 If h and k are positive numbers, how do you think the graphs of the following functions will be related to the graph of $y = x^2$? Test your ideas using a computer algebra system that allows you to define functions like $f(x) = -(x + h)^2 + k$ and dynamically adjust values of h and k.

   a. $y = -(x + h)^2 + k$

   b. $y = -(x - h)^2 + k$

   c. $y = -(x - h)^2 - k$

   d. $y = -(x + h)^2 - k$

6 The next diagram shows graphs of four other variations on the basic quadratic function $q(x) = x^2$. Use ideas that you have developed from your work on Problems 2–5 to write symbolic rules for each new function. Be prepared to explain how you could develop each rule using reasoning alone, without trial-and-error graphing or a curve-fitting program.

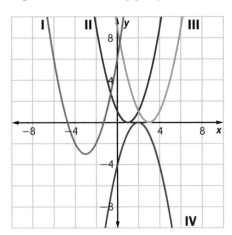

7 Assume h and k are positive numbers. Based on your work in Problems 1–6, how do you think the graphs of the following functions will be related to that of $f(x)$?

a. $g(x) = f(x + h) + k$

b. $g(x) = f(x - h) + k$

c. $g(x) = f(x - h) - k$

d. $g(x) = f(x + h) - k$

8 Explain why the patterns you have observed in examples involving $y = |x|$ and $y = x^2$ seem likely to work with other functions as well.

9 Test your ideas about the graphical relationships of $f(x)$ and $f(x \pm h) \pm k$ on these pairs of functions. In each case:

- make a sketch of what you think the two graphs will look like.

- check your prediction using technology.

a. $f(x) = x^3$ and $g(x) = f(x + 3) + 1$ graphed on $[-6, 6]$

b. $f(x) = x^3$ and $g(x) = f(x - 3) - 2$ graphed on $[-6, 6]$

10 The following is a graph of $y = \cos t$ for values of t from -2π to 2π. On a copy of this graph, sketch the graphs of the variations of the $\cos t$ function in Parts a–c.

a. $y = -\cos t$

b. $y = 1 - \cos t$

c. $y = \cos (t - 2)$

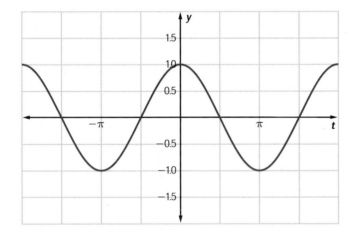

11 **Properties of Horizontal Translations** Suppose you are told that functions $f(x)$ and $g(x)$ have graphs that are images of each other under horizontal translation $(x, y) \rightarrow (x \pm h, y)$ with $h > 0$.

a. Describe the relationship between the local maximum and minimum points of the two functions.

b. Describe the relationship between the zeroes of the two functions.

c. Describe the relationship between the y-intercepts of the two functions.

d. Illustrate your answers to Parts a–c by sketching graphs of
$f(x) = x^3 - x^2 - 2x$ and the function
$g(x) = (x - 1)^3 - (x - 1)^2 - 2(x - 1)$.
Label key points on each with their approximate coordinates.

12 The graph shown below is of a piecewise-defined function with the given rule:

$$f(x) = \begin{cases} 3x + 12 & \text{for } x \leq -2 \\ x^2 + 2 & \text{for } -2 < x \leq 1 \\ -x + 2 & \text{for } x > 1 \end{cases}$$

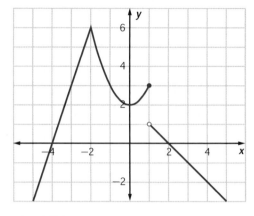

a. On a coordinate grid, copy the graph of $f(x)$.

b. On the same coordinate grid, sketch the graph of $g(x) = f(x + 1)$.

c. Write the symbolic rule for $g(x)$.

SUMMARIZE THE MATHEMATICS

In this investigation, you discovered connections between the rules of functions when their graphs are related by horizontal translations.

a Suppose the graph of a function $g(x)$ is the image of the graph of $f(x)$ after translation of h units to the right. How can the rule for $g(x)$ be derived from the rule for $f(x)$?

b Suppose the graph of a function $j(x)$ is the image of the graph of $f(x)$ after translation of h units to the left. How can the rule for $j(x)$ be derived from the rule for $f(x)$?

c Suppose the graph of a function $g(x)$ is the image of the graph of $f(x)$ after translation of h units to the right. How are locations of maximum and minimum points and zeroes of the two functions related? What if the translation is h units to the left?

Be prepared to share your ideas and reasoning with the class.

 CHECK YOUR UNDERSTANDING

For each function in Parts a–c, without using technology:

- sketch a graph of the function and the specified translation of that graph.

- give the rule for the function whose graph is the translation image of the original function.

a. $p(x) = -|x|$ graphed on the interval $[-8, 8]$ and then translated 4 units to the right

b. $h(x) = x^2$ graphed on the interval $[-6, 6]$ and then translated 3 units to the left

c. $d(x) = -1.5x$ graphed on the interval $[-6, 6]$ and then translated 2 units to the right

1 For each function below, find a rule for the function $g(x)$ whose graph is the image of the graph of $f(x)$ under the indicated transformation.

a. $f(x) = x^2 + 4x - 5$; transformation $(x, y) \rightarrow (x, y + 3)$

b. $f(x) = 4x - 3$; transformation $(x, y) \rightarrow (x, y - 4)$

c. $f(x) = 30(2.5^x)$; transformation $(x, y) \rightarrow (x, y - 7)$

2 For each function below, find the transformation $(x, y) \rightarrow (?, ?)$ that maps the graph of $f(x)$ onto the graph of $g(x)$.

a. $f(x) = -x^2 + 4x - 5$ and $g(x) = -x^2 + 4x$

b. $f(x) = |x| - 5$ and $g(x) = |x| + 4$

c. $f(x) = \frac{1}{x} + 2$ and $g(x) = \frac{1}{x}$

d. $f(x) = 7(0.5^x)$ and $g(x) = 3 + 7(0.5^x)$

3 The diagram below shows the graph of $j(t) = \sin t$ and two related functions $k(t)$ and $m(t)$.

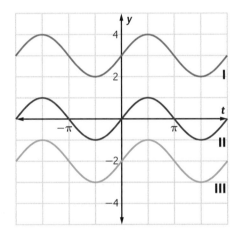

a. Which is the graph of $j(t)$ and how do you know?

b. What rules for $k(t)$ and $m(t)$ would produce the other two graphs?

c. What coordinate rules $(t, y) \rightarrow (?, ?)$ for geometric transformations would map the graph of $j(t)$ onto:

   i. the graph of $k(t)$?

   ii. the graph of $m(t)$?

4 Suppose that $f(x)$ and $g(x)$ are functions for which $g(x) = f(x) - 3$ for all values of x.

a. How are the graphs of $f(x)$ and $g(x)$ related geometrically?

b. If $f(x)$ has a local minimum at $(-4, -2)$ and a local maximum at $(3, 7)$, what (if anything) can be said about local minimum and maximum points for $g(x)$?

c. If $g(x)$ has y-intercept $(0, -1)$, what is the y-intercept of $f(x)$?

d. If $f(x)$ has zeroes at $x = 7$ and $x = -1$, what (if anything) can be said about the zeroes of $g(x)$?

5 Suppose that an amusement park predicts that daily net income (in dollars) from ticket, food, and gift shop sales will be related to the price p (in dollars) charged for admission by

$$I(p) = -100p^2 + 3{,}000p + 100{,}000.$$

a. Sketch a graph of $I(p)$ for $0 \le p \le 50$. Label the maximum income point, the y-intercept, and any zeroes with their coordinates.

b. Suppose that the park has fixed daily expenses of $45,000. How does that information and the income function $I(p)$ combine to give a function $P(p)$ that tells the daily profit of the park as a function of admission price?

c. Add the graph of $P(p)$ to your sketch in Part a. Label the maximum profit point, the y-intercept, and any zeroes with their coordinates. Then explain how, if at all, those key points on the profit function graph are related to corresponding points on the net income graph.

6 Restaurants cook pizza in very hot ovens, but as soon as the pizza comes out of the oven, it begins cooling toward room temperature. Data in the following table illustrate the kind of cooling pattern that might be expected.

Cooling Time (in minutes)	0	2	4	6	8
Temperature (in °F)	575	325	200	140	110

a. Suppose that the temperature is 80°F in the room where the pizza is cooling. What transformation of the given temperature data seems likely to yield a pattern that could be modeled by an exponential decay function?

b. What exponential function matches the pattern in the transformed data?

c. What function is a good model for the pattern in change over time of the original pizza temperature data?

d. What temperature does your model from Part c predict for a time 15 minutes after the pizza comes out of the oven?

e. At approximately what time will the pizza reach a temperature of 212°F, the boiling point of water?

7 In each part, find a rule for the function $g(x)$ whose graph is the image of the graph of $f(x)$ under the indicated transformation.

a. $f(x) = x^2 + 4x - 5$; transformation $(x, y) \rightarrow (x, -y)$

b. $f(x) = 4x - 3$; transformation $(x, y) \rightarrow (x, -y)$

c. $f(x) = 30(2.5^x)$; transformation $(x, y) \rightarrow (x, 7 - y)$

d. $f(x) = -x^2 + 4x + 5$; transformation $(x, y) \rightarrow (x, -y - 4)$

8 In each part, find the transformation $(x, y) \rightarrow (?, ?)$ that maps the graph of $f(x)$ onto the graph of $g(x)$.

a. $f(x) = -x^2 + 4x$ and $g(x) = x^2 - 4x$

b. $f(x) = |x| - 5$ and $g(x) = -|x| + 5$

c. $f(x) = \frac{1}{x} + 2$ and $g(x) = -\frac{1}{x}$

d. $f(x) = 7(0.5^x)$ and $g(x) = 10 - 7(0.5^x)$

9 Suppose that $f(x)$ and $g(x)$ are functions for which $g(x) = -f(x)$ for all values of x.

a. How are the graphs of $f(x)$ and $g(x)$ related geometrically?

b. Suppose $f(x)$ has a local minimum at $(-4, -2)$ and a local maximum at $(3, 1)$. What (if anything) can be said about local minimum and maximum points for $g(x)$?

c. Suppose $f(x)$ has y-intercept $(0, -4)$. What is the y-intercept of $g(x)$?

d. Suppose $f(x)$ has zeroes at $x = 7$ and $x = -1$. What (if anything) can be said about the zeroes of $g(x)$?

10 A Summit Street snack vendor takes an ice cream bar out of the freezer which keeps frozen snacks at 0°F and forgets to put it back into the freezer when the customer changes his mind. If the air temperature in the vendor's truck is about 85°F, the temperature of the melting ice cream bar will increase in a pattern like that shown in the next table.

Warming Time (in minutes)	0	5	10	15	20	25	30	35
Temperature (in °F)	0	35	55	67	75	79	81	83
Temperature Difference (in °F)								

a. Use a regression routine on your calculator or computer software to find a good-fitting model for the relationship between time and the difference between the air temperature and ice cream bar temperature.

b. Use the model from Part a and what you know about translation and reflection of graphs to devise a rule for the function that gives actual temperature of the ice cream bar at any time after it has been taken from the freezer.

c. Compare the shape of the graph for your function model in Part b and a plot of the given (*warming time, temperature*) data. Explain why any observed differences between the two patterns occur.

11 In each part, find a rule for the function $g(x)$ whose graph is the image of the graph of $f(x)$ under the indicated transformation.

a. $f(x) = 4x - 3$; transformation $(x, y) \rightarrow (x - 5, y)$

b. $f(x) = x^2 + 4x$; transformation $(x, y) \rightarrow (x + 3, y)$

c. $f(x) = \sin x$; transformation $(x, y) \rightarrow (x + 2, y - 1)$

d. $f(x) = |x|$; transformation $(x, y) \rightarrow (x - 2, 4 - y)$

12 Suppose that $f(x)$ and $g(x)$ are functions for which $g(x) = f(x - 5)$ for all values of x.

a. How are the graphs of $f(x)$ and $g(x)$ related geometrically?

b. If $f(x)$ has a local minimum at $(-4, -2)$ and a local maximum at $(3, 1)$, what can be said about local minimum and maximum points for $g(x)$?

c. If $f(x)$ has y-intercept $(0, -4)$, what is the y-intercept of $g(x)$?

d. If $f(x)$ has zeroes at $x = 7$ and $x = -1$, what (if anything) can be said about the zeroes of $g(x)$?

13 To win a prize for the most creative Punkin' Chunker, one group designed a rapid-fire machine that could fire 5 pumpkins at one-second time intervals. The trajectory of each pumpkin is the same, with height (in feet) at any time t seconds after firing given by the quadratic function

$$h(t) = -16t^2 + 144t + 20.$$

a. Sketch a graph of $h(t)$ and label with coordinates the y-intercept, the maximum height point, and the point where the pumpkin hits the ground.

b. What rule for function $h_2(t)$ will describe the height of the *second* pumpkin at any time t seconds after the first pumpkin is shot? Add a sketch of the graph of that function to the diagram of Part a.

c. What rule for function $h_5(t)$ will describe the height of the *fifth* pumpkin at any time t seconds after the first pumpkin is shot? Add a sketch of the graph of that function to the diagram of Part a.

14 Suppose that the mean of test scores on a mathematics exam is 75.6 and the standard deviation is 5.9. If the teacher adds 10 points to every score to correct for a problem all students missed because the wording was unclear, how will that rescoring change:

a. the class mean?

b. the class standard deviation?

15 The diagram below shows a geometric figure on a coordinate grid. Explain how each of the following attributes of that figure will or will not change under the transformations described in Parts a–c.

- perimeter

- area

- sum of angle measurements

a. $(x, y) \rightarrow (x, y + 3)$

b. $(x, y) \rightarrow (x + 3, y)$

c. $(x, y) \rightarrow (x, -y)$

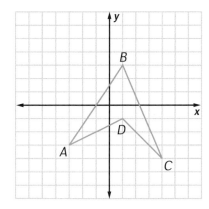

16 What are the rules for the inverses of each of these geometric transformations?

Transformation	Inverse
a. $(x, y) \rightarrow (x, y + 3)$	$(x, y) \rightarrow (?, ?)$
b. $(x, y) \rightarrow (x + 3, y)$	$(x, y) \rightarrow (?, ?)$
c. $(x, y) \rightarrow (x, -y)$	$(x, y) \rightarrow (?, ?)$

17 A circle with center O and radius r is the set of all points at distance r from O.

a. Using the sketch at the right, explain why the equation for a circle of radius r centered at the origin of a coordinate grid must be $x^2 + y^2 = r^2$.

b. How does the reasoning developed in Investigation 3 for horizontal translation of graphs explain why the equation for a circle of radius r and center $(h, 0)$ must be $(x - h)^2 + y^2 = r^2$?

c. How does similar reasoning explain why the equation for a circle of radius r and center (h, k) must be $(x - h)^2 + (y - k)^2 = r^2$?

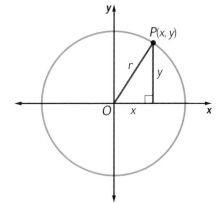

18 The diagram below shows graphs of the circular functions $f(t) = \cos t$ and $g(t) = \sin t$ on an interval from -2π to 2π. You can see that the graphs have the same overall shape.

   a. Which graph is $f(t) = \cos t$ and which is $g(t) = \sin t$ and how do you know?

   b. For what values of k is $\cos(t + k) = \sin t$ for all values of t?

   c. For what values of j is $\cos t = \sin(t + j)$ for all t?

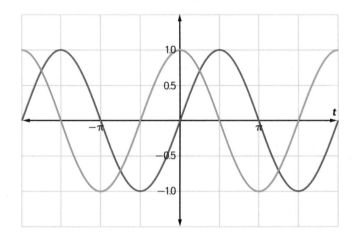

19 For each of the following functions:

   • use your skill in completing the square of a quadratic polynomial to write an equivalent function rule in vertex form, $y = (x - h)^2 + k$.

   • find the coordinates of the maximum or minimum points on the graphs of the functions.

   • explain how the maximum and minimum coordinates could be found by thinking about translating (and possibly reflecting) the graph of $y = x^2$.

   a. $y = x^2 - 4x + 7$

   b. $y = x^2 + 6x + 8$

20 Combine your knowledge about function transformations and solving inequalities. For each inequality below:

   • sketch the graphs of the functions involved in the inequality.

   • locate the intersection points.

   • record the solution using inequality symbols and interval notation.

   a. $|x - 6| \le 4$

   b. $|x + 6| \ge 4$

   c. $|x - 2| < 0.5$

   d. $|x - 2| \le \frac{1}{2}x + 2$

21 Shown below is the graph of the *square root function,* $f(x) = \sqrt{x}$, for $0 \le x \le 20$. On a copy of the diagram, sketch graphs of the following variations of the square root function, and then check your ideas. With another color pen or pencil, note any needed corrections.

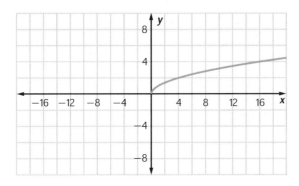

a. $g(x) = -\sqrt{x}$

b. $h(x) = \sqrt{x} + 5$

c. $j(x) = -\sqrt{x} + 5$

d. $k(x) = \sqrt{x + 4}$

REFLECTIONS

22 The diagram at the right shows graphs of $y = x^2$ and $y = x^2 - 4$. You learned in this lesson that those graphs are related by a vertical translation. The difference between corresponding y values should be 4 for every x. But somehow the graphs seem to get closer together as $|x|$ increases. How can you explain this apparent contradiction?

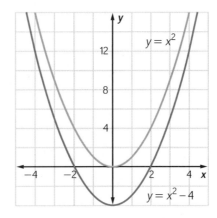

23 Look back at Connections Task 20. How could you use similar reasoning to solve the following inequalities? Record the solutions using inequality symbols and interval notation.

a. $\sqrt{x} - 4 \le 0$

b. $\sqrt{x + 5} < 3$

24 Why is it reasonable to say that the graphs of $f(x) = |x|$ and $g(x) = -|x| + 7$ are congruent?

25 Look back at your work in this lesson. Give specific examples where you found the following mathematical practices helpful and explain how they were helpful.

a. Making sense of problems and persevering in solving them

b. Reasoning both quantitatively and algebraically

c. Searching for and making use of patterns or structure in mathematical situations

26 Many students find it hard to believe (or remember) that the graph of $y = (x - 3)^2$ can be found by translating the graph of $y = x^2$ three units to the *right*. Or that the graph of $y = (x + 3)^2$ can be found by translating the graph of $y = x^2$ three units to the *left*.

How do you think about the relationship of rules and graphs in the general case where $g(x) = f(x \pm h)$ with $h > 0$ so that you can come up with the correct connection between rules and graphs? How can you check the correctness of your thinking?

EXTENSIONS

27 The *Custom Gear* company designs and makes baseball caps and shirts with clever messages and sketches for groups that want personalized items. For each personalized baseball cap, they charge $250 for creating the pattern and then an additional $15 to produce each individual cap.

a. What function gives the cost of an order for x caps?

b. What function gives the cost per cap of an order for x caps?

c. How can the function in Part b be expressed with an equivalent rule in the form $y = m + \dfrac{b}{x}$?

d. Sketch a graph of the cost per cap function and describe the horizontal asymptote for that graph.

e. Explain what the horizontal asymptote tells about the cost per cap as the number of caps ordered increases.

f. Explain how you can identify the horizontal asymptote by analyzing the expression $m + \dfrac{b}{x}$.

28 From your many encounters with quadratic functions, you know that the graph of $y = ax^2 + bx + c$, $a > 0$, has a minimum point when $x = -\dfrac{b}{2a}$.

a. Why must the graph of $y = x^2$ have its minimum point at $(0, 0)$?

b. If h is a positive number, what is the minimum point on the graph of $y = (x - h)^2$? How does what you learned in Investigation 3 justify your answer?

c. Expand the expression $(x - h)^2$ to standard polynomial form and explain how it and the expression $-\dfrac{b}{2a}$ confirm your answer to Part b.

d. If h and k are positive numbers, what is the minimum point on the graph of $y = (x - h)^2 + k$? How do results from your work in Investigations 1 and 3 justify your answer?

e. Expand the expression $(x - h)^2 + k$ to standard polynomial form. Then explain how it and the $-\dfrac{b}{2a}$ rule confirm your answer to Part d.

29 In this unit, you have seen how simple geometric transformations of a function graph are connected to corresponding transformations of the function rule. In this task, you will investigate how horizontal shifts are related to the zeroes of quadratic functions and, more generally, to the quadratic formula.

a. First examine each of the following pairs of quadratic functions and their graphs. For each pair of functions, do the following.

- Write rules for the lines of symmetry.

- Describe geometrically how the graphs are related.

- Explain how the zeroes of the two functions are related.

- Solve $f(x) = 0$ using algebraic reasoning. What can you conclude about the solutions of $g(x) = 0$?

i. $g(x) = x^2 - 10x + 16$ $f(x) = x^2 - 9$

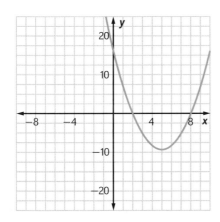 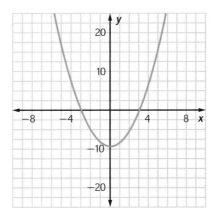

ii. $g(x) = 4x^2 + 24x - 45$ $f(x) = 4x^2 - 81$

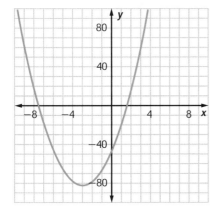 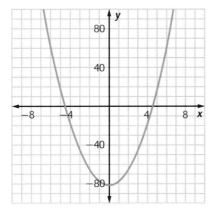

b. Just as you are able to transform the graph of $g(x)$ to the graph of $f(x)$ for each pair of functions in Part a, you can similarly transform the function rule. Consider the function:

$$g(x) = x^2 - 10x + 16 = 0$$

i. Write a coordinate rule $(x, y) \rightarrow (?, ?)$ that will transform the graph of g so that its image f has the y-axis as its line of symmetry.

ii. Find a number a so that $g(x + a) = f(x)$ for all x.

iii. Solve $g(x + a) = 0$.

iv. How can your answer from part iii be used to solve $g(x) = 0$? What are the solutions?

v. How is your work in parts i–iv related to the "translate-transform-translate" method you used in Course 2?

c. Use algebraic reasoning similar to what you used in Part b to find the zeroes of $g(x) = 4x^2 + 24x - 45$.

30 You can use reasoning similar to that in Extensions Task 29 to provide another proof of the quadratic formula. To begin, use the fact that $x = -\dfrac{b}{2a}$ is the line of symmetry for the parabola $g(x) = ax^2 + bx + c$. Transform the graph of g so that its image f has the y-axis as its line of symmetry. Use the transformed graph f and a combination of algebraic and geometric reasoning to show that if $ax^2 + bx + c = 0$ and $a \neq 0$, then

$$x = -\frac{b}{2a} \pm \frac{\sqrt{b^2 - 4ac}}{2a}.$$

31 A patch of snow is melting on a warm spring day. Suppose the initial temperature of the snow is $-5°C$ and that over the course of a half hour, the snow increases in temperature at a linear rate to the melting point of $0°C$. During the next eight hours, the snow slowly melts with the water collecting in a puddle. Suppose also that, once it is melted, the temperature of the water increases at half the rate as that of the snow for one half hour.

a. Write a piecewise linear function for the temperature of the water as a function of time over the 9-hour span.

b. Sketch a graph of your function and explain what is happening during each portion of the graph.

REVIEW

32 Consider the function $h(x) = x^2 + 5x - 14$.

a. Evaluate $h(4)$ and $h(-5)$.

b. Find the zeroes of $h(x)$.

c. For what values of x is $h(x) > 0$?

d. For what values of x is $h(x) = 10$?

33 Find the coordinates of the vertex, the x-intercepts, and the y-intercept of the graph of each quadratic function.

a. $y = x^2 - 6x$

b. $y = x^2 - 8x + 15$

c. $y = (x + 5)^2 - 3$

34 Solve each equation.

a. $3(x + 7) - 2x = 5x - 7$

b. $x(x + 4) = x^2 - 12$

c. $(2x - 3)^2 = 25$

d. $3x^2 - 5x + 1 = 2$

35 Rewrite each expression as a product of linear factors.

a. $x^2 + 12x + 20$

b. $25x^2 - 16$

c. $x^2 - 6x + 9$

d. $x^2 + 7x - 30$

e. $9 - 4x^2$

f. $3x^3 + 15x^2 + 12x$

36 In Course 3, you learned that angles of rotation can be measured in degrees, radians, and revolutions. Fill in each blank with the appropriate value.

a. $270° = $ _____ radians $= $ _____ revolutions

b. _____ $° = \frac{\pi}{4}$ radians $= $ _____ revolutions

c. _____ $° = $ _____ radians $= \frac{1}{6}$ revolution

d. _____ $° = 3\pi$ radians $= $ _____ revolutions

37 Write each expression in a simpler equivalent exponential form using only positive exponents.

a. $3x^{-2}$

b. $\dfrac{x^4(2x^3)^2}{x^{-2}}$

c. $-8x^4y(4x^{-2}y)$

d. $3(5x)^{-1}$

38 It is 6 miles from Uma's house to the swimming beach at the lake.

a. If Uma walks an average of 4 miles per hour, how long will it take her to walk to the swimming beach?

b. Erina took 40 minutes to run to the beach from Uma's house. What was Erina's average running speed in miles per hour?

c. Claudio lives further from the swimming beach than does Uma. He rides his bike at an average speed of 12 miles per hour and arrives at the beach in 50 minutes. How far does Claudio live from the swimming beach?

39 Consider the following matrix representation of $\triangle ABC$.

$$\triangle ABC = \begin{bmatrix} -4 & 2 & 2 \\ 2 & 2 & -3 \end{bmatrix}$$

a. On separate grids, sketch and label $\triangle ABC$ and its image $\triangle A'B'C'$ under each of the following transformations.

  i. $(x, y) \rightarrow \left(\frac{1}{2}x, y\right)$

  ii. $(x, y) \rightarrow \left(x, \frac{1}{3}y\right)$

  iii. $(x, y) \rightarrow (-2x, -2y)$

b. Determine whether each of the image triangles in Part a is similar, congruent, or neither similar nor congruent to $\triangle ABC$. Explain your reasoning.

Customizing Models by Stretching and Compressing

In the 21st century, we measure time to the nanosecond and we can tell the time of day at any place in the world using clocks that communicate with satellites in Earth's orbit. But, in the 17th and 18th centuries, development of accurate clocks was a scientific problem that attracted the attention of famous scientists like Huygens and Galileo.

One of the most effective solutions to the time-keeping problem relies on the periodic motion of pendulums like those that you now see in grandfather clocks. Power from falling weights causes the pendulum to swing from side to side in a pattern like that shown in the following graph. This motion is then transferred to the clock hands.

Displacement of Pendulum Over Time

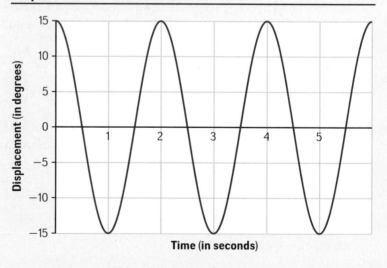

THINK ABOUT THIS SITUATION

Study the graph and use the information it reveals to answer these questions.

a What does the graph tell about the pattern of change over time in pendulum position?

b What familiar functions have graphs similar to that showing pendulum motion?

c How is the pendulum graph different from graphs of the functions it most resembles?

d How could you modify the rule for a familiar function to fit this somewhat different pattern?

In this lesson, you will continue study of strategies for using familiar functions and transformations of their graphs to develop mathematical models. The focus will be on techniques for modeling patterns that are produced by vertical or horizontal stretching or compressing of familiar graphs.

INVESTIGATION 1

Vertical Stretching and Compressing

The graph of pendulum motion looks a lot like that of the circular function $y = \cos x$. But the pendulum graph shows variation between -15 and 15 (not between -1 and 1) and the pattern of variation repeats in cycles of 2 seconds (not $2\pi \approx 6.28$ seconds). As you work on the problems of this investigation, look for answers to these questions:

What are the connections between rules for functions whose graphs are related by vertical stretching or compressing?

How can those connections be used to build related models for important data patterns?

 The diagram below shows graphs of $y = \cos x$ and two functions $f(x)$ and $g(x)$ with graphs related to $\cos x$ by *vertical stretching*—away from the x-axis.

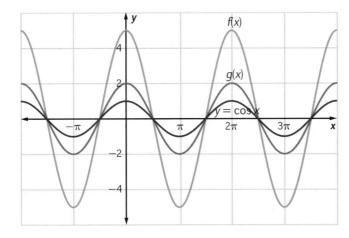

a. Complete a copy of the following table showing a sample of approximate (x, y) values for the three functions. (Remember: $\frac{\pi}{2} \approx 1.57$, $\pi \approx 3.14$, $\frac{3\pi}{2} \approx 4.71$, and $2\pi \approx 6.28$.)

x	$-\frac{3\pi}{2}$	$-\pi$	$-\frac{\pi}{2}$	0	2π	3π	$\frac{7\pi}{2}$
$f(x)$							
$g(x)$							
$\cos x$							

b. What geometric transformations $(x, y) \rightarrow (?, ?)$ do you think would map the graph of:

   i. $y = \cos x$ onto $f(x)$?

   ii. $y = \cos x$ onto $g(x)$?

   iii. $g(x)$ onto $y = \cos x$?

c. What rules would you expect to model the patterns shown in the graphs of $f(x)$ and $g(x)$? Compare your rules with those of your classmates and resolve any differences. Then check your agreed-upon ideas using technology.

2 The next diagram shows the graph of $y = \sin x$ and another function $h(x)$ that is related to it by *vertical compressing*—toward the x-axis.

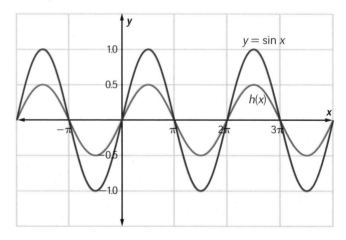

a. What geometric transformation $(x, y) \rightarrow (?, ?)$ do you think would map the graph of $y = \sin x$ onto the graph of $h(x)$?

b. Write a rule for $h(x)$.

3 Now reconsider the pendulum graph shown at the start of this lesson. What variation of the rule $y = \cos x$ will define a function that oscillates between -15 and 15, giving a range of values comparable to those of the pendulum displacement?

Applications of Vertical Stretching and Compressing The ideas you have developed about vertical stretching and compressing of $y = \cos x$ and $y = \sin x$ graphs can be used in work with other families of functions.

4 In several earlier *Core-Plus Mathematics* courses, you have used quadratic functions to model the relationship between time and distance of moving objects (soccer balls, pumpkins, skate boarders, luge riders, and others) as they respond to the pull of gravity. The rules for those functions were all derived from the simplest quadratic $y = x^2$. For instance, if you drop a ball from the top of a 60-foot tall building, its position at any time t seconds later will be given by $h(t) = 60 - 16t^2$.

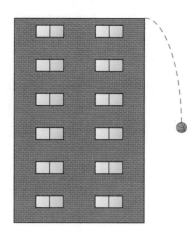

The diagram below shows a graph of $y = x^2$.

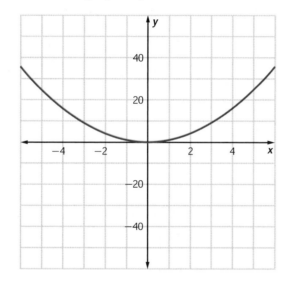

a. On a copy of this diagram, sketch and label the graphs for these functions.

$$y = -x^2 \qquad y = -16x^2 \qquad y = 60 - 16x^2$$

b. Study the following sequences of transformations to see which would lead to a correct graph for $h(x) = 60 - 16x^2$. In each case, provide algebraic reasoning that justifies your answer.

 i. Stretch the graph of $y = x^2$ vertically by a factor of 16, then reflect across the x-axis, and then translate up 60 units.

 ii. Stretch the graph of $y = x^2$ vertically by a factor of 16, then translate up 60 units, and then reflect across the x-axis.

 iii. Reflect $y = x^2$ across the x-axis, translate up 60 units, and then stretch vertically by a factor of 16.

5 In many science museums, one of the most popular exhibits is a Foucault pendulum. With a large mass attached to a long cable, it swings in a way that demonstrates the rotation of Earth.

The picture above shows a Foucault pendulum at the Reuben H. Fleet Science Center in Indianapolis. The functions in Parts a and b describe possible patterns of displacement from vertical (in degrees) as such a pendulum swings.

Draw graphs of each function on $[0, 4\pi]$. Then explain what each rule and graph tell about motion of the pendulum with passage of time t (in seconds).

a. $f(t) = 10 \cos t$

b. $g(t) = 25 \cos t$

6 **Properties of Vertical Stretching and Compressing** Consider the following graph of a function $f(x)$.

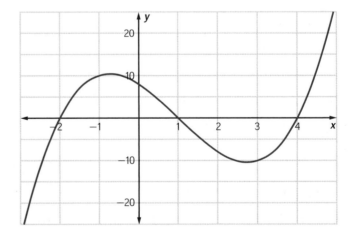

On a copy of the diagram:

a. sketch and label the graphs of $g(x) = 1.5f(x)$ and $h(x) = -2f(x)$.

b. locate all local maximum and minimum points of the three functions and label those points on the graphs with their approximate coordinates.

c. locate all zeroes of the three functions and label those points on the graphs with their approximate coordinates.

d. locate the y-intercepts of the three functions and label those points on the graphs with their approximate coordinates.

e. Look over the results of your work in Parts a–d. Formulate conjectures that answer these questions about vertical stretching and compressing of graphs for any function $f(x)$. Be prepared to explain your reasoning.

   i. How are local maximum and minimum points of functions $f(x)$ and $g(x)$ related if $g(x) = kf(x)$?

   ii. How are zeroes of $f(x)$ and $g(x)$ related if $g(x) = kf(x)$?

   iii. How are y-intercepts of $f(x)$ and $g(x)$ related if $g(x) = kf(x)$?

f. Test your ideas in Part e by comparing $f(x) = x^2 + x - 6$ and $g(x) = 3f(x)$.

   i. What is the standard polynomial form of the rule for $g(x)$?

   ii. How are the local maximum and minimum points, zeroes, and y-intercepts of the two functions related?

7 One of the simplest strategies for individuals and organizations to earn income on savings is to invest in mutual funds that spread the individual contributions over many different common stocks, bonds, or other assets. At the start of one year, a club invested $5,000 in such a fund and the profit on that investment varied over the next 12 months as shown in the following graph.

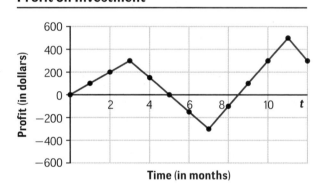

Profit on Investment

Suppose that the club had instead invested $10,000 in the same mutual fund.

a. What would the maximum and minimum profit figures have been during the year and when would they have occurred?

b. At what times during the year would the profit have been $0?

c. What would the profit have been at the end of the year?

d. How would the profit graph differ from the graph tracking profit for the $5,000 investment?

SUMMARIZE THE MATHEMATICS

In this investigation, you explored connections between rules of functions whose graphs are related by vertical stretching or compressing.

a Suppose the y-coordinate of each point on the graph of $g(x)$ is exactly k times the y-coordinate of the point just below or above it on the graph of $f(x)$. What is the connection between the rules of the two functions?

b What is the coordinate rule $(x, y) \rightarrow (?, ?)$ that will map the graph of $f(x)$ onto the graph of $kf(x)$ by vertical stretching or compressing with a scale factor of k?

c If $g(x) = kf(x)$, how are the graphs of $g(x)$ and $f(x)$ related when $|k| > 1$? When $|k| < 1$?

d How are the local maximum and minimum points, zeroes, and y-intercepts of a function $g(x)$ related to those of $f(x)$ if $g(x) = kf(x)$ for all x when $k > 0$? When $k < 0$?

Be prepared to explain your ideas to the class.

 CHECK YOUR UNDERSTANDING

For each function in Parts a and b, without using technology:

- sketch a graph of the function and the specified stretching or compressing of that graph.

- give a rule for the function whose graph is the image after stretching or compressing of the original graph.

a. $p(x) = -|x|$ graphed on $[-5, 5]$ and then stretched by a factor of 1.5 away from the x-axis

b. $h(x) = x^2$ graphed on $[-5, 5]$ and then compressed by a factor of 0.5 toward the x-axis

Horizontal Stretching and Compressing

Recall from Investigation 1 that the motion of a pendulum swinging from 15° right of center to 15° left of center and back in 2-second intervals can be represented by a graph like that of the circular function $y = \cos x$. The function $a(t) = 15 \cos t$ has a graph with *amplitude* 15—half the absolute value of the difference between minimum and maximum values. This amplitude is the same as that of the pendulum displacement function. But $a(t) = 15 \cos t$ has *period* 2π (taking approximately 6.28 seconds to complete each swing from maximum to minimum and back), while the clock pendulum has period 2 seconds.

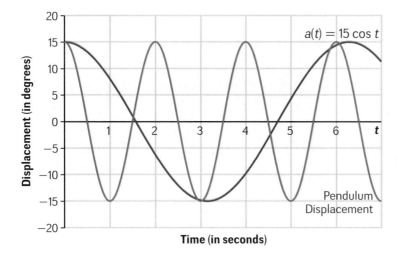

Time (in seconds)

Finding the rule for a function that accurately represents the displacement of the pendulum at any time t requires a modification of $a(t) = 15 \cos t$ that corresponds to horizontal compression of its graph toward the y-axis. As you work on the problems of this investigation, look for answers to these questions:

> *What are the connections between rules for functions whose graphs are related by horizontal stretching or compressing?*

> *How can those connections be used to build related models for important data patterns?*

As is often the case, it helps to develop a strategy for answering these questions by working on somewhat simpler problems first.

The next diagram shows $y = \cos x$ and a function $f(x)$ whose graph was obtained by horizontal compressing of $y = \cos x$ by a scale factor 0.5. Think about what is a suitable rule for $f(x)$.

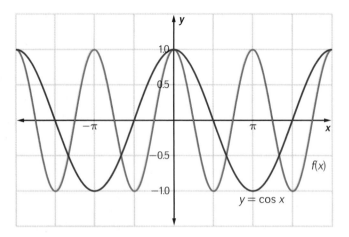

1 Test some of your ideas about $f(x)$ by comparing the graphs of $y = \cos x$ and the rule(s) you came up with to the given diagram. Answer the following questions to develop explanations for why the rule you came up with works and to find a strategy for finding the rule for $f(x)$ in this and similar cases.

a. When one group of students in a Sitka, Alaska class started work on this problem, they noticed that $f(x)$ seems to track through the same pattern of values as $y = \cos x$, but twice as fast. They wondered how to make a rule for $f(x)$ so that when x increases from 0 to π, the function $f(x)$ actually operates on inputs from 0 to 2π. Can you see a way to do that?

b. Write the geometric transformations $(x, y) \rightarrow (?, ?)$ that map:

   • the graph of $y = \cos x$ onto the graph of $f(x)$.

   • the graph of $f(x)$ onto the graph of $y = \cos x$.

   How do these answers suggest a way to modify the rule for $y = \cos x$ to get a rule for $f(x)$?

c. Examine how graphs of the functions below are related to the graph of $y = \cos x$ for various values of k.

   • $y = \cos x + k$ • $y = \cos (x + k)$ • $y = k \cos x$

   What do the patterns in these cases suggest about developing rules for functions whose graphs are related by horizontal stretching or shrinking?

d. What do your answers to Parts a–c suggest as a way to modify the function $y = \sin x$ so that its graph is the sine graph stretched or compressed horizontally by a scale factor 0.5? By other scale factors?

2 Find the rules, amplitude, and period of functions whose graphs are related to $y = \cos x$ by:

a. horizontal stretching with scale factor 2.

b. horizontal compressing with scale factor $\frac{1}{3}$.

c. horizontal stretching with scale factor 3.

d. horizontal compressing with scale factor $\frac{2}{3}$ and vertical stretching with scale factor 5.

3 Finding a rule for the function $d(t)$ giving displacement of the clock pendulum at any time t requires modification of the basic cosine function so that the new function is periodic with amplitude 15 and period 2 seconds. Use ideas from your work on Problems 1 and 2 to find the required rule. Compare your rule with those of others and resolve any differences.

Modeling Tidal Patterns The graph representing pendulum motion is similar to graphs of many other physical phenomena. For example, if you live in or visit an ocean seaport city or beach resort town, you will have a chance to observe the tidal patterns of change in water depth that are caused by the gravitational pull of the Moon. A 24-hour graph of water depth at a point near the shore of the harbor in Boston, New York, Miami, San Diego, San Francisco, or Seattle might produce a graph like that in the next diagram.

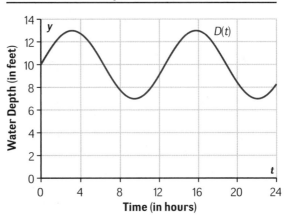

Harbor Water Depth Over Time

4 Suppose that this tidal pattern has period $4\pi \approx 12.57$ hours. What function $D(t)$ will model the pattern of change in water depth over time? You might find it helpful to think about the following questions as you build a function rule.

- What variation on $y = \sin t$ has amplitude 3 feet?

- What variation on the result in Part a will have period $4\pi \approx 12.57$ hours?

- What variation on the result in Part b will have range [7, 13] feet?

Modeling Wave Motion Transmission of energy in the form of light, electrical current, sound, radio and television signals, radar, or earthquake vibrations occurs in patterns that scientists call *wave motion*. This name makes sense because when the intensity of light, sound, or earthquake vibration is measured and graphed over time, the result will be a sinusoidal pattern that resembles a profile of water waves.

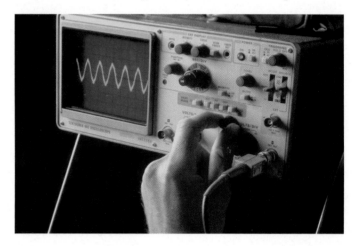

For example, if you use an oscilloscope to monitor voltage in a standard U.S. household electrical circuit, the display will look something like the graph below. The amplitude of this periodic function is 150 volts and its period is $\frac{1}{60} \approx 0.0167$ seconds. Another common way to describe the period is to say that the current alternates with *frequency* of 60 cycles per second.

Alternating Current in an Electrical Circuit

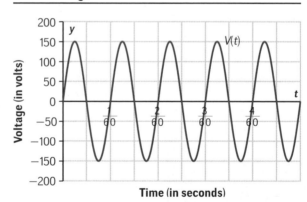

5 Find a rule for the function $V(t)$ giving voltage in a household electrical circuit at any time t by modifying the basic sine function. You might think about the following questions as you build a function rule.

 a. What variation on $y = \sin t$ has amplitude 150 volts?

 b. What variation on the result in Part a will have period $\frac{1}{60} \approx 0.0167$ seconds?

Properties of Horizontal Stretching and Compressing In previous investigations, you have seen that when the graphs of two functions are related by a horizontal or vertical translation or by vertical stretching or compressing, it is often possible to infer locations of maximum and minimum points, zeroes, and the y-intercept of one function from those of the other. In Problems 6 and 7, you will investigate if similar inferences can be made when two functions have graphs related by horizontal stretching or compressing.

6 Consider the cubic polynomial function $f(x) = x^3 + x^2 - 9x - 9$, with graph shown here.

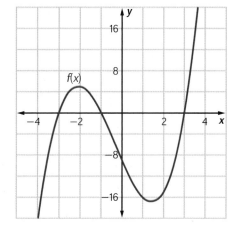

a. Use the graph to estimate:

   i. coordinates for the local maximum and minimum points on the graph of $f(x)$.

   ii. the zeroes of $f(x)$.

   iii. coordinates for the y-intercept on the graph of $f(x)$.

b. Now consider the function $g(x) = f(2x)$. Use the relationship between $g(x)$ and $f(x)$ to estimate each of the following features of $g(x)$ and check your ideas using technology.

   i. Coordinates for the local maximum and minimum points on the graph of $g(x)$

   ii. The zeroes of $g(x)$

   iii. Coordinates for the y-intercept on the graph of $g(x)$

c. Next consider the function $h(x) = f(0.5x)$. Use the relationship between $h(x)$ and $f(x)$ to estimate each of the following features of $h(x)$ and check your ideas using technology.

   i. Coordinates for the local maximum and minimum points on the graph of $h(x)$

   ii. The zeroes of $h(x)$

   iii. Coordinates for the y-intercept on the graph of $h(x)$

7 Suppose $c(x)$ and $d(x)$ are two functions such that $d(x) = c(kx)$ for some positive number k. Use results of your work in Problem 6 and your earlier exploration of horizontal stretching and compressing of sine and cosine functions to answer these questions.

a. How will coordinates of the local maximum and minimum points on graphs of $c(x)$ and $d(x)$ be related?

b. How will the zeroes of the two functions be related?

c. How will coordinates of the y-intercepts of the two functions be related?

Be prepared to explain why your answers make sense based on the effects of horizontal stretching and compressing of the function graphs.

In this investigation, you explored connections between rules of functions whose graphs are related by horizontal stretching or compressing.

a How will the graph of $g(x) = \cos kx$ be related to the graph of $y = \cos x$ when $k > 1$? When $0 < k < 1$?

b What is the coordinate rule $(x, y) \rightarrow (?, ?)$ that will map the graph of $f(x)$ onto the graph of $g(x) = kf(x)$ by horizontal stretching or compressing with a scale factor of k?

c What do the values of a, b, and c tell about the relationship of $y = \cos x$ to $f(x) = a \cos bx + c$?

d Suppose $f(x)$ and $g(x)$ are functions such that $g(x) = f(kx)$ for all x. How are the local maximum and minimum points, zeroes, and y-intercepts of $g(x)$ related to those of $f(x)$?

Be prepared to explain your ideas to the class.

 CHECK YOUR UNDERSTANDING

For each function in Parts a and b, without using technology:

- sketch a graph of the function and the specified stretching or compressing of that graph.

- give a rule for the function whose graph is the image after stretching or compressing of the original graph.

a. $p(x) = -\cos x$ graphed on $[-4\pi, 4\pi]$ and then stretched by a factor of 1.5 away from the y-axis

b. $h(x) = -\sin x$ graphed on $[-2\pi, 2\pi]$ and then compressed by a factor of 0.5 toward the y-axis

APPLICATIONS

1. In each part, find a rule for the function $g(x)$ whose graph is the image of the graph of $f(x)$ under the indicated transformation.

   a. $f(x) = 3x^2 + 4$; transformation $(x, y) \rightarrow (x, 0.5y)$

   b. $f(x) = 3 \cos x$; transformation $(x, y) \rightarrow (x, 2.5y)$

   c. $f(x) = 5 \sin x$; transformation $(x, y) \rightarrow (x, 4y + 1)$

   d. $f(x) = |x|$; transformation $(x, y) \rightarrow (x + 2, 0.5y)$

2. In each part, find the transformation $(x, y) \rightarrow (?, ?)$ that maps the graph of $f(x)$ onto the graph of $g(x)$.

   a. $f(x) = 3x^2 + 4$ and $g(x) = 9x^2 + 12$

   b. $f(x) = 3x^2 + 4$ and $g(x) = 9x^2 + 15$

   c. $f(x) = \sin x$ and $g(x) = 4 \sin(x - 3)$

   d. $f(x) = |x|$ and $g(x) = 7 - 2|x + 4|$

3. Suppose $h(x)$ and $j(x)$ are two functions such that $h(x) = 7j(x)$ for all values of x.

   a. If $j(x)$ has a local maximum point at $(-3, 2)$ and a local minimum point at $(5, -3)$, what can be said about local maximum and minimum points of $h(x)$?

   b. If $h(x)$ has y-intercept $(0, -2)$, what can be said about the y-intercept of $j(x)$?

   c. If $j(x)$ has zeroes -5, -1, and 8, what can be said about the zeroes of $h(x)$?

4. When distance is measured in meters and time in seconds, the function $d(t) = 4.9t^2$ tells the approximate distance fallen in t seconds by an object dropped from a high place anywhere on Earth.

   a. Gravitational force near the surface of the Moon is one-sixth that near the surface of Earth. What distance function $d_M(t)$ is implied for falling objects on the Moon?

   b. If an object is dropped from a point 50 meters above the surface of the Moon, what function tells its altitude at any time t after it is dropped?

   c. When will the dropped object described in Part b reach the Moon's surface?

5 If the temperature in a restaurant's pizza oven is set at 500°F, it will actually vary above and below that setting as time passes. Suppose that the actual temperature is a function of time in minutes with rule in the form $D(t) = a \sin t + b$.

  a. What are the values of a and b if the oven temperature varies from 490°F to 510°F?

  b. What are the values of a and b if the oven temperature varies from 500°F to 510°F?

6 On a hilltop overlooking the city of Prague in the Czech Republic, there is a giant *metronome* that swings back and forth during daylight hours. The metronome was built in 1991 to celebrate freedom from communist rule and Prague's proud musical heritage.

  Suppose that the metronome is set to swing 30° either side of vertical, making one complete swing every $6\pi \approx 18.85$ seconds.

  a. What function tells the angle from vertical of the metronome at any time t seconds after it passes the vertical position?

  b. What function tells the angle from vertical of the metronome at any time t seconds after it leaves its extreme left position?

7 The Santa Monica Pier in California reaches out into the Pacific Ocean. There are restaurants, arcade games, and amusement park rides on the pier, making it a popular spot for weekend entertainment.

  Suppose that the water halfway out along the pier is 20 feet deep at high tide and that water depth varies 4 feet from high to low tide.

  a. The depth of water (in feet) at that point off the Santa Monica Pier varies according to a function in the form $d(t) = a \cos 0.5t + b$, where t represents time (in hours) after high tide. What values of a and b will give $d(t)$ that matches the given information about water depth and tide range?

b. Suppose that at the end of the pier, typical water depth is 25 feet at high tide, but the range from high to low is the same 4 feet. What function $d_1(t)$ gives the typical depth of water at the end of the Santa Monica Pier at any time t hours after high tide?

c. Tidal variation tends to be greater as one moves away from the equator, but the time from high to low is the same around the world. Suppose that at the end of a long pier extending into Cook Inlet in Alaska, the water is typically 45 feet deep at high tide, and that the range from high to low tide is 15 feet. What function $d_2(t)$ gives the depth of water at the end of the Cook Inlet pier at any time t hours after high tide?

d. What function $d_3(t)$ would give the depth of water as a function of time since high tide at a point closer to shore along the Cook Inlet pier where the mean high tide depth is only 25 feet?

8 The following diagrams show three periodic relations that are different from the sine and cosine functions but useful in modern electronics. For each graph:

- identify the period and amplitude of the relation $r(t)$.

- sketch and label the graphs of two variations $v(t) = 2r(t)$ and $w(t) = r(2t)$.

- identify the period and amplitude of the two variations.

a. Square Wave

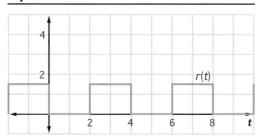

b. Triangle Wave

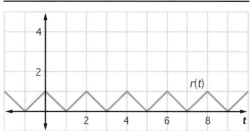

c. Saw Tooth Wave

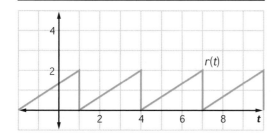

9 Find values of k so that each of the following functions has the indicated period.

a. $f(t) = \cos kt$; period π

b. $g(t) = \cos kt$; period $\frac{\pi}{2}$

c. $h(t) = \cos kt$; period 6π

d. $j(t) = \sin kt$; period 10

e. $p(t) = \cos kt$; period 0.1

f. $s(t) = 4\cos kt + 15$; period $\frac{1}{60}$

10 The London Eye, also called the Millennium Wheel, is one of the largest observation wheels in the world. It stands 135 meters (443 feet) tall in the Jubilee Gardens of London, England. Passengers ride in 32 air-conditioned sealed capsules that are mounted on the wheel's exterior.

It takes about 30 minutes for each complete rotation of the wheel. Assume that the wheel rotates at a constant speed. As you ride in one of the capsules, your position changes in two directions—horizontally and vertically.

a. If you step into a capsule at the bottom of the Millennium Wheel, your distance from the vertical axis of symmetry will be given at any time t (in minutes) by a function $d(t) = a \sin kt$.

    i. What value of k will guarantee that the distance function has period 30 minutes?

    ii. What value of a is implied by the fact that the wheel has diameter 135 meters?

b. If you step into a capsule at the bottom of the wheel, your height above the bottom of the wheel at any time t will be given by a function $h(t) = a(1 - \cos kt)$.

    i. What value of k will guarantee that the height function has period 30 minutes?

    ii. What value of a is implied by the fact that the wheel has diameter 135 meters?

c. Use results of your work on Parts a and b to find the location of your capsule (distance from vertical and height above the bottom of the wheel) at the following times after boarding at the bottom of the wheel. Check each answer by reasoning with principles from geometry and trigonometry and the fact that the capsule turns through an angle of about 12° each minute.

    i. 30 minutes   ii. 10 minutes

    iii. 15 minutes   iv. 22.5 minutes

11 The periodic patterns of change in many physical and biological situations are often monitored by instruments that produce electronic displays like those seen on oscilloscopes. For example, an instrument monitoring pressure created by the beating of your heart might produce a readout that looks like the diagram at the right.

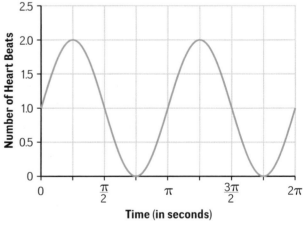

Heart Beat Monitor

a. The graph has period of π seconds.

    i. What values of a, b, and c will produce a model in the form $f(t) = a \sin bt + c$ that matches the graph on the readout shown?

    ii. What pulse rate in beats per minute is indicated by the graph and rule?

b. Suppose that the pulse rate doubled.

    i. How would the displayed graph change?

    ii. How would you modify the rule for $f(t)$?

c. Suppose that the pulse rate slowed to half the rate shown in the graph.

    i. How would the displayed graph change?

    ii. How would you modify the rule for $f(t)$?

12 For a simple pendulum, the time it takes to swing from side to side depends only on the length of the pendulum—not the weight attached or the point from which it is released. Since we usually set a pendulum in motion from its maximum displacement point, it makes sense to model pendulum motion with variations of $y = \cos t$. Displacement is naturally measured by degrees left or right of vertical and time is naturally measured in seconds.

    What variation of $y = \cos t$ will describe motion of a pendulum with maximum displacement 20° on both sides of vertical and period of 10 seconds (i.e., one complete swing from right to left and back again every 10 seconds)?

CONNECTIONS

13 The diagram below shows a graph of $v(x) = |x|$ and two variations $r(x)$ and $s(x)$ of that basic function.

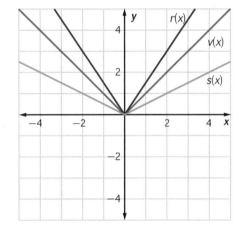

a. What are the rules for $r(x)$ and $s(x)$?

b. What coordinate rule $(x, y) \rightarrow (?, ?)$ will map the graph of $v(x)$ onto the graph of $r(x)$?

c. What coordinate rule $(x, y) \rightarrow (?, ?)$ will map the graph of $v(x)$ onto the graph of $s(x)$?

d. When asked to describe the relationship of the graph of $r(x)$ to those of $v(x)$ and $s(x)$, most people say that the graph of $r(x)$ is "narrower." Why does the visual descriptor "narrower" imply "increasing at a faster rate" as values of x move away from 0 in both directions?

14 Suppose that scores on a 40-question multiple-choice test are recorded as number correct and that the mean number correct for a class is 28, with standard deviation 4.3. Suppose that each student's number correct score is converted to a percent score.

a. Why can that transformation of scores be accomplished by multiplying each score by 2.5?

b. What is the mean percent correct?

c. What is the standard deviation of percent correct scores?

15 Imagine a physics class in which the majority of students failed a unit test. The grading scale was from 0 to 100. Rather than have the students repeat the test, the teacher used a correction factor $f(x) = 10\sqrt{x}$ where x is the original grade and $f(x)$ is the improved grade.

a. What is the improved grade for a student who scored 49 on the original test?

b. Will this correction factor improve the grades of all students? Explain your answer:

   i. using graphs.

   ii. using algebraic reasoning.

c. Will some students gain more points than others? Explain your reasoning.

d. Why do you think the parameter 10 was chosen in defining the correction factor?

16 The diagram below shows a geometric shape on a coordinate grid. Explain how each of the following attributes of that figure will or will not change under the transformations described in Parts a–c.

- perimeter

- area

- sum of angle measurements

a. $(x, y) \rightarrow (x, 3y)$

b. $(x, y) \rightarrow (3x, y)$

c. $(x, y) \rightarrow (3x, 3y)$

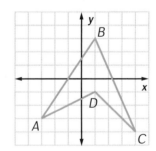

17 For what values of k will the following functions $g(x)$ have graphs that are congruent to the graph of $f(x)$?

a. $g(x) = kf(x)$

b. $g(x) = f(kx)$

c. $g(x) = f(x) + k$

d. $g(x) = f(x + k)$

18 Find matrices $\begin{bmatrix} a & b \\ c & d \end{bmatrix}$ that could be used to find images of points $\begin{bmatrix} x \\ y \end{bmatrix}$ by left multiplication to accomplish the following transformations of the coordinate plane.

a. $(x, y) \rightarrow (x, 3y)$

b. $(x, y) \rightarrow (3x, y)$

c. $(x, y) \rightarrow (3x, 3y)$

d. $(x, y) \rightarrow (2x + y, 5x + 3y)$

REFLECTIONS

19 Marie graphed the function $f(x) = x(x + 2)(x - 3)$ as shown on the left below. Dario graphed $g(x) = 3f(x)$ as shown on the right. Then he looked at Marie's graph and decided since they were the same that his graph was incorrect. Assuming that the graph of $g(x) = 3f(x)$ is correct, what should Marie say to Dario?

Marie's Graph

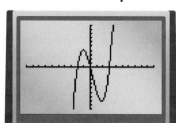

Dario's Graph

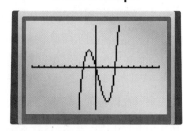

20 As was the case with horizontal translation of graphs, the connection between geometric action and function rule for graphs that are horizontal stretches or compressions of each other does not seem natural. Stretching the graph horizontally means multiplying the independent variable by a number less than 1; compressing the graph horizontally means multiplying by a number greater than 1.

a. How do you think about the problem of matching transformed rules to transformed graphs so that you can get the right connections?

b. Why do you think this task proves to be so challenging?

21 Look back at the voltage modeling problem in Investigation 2 (page 62). Would it be possible to model the voltage in a household electrical circuit at any time t using transformations of the basic cosine function? Explain your reasoning.

22 In previous work, you have used linear (exponential) functions to model patterns of change in which the dependent variable changes by a constant amount (by a constant factor) for each unit change in the independent variable. What connections do you see between that work and the use of translations versus stretches in Lessons 2 and 3?

EXTENSIONS

23 The number of hours between sunrise and sunset at any location on Earth varies in a predictable pattern with a period of 365 days. In the Northern Hemisphere, the date of maximum sunlight hours is on or around June 21 (the Summer solstice) and the date of minimum sunlight hours is on or around December 21 (the Winter solstice). The number of hours of sunlight is midway in the range at the Spring and Fall equinoxes—around March 21 and September 21, respectively.

In Chicago, Illinois, the "longest day" has about 15 hours of sunlight and the "shortest day" has about 9 hours of sunlight.

a. Using the Spring equinox as a starting point, one can model the number of hours of sunlight in Chicago with a function in the form $S_1(t) = a \sin bt + c$, where t is time in days after March 21. What values of a, b, and c will produce a model that meets the conditions described above?

b. What variation $S_2(d)$ of the function from Part a will model the change in sunlight hours in a way that d represents *day of the year* beginning with January 1 as day 1? Here are some hints.

   i. On what day t will your function in Part a reach its maximum value?

   ii. On what day d do you want your new function $S_2(d)$ to reach its maximum value (i.e., what day of the calendar year is June 21)?

   iii. What transformation of the graph of $S_1(t)$ will place its maximum point at the desired day for the new function $S_2(d)$?

   iv. What rule for $S_2(d)$ is implied by the answer to part iii?

24 In many situations involving periodic patterns of change, the defining conditions are given as amplitude and *frequency* rather than period. The **frequency** of a wave motion is the number of complete periods that occur in a given time period. For electronic phenomena, the standard unit of frequency is the *hertz* (Hz) which is one cycle per second.

a. Describe the relationship between the period p and frequency f for a periodic function in each of the following cases.

   i. If the frequency is 30 cycles per second, what is the period?

   ii. If the period for one cycle is 0.2 seconds, what is the frequency in cycles per second?

   iii. If the frequency is f, what is the period p?

   iv. If the period is p, what is the frequency f?

b. What variations of $y = \sin x$ and $y = \cos x$ have period 1 (rather than 2π)?

c. What variations of $y = \sin x$ and $y = \cos x$ have period p (rather than 2π)?

d. What variations of $y = \sin x$ and $y = \cos x$ have frequency f?

25 Use your results from Extensions Task 24 to find variations of $y = \sin x$ that meet these modeling conditions.

a. Frequency 4 and range $[-5, 5]$

b. Frequency 0.1 and range $[-6, 6]$

c. Period 20 and range $[0, 12]$

d. Period 0.05 and range $[-10, -5]$

REVIEW

26 Solve each inequality and represent the solution using interval notation and a number line.

a. $12 - 8x > 3(6 - 2x)$

b. $-12 + 6(x - 5) \le -10 - 2x$

c. $(x - 5)(x + 8) > 0$

d. $|x - 3| \ge 4$

27 For each of the following functions, determine if the function has an inverse. If so, write a function rule for the inverse of the function.

a. $f(x) = 9x + 3$

b. $g(x) = \dfrac{5x - 4}{3}$

c. $h(x) = \dfrac{3}{x} - 7$

28 Find the indicated angle measure or side length(s).

a. Find m$\angle B$.

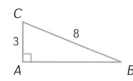

b. Find AC.

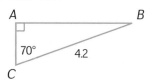

c. Find BC and AC.

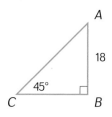

d. Find AB.

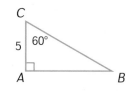

29 The table below indicates the number of minutes Regina spent practicing the piano each day for the last several weeks. Find the mean and median number of minutes she practiced during this time period.

Minutes of Practice	Number of Days
0	5
10	2
20	4
30	6
45	2
60	1

30 Write each sum of rational expressions in equivalent form as a single algebraic fraction. Then simplify the result as much as possible.

a. $\dfrac{x}{3} + \dfrac{2x + 3}{3}$

b. $\dfrac{6x}{5} + \dfrac{x + 1}{2}$

c. $\dfrac{3x + 5}{x} + 2x$

31 Points A, B, and C are on the circle with center O and radius 8 cm. In addition, $m\angle AOC = 100°$ and $AB = BC$. Find each of the following.

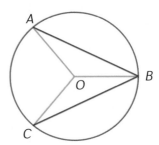

a. $m\widehat{AC}$ b. $m\angle ABC$ c. $m\widehat{AB}$ d. BC

LESSON

Combining Functions

In building models for relationships among variables, you often start with a few basic functions and assemble them into functions with more complex rules. For example,

The height h (in feet) of a tennis lob is likely to change over time t (in seconds) according to a function like
$$h(t) = -16t^2 + 32t + 8.$$

The profit per sale p (in dollars) of a popular song at an Internet music site might depend on the price per copy x (in dollars) according to a function like
$$p(x) = \frac{-2{,}000x^2 + 11{,}000x - 9{,}000}{10{,}000 - 2{,}000x}.$$

The height above ground h (in meters) of a rider on a large Ferris wheel is likely to change over time t (in minutes) according to a function like $h(t) = 50 \sin 0.2t + 60$.

The temperature T (in °F) of a bottle of water placed in an ice chest cools over time t (in minutes) according to a function like $T(t) = 45(0.9^t) + 35$.

In this lesson, you will study properties of functions that result from combining simpler functions by arithmetic operations of addition, subtraction, multiplication, and division and by a new operation called *function composition*.

INVESTIGATION 1

Arithmetic with Functions

The most common way that function models are built by combining simple pieces is using the arithmetic operations of addition, subtraction, multiplication, and division. As you work on the problems of this investigation, look for answers to these questions:

> *When do the sum, difference, product, and quotient of two given functions provide useful models for patterns of change in other variables?*
>
> *How can you determine properties of functions that have been constructed from two or more simpler functions by arithmetic operations?*

1 For each of the following situations, describe an arithmetic combination of the given functions that will model the pattern of change in the indicated variable. Be prepared to explain your reasoning.

a. If $I(x)$ gives income from Internet sales of x copies of a song and $c(x)$ gives cost of providing those downloads, what function $P(x)$ will give profit to the music sales company?

b. If $p(x)$ gives the profit per sale of a movie sold by Internet download when price is x dollars and $s(x)$ gives the number of sales at that price, what function $T_p(x)$ will give total profit from sales of the movie?

Photodisc/SuperStock

2 Data sensors in modern automobiles measure gallons of gasoline in the car's fuel tank $f(t)$ and miles traveled $d(t)$ at any time t.

a. What will the following functions tell about a car's fuel economy over any one-minute period?

   i. $d_1(t) = d(t + 1) - d(t)$

   ii. $f_1(t) = f(t + 1) - f(t)$

   iii. $E(t) = \dfrac{d_1(t)}{f_1(t)}$

b. If the function $f(t)$ gives gallons of gasoline in a car's fuel tank and the function $E(t)$ gives fuel economy in miles per gallon at any time t, what combination of those functions will predict miles until empty?

3 The examples in Problems 1 and 2 should have suggested ways to define and evaluate the sum, difference, product, and quotient of two functions. If $f(x) = 3x^2 - 2$ and $g(x) = x + 5$, use definitions that make sense to you to find specific values and general rules for these combinations of functions.

a. $[f + g](-2) =$
   $[f + g](x) =$

b. $[f - g](5) =$
   $[f - g](x) =$

c. $[f \cdot g](-4) =$
   $[f \cdot g](x) =$

d. $[f \div g](7) =$
   $[f \div g](x) =$

4 The diagram below shows graphs of two functions f and g. Use the graphs to study the difference of the two functions, $f - g$.

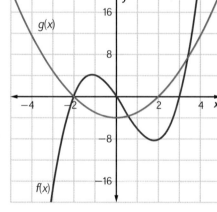

a. Estimate the coordinates of the y-intercept of $f - g$.

b. Estimate the zeroes of $f - g$.

c. On a copy of the grid, sketch a graph of $f(x) - g(x)$.

d. Describe intervals on which:

   i. $f - g > 0$

   ii. $f - g < 0$

5 The rules used to produce the graphs in Problem 4 are $f(x) = x^3 - x^2 - 6x$ and $g(x) = x^2 - 4$. Use those rules, your skill in work with algebraic expressions, and a CAS (where useful) to further study the difference function $f - g$.

a. Find a rule for $f - g$.

b. Find the y-intercept of $f - g$.

c. Find the zeroes of $f - g$.

d. Find the intervals on which:

   i. $f - g > 0$

   ii. $f - g < 0$

e. Estimate coordinates of the local maximum and local minimum points of $f - g$.

f. Check your sketched graph from Problem 4 Part c against that produced using technology.

SUMMARIZE THE MATHEMATICS

In this investigation, you explored the use of sums, differences, products, and quotients of functions to model problem situations.

 a How can $f + g$, $f - g$, $f \cdot g$, and $f \div g$ be evaluated for specific values of the independent variable?

b If you are given rules for functions $f(x)$ and $g(x)$, how can you use those rules to find rules for $[f + g](x)$, $[f - g](x)$, $[f \cdot g](x)$, and $[f \div g](x)$?

c How can the graphs of f and g give information about the y-intercept, zeroes, and local maximum and minimum points of the function $f - g$?

Be prepared to share your methods and reasoning with the class.

✔ CHECK YOUR UNDERSTANDING

Study the diagram that follows showing the graphs of functions $f(x)$ and $g(x)$.

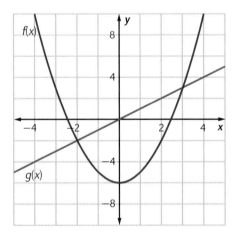

Using information from the graphs:

a. estimate coordinates of the y-intercept of $f - g$.

b. estimate the zeroes of $f - g$.

c. estimate coordinates of any local maximum and minimum points of $f - g$.

d. estimate the values of:

    i. $[f \cdot g](-1)$ **ii.** $[f + g](2)$ **iii.** $[f \cdot g](2.5)$

Composition of Functions

Combining functions by addition, subtraction, multiplication, and division is a natural extension of arithmetic operations with numbers. But there are other modeling tasks in which two or more functions are naturally combined by an operation called *function composition*. As you work on the problems of this investigation, look for answers to these questions:

What does it mean to combine functions by composition?

When and how can rules for given functions be combined to produce rules for composite functions?

How can rules for given functions be analyzed to find simpler functions that are composed to give those rules?

1 Nearly every American state now runs some sort of daily or weekly lottery to raise funds for support of government operations. To make budget plans, those governments need to predict future net income from lottery operations. Suppose that in one state, the key relationships are given by two functions:

- Annual net income or profit from lottery operation (in millions of dollars) depends on the state's adult population *p* (in millions) according to the function:

$$I(p) = 25p - 7.5$$

- State population *t* years from now is predicted by the function:

$$p(t) = 5(1.02^t)$$

a. What conditions in the lottery operation could explain the occurrence of the numbers 25 and -7.5 in the rule for $I(p)$?

b. What do the numbers 5 and 1.02 tell about the pattern of change in state population that is expressed by the rule for $p(t)$?

c. What are the predicted state population and the lottery net income at a point 5 years in the future?

d. How would you use the rules for $p(t)$ and $I(p)$ to calculate the predicted annual lottery net income at any time *t* in the future?

e. What single algebraic rule can be used to calculate the predicted annual lottery net income at any point *t* years in the future? That is, how can net income be expressed as a function of *t*?

The problem of predicting future annual lottery income is an example of the kind of situation in which it is helpful to combine two functions by the operation of **function composition**. Finding predicted net income at any time t is a two-step process. First, the function $p(t)$ is used to find the predicted adult population t years in the future; then the function $I(p)$ is used to find the predicted annual lottery income at that time. The *composite* of those two functions is expressed using the notation $I(t) = I(p(t))$. In general,

For any given functions f and g, values of the composite function $f \circ g$ are calculated by finding $f(g(x))$.

The notation $f \circ g$ indicates "function f following function g." The notation $f(g(x))$ is read as "f of g of x." To form the composite of two functions, the range of the first function (in this case g) must fall within the domain of the second function (in this case f).

2 When retail businesses rent space in shopping centers, their monthly rental fee often depends, at least in part, on the income earned by the business. For example, if you owned a barber or beauty shop, the monthly rent for your space might be given by a function like

$$R(I) = 0.02I + 500$$

where I is your income in dollars. Your income depends, in turn, on the number of customers n who come to your shop in a month, perhaps according to a function like

$$I(n) = -0.0125n^2 + 25n.$$

a. What is the rent for your shop in a month when you have:

  i. 750 customers?

  ii. 1,000 customers?

  iii. 1,500 customers?

b. What is the rule, in simplest form, of the composite function $R(n) = R(I(n))$?

Properties of Composite Functions Since composition of functions occurs often and in many different contexts, it is helpful to develop ability in working with that operation without depending on the guidance of clues from problem situations. Rules for functions are most commonly expressed in terms of the generic variable x, so forming rules for composite functions requires some care.

For example, if $f(x) = 3.5x$ and $g(x) = 5 - x$, the rule for the composite $f \circ g$ is calculated by finding $f(g(x)) = 3.5(5 - x)$. The rule for f operates on the result from g operating on an input value x.

3 Find specific outputs and algebraic rules for composite functions in Parts a–e. Express each composite function rule in simplest algebraic form.

a. If $f(x) = 3x + 5$ and $g(x) = 4 - 2x$, find:

   i. $f(g(7))$　　　　　　　　　ii. $g(f(7))$

   iii. $f(g(-3))$　　　　　　　iv. $g(f(-3))$

   v. a rule for $f(g(x))$　　　vi. a rule for $g(f(x))$

b. If $f(x) = x^2$ and $g(x) = 4x - 3$, find:

   i. $f(g(5))$　　　　　　　　　ii. $g(f(5))$

   iii. $f(g(-2))$　　　　　　　iv. $g(f(-2))$

   v. a rule for $f(g(x))$　　　vi. a rule for $g(f(x))$

c. If $h(n) = n^2 + 4n + 3$ and $j(n) = 2n + 1$, find rules for:

   i. $h(j(n))$　　　　　　　　　ii. $j(h(n))$

d. If $r(x) = \dfrac{1}{x}$ and $s(x) = 2x + 3$, find rules for:

   i. $r(s(x))$　　　　　　　　　ii. $s(r(x))$

e. If $s(t) = t^2$ and $r(t) = \sqrt{t}$, find rules for:

   i. $s(r(t))$　　　　　　　　　ii. $r(s(t))$

f. Based on the results of your work in Parts a–e, would you say that function composition is or is not a commutative operation? That is, will $f \circ g = g \circ f$ in all cases? Explain your reasoning.

4 Sometimes it is helpful to decompose a given function into the composite of two simpler functions. For example, suppose you are planning an after-prom party at a local club.

You are provided this information about club costs.

• There is a charge for rental of the club party room.

• There is a charge per person for food and drinks.

• There is a service charge that is a percent of the total bill for room rent, food, and drinks.

Suppose also that you were not told the details of those separate charges, but only that the function $C(n) = 1.18(35n + 400)$ gives total cost of the event as a function of the number of people attending.

a. Looking at the rule for $C(n)$ what would you suspect to be:

    i. the cost for party room rental?

    ii. the cost per person for food and drinks?

    iii. the percent rate of the service charge?

b. What function $P(n)$ probably gives party costs for room rental, food, and drinks for n guests (before the service charge is applied)?

c. What function $C(P)$ probably shows how to calculate total party cost, including service charge, as a function of P, the bill for room rental, food, and drinks?

d. What is the rule for $C(n) = C(P(n))$?

5 For each of the following functions $f(x)$, give rules for two functions $g(x)$ and $h(x)$ so that $f(x) = g(h(x))$. Do not use the simple choices such as $h(x) = x$ and $g(x) = f(x)$. There are several possible combinations of functions in each case.

a. $f(x) = 7x - 3$ **b.** $f(x) = (4x + 9)^2$

c. $f(x) = \dfrac{3}{x - 5}$ **d.** $f(x) = \sqrt{3x - 5}$

6 Find simplest possible algebraic rules for each of the composite functions in Parts a–d. State any restrictions on the domains of the composite functions.

a. If $f(x) = x + 5$ and $g(x) = x - 5$, what are the rules for $f(g(x))$ and $g(f(x))$?

b. If $f(x) = 2x + 5$ and $g(x) = 0.5x - 2.5$, what are the rules for $f(g(x))$ and $g(f(x))$?

c. If $s(x) = (x - 3)^3$ and $r(x) = \sqrt[3]{x} + 3$, what are the rules for $s(r(x))$ and $r(s(x))$?

d. If $f(x) = \dfrac{1}{x} + 5$ and $g(x) = \dfrac{1}{x - 5}$, what are the rules for $f(g(x))$ and $g(f(x))$?

e. What pattern occurs in each of the composite functions calculated in Parts a–d, and what does that pattern say about the special relationship between the pairs of functions involved?

Applying Function Composition Composition of functions gives a useful way of thinking about some problems that you have met before in work on linear and exponential functions.

7 One of the standard ways of describing exponential growth is to find **doubling time**—the constant interval of time that it takes a population or a financial investment to increase to twice its starting value. Suppose, for example, that an experimental bacteria population starts at 25 and grows exponentially with doubling time 20 minutes.

a. What bacteria counts would you expect after 20, 40, 60, and 80 minutes?

b. The bacteria do not wait until the end of each 20-minute period to suddenly all split in two at once. Use the population estimates in Part a to find the rule of an exponential function $p_1(t)$ that predicts the population at any time t minutes after the start of the experiment.

c. Compare tables of values for the modeling function derived in Part b to those of the function $p_2(t) = 25\left(2^{\frac{t}{20}}\right)$ and describe similarities and differences of the two functions.

d. Explain how properties of fractions and exponents justify each step in the following algebraic reasoning and how the result explains similarity of values for $p_1(t)$ and $p_2(t)$.

$$
\begin{aligned}
p_2(t) &= 25\left(2^{\frac{t}{20}}\right) &&(1)\\
&= 25(2^{0.05t}) &&(2)\\
&= 25(2^{0.05})^t &&(3)\\
&\approx 25(1.035)^t &&(4)
\end{aligned}
$$

e. What separate simpler functions could be composed to produce the rule $p_2(t) = 25\left(2^{\frac{t}{20}}\right)$?

SUMMARIZE THE MATHEMATICS

In this investigation, you explored strategies and uses for the operation of function composition.

a If you are given rules for two functions $f(x)$ and $g(x)$, how can you use those rules to find the value of $f \circ g$ for specific values of x?

b If you are given rules for functions $f(x)$ and $g(x)$, how can you use those rules to find a rule for $f(g(x))$?

c What special relationship between functions f and g is indicated when $f(g(x)) = g(f(x)) = x$?

d Is composition of functions a commutative operation? Give a justification or a counterexample.

Be prepared to share your thinking with the class.

 CHECK YOUR UNDERSTANDING

Think about the meaning of function composition and its relation to function inverses as you complete the following tasks. Suppose that $f(x) = 3x + 5$ and $g(x) = x^2 + 1$.

a. Find rules for $f \circ g$ and $g \circ f$.

b. If possible, find the rules for functions $h(x)$ and $j(x)$ so that:

    i. $h \circ f = x$ for all x.   **ii.** $j \circ g = x$ for all x.

c. If a landlord owns an apartment building worth 2.5 million dollars and expects the value of that property to increase exponentially with doubling time 10 years, what function will predict the value of the investment at any time t years in the future?

APPLICATIONS

1 Consider the functions $f(x) = 4x + 3$ and $g(x) = \dfrac{8}{x + 2}$. Evaluate the following combinations of those functions.

    a. $[f + g](3)$ **b.** $[f - g](3)$

    c. $[f \cdot g](3)$ **d.** $[f \div g](3)$

2 The diagram below shows graphs of two functions h and j. Use information from the graphs to study the arithmetic combinations of the functions listed in Parts a–e.

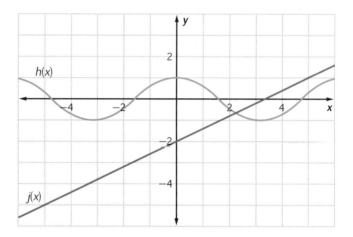

    a. Estimate coordinates of the y-intercept of $h - j$.

    b. Estimate coordinates of the zeroes of $h - j$.

    c. Estimate the value of $[h \cdot j](0)$.

    d. Estimate the value of $[h \div j](0)$.

    e. Estimate the value of $[h + j](0)$.

3 The rules used to produce the graphs in Applications Task 2 are $h(x) = \cos x$ and $j(x) = 0.6x - 2$. Use those rules, your skill in work with trigonometric and algebraic expressions, and a CAS as needed to find more accurate estimates or exact answers (where possible) for that task.

4 Major League Lacrosse seasons run from April through August. Each team has income and expenses related to ticket sales, concession sales, and parking fees.

    a. If the average ticket price is $15 for a Chesapeake Bayhawks game, what function $T(x)$ shows how the income from ticket sales depends on the attendance at each game?

    b. If net income from concession sales averages $5 per game customer, what function $C(x)$ shows how net income from that source depends on the number of ticket sales?

c. If all game customers arrive by car, with an average of 4 people per car, and the fee for parking is $5, what function $P(x)$ shows how income from that source depends on the number of ticket sales?

d. What function $I(x)$ shows how total net income for a game depends on the number of ticket sales?

e. Suppose that market research suggests that the number of tickets sold x depends on the average ticket price p according to the rule $x = 10,000 - 200p$. Find rules that express:

    i. income from ticket sales as a function of ticket price.

    ii. net income from concession sales as a function of ticket price.

    iii. income from parking fees as a function of ticket price.

    iv. total net income as a function of ticket price.

    v. net income per ticket sale as a function of ticket price.

5 If $f(x) = 2x^2$ and $g(x) = 5 - x$, find:

a. $f(g(2))$ **b.** $g(f(1))$

c. $f(g(-3))$ **d.** $g(f(-1))$

e. a rule for $f(g(x))$ **f.** a rule for $g(f(x))$

6 For each of the following functions $f(x)$, give rules for two functions $g(x)$ and $h(x)$ so that $f(x) = g(h(x))$.

a. $f(x) = 3x + 5$ **b.** $f(x) = (x - 9)^2$

c. $f(x) = \cos(2x + 3)$ **d.** $f(x) = \dfrac{1}{3x + 2}$

7 When you eat in a restaurant that has table service, the menu prices do not usually include gratuity for the server or sales tax on the food and drink.

a. What is the total cost of a meal with a food and drink charge of $25, gratuity of 20%, and tax of 6%, if the tax is added before the gratuity?

b. What is the total cost of the meal in Part a if the tax is applied to the food and drink *and* gratuity?

c. Does $(1.06)(1.20)(25) = (1.20)(1.06)(25)$? How does the answer explain the results of Parts a and b?

8 In your work on Problem 2 of Investigation 2, you saw that monthly rent for a barber or beauty shop could depend on income according to a function like $R(I) = 0.02I + 500$ and that income could depend on the number of customers according to a function like $I(n) = -0.0125n^2 + 25n$.

a. The number of customers at your shop in a month will depend on the average price p in dollars charged for a haircut. Suppose that the number of customers function is $n(p) = 2,000 - 80p$. Find the number of customers, the shop income, and the monthly rent when the average price charged for a haircut is:

    i. $12   **ii.** $15   **iii.** $20

b. Find rules that express shop income and rent as functions of haircut price. Then explain how each function is the result of composing two or more other functions.

    i. $I(p)$ **ii.** $R(p)$

9 When you receive a shot of medicine like penicillin, it begins decaying exponentially. Suppose that a 200-mg injection of one such medicine has a half-life of 12 hours.

a. What amounts of active medicine would you expect after 12, 24, 36, 48, and 60 hours?

b. What exponential regression model matches the pattern of (*time*, *medicine*) data in Part a?

c. Use what you know about properties of exponents to show that the expression $200\left(0.5^{\frac{t}{12}}\right)$ is equivalent to the expression in the regression equation of Part b.

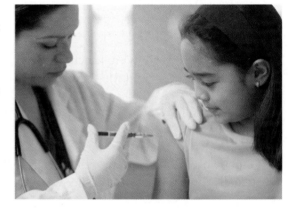

CONNECTIONS

10 A matrix like $M = \begin{bmatrix} 5 & 2 \\ 2 & 1 \end{bmatrix}$ can be used to define a function of two variables (x, y) with rule of assignment $\begin{bmatrix} 5 & 2 \\ 2 & 1 \end{bmatrix}\begin{bmatrix} x \\ y \end{bmatrix} = \begin{bmatrix} 5x + 2y \\ 2x + y \end{bmatrix}$ or $(5x + 2y, 2x + y)$.

Use the particular matrices $M = \begin{bmatrix} 5 & 2 \\ 2 & 1 \end{bmatrix}$ and $N = \begin{bmatrix} 4 & 1 \\ 7 & 2 \end{bmatrix}$ to explore properties of arithmetic operations with such matrix functions.

a. Does $(M + N)\begin{bmatrix} x \\ y \end{bmatrix} = M\begin{bmatrix} x \\ y \end{bmatrix} + N\begin{bmatrix} x \\ y \end{bmatrix}$? If so, will that relationship hold for any pair of 2 × 2 matrices M and N? If not, why not?

b. Does $(M - N)\begin{bmatrix} x \\ y \end{bmatrix} = M\begin{bmatrix} x \\ y \end{bmatrix} - N\begin{bmatrix} x \\ y \end{bmatrix}$? If so, will that relationship hold for any pair of 2×2 matrices M and N? If not, why not?

c. Does $(M \cdot N)\begin{bmatrix} x \\ y \end{bmatrix} = M\begin{bmatrix} x \\ y \end{bmatrix} \cdot N\begin{bmatrix} x \\ y \end{bmatrix}$? If so, will that relationship hold for any pair of 2×2 matrices M and N? If not, why not?

11 Determine which of the following properties hold for operations with 2×2 matrices M, N, and P *and* for the comparable operations for functions f, g, and h.

  a. Commutative Property of Addition:

    i. $M + N = N + M$

    ii. $f + g = g + f$

  b. Commutative Property of Multiplication:

    i. $M \cdot N = N \cdot M$

    ii. $f \cdot g = g \cdot f$

  c. Distributive Property of Multiplication over Addition:

    i. $M(N + P) = MN + MP$

    ii. $f \cdot (g + h) = f \cdot g + f \cdot h$

12 The diagram below shows the graphs of $c(x) = \cos x$ and $s(x) = \sin x$ over the interval $[-2\pi, 2\pi]$. Use the graphs to think about the quotient function $f(x) = \dfrac{s(x)}{c(x)}$.

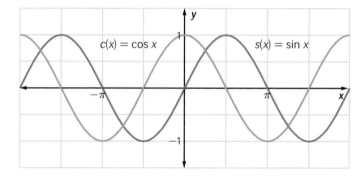

  a. What is the domain of $f(x)$?

  b. What is the y-intercept of $f(x)$?

  c. What are the zeroes of $f(x)$?

  d. What are the asymptotes, if any, of $f(x)$?

  e. Produce a graph of $f(x)$. What familiar function is $f(x)$?

13 In the Course 3 *Inverse Functions* unit, you learned that if functions $f(x)$ and $g(x)$ are inverses of each other and $f(a) = b$, then $g(b) = a$. In Problem 6 on page 82, you may have noticed, in a few cases, for inverse functions $f(x)$ and $g(x)$, that $f(g(x)) = x$ and $g(f(x)) = x$. By definition, functions f and g are **inverses** of each other if and only if $f(g(x)) = x$ for every x in the domain of g and $g(f(x)) = x$ for every x in the domain of f. The convention in mathematics is to write the inverse of $f(x)$ as $f^{-1}(x)$.

a. For each of the functions below, find the inverse function. Then verify by composition that $f(f^{-1}(x)) = x$ and $f^{-1}(f(x)) = x$.

i.

x	−4	−1	2	3
y	3	−5	4	2

ii. $f(x) = x - 3$

b. One strategy for finding an inverse function rule is to *swap roles of y and x* as outlined below. Study the strategy and explain why it makes sense in terms of the definition of inverse functions.

$$\text{Suppose} \quad f(x) = \frac{4}{x - 3}.$$

$$\text{Let} \quad y = \frac{4}{x - 3}.$$

Swap the roles of y and x to get

$$x = \frac{4}{y - 3}.$$

$$\text{Then} \quad y - 3 = \frac{4}{x}.$$

$$\text{So,} \quad f^{-1}(x) = \frac{4}{x} + 3.$$

c. Use the strategy outlined in Part b to find the rule for the inverse of $f(x) = x^3 + 2$. Then verify by function composition that you have found the inverse function for $f(x)$.

d. Explain why you need to restrict the domain to find an inverse for $f(x) = x^2$, but not for $f(x) = x^3$.

14 Suppose that when a single function f is composed with itself many times, the number of iterations is indicated by an exponent. For example, $f \circ f = f^2$, $f \circ f \circ f = f^3$, and in general f^n means a composite function with f used n times.

a. If $f(x) = 3x$, find these values.

   i. $f^2(4)$ **ii.** $f^3(-2)$ **iii.** $f^5(4)$

b. If $g(x) = x + 5$, find these values.

   i. $g^2(4)$ **ii.** $g^3(-2)$ **iii.** $g^5(4)$

c. What simple algebraic rules give results of:

   i. $f^n(a)$ **ii.** $g^n(a)$

d. What recursive formulas of the form "$y_n = \ldots$" can be used to produce the same results as the function calculations indicated in each item of Part c?

15 In Parts a–d, describe in words the composite of geometric transformations that will map the graph of $f(x)$ onto the graph of $g(x)$.

  a. $f(x) = x^2$ and $g(x) = -5x^2$

  b. $f(x) = \cos x$ and $g(x) = -\cos x + 3$

  c. $f(x) = x^2$ and $g(x) = (x + 7)^2 - 3$

  d. $f(x) = \cos x$ and $g(x) = 10 \cos 0.5x$

REFLECTIONS

16 In making plans for a benefit concert, organizers expect to earn \$45 from each ticket sale, \$10 from each sale of a souvenir poster or T-shirt, and \$15 from parking of each car. One person on the planning committee said, "This means the total income will be given by the function $I(x) = 45x + 10x + 15x$, or $I(x) = 70x$. In other words, the concert will earn \$70 per customer.

  a. Do you agree? Why or why not?

  b. What does your answer tell you about the uses and limitations of combining functions by arithmetic operations of addition, subtraction, multiplication, and division?

17 The notation for composition of functions "$f(g(x))$" looks a bit like that for multiplication. But composition of functions is *not* in general commutative while multiplication of functions is.

  a. How could you use the two functions $f(x) = x^2$ and $g(x) = x + 3$ to demonstrate that multiplication of functions is commutative but composition of functions is not commutative?

  b. How would you respond to someone who points out these cases where composition of functions *is* commutative?

  Case I $f(x) = x - 5$ and $g(x) = x + 3$

  Case II $f(x) = 5x$ and $g(x) = 3x$

18 What features of a problem situation give you a hint that the relationships among variables can be modeled by composition of functions?

19 You might have noticed that composition of functions is similar in some ways to your work in Course 3 with recursion. Write rules in the form $f(n)$ for functions that are being composed with themselves in these recursive formulas.

  a. $y_n = y_{n-1} + 7.5$, starting at 4

  b. $y_n = 7.5y_{n-1}$, starting at 4

20 Consider again the function $g(x) = x + 3$ in Reflections Task 17.

a. What is $g^{-1}(-1)$? What is $g^{-1}(5)$? What is $g^{-1}(0)$?

b. Write a rule for $g^{-1}(x)$.

c. Show that $g \circ g^{-1}$ and $g^{-1} \circ g$ are each the **identity function** $i(x) = x$.

EXTENSIONS

21 Quadratic functions with rules in the form $f(x) = ax^2 + bx + c$ can be written as the sum of two functions $g(x) = ax^2$ and $h(x) = bx + c$. Thinking about quadratic functions in this way reveals interesting properties of the way the two simple components combine to produce the family of all quadratics.

a. Graph $f(x) = ax^2 + bx + c$ and $h(x) = bx + c$ for several different combinations of values for the parameters a, b, and c. Describe the relationship between the two graphs that occurs in every case. (If you have a calculator or computer software that will allow entry of generic function rules and slider manipulation of parameters, that tool will be very useful in this investigation.)

b. Use the fact that $ax^2 + bx + c = (ax^2) + (bx + c)$ to explain the pattern you discovered in Part a.

22 In earlier geometry study, you explored the role of matrices in defining transformations of points on the coordinate plane. You saw that matrices like $M = \begin{bmatrix} 5 & 2 \\ 2 & 1 \end{bmatrix}$ and $N = \begin{bmatrix} 4 & 1 \\ 7 & 2 \end{bmatrix}$ can be used to define geometric transformations that map any point (x, y) to the points with coordinates determined by

$$\begin{bmatrix} 5 & 2 \\ 2 & 1 \end{bmatrix}\begin{bmatrix} x \\ y \end{bmatrix} = \begin{bmatrix} 5x + 2y \\ 2x + y \end{bmatrix} \text{ and } \begin{bmatrix} 4 & 1 \\ 7 & 2 \end{bmatrix}\begin{bmatrix} x \\ y \end{bmatrix} = \begin{bmatrix} 4x + y \\ 7x + 2y \end{bmatrix}.$$

a. Find the images of the following points under the composite transformation $M \circ N$; that is, M *following* N.

   i. $(2, 5)$

   ii. $(-2, 3)$

   iii. $(4, -6)$

b. Find the product matrix MN and apply it to each of the points in Part a.

c. Compare the results of the matrix operations in Parts a and b and see if you can prove that the same thing will happen when matrices M and N are applied to any point (x, y).

d. Now see if you can prove that the observation in Part c will occur for any two matrices $P = \begin{bmatrix} a & b \\ c & d \end{bmatrix}$ and $Q = \begin{bmatrix} e & f \\ g & h \end{bmatrix}$ and any point (x, y).

23 When two functions are combined by the operation of composition, it is important to be careful about the domains and ranges of the two functions involved.

   a. Suppose that $f(x) = \sqrt{x}$ and $g(x) = 3x - 4$ and you are asked to evaluate the composite function $f(g(0))$.

   i. What is the complication that occurs in trying to find that composite function value?

   ii. For what value(s) of x is it possible to evaluate $f(g(x))$?

   b. If $f(x) = \dfrac{3}{x-4}$ and $g(x) = 2x$, there is at least one value of x for which the composite function $f(g(x))$ cannot be evaluated.

   i. What is that value of x?

   ii. For what value(s) of x is it impossible to evaluate $g(f(x))$?

   c. If $f(x) = |x|$ and $g(x) = \sqrt{x}$, what are the domain and range of:

   i. $f(g(x))$

   ii. $g(f(x))$

   d. Based on results of your work in Parts a, b, and c, how would you complete the sentence that begins, "For any functions $f(x)$ and $g(x)$, the domain of the composite function $f(g(x))$ consists of all x for which …"?

REVIEW

24 Suppose that a softball is hit from a height of 1.25 meters with an initial upward velocity of 25 meters per second. Recall that the function rule for the height of a ball is $h(t) = -4.9t^2 + v_0 t + h_0$, where v_0 is the initial upward velocity in meters per second and h_0 is the initial height in meters.

   a. Write a function rule that will give the height of the softball after any number of seconds.

   b. How high will the ball be after 2 seconds?

   c. At what times will the ball be more than 20 meters above the ground?

   d. If the ball is not caught, when will it hit the ground?

25 Without using technology, decide if each statement is true or false. Then check your answer using technology.

   a. $\log 1{,}000 = 3$

   b. $\log 0.01 = -3$

   c. $13 < \log 1{,}489 < 14$

   d. $-2 < \log 0.78 < -1$

26 Recall that the Law of Cosines and the Law of Sines can be used to find angle measures and side lengths of any triangle. The Law of Cosines and the Law of Sines are reviewed below.

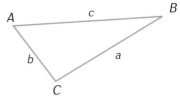

Law of Cosines: $c^2 = a^2 + b^2 - 2ab \cos C$

Law of Sines: $\dfrac{a}{\sin A} = \dfrac{b}{\sin B} = \dfrac{c}{\sin C}$

Use the Law of Cosines or the Law of Sines to find the indicated angle measure or side length.

a. Find AC.

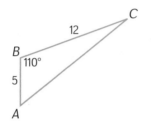

b. Find AB.

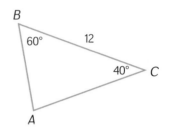

c. Find $m\angle B$.

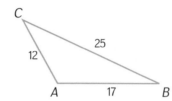

27 Many car rental plans have price rules that charge a fixed minimum fee that allows perhaps 100 miles usage without additional charge. Miles beyond the daily allowance might cost $0.50 each. Similar *piecewise* rules are common with cellular telephone plans.

a. What rule expresses the charging plan with a fixed minimum of $35 per day and $0.50 for each mile beyond 100 per day? Write the rule in two forms—one in piecewise form that shows all three relevant parameters 35, 0.50, and 100 and another that is in standard linear function form $ax + b$.

b. What rule expresses the cell phone plan that charges $75 per month plus $0.65 for each daytime minute beyond 1,000? Write the rule in two forms— one in piecewise form that shows all three relevant parameters 75, 0.65, and 1,000 and another that is in standard linear function form $ax + b$.

c. For each pair of rules you wrote for the cell phone charging plans in Part b, which rule is more efficient for calculations? Which rule better shows how the plan works?

28 Each of the following shows the start of an arithmetic or geometric sequence. Determine the 12th term of each sequence. Then find a formula for the nth term of each sequence.

  a. 12, 21, 30, 39, 48, …

  b. 2,000, 1,000, 500, 250, 125, …

  c. 2, 6, 18, 54, 162, …

  d. 428, 408, 388, 368, 348, …

29 At the beginning of the school year, the 422 seniors at Aztec High School were surveyed and asked about whether they have their own car and if they had a summer job. Some of the results are shown in the chart below.

	Had Summer Job	No Summer Job	Total
Has Own Car			
Does Not Have Own Car	31		92
Total		75	422

  a. Fill in the rest of the chart.

  Suppose one of the seniors at Aztec High School is randomly chosen. Determine each of the following probabilities.

  b. P(*has own car*)

  c. P(*had summer job*)

  d. P(*has own car* and *had summer job*)

  e. P(*has own car* or *had summer job*)

  f. P(*had summer job | has own car*)

30 Consider a circle with center at the origin O and radius 1. Point A on the circle has coordinates $(1, 0)$.

  a. Write an equation for this circle.

  b. Consider point B on the circle so that $m\angle AOB = 60°$. Find the coordinates of point B.

  c. Consider point C on the circle in the second quadrant so that $m\angle AOC = 115°$. Find the coordinates of point C.

Looking Back

In this unit, you revisited a variety of basic types of functions and developed strategies for combining those functions to build models for more complex relationships. As a result of that work, you have developed greater skill in using linear, quadratic, power, exponential, and circular functions to model data patterns and problem conditions. You also have learned how to modify the rules of basic function types to model and analyze data patterns that are related to familiar functions by vertical and horizontal translation, by reflection across the x-axis, and by vertical and horizontal stretching or compressing. Finally, you have increased your skill and understanding of ways to build and analyze functions by the arithmetic operations of addition, subtraction, multiplication, and division and by the new operation of function composition.

The tasks in this final lesson give you a chance to review your skill and understanding of function families, transformations, and function operations.

Thinking in Millennia New Year's Day is celebrated in cultures and countries around the world. But January 1, 2000 was a very special date, because it marked the beginning of a new millennium, or thousand-year time period. The occasion prompted many comparisons with life at the start of the previous millennium in the year 1000. At that time:

- Earth's human population was about 250 million and growing at a rate of 0.1% per year.

- one fourth of the population lived in China, and the world's largest city was Cordoba, Spain with a population of 450,000.

- half of all children died before the age of five.

By the year 2000, the world's population had increased to 6 billion (6,000 million) and it was growing at an annual rate of 1.7%.

1 About how many people were added to the world population in the year 1000? In the year 2000?

2 The relatively low world population growth rate continued until the 1700s, when more modern medicine and improved water and sewage systems emerged.

   a. Suppose that world population growth had continued at an annual rate of 0.1% from 1000 to 1700. What function would this condition imply as a model for estimating world population $P(t)$ in year $1000 + t$?

   b. What world population does your model in Part a predict for 1700? Compare your prediction with the actual world population in 1700 that is estimated to have been about 640 million.

   c. Suppose that world population had continued to increase by the same number of people in each year after 1000. What function would this condition imply as a model for estimating world population in year $1000 + t$?

   d. What world population does your model in Part c predict for 1700?

   e. If world population growth had continued at the rates in the year 1000 until the year 2000, what would the population models have predicted for the year 2000:

      i. using the 0.1% growth rate condition in Part a?

      ii. using the constant number-of-people-per-year growth rate condition in Part c?

   f. If world population grows beyond the year-2000 figure of 6 billion at the rate of 1.7% per year:

      i. what function would this condition imply as a model for predicting world population in year $2000 + t$?

      ii. what world population does the model in part i predict for the year 2050?

      iii. what reasons can you imagine for doubting that the prediction in part ii will actually occur?

Planetary Motion Whenever scientists report an unusual astronomical event, we are reminded that our Earth is a very small planet in a very large universe. For example, when the Hale-Bopp comet flew within sight of Earth during 1996 and 1997, there was considerable discussion about the chances that other comets and asteroids might actually enter Earth's atmosphere. Some scientists even made estimates of the damage that would result from such an event.

One theory predicts that if an asteroid with a diameter of only 3 miles were to land in the middle of the North Atlantic Ocean, it would send a 300-foot tsunami crashing on the shores of North America and Europe. Fortunately, such an event is estimated to occur only once every 10,000,000 years!

3 Comets and asteroids have irregular shapes, but most can be approximated as spheres.

a. If an asteroid has average diameter d miles, what formulas give the:

   i. area of the cross section at a diameter of the asteroid?

   ii. total surface area of the approximately spherical body?

   iii. volume of the approximately spherical body?

b. The diagram below shows graphs of the three measurement functions in Part a on the interval $0 \leq d \leq 10$.

   i. Match the functions and graphs and explain how you know you are correct.

   ii. Explain what the relative shape of the three graphs says about the rates at which disk area, surface area, and volume change as asteroid diameter increases.

Asteroid Measurements

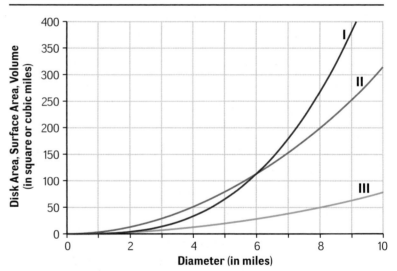

c. Comet Hale-Bopp appeared recently near Earth. It last came near Earth over 4,200 years ago, but the gravitational pull of the planet Jupiter will cause it to return near Earth again in only 2,400 years. The gravitational attraction of two large bodies is directly proportional to the product of their masses and inversely proportional to the square of the distance between their centers.

   i. If m_J and m_H represent the masses of Jupiter and comet Hale-Bopp respectively and d represents the distance between the centers of those masses, what is the form of the rule for the function $g(d)$ giving the gravitational attraction of those bodies at any distance d?

   ii. What will a graph of $g(d)$ look like?

4 For residents of Earth, the most important planetary body in our solar system is the Moon. The visible Moon varies in size from a full moon to a new moon (not visible at all) and back to a full moon in a cycle that takes roughly 30 days. Dates of many important religious and cultural events are set by reference to lunar calendars.

a. What function family seems likely to be the best starting point in building a model that gives visible area at any time during its 30-day cycle of phases:

   i. if you assume that the cycle starts with a full moon?

   ii. if you assume that the cycle starts with a half-moon on its way toward a full moon?

b. What particular members of the function families described in your response to Part a are likely to be good models of change in the visible moon if we assume that a full moon is 100%, a new moon is 0%, the cycle is 30 days long, and

   i. the cycle starts with a full moon?

   ii. the cycle starts with a half-moon on its way toward a full moon?

Matching Function Rules and Graphs In the investigations of this unit, you discovered that when building models of data patterns, it helps if you can identify a likely function rule by inspecting the graph.

5 Match each of the functions given in Parts a–j with their graphs in the diagrams below without using technology.

a. $y = x - 2$

b. $y = (x + 2)^2 - 5$

c. $y = 1.5^x - 2$

d. $y = 1.5x - 2$

e. $y = 3 \cos x + 1$

f. $y = 3 \cos x - 1$

g. $y = (x - 2)^2 - 5$

h. $y = 3 \cos 2x - 1$

i. $y = -2(x + 1)$

j. $y = -2|x| + 2$

Graph I

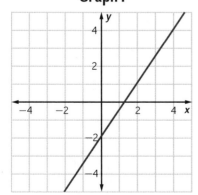

Graph II

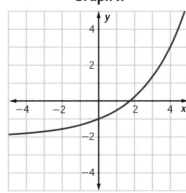

Graph III

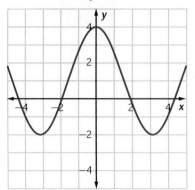

Graph IV

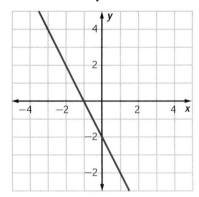

Graph V

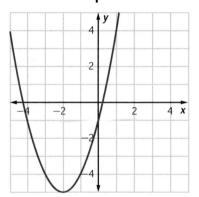

Graph VI

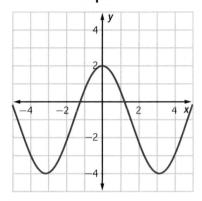

Graph VII

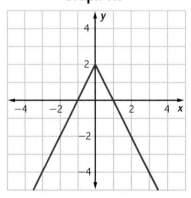

Graph VIII

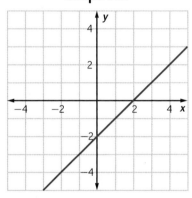

Graph IX

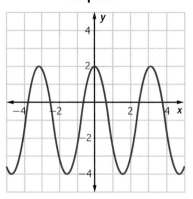

Graph X

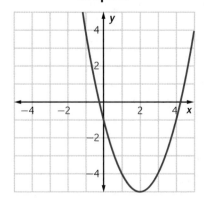

Operations that Combine Functions In this unit, you also learned how to build more complex models for patterns of change by combining basic functions through arithmetic operations and function composition.

6. Suppose that $f(x) = 5x$, $g(x) = \sin x$, and $h(x) = x + 3$. Use these functions to evaluate or find rules for:

   a. $f(-1) \div g\left(\frac{\pi}{2}\right)$

   b. $g(f(2)) + h(-5)$

   c. $h(g(f(x)))$

   d. $f(g(h(x)))$

7. At the Wild Water Amusement Park, managers use data from previous years to predict:

   • daily attendance as a function $a(T)$ of predicted high temperature T,

   • daily operating cost as a function $c(a)$ of daily attendance, and

   • daily income as a function $I(a)$ of daily attendance.

How can these functions be combined to give functions that predict:

a. daily profit as a function of daily attendance?

b. daily profit per customer as a function of daily attendance?

c. daily profit as a function of predicted high temperature?

d. daily profit per customer as a function of predicted high temperature?

SUMMARIZE THE MATHEMATICS

In this unit, you investigated a variety of situations in which rules for familiar functions had to be modified to model patterns in data plots and conditions in particular problems. You also explored how to build and analyze functions using arithmetic operations and function composition.

a What table and graph patterns and problem conditions are clues to use the following families of functions as models?

   i. Linear   **ii.** Quadratic

  iii. Exponential   **iv.** Direct power

   v. Inverse power   **vi.** Absolute value

  vii. Sine   **viii.** Cosine

   ix. Square root   **x.** Common logarithm

b What are the general forms of rules for each of the types of functions listed in Part a? What do the values of the parameters in those rules tell you about the patterns in tables and graphs of particular functions?

c How can you adjust the rule of $f(x)$ so that its graph matches graphs of $f(x)$ that are related by these transformations?

   i. Vertical translation   **ii.** Horizontal translation

  iii. Vertical stretching/compressing   **iv.** Horizontal stretching/compressing

   v. Reflection across the x-axis

d How can the basic sine and cosine functions be modified to give functions with amplitude A and period p?

e How do you use the rules for two functions $f(x)$ and $g(x)$ to calculate values for these functions?

   i. $f(x) + g(x)$ **ii.** $f(x) - g(x)$

  iii. $f(x) \cdot g(x)$ **iv.** $f(x) \div g(x)$

   v. $f(g(x))$

Be prepared to share your responses and reasoning with the class.

 CHECK YOUR UNDERSTANDING

Write, in outline form, a summary of the important mathematical concepts and methods developed in this unit. Organize your summary so that it can be used as a quick reference in future units.

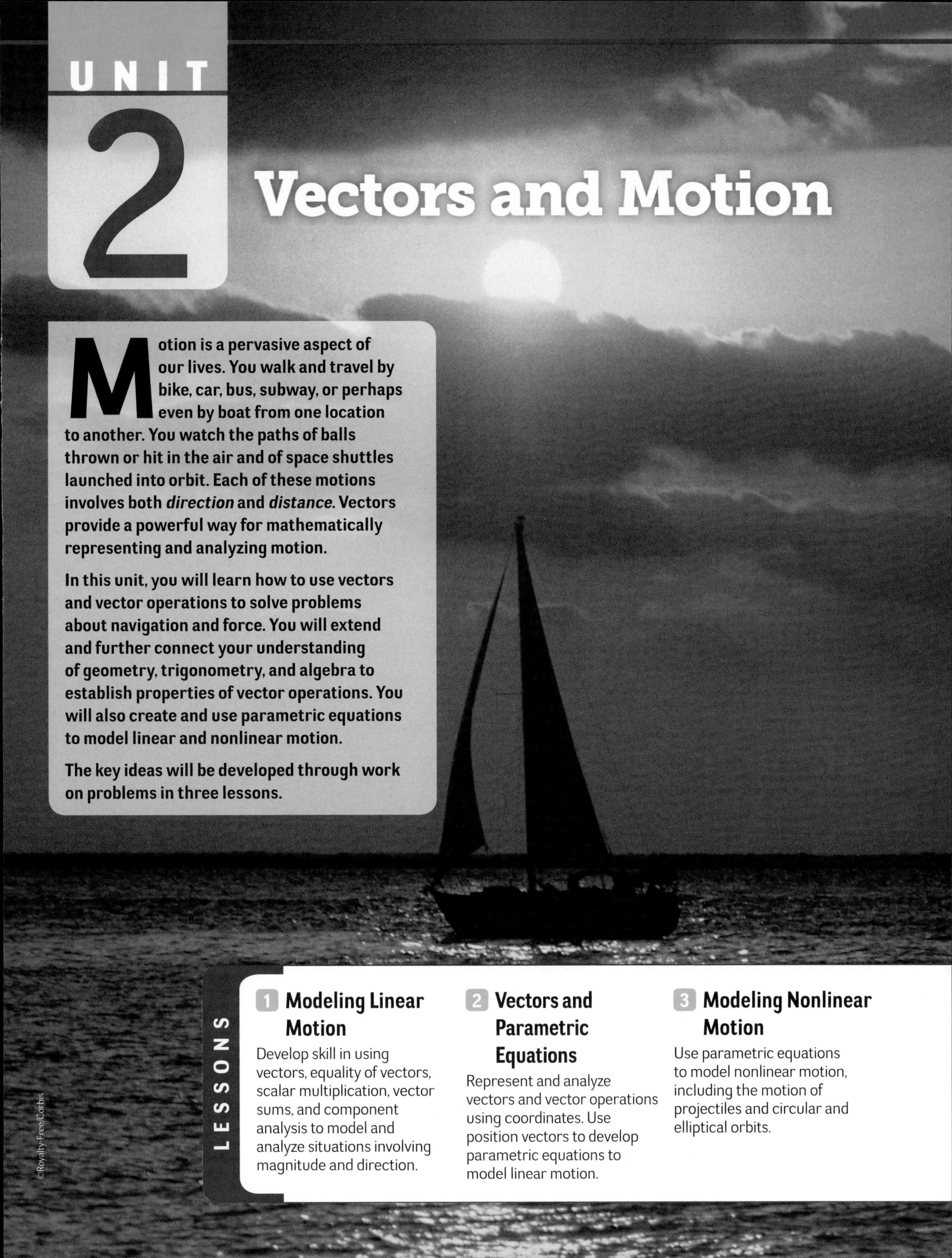

Vectors and Motion

Motion is a pervasive aspect of our lives. You walk and travel by bike, car, bus, subway, or perhaps even by boat from one location to another. You watch the paths of balls thrown or hit in the air and of space shuttles launched into orbit. Each of these motions involves both *direction* and *distance*. Vectors provide a powerful way for mathematically representing and analyzing motion.

In this unit, you will learn how to use vectors and vector operations to solve problems about navigation and force. You will extend and further connect your understanding of geometry, trigonometry, and algebra to establish properties of vector operations. You will also create and use parametric equations to model linear and nonlinear motion.

The key ideas will be developed through work on problems in three lessons.

©Royalty-Free/Corbis

LESSONS

1 Modeling Linear Motion

Develop skill in using vectors, equality of vectors, scalar multiplication, vector sums, and component analysis to model and analyze situations involving magnitude and direction.

2 Vectors and Parametric Equations

Represent and analyze vectors and vector operations using coordinates. Use position vectors to develop parametric equations to model linear motion.

3 Modeling Nonlinear Motion

Use parametric equations to model nonlinear motion, including the motion of projectiles and circular and elliptical orbits.

Modeling Linear Motion

Each day you confront motion in nearly everything you do. You may walk, ride a bicycle, or ride in a car or bus to school. You may take a subway train to meet friends at a shopping mall. You see aircraft fly overhead and you see the position of the Sun in the sky move, from morning when it rises in the east to evening when it sets in the west. You might run in a race or throw, kick, or hit a ball.

In this unit, you will learn to use an important tool for modeling motion—*vectors*. Vectors are useful in situations that involve *magnitude* (such as distance) and *direction*. These are important descriptors of motion. The simplest motion is linear—movement along a straight line.

Linear motion is used to plan and guide hiking routes and courses of boats and ships. Think about how you might describe or represent a planned route on a map. Also think about conditions that might affect a planned route and how you might incorporate that information in the planning process.

THINK ABOUT THIS SITUATION

Suppose you wanted to map out a route that involved sailing 3 km west from Bayview Harbor to Presque Island, then 6 km south to Rudy Point, and then 5 km southeast to Pleasant Bay.

a How could you represent the planned route geometrically?

b How could you represent a direct sailing route from Bayview Harbor to Pleasant Bay?

c How could you estimate the length of the route in Part b? How would you describe its direction?

d How would a northeast water current affect the path along which you would steer the boat to maintain the route in Part b?

In this lesson, you will learn how to represent vectors geometrically, how to scale vectors, and how to combine vectors by addition in the context of solving applied problems.

INVESTIGATION 1

Navigation: What Direction and How Far?

Vectors and vector operations are used extensively in navigation on water and in the air. As you work on the problems of this investigation, look for answers to these questions:

How can vectors be represented geometrically with directed line segments?

How can vectors and scalar multiples of vectors be used to model navigation routes?

Charting a Boat's Course

Imagine that you are navigating a boat along the small portion of the Massachusetts coast shown in the nautical chart at the right. Note that within the chart itself, there are several aids to navigation such as buoys, landmarks, and scales. The buoys are painted red or green and may have a red or green flashing light. A circle (on land) with a dot at its center indicates an easily recognized landmark such as a stone tower or a tank.

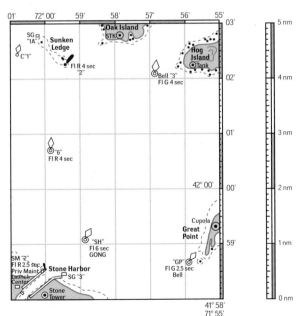

Adapted from Frank J. Larkin. *Basic Coastal Navigation.* Sheridan House Inc. 1998.

1. As a class, examine a copy of the nautical chart.

   a. At the right of the chart is a nautical mile (nm) scale. Use this scale to find the distance from the "SH" buoy to the "GP" buoy. Measure between the centers of the circles that mark the buoys.

   b. There are other scales at the top and along the right edge of the chart. What do you think these scales represent? Share your ideas with classmates.

   c. What other scale on this chart can be used to measure nautical miles? What does a nautical mile represent based on this scale? Share your ideas with your classmates.

   d. A nautical mile is 6,076.1033 feet. How does a nautical mile compare to a statute mile (regular mile)?

Coastal water nautical charts are designed so that the top is due north and the right side is due east. You can use your knowledge of directed angles measured counterclockwise from the horizontal (due east) to describe the direction of a craft. Thus, you can say due east is 0°, due north is 90°, due west is 180°, and due south is 270°.

2. The course of a boat starting at Buoy 6 and moving 30° north of east is shown in the chart below. Use a copy of the nautical chart to complete this problem.

   Using a ruler made from the nautical mile scale, measure distances to the nearest $\frac{1}{10}$ nm. Measure angles to the nearest degree using a protractor.

   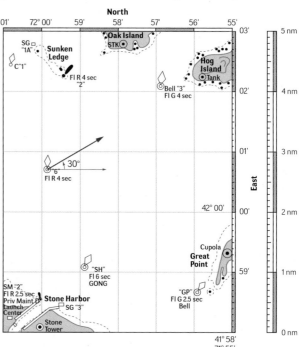

   a. Mark and label a point P on a copy of the chart to represent a boat that is 3 nautical miles from the "3" bell and is headed at an angle of 290°. What buoy is nearest to P?

b. Draw an arrow from the "SH" buoy to the "6" buoy. What is the direction in degrees? What is the distance in nautical miles?

c. What are the direction and distance of the path from the "6" buoy to the center of the mouth of the channel at Stone Harbor?

d. Why are arrows particularly useful representations for nautical paths?

3 The arrows that indicate boating routes are *directed line segments*. They have both a *magnitude* (length) and a *direction*. Thus, an arrow is a geometric representation of a **vector**—a quantity with magnitude and direction. A vector with a length of 1" and direction of 45° is shown at the right.

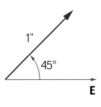

a. Accurately draw arrows representing vectors with the following characteristics.

  i. Length: 2 nm; direction: 70° (Use your nautical ruler and a protractor.)

  ii. Length: 5 cm; direction: 110°

  iii. Magnitude: 7 cm; direction: 300°

b. Draw an arrow for each vector described. State what length you chose to represent 1 knot and what length you chose to represent 1 mph.

  i. A boat with a speed of 2 knots (nautical miles per hour) at a direction of 180°

  ii. Speed of 40 mph at a direction of 240°

  iii. Force of a 15 mph wind blowing *from* the northeast

c. Compare the arrows you drew in Parts a and b with your classmates. Resolve any differences.

Denoting Vectors Vectors can be denoted in various ways. One way is to use italicized letters with arrows over them, such as $\vec{a}$ or $\vec{v}$. When the **initial point**, or **tail**, A and **terminal point**, or **head**, B are labeled, the notation $\overrightarrow{AB}$ can be used.

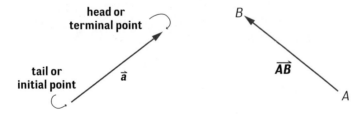

Since a vector $\vec{v}$ is determined by its magnitude r, and its direction θ, $\vec{v}$ can also be represented as a pair, $[r, \theta]$, called the **polar form** or **polar representation** of the vector.

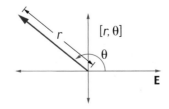

4 Since arrows representing vectors, as in Problem 3, can be drawn anywhere in a plane, it is important to know that two arrows drawn using the same scale represent **equal vectors** when they have the same magnitude and the same direction.

Explain why the following method provides a geometric test for the equality of the vectors $\overrightarrow{PQ}$ and $\overrightarrow{RS}$.

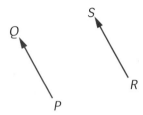

Step 1. Connect the *heads* Q and S and connect the *tails* P and R.

Step 2. If $PQSR$ is a parallelogram, then $\overrightarrow{PQ} = \overrightarrow{RS}$.

Scalar Multiples of a Vector In the problems that follow, use either a ruler, protractor and graph paper, or interactive geometry software with vector drawing and analysis capabilities. It may also be helpful to use the "Stone Harbor" custom app in *CPMP-Tools*. When the instructions ask you to make an "accurate drawing" or an "accurate sketch" of a vector, you can do so with geometry software or carefully draw an arrow on graph paper using a ruler and protractor to measure. If, however, the instructions are simply "sketch" or "draw" a vector, you may make a freehand sketch of an arrow that approximates the characteristics of an accurately drawn vector in order to guide your thinking. Note that "draw a vector" actually means "draw a geometric representation of the vector" (an arrow).

5 A fishing boat leaves the mouth of the Stone Harbor channel trolling on a heading due north at a speed of 1.5 knots (nautical miles per hour).

a. On your copy of the nautical chart, sketch the vector $\vec{v}$ representing the distance and direction traveled from the middle of the channel opening during the first hour.

b. Use the vector $\vec{v}$ in Part a to determine the vector for a 2-hour trip at the same speed and in the same direction. Sketch this vector. Label it $2\vec{v}$.

c. Sketch and label a vector that locates the fishing boat at the end of 20 minutes and another that locates it at the end of 2.5 hours.

d. Now sketch another vector that has the same length as $2\vec{v}$ but is not equal to $2\vec{v}$, and another vector that is equal to $2\vec{v}$. Compare your vectors with those of your classmates.

e. In general, how would you sketch a vector that was a positive number k times a given vector? How are the lengths and directions of these two vectors related?

6 Suppose another boat begins a trip at the same point at the mouth of the channel at Stone Harbor headed at a direction of 20° and at a speed of 2 knots.

a. Sketch the vector $\vec{v}$ showing the approximate position at the end of the first hour.

b. Suppose the boat returns to the harbor along the same route at the same speed. Sketch the return vector and give its magnitude and direction.

c. The word "opposites" can be used to denote the vectors in Parts a and b. How is the word "opposite" descriptive of the relationship between the two vectors?

d. Sketch a vector opposite to the vector $\vec{v}$ in Part a with initial point at the "3" bell. Give its magnitude and direction.

When a vector $\vec{a}$ is multiplied by a real number k, the number is called a **scalar** and the product, $k\vec{a}$, is a **scalar multiple** of $\vec{a}$. (In a similar manner, $k\overrightarrow{AB}$ is a scalar multiple of the vector $\overrightarrow{AB}$.) When $k > 0$, $k\vec{a}$ is the vector whose length is k times the length of $\vec{a}$ and has the same direction as $\vec{a}$ as shown at the right.

$\vec{a}$ $\qquad$ $k\vec{a}, k > 1$ $\qquad$ $k\vec{a}, 0 < k <$

7 For vector $\vec{a}$ shown above, the **opposite of vector** $\vec{a}$, denoted $-\vec{a}$, is shown at the right. The scalar multiple $k\vec{a}$ when $k < 0$ is shown at the far right.

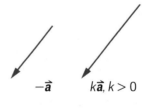

$-\vec{a}$ $\qquad$ $k\vec{a}, k > 0$

a. Compare the relationship between $\vec{a}$ and $k\vec{a}$ when $k > 0$ to the relationship between $-\vec{a}$ and $k\vec{a}$ when $k < 0$.

b. Suppose $\vec{a} = [10, 50°]$ and $\vec{v} = [8, 20°]$. Write each of the following vectors in polar form $[r, \theta]$, where r is the vector's magnitude and its direction θ satisfies $0° \le \theta < 360°$.

    i. $5\vec{a}$ $\qquad\qquad\qquad$ **ii.** $0.2\vec{v}$

    iii. $-\vec{a}$ $\qquad\qquad\qquad$ **iv.** $-3\vec{v}$

c. Suppose $k < 0$ and $\vec{a} = [r, \theta]$. Write $k\vec{a}$ in polar form.

Janet S. Robbins

SUMMARIZE THE MATHEMATICS

In this investigation, you explored how vectors—quantities with magnitude and direction—can be represented geometrically by arrows.

a Describe how you know when two arrows represent equal vectors.

b How are a vector and a scalar multiple of that vector similar? How are they different?

c How are vectors $\overrightarrow{AB}$ and $\overrightarrow{BA}$ alike? How are they different? What is another way to write $\overrightarrow{BA}$ using A and B?

d What is always true about the magnitudes and directions of two opposite vectors?

Be prepared to explain your ideas to the class.

 CHECK YOUR UNDERSTANDING

Daily ferries shuttle people and cars between Manitowoc, Wisconsin, and Ludington, Michigan. Use a copy of this map of Lake Michigan to complete the following tasks.

a. Draw the vector for the ferry route from Manitowoc to Ludington. Label it $\vec{v}$. Measure its magnitude and direction.

b. Find the magnitude and direction of $-\vec{v}$. Draw $-\vec{v}$ beginning at Charlevoix, Michigan.

c. Sketch $0.5\vec{v}$ from Milwaukee, Wisconsin. Find its magnitude and direction.

d. Are the vectors representing the route from Charlevoix to Escanaba and the route from South Haven to Milwaukee approximately the same? Explain.

Changing Course

In the previous investigation, you used vectors to model straight-line paths. But hiking paths or the course of a ship often involve changes in direction.

As you complete the problems in this investigation, look for answers to the following question:

How can vectors be used to model routes when there is a change of course during the trip?

1 **Vector Sums** Roberta, the skipper of the fishing boat *High Hopes*, leaves the mouth of the Stone Harbor channel at a speed of 6 knots at 25°. She travels for 20 minutes, then turns so that she is heading in a direction of 100° at the same speed and travels for 30 minutes before deciding to drop anchor and begin fishing.

a. Using the "Stone Harbor" custom app or a copy of the nautical chart, draw a vector diagram showing the paths taken and the position of the *High Hopes* at the end of 50 minutes. What is the magnitude of each of these two vectors?

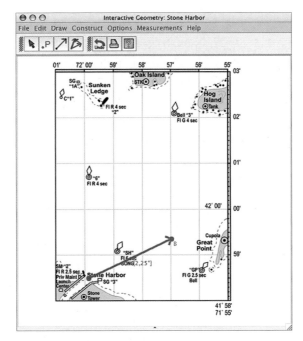

b. Suppose the fish are biting and Roberta wants to inform Clarissa, the skipper of the *Salmon King*, of where she is located so Clarissa can join her. Draw a vector representing the path Clarissa should take from the mouth of Stone Harbor channel directly to the *High Hopes*. What direction should Roberta advise her to take? How far will she need to travel?

c. The vector representing the path that Clarissa should travel to the good fishing spot is called the **sum** or **resultant** of the two vectors that describe the route taken by the *High Hopes*. How are the initial and terminal points of the resultant vector in Part b related to the two vectors that represent the route taken by the *High Hopes*?

d. Suppose Roberta had left the harbor at a speed of 6 knots in a direction of 100° for 30 minutes, and then turned to a direction of 25° and traveled for 20 minutes at the same speed. Draw a sketch of Roberta's path and the resultant vector. What are the direction and the magnitude of the resultant vector?

e. Now sketch two additional two-leg routes to this good fishing spot.

2 Based on your work in Problem 1, decide if each of the following generalizations is true or false. In each case, explain your reasoning.

a. The vector sum of any two given vectors is unique.

b. If a vector is the sum of two given vectors, it cannot be the sum of two other vectors.

c. If $\vec{v} = [r, \theta_1]$ and $\vec{w} = [s, \theta_2]$, then $\vec{v} + \vec{w} = [r + s, \theta_1 + \theta_2]$.

3 For the following vectors, the magnitude is in centimeters and the given angle measure is the vector's direction: $\vec{a} = [5, 70°]$, $\vec{b} = [4, 30°]$, $\vec{c} = [4, 350°]$, and $\vec{d} = [3, 250°]$. Make accurate drawings of each vector sum and measure to find the magnitude (to the nearest 0.1 cm) and direction (to the nearest 5°) for each resultant vector.

a. $\vec{a} + \vec{b}$

b. $\vec{a} + \vec{d}$

c. $\vec{a} + \vec{b} + \vec{c}$

4 Now investigate some general properties of vector addition. Begin by sketching any two vectors $\vec{a}$ and $\vec{b}$ as arrows that have different directions but no points in common.

a. Draw a diagram showing how to place $\vec{a}$ and $\vec{b}$ to find $\vec{a} + \vec{b}$. Do the same for $\vec{b} + \vec{a}$. What do you notice about the two vector sums? Compare your observations to those of others and resolve any differences.

b. To which property of real number operations is this similar?

c. Choose a point in the plane. Place $\vec{a}$ and $\vec{b}$ so their initial points are at this point. Then draw a single diagram showing how to find $\vec{a} + \vec{b}$ and $\vec{b} + \vec{a}$. What shape is formed? Prove your conjecture.

5 On a sheet of plain paper or graph paper, make an accurate drawing of vector $\vec{u}$ with magnitude 4 cm and direction 200° and vector $\vec{v}$ with magnitude 5 cm and direction 70°.

a. Without drawing or measuring, find the magnitude and direction of as many of the following vectors as possible. Explain your reasoning in each case.

i. $2\vec{u}$	ii. $\vec{v} + \vec{u}$
iii. $\vec{u} + \vec{v}$	iv. $3(\vec{u} + \vec{v})$
v. $3\vec{u} + 3\vec{v}$	vi. $-2\vec{v}$
vii. $2\vec{v} + (-2\vec{u})$	viii. $-2\vec{v} + (-2\vec{u})$

b. For the remaining vectors in Part a, find the magnitude and direction by measuring. Use as few drawings as possible. Look for possible connections between pairs of vectors that might reduce your work.

c. What general rule is suggested by parts iv and v in Part a? Test your conjecture.

d. The sum of two vectors is always a vector. Describe the resultant vector for $\vec{u} + (-\vec{u})$.

6 **Horizontal and Vertical Components** The chart below shows a vector $\vec{v}$ with magnitude 1.3 nm and direction 0° and a vector $\vec{w}$ with magnitude 1.8 nm and direction 90° that represent one route to a good fishing area.

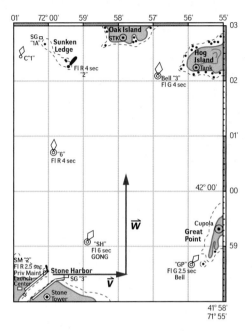

a. *Calculate* (do not measure) the magnitude of the resultant vector $\vec{v} + \vec{w}$.

b. Use trigonometric ratios to compute the direction of the direct route $\vec{v} + \vec{w}$ to the good fishing spot.

c. Starting at the harbor, is it possible to find another pair of vectors with directions 0° and 90° that have the same vector sum as in Part a? Explain your reasoning.

7 Now investigate further how a vector can be thought of in terms of the sum of horizontal and vertical vectors called its **components**.

a. Suppose a vector represents a 2-nautical mile route with a direction of 78°. Use trigonometric ratios to compute the lengths of the east (0°) and north (90°) legs of a route to the same location.

b. Suppose a vector $\vec{v}$ represents a 2-nm route with a direction of 125°. Make a sketch of the vector $\vec{v}$ and include the west and north vectors that would give the resultant vector $\vec{v}$. Compute the magnitudes of the west and north vectors.

c. Now think more generally. How would you compute the magnitudes of the horizontal and vertical components of the vectors described below? Compare your methods with those of your classmates and resolve any differences.

   i. Any 2-nm vector with a direction θ between 180° and 270°

   ii. Any 5-nm vector with a direction θ between 270° and 360°

   iii. Any 10-nm vector that points due north or due south

   iv. Any 10-nm vector that points due east or due west

SUMMARIZE THE MATHEMATICS

In this investigation, you explored the geometry of addition of vectors.

a Describe geometrically how you can find the resultant, or sum, of two vectors.

b Any nonzero vector can be represented as the sum of a horizontal vector and a vertical vector. Illustrate and explain how this can be done for a given vector.

c In the vector diagram below, $\overrightarrow{AC}$ and $\overrightarrow{CB}$ are the horizontal and vertical components, respectively, of $\overrightarrow{AB}$.

> **i.** If you know the magnitudes of $\overrightarrow{AC}$ and $\overrightarrow{CB}$, how would you calculate the magnitude and direction of $\overrightarrow{AB}$?
>
> **ii.** If you know the magnitude and direction of $\overrightarrow{AB}$, how would you calculate the magnitudes of $\overrightarrow{AC}$ and $\overrightarrow{CB}$?

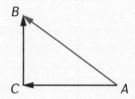

Be prepared to share your ideas and reasoning with the class.

✔ CHECK YOUR UNDERSTANDING

Use what you have learned about adding vectors and horizontal and vertical components of a vector to compute (not measure) answers to the questions below. Check that your answers are reasonable by measuring.

a. Suppose Clarissa wants to fish in the secluded bay behind Great Point, as shown in the chart. The vector [3.1 nm, 10°] represents a direct route to the bay. Since this route crosses land, Clarissa decides to head east and then due north to the fishing spot. How many nautical miles should she travel east before turning north? How far north from there is the fishing spot?

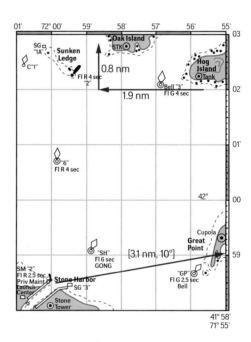

b. Suppose Roberta needs to travel from her location south of Hog Island to the west side of Oak Island before nightfall. The west and north vectors for one route are shown on the chart. If Roberta decides to take a direct route (across the rocky area) rather than the west/north route, how many nautical miles can she shave off the trip? In what direction should she head?

Go with the Flow

The vector models you have been using for navigation assume that the force moving a boat is the only one acting on the craft. When this is the case, the craft moves in a straight line in the direction of the force. In reality though, two (or more) forces often act simultaneously on an object. For example, currents in the ocean are forces that move the boats in the direction of the current. Sailing ships without motors use water currents to help them enter and leave port. The wind, too, is a force that affects the path that a boat or an airplane follows. A fundamental principle of physics is that the effect of two forces acting on a body is the sum of the forces.

As you work on problems in this investigation, look for answers to this question:

How can vectors be used to analyze the effect of two or more forces simultaneously acting on an object?

1. Suppose a boat leaves port *P* headed in a direction of 60° with the automatic pilot set for 10 knots. On this particular day, there is a 4-knot ocean current with a direction of 30°. The vector diagram at the right shows the effect of the current on the position of the boat at the end of one hour.

   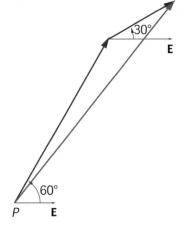

   a. Assuming a scale of 1 cm = 2 nm, verify the accuracy of the diagram.

   b. The sum of the original course and current vectors gives the position of the boat in one hour. Determine how far the boat will actually travel in one hour:

      i. using the scale diagram.

      ii. using the Law of Cosines. (*Hint:* The obtuse angle of the triangle is 150°. Why?)

   c. At what speed and in what direction will the boat actually travel during the first hour? Will it continue to travel similarly during the next hour if all conditions remain the same? Explain.

Steve Mason/Getty Images

2 In Problem 1, you were able to determine the course of the boat using either a scale drawing and measuring or using the Law of Cosines. Now examine the situation in terms of component vectors.

  a. Make a sketch similar to the one in Problem 1 that shows the planned-course vector and the current vector without the resultant vector. Then sketch the horizontal and vertical components of each vector.

  b. Compute the lengths of the horizontal and vertical components of each vector.

  c. Using the component vectors found in Parts a and b, find the magnitude and direction of the resultant vector representing the actual route.

  d. Compare the results of this problem with those of Problem 1.

3 Now consider vector $\vec{v}$, with length 4 cm and direction 35°, and vector $\vec{u}$ with length 5 cm and direction 70°.

  a. On graph paper, draw $\vec{v} + \vec{u}$.

  b. Draw the horizontal and vertical components of $\vec{v}$ and of $\vec{u}$.

  c. Draw the resultant of the horizontal components and the resultant of the vertical components.

  d. Draw the sum of the two resultant vectors found in Part c. How is this sum related to $\vec{v} + \vec{u}$?

  e. Write a summary statement describing how the components of two vectors can be used to find the sum of the two vectors.

4 Next make a sketch of a vector showing the location of an airplane traveling at a fixed altitude after one hour if it is headed in a direction of 40° and its speed in still air is 500 mph, but the wind is blowing at 40 mph from the northwest.

  a. Augment your sketch to show the horizontal and vertical components $\vec{h}_v$ and $\vec{v}_v$ of the planned-course velocity vector $\vec{v}$. On the sketch, represent the horizontal and vertical components $\vec{h}_w$ and $\vec{v}_w$ of the wind velocity vector $\vec{w}$.

  b. Use the component vectors in Part a to determine the direction and distance the airplane traveled in one hour.

  c. What was the effective speed of the airplane during that one hour of flying? (This is called the *ground speed*.)

Combining Forces The process illustrated in Problems 2–4, called **component analysis of vectors**, is a very powerful tool for analyzing linear motion problems. It reduces a complex situation to one in which only component vectors with the same direction are added. The next two problems will provide you with further practice in using component analysis to solve applied problems.

5 Two men have to move a doghouse on skids to a new position due east of its present location. They tie ropes to the doghouse and pull as follows: Thad pulls with a force of 100 pounds at a direction of 35°, while Jerame pulls with a force of 120 pounds at a direction of 315°.

a. Make a sketch showing the force vectors and resultant vector. Ignore the force of friction.

b. Find the direction at which the doghouse should move under these conditions.

c. If friction requires 150 lb of force to move the doghouse, will it move with the given effort of Thad and Jerame? Explain.

d. What would be the effect of Jerame changing the direction at which he pulls, so that Jerame's south force cancels out Thad's north force?

e. If Thad pulls as before, in what direction should Jerame pull to slide the doghouse due east when they both pull it?

6 In Problem 1, you found that an ocean current can cause a boat to travel off the desired course. As shown in the first diagram at the right, the resultant vector $\vec{v}$ is the sum of the desired course of the boat $\vec{b}$ and the current vector $\vec{c}$.

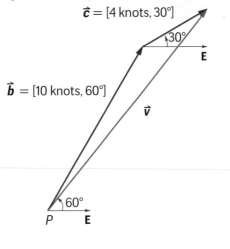

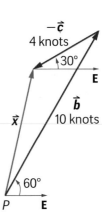

Suppose the captain of the boat wants to stay on the desired course $\vec{b} = [10 \text{ knots}, 60°]$ by adjusting for the current vector $\vec{c} = [4 \text{ knots}, 30°]$. In the second diagram at the right, vector $\vec{x}$ represents this adjusted course setting.

a. How can $\vec{x}$ be expressed in terms of $\vec{b}$ and $\vec{c}$?

b. Explain why this adjusted course path will give the desired course [10 knots, 60°].

c. What are the magnitude and direction of $\vec{x}$?

d. Using *CPMP-Tools* Vector Geometry, check your answer to Part c.

In this investigation, you examined how vectors can be used to analyze situations in which more than one force is acting on an object.

 a Describe how vectors can be used to model linear motion in moving air or water.

b Explain how the horizontal and vertical components of vectors can be used to determine the direction and speed of a boat or airplane that is moving at a fixed speed along a linear path in water or air that is also moving at a fixed speed along a linear path.

Be prepared to share your descriptions and thinking with the class.

✔ CHECK YOUR UNDERSTANDING

Flight plans for a commercial jet airplane indicate a flight in the direction of 70° and an average speed of 600 mph. Today, atmospheric winds are blowing from the northwest at an average of 50 mph.

Determine how to adjust the flight plan by:

a. drawing a vector model of the effect of the wind on the jet.

b. drawing a vector model showing the direction needed to keep the jet on the planned course and computing its direction.

c. computing the speed that the jet needs to maintain the desired average of 600 mph.

1. Tony Hillerman was a mystery writer whose books are often based on the Native American cultures of New Mexico, Utah, Colorado, and Arizona. The map below shows Hillerman Country in which Navajo Tribal Police Officers Joe Leaphorn and Jim Chee solve mysteries. In Hillerman's novels, they travel mostly by car throughout the reservations, but for this task assume they have a helicopter. Use careful sketches and measure the desired magnitudes and angles.

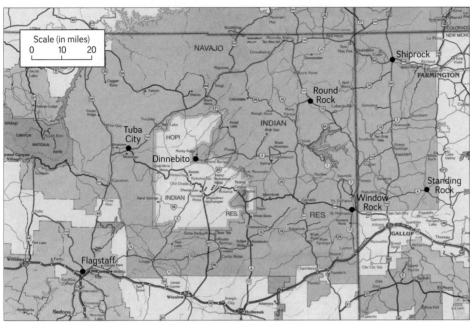

Source: Adapted from "Recreation Map of Arizona and the Four Corners Region," North Star Mapping.

   a. Suppose Jim and Joe are stationed at Shiprock. What direction should Jim chart to go to Tuba City to investigate a car accident? What is the distance he must fly by helicopter?

   b. Jim is to fly from Tuba City to Flagstaff to meet with FBI officials. In what direction is he headed? At 100 mph, what is his flying time?

   c. Plot the round trip from Shiprock to Round Rock to Window Rock to Standing Rock and back to Shiprock. Give the direction and distance of each part of the trip.

2. Suppose a boat leaves harbor at noon in still water at a direction of 310° and a speed of 4.4 knots, and vector $\vec{v}$ represents the distance and direction traveled during the first hour of the trip.

   a. Label a point on your paper as the harbor. Use a scale of 1 cm = 1 knot. Sketch vector $\vec{v}$, and write it in polar form [*magnitude, direction*].

   b. Sketch the vector that represents the position of the boat at 1:30 P.M. Write this vector in terms of $\vec{v}$ and in polar form.

   c. At 3:00 P.M., the skipper turns the boat back toward the harbor, maintaining a speed of 4.4 knots. Make a sketch of vector $\vec{w}$ that represents the distance and direction that the boat travels from 3:00 P.M. until 5:00 P.M. Then write this vector in polar form.

3 A Coast Guard cutter is located in a harbor in Ludington when an SOS comes in from a boat located at a point due west of Grand Haven and due south of Manitowoc. Grand Haven is about 62 miles from Ludington at a direction of 280°, and Manitowoc is about 60 miles from Ludington at a direction of 170°.

a. Find the components of the vector from Ludington to Grand Haven.

b. Find the components of the vector from Ludington to Manitowoc.

c. Use a copy of the map above as a guide to sketch the components that you found in Parts a and b and the location of the boat that is in trouble.

d. Using your sketch as a guide, calculate the magnitude and direction of the vector from Ludington to the location of the boat.

e. Suppose a second Coast Guard cutter is located at Manitowoc. Assuming this cutter can travel at the same speed as the first, which boat should respond to the SOS?

4 Any vector can be thought of in terms of its horizontal and vertical components. Similarly, if you know the components of a vector, you can determine the vector's magnitude and direction.

a. Calculate the directed lengths of the horizontal and vertical components of the following vectors written in polar form.

   i. [3 nm, 86°]   **ii.** [5 nm, 285°]

   iii. [2.5 nm, 120°]   **iv.** [6 nm, 315°]

b. Calculate the magnitude and direction of the vectors with the following components.

   i. 5 nm east, 3 nm north

   ii. 12 nm east, 7 nm south

   iii. 7 nm west, 1 nm north

   iv. 21 nm west, 6 nm south

5 The nautical chart used in Investigation 1 is reproduced below. The boat *Open C* is located at the flashing red light at Sunken Ledge when its skipper learns that fishing action has begun near the "GP" buoy.

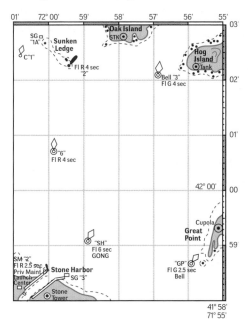

a. In what direction should the skipper head for the "GP" buoy?

b. At 6 knots, how long would the trip take in still water?

c. Now suppose there is a heavy northeast wind that will move boats at a rate of about 2 knots. Make a vector diagram showing the effect of the wind on the course of the *Open C*.

d. In the wind, what is the direction of the route the *Open C* actually travels?

e. What direction should the skipper plot to account for the wind and follow a direct route to the "GP" buoy?

6 A balloon ride can be a very beautiful and peaceful experience, but balloon operators must always be on guard for the effects of the wind. Suppose a balloon rises at a constant rate of 2.4 meters per second, but there is a wind blowing at 1 meter per second from the west.

a. Taking the effect of the wind into account, what are the speed, direction, and components of the balloon's velocity vector?

b. How high above the ground will the balloon be after 5 minutes?

c. After 5 minutes, the balloon is directly above a monument on the ground. How far is the monument from the point at which the balloon originally ascended?

Medioimages/Superstock

7 Maria and Kim are volleyball players on their respective school teams. Suppose that in a conference match, at the same time, they each block the ball when it is directly over the net. Maria's hit has a force of 50 pounds at a direction of 325°. Kim's hit has a force of 40 pounds at a direction of 60°.

a. Sketch the vectors involved if the net is on the east-west line.

b. Assuming that the ball moves in the direction of the resultant force, on whose side of the net will the ball land? How can component vectors be used to verify this?

c. At what angle should Maria hit the ball so that it follows the top of the net or goes onto Kim's side?

CONNECTIONS

8 In each of the diagrams below, a figure *F* and its image *G* under a translation are shown.

Translation I

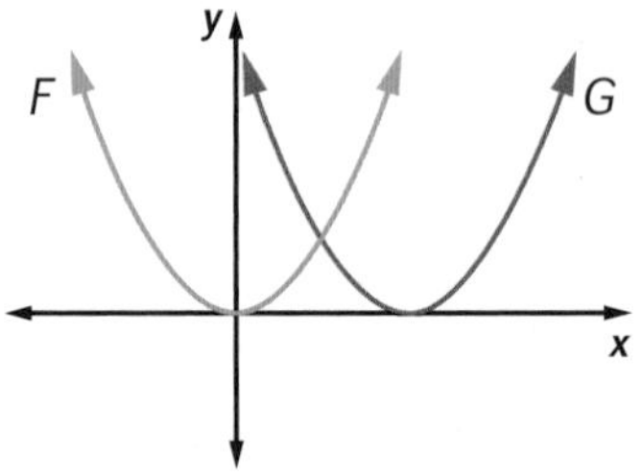

Translation II

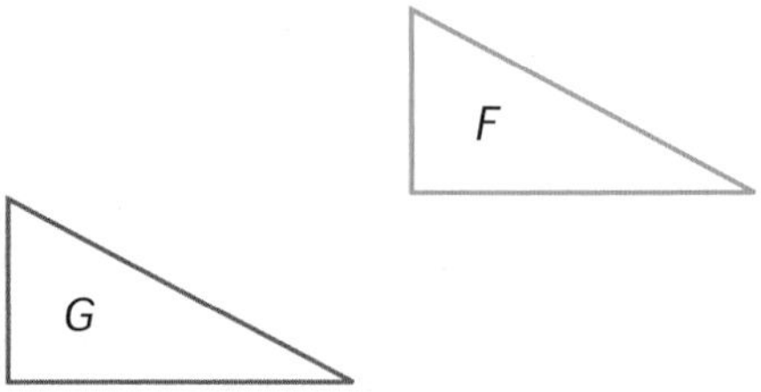

a. How could you use a vector to describe each translation?

b. Can every translation be described by a vector? Explain your reasoning.

c. Sketch any triangle $\triangle RST$ and its image under the composition of two translations, the first with vector $\vec{u} = [3 \text{ in.}, 90°]$ and the second with vector $\vec{w} = [3 \text{ in.}, 180°]$. In your drawing, also sketch the single vector that represents the composite of the two translations.

d. Describe how the vector representing the composite transformation in Part c is related to the two translation vectors, $\vec{u}$ and $\vec{w}$.

9 In Investigation 2, you found the sum $\vec{u} + \vec{v}$ by placing the vectors head-to-tail as shown below. If $\overrightarrow{AB}$ represents $\vec{u}$ and $\overrightarrow{BC}$ represents $\vec{v}$, then $\vec{u} + \vec{v} = \overrightarrow{AC}$.

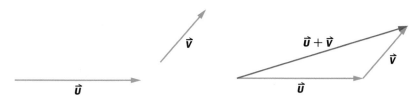

a. The *parallelogram law* provides a second way to add the two vectors $\vec{u}$ and $\vec{v}$.

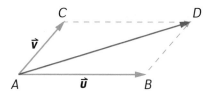

Explain as precisely as you can why by completing the parallelogram *ACDB*, the diagonal vector $\overrightarrow{AD}$ of □*ACDB* equals $\vec{u} + \vec{v}$.

b. Suppose $\vec{a} = [3 \text{ cm}, 20°]$ and $\vec{b} = [5 \text{ cm}, 50°]$. Use the parallelogram law to find $\vec{a} + \vec{b}$. Express the sum in polar form.

10 Vector addition and scalar multiplication have some of the same properties as real number addition and multiplication.

a. Using the vectors shown at the right, make sketches illustrating the following properties.

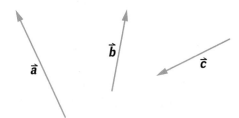

   i. $\vec{a} + \vec{b} = \vec{b} + \vec{a}$

   ii. $(\vec{a} + \vec{b}) + \vec{c} = \vec{a} + (\vec{b} + \vec{c})$

   iii. $2(\vec{b} + \vec{c}) = 2\vec{b} + 2\vec{c}$

b. For any two real numbers, s and t, $s - t = s + (-t)$. The expression $\vec{a} - \vec{b}$, by definition, is the vector that when added to $\vec{b}$ gives the resultant $\vec{a}$, that is, $(\vec{a} - \vec{b}) + \vec{b} = \vec{a}$. The expression $\vec{a} + (-\vec{b})$ represents the sum of vectors $\vec{a}$ and $-\vec{b}$. Using the vectors above, sketch $\vec{a} - \vec{b}$ and $\vec{a} + (-\vec{b})$. Show that the processes you use to make the sketches are different, but the vectors that result are equal vectors.

11 Consider the partial tiling of equilateral triangles shown at the right. Write expressions for each of the following vectors in terms of $\vec{u}$ and $\vec{v}$.

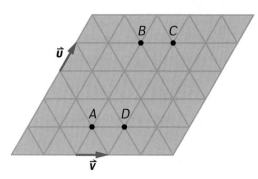

a. $\overrightarrow{AB}$ **b.** $\overrightarrow{BC}$

c. $\overrightarrow{AC}$ **d.** $\overrightarrow{BD}$

12 Using the diagram below and properties in Connections Task 10, write each expression in a simpler form.

a. $\vec{b} - \vec{u}$

b. $\vec{w} - \vec{a}$

c. $\vec{v} + \vec{u} - \vec{b}$

d. $\vec{b} + \vec{a} - \vec{v} - \vec{w}$

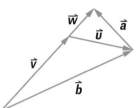

13 On a piece of paper, mark a point O in the center. Using O as the beginning point, accurately draw a 1-inch vector, $\vec{a}$, pointing horizontally to the right and accurately draw a 1.5-inch vector, $\vec{b}$, pointing straight upward.

a. What is the measure of the angle between these vectors?

b. Choose a point P on your paper so that the length of $\overrightarrow{OP}$ is 4 inches. Find scalars m and n so that $\overrightarrow{OP} = m\vec{a} + n\vec{b}$.

c. If Q is any other point, can you always find scalars m and n such that $m\vec{a} + n\vec{b} = \overrightarrow{OQ}$? Explain your reasoning.

d. Revise your answer for Part c when $\overrightarrow{OQ}$ has the same direction as $\vec{a}$ or $\vec{b}$. When Q coincides with point O.

REFLECTIONS

14 Wind patterns over a region are sometimes plotted using vectors that show the direction and force of the winds at various locations. The wind patterns over the San Francisco Bay on a recent day are shown on the chart at the right.

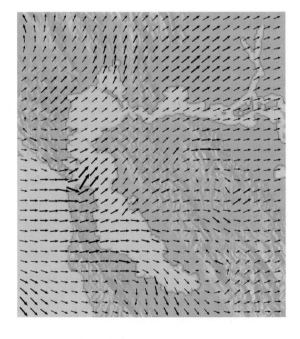

a. Describe where the wind is the strongest. What is the direction of the strongest winds?

b. Describe locations where the wind is blowing from the west. From the south. From the north.

c. Describe two separated locations in which the wind vectors are approximately equal. Approximately opposite. What does this tell you about the wind in those locations?

15 The following questions will help you refine your thinking about the sum, or resultant, of two vectors.

a. If two vectors have the same direction, what is the magnitude of the resultant? What is the direction of the resultant?

b. If two vectors of different lengths have opposite directions, what is the magnitude of the resultant? What is the direction of the resultant?

c. What is the resultant of a vector and its opposite vector?

d. What are the components of a vector with magnitude 5 units and direction east?

e. How would the zero vector be interpreted if it represented a velocity? A displacement? A force? A translation?

16 Mia claimed that the magnitude of the sum of two vectors is always equal to the sum of the magnitudes of the vectors.

a. Give a counterexample to show that Mia's claim is not true for all pairs of vectors.

b. Under what conditions is the magnitude of the sum of two vectors equal to the sum of the magnitudes of the vectors?

c. Your answers in Parts a and b are related to an important property of triangles. What is the property? Explain your reasoning.

17 In problems involving navigation, the direction a vector points can also be described by specifying its **heading** in degrees measured *clockwise* from the north.

a. What is the heading of a ship sailing:

   i. due north?   **ii.** due east?

  iii. due south?   **iv.** due west?

b. Sketch vectors satisfying the given criteria.

   i. $\vec{v}$ has length 5 cm and heading 80°.

  ii. $\vec{n}$ has length 2 cm and heading 130°.

c. For what angles are the direction and heading of a vector identical?

d. What relationship exists between the heading H and direction D of a vector?

18 In this task, you are asked to reflect on some of the important mathematical practices that you used in this lesson.

a. *Practice: Model with mathematics* Vectors were used to model several different types of physical problem situations. Describe at least five different examples of such modeling.

b. *Practice: Use appropriate tools strategically* Various drawing/measuring tools and mathematical tools were used to operate on vectors and to solve problems. Describe the use of at least two different tools of each type.

c. *Practice: Look for and make use of structure* In earlier courses involving arithmetic and algebra, you studied the structure of the real number system under addition. Addition of vectors has many properties in common with addition of real numbers. Name or describe at least five of them. How does understanding common structures help you in reasoning?

EXTENSIONS

19 In Course 3, Unit 3, *Similarity and Congruence*, you used properties of similar triangles to prove the Midpoint Connector Theorem:

If a line segment joins the midpoints of two sides of a triangle, then it is parallel to the third side and its length is one-half the length of the third side.

a. Alonzo attempted to prove this theorem using what he learned about vectors. Check the correctness of Alonzo's argument. Supply a reason for each statement or correct any misstep.

Suppose X and Y are the midpoints of $\overline{AC}$ and $\overline{BC}$, respectively. Orient the vectors as shown in the diagram.

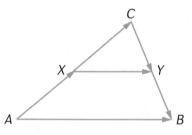

Then $\overrightarrow{AX} = \overrightarrow{XC} = \frac{1}{2}\overrightarrow{AC}$

and $\overrightarrow{CY} = \overrightarrow{YB} = \frac{1}{2}\overrightarrow{CB}$.

Also, $\overrightarrow{AC} + \overrightarrow{CB} = \overrightarrow{AB}$

and $\overrightarrow{XC} + \overrightarrow{CY} = \overrightarrow{XY}$

So, $\frac{1}{2}\overrightarrow{AC} + \frac{1}{2}\overrightarrow{CB} = \overrightarrow{XC} + \overrightarrow{CY}$ or

$\frac{1}{2}(\overrightarrow{AC} + \overrightarrow{CB}) = \overrightarrow{XY}$.

Therefore, $\frac{1}{2}\overrightarrow{AB} = \overrightarrow{XY}$.

$\overrightarrow{XY} \parallel \overrightarrow{AB}$.

It follows that $XY = \frac{1}{2}AB$ and $\overline{XY} \parallel \overline{AB}$.

b. How, if at all, would the above **vector proof** need to be modified if $\angle C$ in $\triangle ABC$ was an obtuse angle?

20 Quadrilaterals $ABCD$ and $AEFG$ are parallelograms with $\overrightarrow{BC} = 4\overrightarrow{AG}$ and $\overrightarrow{DC} = 4\overrightarrow{AE}$. Write a vector proof for each of the following statements.

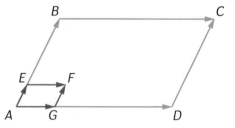

a. $\overrightarrow{AC}$ is a scalar multiple of $\overrightarrow{AF}$.

b. Points A, F, and C are collinear.

21 Jim Chee, a helicopter pilot, wants to fly from Shiprock to Dinnebito in the Hopi-Navajo joint-use region in Hillerman Country. There is a 20-mph northwest wind.

Use the map of Hillerman Country that was provided in Applications Task 1, to help answer the following questions.

a. Suppose Jim leaves Shiprock at 10:00 A.M. and travels at 100 mph heading directly for Dinnebito, with no correction for wind. How far is Jim from Dinnebito? Where will Jim be located at 10:45 A.M.?

b. What course should Jim follow that accounts for the wind and ensures arriving at Dinnebito at 10:45 A.M.?

22 Refer to the nautical chart of the Stone Harbor, Massachusetts region below. Suppose the *Angler* and *Free Spirit* leave the mouth of the channel at Stone Harbor together. Their directions are 55° and 70°, respectively. The *Angler* travels at 4 knots and after 30 minutes sights the *Free Spirit* to the north and west. The line of sight makes an angle of 110° with the traveled path of the *Angler* from the harbor.

a. Draw the situation to scale.

b. Estimate the distance between the boats using the scale drawing.

c. Can vector component analysis be used to determine the distance between the boats? Explain your reasoning.

d. Calculate the distance between the two boats using a method other than component analysis. Compare this distance to your estimate in Part b.

e. At what average speed has the *Free Spirit* traveled?

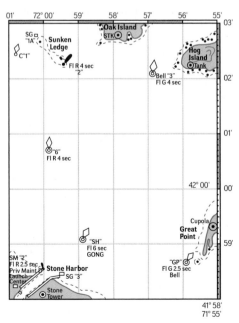

Getty Images/Steve Allen

23 In landscaping an industrial park, a large boulder was to be moved by attaching chains to two tractors that would pull at an angle of 75° between the chains. If one tractor can pull with 1.5 times the force of the other, and the boulder requires a force of 10,000 newtons to be moved, what force is required from each tractor?

REVIEW

24 Consider the circle with radius 5 that is centered at the origin.

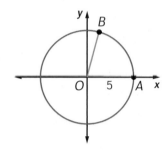

a. If m∠AOB = 75°, find the coordinates of point B.

b. Describe the location of another point B_1 on the circle that has the same x-coordinate as point B. What is the relationship between the y-coordinates of the two points?

c. Describe the location of another point B_2 on the circle that has the same y-coordinate as point B. What is the relationship between the x-coordinates of the two points?

d. Use the circle above to verify that cos 110° = cos 250°.

25 Solve each equation or inequality.

a. $(2x + 6)^2 = 6$

b. $(x - 3)(x + 8)(3x - 5) = 0$

c. $(2x + 1)(x - 5) = 3$

d. $\frac{8}{x} = x + 7$

e. $(x + 4)(3 - 5x) \geq 0$

f. $4x^2 + 8x - 21 > 0$

26 Determine the measures of the remaining sides and angle in each right triangle.

a.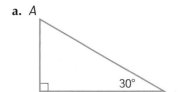

b.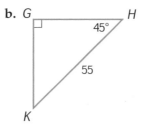

c.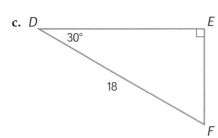

27 Write an equation for each line described below.

a. The line containing the points $(-2, 5)$ and $(-7, -3)$

b. The line parallel to $3x + y = 10$ with x-intercept $(10, 0)$

28 Consider the statement: If $x < 0$ and $y < 0$, then $-(x + y) > 0$.

a. Is the statement true or false? Provide reasoning to support your answer.

b. Write the converse of this statement.

c. Is the converse true or false? Provide reasoning to support your answer.

29 In 2007, 82% of New Hampshire public school seniors indicated that they planned to pursue post-secondary education. (**Source:** www.education.nh.gov/career/guidance/documents/srsurvey2007.pdf) Suppose that you randomly chose two of those seniors.

a. What is the probability that they both indicated that they planned to pursue post-secondary education?

b. What is the probability that neither of them planned to pursue post-secondary education?

c. What is the probability that at least one of them planned to pursue post-secondary education?

30 Rewrite each product in standard polynomial form.

a. $(2x - 5)^2$

b. $(7 + 3x)^2$

c. $(12x - y)^2$

d. $(x^2 + 3y^2)^2$

e. $(x + 1)(x^2 - 8x + 5)$

f. $x^2(x + 1)(3x - 4)$

31 For each graph, write a function rule that matches the graph.

a.

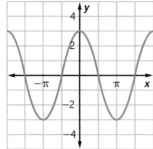

b.

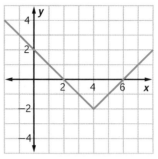

c.

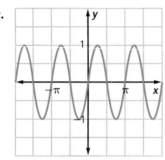

d.

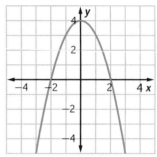

32 Recall that if functions $f(x)$ and $g(x)$ are inverses of each other, then $f(g(x)) = x$ and $g(f(x)) = x$. For each part below, use this relationship to determine whether or not $f(x)$ and $g(x)$ are inverses of each other.

a. $f(x) = 3x - 4$

$g(x) = \frac{1}{3}x + 4$

b. $f(x) = \frac{1}{x} + 2$

$g(x) = \frac{1}{x - 2}$

c. $f(x) = 10^x$

$g(x) = \log x$

33 Three vertices of a parallelogram are $P(0, 0)$, $Q(1, 5)$, and $R(8, 2)$. The fourth vertex S, is a point in the first quadrant.

a. Find the coordinates of vertex S.

b. Find the lengths of each side of the parallelogram.

Vectors and Parametric Equations

The vectors you used in Lesson 1 were located in an east-north coordinate system. Mathematicians and many scientists use a similar system to describe the direction of a vector, namely, by using an x-y coordinate system in which angles are measured counterclockwise from the positive x-axis. This approach allows vectors to be analyzed using coordinate methods.

Suppose a rectangular coordinate system is placed on a nautical map so that the Milwaukee North Shore Marina is located at the origin with the positive *x*-axis pointing east and the positive *y*-axis pointing north. A speedboat leaves the marina in a direction of 30° and proceeds at 16 knots.

a How could the distance and direction traveled after $\frac{1}{4}$ hour be represented by a vector on the coordinate system? After $\frac{1}{2}$ hour? After 1 hour?

b What are the coordinates of the boat's position after $\frac{1}{4}$ hour? After $\frac{1}{2}$ hour? After 1 hour?

c What rules would give the coordinates (*x*, *y*) of the position of the boat at any time *t* (in hours)?

d What rules would give the coordinates (*x*, *y*) of the position of the boat at any time *t* (in hours) if its direction was 40° instead of 30°?

In this lesson, you will learn how to analyze vectors in a coordinate system, how to use coordinate vectors to prove geometric relationships, and how to use vector components to algebraically represent linear motion.

INVESTIGATION **1**

Coordinates and Vectors

In your previous work in *Core-Plus Mathematics*, you saw that there are some strong ties among algebra, geometry, and trigonometry. Flexible and coordinated use of ideas in each of those domains was often helpful in solving problems. As you work on the problems of this investigation, look for further connections among those domains. Begin by looking for answers to this question:

What are some of the advantages of representing vectors in a standard (x, y) coordinate system?

1 In Lesson 1, you represented a vector $\vec{v}$ geometrically as an arrow and in polar form $\vec{v} = [r, \theta]$, where *r* is the magnitude and θ is the direction. The [6, 80°] vector with magnitude 6 and direction 80° is represented on the coordinate system below.

a. Sketch the following vectors on a coordinate grid.

  i. [4, 145°]

  ii. [3, 240°]

  iii. [5, 315°]

b. Sketch the horizontal and vertical components of each of the three "parent" vectors in Part a.

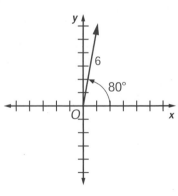

c. Estimate the coordinates of the terminal point of each component vector. In each case, how are the coordinates of the terminal points of the component vectors related to the coordinates of the terminal point of the given "parent" vector? Check your conjecture in the case of the [6, 80°] vector.

d. How can you find the coordinates (x, y) of the terminal point of a vector if you know only its magnitude and its direction? Compare your method with those of your classmates and resolve any differences.

2 Suppose the speedboat in the Think About This Situation traveled at a direction of 130° rather than 30°.

a. Sketch the path of the boat traveling at a speed of 16 knots on a coordinate system and identify its position at $\frac{1}{2}$ hour, 1 hour, and 2 hours into the trip.

b. What are the vector components of each of the boat's positions in Part a?

c. Write rules that give the coordinates of the position of the boat for any time t hours later.

d. What rules represent the coordinates of the position of the boat for any time t hours later and any direction θ?

Vectors in Standard Position A vector with its initial point at the origin of a coordinate system is said to be in **standard position** and is called a **position vector**. Since the terminal point of every position vector has unique rectangular coordinates (x, y), the ordered pair is often identified with the vector. That is, if $\vec{v}$ is a position vector where the terminal point has coordinates (a, b), then we write $\vec{v} = (a, b)$.

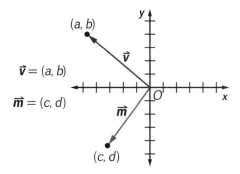

3 Now consider how a standard coordinate system can be used to model linear motion of an aircraft. Suppose after departing New York City, a commercial jet flies at a direction of 190° towards the West Coast at 600 mph. (Assume that this direction and average speed take into account the force exerted by headwinds.)

a. Model this situation by placing New York City at the origin of a coordinate system and sketch the aircraft's path.

b. Find rules for the rectangular coordinates of the position of the aircraft t hours into the flight, assuming no deviation from the course. Then find the coordinates of the aircraft's position after 0.25 hours and 3.2 hours.

c. What are the coordinates of the aircraft's position when it has flown 2,000 miles?

4 Every vector in a coordinate system is equal to some position vector in that system.

    a. Vector $\vec{u}$ has initial point $(-10, 4)$ and terminal point $(-2, 9)$ as shown at the right. What are the coordinates of the position vector that is equal to $\vec{u}$? What is the polar representation of the position vector?

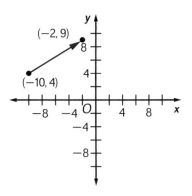

    b. Vector $\vec{v} \neq \vec{0}$ has initial point (a, b) and terminal point (c, d). What are the coordinates of the position vector that is equal to $\vec{v}$? What is the polar form of the position vector?

SUMMARIZE THE MATHEMATICS

In this investigation, you explored representations of vectors in a rectangular coordinate system.

  a Describe the relationships among a vector, its component vectors, and the coordinates of the terminal point when the initial point of the vector is at the origin.

  b Suppose $\overrightarrow{OB}$ is a position vector, and B has coordinates $(r \cos \theta, r \sin \theta)$.

    i. What is the length of $\overrightarrow{OB}$?

    ii. How would you draw $\overrightarrow{OB}$?

    iii. Write $\overrightarrow{OB}$ in polar form.

  c Suppose a position vector $\vec{v}$ is represented in polar form $[r, \theta]$. How can you find its coordinate representation? What does the coordinate representation tell you?

  d Suppose a position vector $\vec{v}$ is represented in coordinate form (x, y). How can you find its $[r, \theta]$ representation?

Be prepared to explain your responses to your classmates.

 CHECK YOUR UNDERSTANDING

Two tugboats are maneuvering a supply barge into a Lake Superior slip. (A slip is a docking place for a boat.) One tugboat exerts a force of 1,500 pounds with direction 340°; the other tugboat exerts a force of 2,000 pounds with direction 70°.

a. Draw the force vectors and the resultant force as position vectors on a coordinate system with the barge at the origin.

b. Determine the coordinate forms of the three vectors. How are they related?

c. Determine the magnitude and direction of the resultant force on the barge.

INVESTIGATION 2

Vector Algebra with Coordinates

In the previous investigation, you explored coordinate representations of vectors. In this investigation, after exploring operations on vectors in a coordinate plane, you will examine how vectors can be used to establish geometric relationships. You will also consider the underlying vector algebra of a video game.

As you work on the following problems, look for answers to these questions:

> *What are some important properties of vectors and their operations?*
>
> *How are these properties similar to, and different from, properties of operations with real numbers?*
>
> *How can vectors and their properties be used to prove geometric statements?*

1 **Vector Algebra** In Lesson 1, you learned about scalar multiples, $k\vec{a}$, of a vector $\vec{a}$ where k is any real number. You also learned about the opposite of vector $\vec{a}$, denoted by $-\vec{a}$.

a. What is meant by a scalar multiple of a vector algebraically? Geometrically?

b. What is meant by the opposite of a vector algebraically? Geometrically?

c. If $\vec{a} = [r, \theta]$ and $0° \le \theta < 180°$, write each vector in polar form.

   i. $3\vec{a}$ ii. $-\vec{a}$

   iii. $-2\vec{a}$ iv. $k\vec{a}$ where $k < 0$

d. Next consider position vector $\vec{b} = (4, -3)$. Write each scalar multiple of $\vec{b}$ below as a position vector in coordinate form.

   i. $5\vec{b}$ ii. $-\vec{b}$

   iii. $-2\vec{b}$ iv. $k\vec{b}$ for any real number k

e. Now generalize your finding in Part d. If position vector $\vec{a} = (x, y)$ and k is a scalar, what are the coordinates of position vector $k\vec{a}$ in terms of $k, x,$ and y? Explain your reasoning.

www.blender.org

2 As you have previously seen, the resultant, or sum, of two vectors can be determined geometrically by using the *head-to-tail* definition of a vector sum or the parallelogram law.

a. Consider position vectors $\vec{a} = (2, -3)$ and $\vec{b} = (-5, 4)$. Draw $\vec{a}$ and $\vec{b}$ on a coordinate system. Then find each of the following vectors in coordinate form.

 i. $\vec{a} + \vec{b}$

 ii. $\vec{b} + (-\vec{a})$

 iii. $-2(\vec{a} - \vec{b})$

 iv. $-2\vec{a} + 2\vec{b}$

b. General position vectors $\vec{a} = (x_1, y_1)$ and $\vec{b} = (x_2, y_2)$ are shown below. Determine the coordinates of the resultant $\vec{a} + \vec{b}$. Use the diagram to help explain your answer.

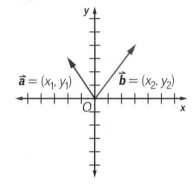

c. Write in words the general principle you discovered in Part b.

d. If $\vec{a} = (x_1, y_1)$, what are the coordinates of $-\vec{a}$ and of $\vec{a} + (-\vec{a})$?

e. How could you interpret $(0, 0)$ as a vector in a motion or force situation? This special vector is called the **zero vector** and is sometimes written $\vec{0}$.

3 In your previous studies, you determined that some properties of real number addition and multiplication are also true for matrices. This is also the case for vectors. Determine which properties below are true for vectors. For each property that you think is true, use general coordinate representations, such as $\vec{a} = (x_1, y_1)$, $\vec{b} = (x_2, y_2)$, and $\vec{c} = (x_3, y_3)$, to write a proof of the statement. For those that you think are false, give a counter example using specific numerical coordinates. Share the work with your classmates and be prepared to explain your proof or counter example.

a. Commutative Property for Addition: $\vec{a} + \vec{b} = \vec{b} + \vec{a}$

b. Associative Property for Addition: $(\vec{a} + \vec{b}) + \vec{c} = \vec{a} + (\vec{b} + \vec{c})$

c. Additive Identity Property: $\vec{a} + \vec{0} = \vec{0} + \vec{a} = \vec{a}$

d. Additive Inverse Property: $\vec{a} + (-\vec{a}) = \vec{0}$

e. Distributive Property for Scalar Multiplication: $k(\vec{a} + \vec{b}) = k\vec{a} + k\vec{b}$

f. Multiplicative Property of Zero: $0\vec{a} = k\vec{0} = \vec{0}$

g. Addition Property of Equality: If $\vec{a} = \vec{b}$, then $\vec{a} + \vec{c} = \vec{b} + \vec{c}$.

h. Scalar Multiplication Property of Equality: If $\vec{a} = \vec{b}$ and k is a scalar, then $k\vec{a} = k\vec{b}$.

4 **Using Vectors to Verify Geometric Properties** You have previously proven geometric statements by reasoning with, and without, the use of coordinates. Vectors provide another method for proving results in geometry.

a. In the vector diagram below, $\vec{a}$ and $\vec{b}$ are the position vectors with terminal points $A(x_1, y_1)$ and $B(x_2, y_2)$, respectively. Explain why $\overrightarrow{AB}$ represents the vector $\vec{b} - \vec{a}$.

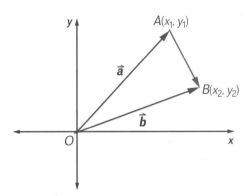

b. On a copy of the diagram, let M be the midpoint of $\overrightarrow{AB}$ and let the position vector with terminal point M be $\vec{m}$. Provide reasons that support the statements in the following proof of the midpoint formula.

$$\overrightarrow{AM} = \overrightarrow{MB} \tag{1}$$

$$\vec{m} - \vec{a} = \vec{b} - \vec{m} \tag{2}$$

$$2\vec{m} = \vec{a} + \vec{b} \tag{3}$$

$$\vec{m} = \frac{1}{2}(\vec{a} + \vec{b}) \tag{4}$$

$$= \left(\frac{x_1 + x_2}{2}, \frac{y_1 + y_2}{2}\right) \tag{5}$$

5 **Designing Video Games** In one video game for cell phones, the protector (player) operates a laser gun located at the center of the bottom of the video screen, the origin of the coordinate system. Meteors fall from the top of the screen straight down. Each meteor drops at a constant rate. But the location of the drop, the drop rate, and the release time vary between meteors. The protector's goal is to fire the laser gun in order to explode each meteor before it causes damage to Earth (the bottom of the screen).

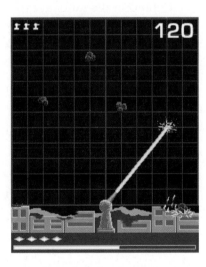

a. Suppose the video screen grid is 12 by 12 units. One meteor drops at a rate of 3 units per second along the line $x = 4$. One second later, a second meteor begins to fall at a rate of 4 units per second along the line $x = -3$.

  i. If $t = 0$ is the time that the first meteor begins to fall, make a table showing the coordinates of the location of each meteor after $t = 0, 1, 2, 3$, and 4 seconds.

  ii. For what time interval is each meteor visible on the screen?

b. Write rules that give the coordinates of each meteor at time *t* in seconds.

c. Suppose the protector aims and fires the laser gun one second after seeing the first meteor begin to fall. Assuming the laser beam hits a target the moment it is fired, at what angle should the gun be aimed to explode that meteor?

d. The protector hits the first meteor after one second, and then takes one more second to turn the laser gun toward the second meteor. Determine the measure of the angle through which the protector should turn the laser gun to hit the second meteor.

6 **Dot Product** The task of finding the measure of the angle between two vectors (as in Parts c and d of Problem 5) occurs frequently in solving applied problems. Look more closely at the mathematics involved.

Let $\vec{a} = (x_1, y_1)$ and $\vec{b} = (x_2, y_2)$ be position vectors as shown in the diagram at the right. Let α (Greek letter "alpha") be the measure of the angle between the two vectors $(0° < \alpha \leq 180°)$.

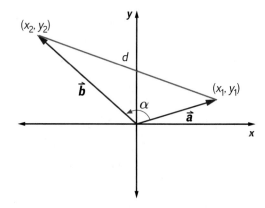

a. Write two expressions for the square of the distance *d* between the terminal points of vectors $\vec{a}$ and $\vec{b}$. One expression should involve cos α.

b. Using the two expressions in Part a, write an expression for cos α in terms of the coordinates $\vec{a}$ and $\vec{b}$.

c. Look at the rational expression you found for cos α. The numerator, $x_1x_2 + y_1y_2$, of the expression is called the **inner product** or **dot product** of the vectors $\vec{a}$ and $\vec{b}$, written $\vec{a} \cdot \vec{b}$ or $(x_1, y_1) \cdot (x_2, y_2)$. Write in words how to calculate the dot product of two nonzero position vectors.

d. Describe the denominator of the rational expression that is equivalent to cos α.

e. Use the expression you found in Part b to help complete the following statement:

> *Two nonzero position vectors are perpendicular if and only if their dot product is _____ .*

Write an argument to support your statement. What two if-then statements must you prove?

7 Explain as precisely as you can why your statement in Part e of Problem 6 is true for *any* two nonzero vectors.

8 Determine the angle α between each pair of vectors where $0° < \alpha \leq 180°$.

a. $(2, 3)$ and $(-3, 2)$

b. $(-2, 1)$ and $(3, -5)$

c. $(-1, -5)$ and $(-3, -2)$

9 What must be true about the measure of the angle between a pair of vectors if their dot product is a positive number? A negative number? Explain your reasoning.

10 Now look back at the video game problem (Problem 5) of Investigation 2.

a. Use the dot product method to determine the measure of the angle needed to turn the laser gun between the first and second shots. Check that your answer is the same as what you found in Problem 5 Part d.

b. Suppose the protector's first shot misses the first meteor, but the protector explodes it with a second shot requiring a total of 1.5 seconds to do so.

    i. What are the coordinates of the first meteor when it explodes?

    ii. Keeping in mind that the protector needs one more second to turn and aim toward the second meteor, at what coordinate location should the protector aim to hit the second meteor?

    iii. Determine the angle through which the protector should turn the laser gun after hitting the first meteor in order to hit the second meteor.

SUMMARIZE THE MATHEMATICS

In this investigation, you learned how to calculate and interpret scalar multiples, sums, and dot products of position vectors. You established useful properties of vector addition and scalar multiplication and determined how to find the angle formed by two position vectors. You also learned how vectors can be used to prove geometric statements.

a Suppose $\vec{a} = (x_1, y_1)$ and $\vec{b} = (x_2, y_2)$. Write $2\vec{a} - 3\vec{b}$ in coordinate form.

b In Problem 3, you proved eight properties of vector operations. Why does it make sense that these real number properties for addition are also true for vectors?

c Explain how the inner product, or dot product, of two position vectors can be used to determine the measure of the angle between the vectors.

d Describe a way to test whether two nonzero position vectors are perpendicular. Why does this method work?

Be prepared to discuss your ideas and reasoning with the class.

✓ CHECK YOUR UNDERSTANDING

Consider the two position vectors $\vec{a} = (4, 2)$ and $\vec{b} = (6, -1)$ graphed on the coordinate system below.

a. Write $4\vec{a} - \vec{b}$ in coordinate form.

b. Let A and B be the terminal points of $\vec{a}$ and $\vec{b}$, respectively. Consider $\angle AOB$, where O is the origin. Using vectors, verify that the midpoint of $\overline{AB}$ is $M\left(5, \frac{1}{2}\right)$.

c. Use the dot product to find the measure of the angle between $\vec{a}$ and $\vec{b}$.

d. Suppose position vector $\vec{c}$ is perpendicular to $\vec{b}$ and the y-coordinate of $\vec{c}$ is -7. Is that enough information to determine a unique x-coordinate of $\vec{c}$? If so, find it. If not, explain why not.

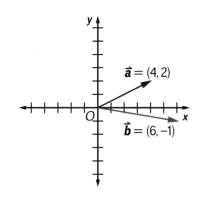

Follow That Dot

In modeling the straight-line motion of a boat, the terminal point of a vector identifies the location of the boat. But motion involves *change* in location over time. A boat, for example, is at different places at different times during a trip.

In this investigation, you will explore how algebraic rules can be used to give the location (x, y) of an object moving in a straight line in terms of the time it has been moving, known as its *elapsed time*. As you complete the following problems, look for answers to this question:

*What are parametric equations and how can
they be used to simulate linear motion?*

1 Suppose the *Wayfarer* begins at the origin of a coordinate system and follows a steady course with a direction of 60° and a speed of 8 knots.

a. Sketch the path of the *Wayfarer* on a coordinate system.

b. Write rules for the horizontal and the vertical components of a point on the ship's path in terms of elapsed time t in hours.

c. A partial table of values for elapsed time and the corresponding horizontal and vertical components is shown at the right. Use an appropriate technology tool to produce a table of values for t up to 2.0 hours.

d. Describe how t changes. How do x and y change as functions of elapsed time t? What are the units of measure for x and y?

Elapsed Time t (in hours)	Horizontal Component x	Vertical Component y
0.0	0.0	0.00
0.1	0.4	0.69282
0.2	0.8	1.3858
0.3	1.2	2.0785
0.4	1.6	2.7713
⋮	⋮	⋮
1.3		
1.4		
1.5		
1.6		
1.7		
1.8		
1.9		
2.0		

e. Make a scatterplot of the (*horizontal component, vertical component*) data. Use the **ZSquare** window on your calculator or an equivalent viewing window with your software.

    i. Describe the pattern in the plot. Explain why this pattern makes sense.

    ii. Does the directed angle determined by the plot make sense?

Parametric Equations Most graphing calculators and computer graphing software have a *parametric function* capability that enables you to quickly construct a table like the one on the previous page. Set your calculator or software to accept angle measures in degrees and parametric equations, and to display multiple graphs simultaneously in dot (not connected) format.

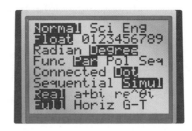

Shown above is the **MODE** screen of one popular graphing calculator that shows the correct settings. You may need to set graph styles differently on your calculator or software.

2 Next choose the [Y=] menu. Notice that the equations are paired. For the first pair, enter your rules for the *x* and *y* components from Problem 1 Part b. Your display should be similar to one of the two screens below.

 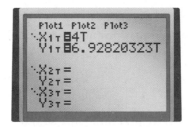

a. Why do both of these displays represent how *x* and *y* change with respect to time *t*?

b. Now use the table-building capability of your technology to generate a table for **T**, **X₁ₜ**, and **Y₁ₜ** for **0 ≤ T ≤ 2** with increments of 0.1 for **T**. Compare this new table with the one you completed in Problem 1. Do the patterns of change you noted for *t*, *x*, and *y* continue in this extended table?

Once you have the rules for **X₁ₜ** and **Y₁ₜ** entered in your calculator or software, you can display a graph of the model for the path of the ship. As with other graphical displays, you need to first set the viewing window. The settings shown at the right do the following:

- Since the independent variable here is **T**, **Tmin = 0** sets the calculator to begin evaluating **X₁ₜ** and **Y₁ₜ** at **T = 0**.

- **Tstep = 0.1** increments **T** by 0.1 at each step until **T** is larger than **Tmax**.

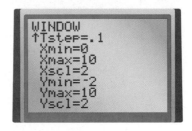

- The **X** and **Y** settings establish the lower and upper bounds of the viewing screen.

3 Now set your calculator or computer software to conform to the conditions at the bottom of page 139. Then plot the (x, y) pairs from Problem 2 using the **ZSquare** (or equivalent) window.

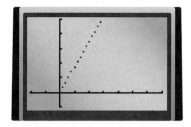

a. Compare your display with the one shown at the right. If they differ, check each menu and your graph style settings.

b. Explore your display by tracing along it. Compare the values of **T**, **X**, and **Y** shown on the screen with those in your table.

c. Experiment with various ranges and step sizes for **T**. Experiment with this new graphing tool.

   i. How many points are displayed? Why?

   ii. What settings will show 21 points?

   iii. How can you produce a graph that begins at a point other than the origin?

   iv. How can you make sure all the points are displayed on the screen?

Modeling Linear Motion The rules you used to generate the coordinates for the terminal point of the vector locating the *Wayfarer* at any time t are called **parametric equations** and the variable t is called a **parameter**. In this case, your parametric equations were:

$$x = 8t \cos 60° \qquad \text{or} \qquad x = 4t$$
$$y = 8t \sin 60° \qquad\qquad y = 6.92820323t$$

In the next several problems, you will explore ways to use parametric equations to model linear motion in different situations.

4 Suppose a commercial jet leaves Los Angeles International Airport on a course with direction 15°; a direction that takes into account the force exerted by tailwinds.

a. Develop a parametric equation model for the location of the plane after t hours if the airport is placed at the origin of the coordinate system, and the plane flies a straight path at a constant speed of 600 mph.

b. How far east has the plane traveled in 2.5 hours? How far north?

c. Simulate the motion of the aircraft by showing its position every half hour until it reaches the East Coast, about 3,100 miles east of Los Angeles. List your viewing window settings.

d. How would you modify your simulation to show the location of the plane every 12 minutes?

5 Now consider how parametric equations could be used to simultaneously model the motion of two boats. The given directions and speeds take into account forces exerted by current or wind. Begin by turning off the axes on your graphing screen.

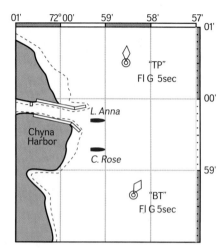

a. Suppose the *Charlotte Rose* leaves its anchorage at noon going due east at a constant speed of 8 knots. Write parametric equations for this motion.

b. Find the time needed to travel 50 nm. Display the 50-nm path on your screen. Trace your graph.

c. Suppose the *Lady Anna* begins at 1:00 P.M. from the harbor 0.4 nm north of *Charlotte Rose*'s starting location, traveling due east at 10 knots. One pair of parametric equations for this motion are $X_T = 10T$ and $Y_T = 0.4$. Display the path.

d. For the parametric model of the *Lady Anna*'s path in Part c, what time of the day is it when $T = 0$? What value of T would represent noon, the time the *Charlotte Rose* leaves her anchorage?

e. To devise a way to display the paths of both the *Charlotte Rose* and the *Lady Anna* so that you can see the boats moving simultaneously, it is helpful to have both parametric models representing the same time of the day for T values. How could you modify the parametric equations in Part c to represent noon when $T = 0$?

f. Use your display to decide which boat travels 50 nm first. Where is the other boat when the first boat has gone 50 nm? At what time will they have traveled the same distance?

g. How would you change your parametric equations if each boat were traveling due north of the original starting point for 50 nm? Make and test the changes.

6 In previous courses, you developed the ability to predict the shape of the graph of various functions by examining their symbolic rules. This problem will help you extend your symbol sense to parametric equations. For each pair of parametric equations:

- predict what the graph will look like including starting and ending locations, then check your prediction using technology.

- make a sketch of the displayed graph and label it with the corresponding pair of equations.

For each case, use **Tmin = 0**, **Tmax = 4**, and **Tstep = 0.1**. The viewing window should be $-40 \leq X \leq 40$ and $-40 \leq Y \leq 40$, with **Xscl = 10, Yscl = 10**.

a. $x = 5t$
$y = 8$

b. $x = -5t$
$y = 8$

c. $x = 6(t - 2)$
$y = -20$

d. $x = 10$
$y = 5t$

e. $x = 10$
$y = -5t$

f. $x = 8$
$y = -4(t - 5)$

7 Look back at the parametric equations and graphs you produced in Problem 6.

a. What general patterns do you see that relate the shape and placement of a graph to the symbolic form of the equations?

b. Write a pair of parametric equations different from those in Problem 6, but with the same shape graph. Trade equations with a partner. Predict what the graph of your partner's equations will look like and then test your prediction. If either your prediction or that of your partner is incorrect, identify the possible cause of the error and then repeat for a different pair of equations.

c. Why do the following parametric equations have the same graph as the parametric equations in Part a of Problem 6?

$$x = 5t \cos 0°$$
$$y = 5t \sin 0° + 8$$

d. Why do the following parametric equations have the same graph as the parametric equations in Part d of Problem 6?

$$x = 5t \cos 90° + 10$$
$$y = 5t \sin 90°$$

e. Determine which of the remaining pairs of parametric equations in Problem 6 can be represented in the form below. Explain your reasoning.

$$x = At \cos θ + B$$
$$y = At \sin θ + D$$

The **general form of parametric equations** for linear motion with a constant velocity is:

$$x = At \cos θ + B$$
$$y = At \sin θ + D$$

As you noted in Problem 7, when $θ = 0°$, these equations simplify to $x = At + B$ and $y = D$. Also, when $θ = 90°$, they simplify to $x = B$ and $y = At + D$.

8 For each part below, sketch a graph of each pair of parametric equations for $0 ≤ t ≤ 2$. Describe how the three graphs in each part are related.

a.
Set 1	**Set 2**	**Set 3**
$x = t \cos 40°$	$x = 2t \cos 40°$	$x = t \cos 40° + 1$
$y = t \sin 40°$	$y = 2t \sin 40°$	$y = t \sin 40° - 2$

b.
Set 1	**Set 2**	**Set 3**
$x = 2t - 1$	$x = 4t - 3$	$x = t$
$y = t$	$y = 2t - 2$	$y = 2t - 1$

9 Now apply your understanding of modeling with parametric equations to the following situation.

Suppose the cabin cruiser *Sawatdee* begins at noon heading due east at 8 knots. The *Delhi Dhaba* begins at noon at a location 60 nm due east of the *Sawatdee* and heads due west at 10 knots.

a. Represent this situation on a coordinate system.

b. Write parametric equations for each motion.

c. At what location and time will they meet?

Motion of a Dropped Object You may recall from your work with power models in previous courses that when an object is dropped from a height above the surface of Earth, the velocity and distance traveled are functions of gravity and time. Neglecting air resistance, the average velocity after t seconds is $-4.9t$ meters per second or $-16t$ feet per second. Thus, the directed distance an object falls in t seconds due to gravity is $(-4.9t \text{ meters/second})(t \text{ seconds}) = -4.9t^2$ meters. In feet, the directed distance traveled is $-16t^2$ feet.

10 Suppose an object is released from a hot air balloon 200 meters above the surface of Earth.

a. Explain why the height in meters of the object above the surface of Earth after **T** seconds can be represented by these parametric equations:

$$X_T = 1 \quad \text{and} \quad Y_T = -4.9T^2 + 200$$

b. For a specific value of **T** (such as 1.34 seconds), what does the corresponding value of **Y**$_T$ tell you about the object?

c. How many seconds after the drop does the object strike the earth; that is, reach the *x*-axis? How did you determine the time?

d. Select appropriate values for **T** and a viewing window to display the motion of the object. Use **Dot** mode. Sketch the display, including the window and **T** values.

e. Describe how the distance the object falls per second changes with increasing time. How can this be observed in the graph? In a table?

f. In this example, **X**$_T$ = 1. Is it important that **X**$_T$ be 1, or could it be another number? Explain your reasoning.

g. Write a pair of parametric equations that describes the height in feet of an object that is dropped from a hot air balloon 150 feet above the surface of Earth.

In this investigation, you explored how parametric equations can be used to model linear motion.

 a How do parametric equations differ from other algebraic equations you have studied?

b How are parametric equations of a point moving along a line related to vectors?

c Describe how you would write parametric equations of:

    i. horizontal linear path at a constant velocity.

    ii. a vertical linear path at a constant velocity.

    iii. an oblique linear path through the origin at a constant velocity.

Be prepared to share your ideas and reasoning with the class.

✔️ CHECK YOUR UNDERSTANDING

Average velocity, after t seconds, of a falling object differs among celestial bodies due to the differing forces of gravity on each body.

Earth	**The Moon**	**Jupiter**

$-4.9t$ m/setc^2	$-0.83t$ m/sec^2	$-11.44t$ m/sec^2

a. Suppose for each of the celestial bodies—Earth, the Moon, and Jupiter—an object is dropped from 100 meters above its surface. Write parametric representations of the falling motion for each case that can be viewed on the same display.

b. Display the motion of the objects in the same viewing window. Describe differences and similarities in the patterns of change.

c. Find the time it would take for the object to strike the surface of Earth. The Moon. The planet Jupiter.

1 Two boaters leave Ludington, Michigan, at 8:00 A.M. One is going to Manitowoc, Wisconsin, at a direction of 170°. The other heads for Milwaukee at a direction of 220°. Ludington is about 60 miles from Manitowoc and about 97 miles from Milwaukee. The radios on the boats are good for distances up to 50 miles.

a. With Ludington as the origin, set up a coordinate system. What are the coordinates of Manitowoc and Milwaukee?

b. Sketch a vector diagram if the boat to Manitowoc travels at 8 mph and the boat to Milwaukee travels at 10 mph. How far from each other are the boats at 9:00 A.M.? At 11:00 A.M.?

c. At about what time will they lose radio contact?

d. How far from their destinations are the boats when they lose radio contact?

2 Suppose $\overrightarrow{OA}$ and $\overrightarrow{OB}$ are position vectors. The length of $\overrightarrow{OA}$ is r and its direction is θ. $\overrightarrow{OB}$ has length s and direction ϕ (Greek letter "phi").

a. What are the components of $\overrightarrow{OA}$ and of $\overrightarrow{OB}$?

b. What are the coordinates of points A and B?

c. Find the coordinates of the position vector $\overrightarrow{OA} + \overrightarrow{OB}$.

d. What are the coordinates of $-\overrightarrow{OA}$?

e. What are the coordinates of the vector $\overrightarrow{OB} - \overrightarrow{OA}$ for $\overrightarrow{OA}$ and $\overrightarrow{OB}$ as described above? Show that these coordinates are equal to those of $\overrightarrow{OB} + (-\overrightarrow{OA})$.

3 Two families of hikers leave a base camp on a mesa traveling in directions of 31° and 42°, respectively. The first family averages about 0.8 mph while the second family averages 1.1 mph.

a. Sketch the hiking paths of the two families on a standard coordinate system. Assume the families continue to hike in the directions they started and at the indicated rates.

b. At the end of one hour, what are the coordinates of their positions? How far apart are they?

c. How far apart are the families after 2 hours? After 3 hours?

d. How does the distance D between the families change as a function of time t?

e. After how much time will they be about 1.5 miles apart?

4 A cell phone game is displayed on the first quadrant of a rectangular coordinate system. The player sights the action on the screen from the origin (lower-left corner). The position vector of an adventurer is (8, 3).

a. What are the magnitude and direction of the adventurer's position vector?

b. Suddenly, the adventurer disappears, but then reappears at (4, 6). Find the measure of the angle between the adventurer's two position vectors.

c. The adventurer disappears a second time, but the player has figured out a pattern for its reappearance. It will reappear at a point on the angle bisector of its previous two position vectors. What will be the angle of the adventurer's position vector when it appears on the screen for a third time?

d. If the magnitude of the adventurer's third position vector is the average of the magnitude of its two previous position vectors, what is the magnitude of the adventurer's third position vector?

e. What are the coordinates of the adventurer's third position vector?

5 Recall that scalar multiplication is a way to multiply a vector by a real number. Which of the following statements about scalars c and d, vector $\vec{a}$, and their products and sums are true? Use position vectors in coordinate form and properties of real numbers to justify your answers. If a statement is false, give a counterexample.

a. $c(d\vec{a}) = (cd)\vec{a}$

b. $(c + d)\vec{a} = c\vec{a} + d\vec{a}$

c. $(c - d)\vec{a} = (d - c)\vec{a}$

d. $c(-\vec{a}) = (-c)\vec{a} = -(c\vec{a})$

6 In Investigation 2 Problem 5 (page 135) of this lesson, you examined a video game that was designed on a 12 by 12 unit rectangular coordinate screen with the origin at the center of the bottom of the screen. Meteors fall from the top of the screen straight down.

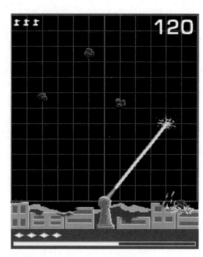

a. In the earlier problem, one meteor fell at a rate of 3 units per second along the line $x = 4$ starting at time $t = 0$ seconds. At time $t = 1$, a second meteor began to fall at a rate of 4 units per second along the line $x = -3$. Use technology to simulate the fall of these meteors.

b. Simulate a meteor shower in which one meteor falls along each of the vertical lines $x = -2, -1, \ldots, 2, 3$ each second in an irregular order that you decide on. A meteor that falls along a vertical line with an even x-coordinate falls at a rate of 3 units per second, and a meteor that falls along a vertical line with an odd x-coordinate falls at a rate of 4 units per second. It should take 5 seconds for all 6 meteors to begin to fall.

c. In reality, a meteor would not fall at a constant rate as it does in this video game; rather its speed would increase at an increasing rate due to gravity. Suppose each coordinate unit on the screen represents 100 meters. Refer to Investigation 3 Problem 10 (page 143) for guidance in simulating a meteor in this video game that is dropped toward earth from 1,200 meters along the line $x = 1$.

  i. How many seconds after the drop does the meteor strike the earth?

  ii. What is the average velocity of the meteor during its 1,200-meter fall?

7 Consider these two sets of parametric equations.

Set 1	Set 2
$x = t$	$x = 2t + 2$
$y = 2t - 3$	$y = 4t + 1$

a. Draw the graph of each pair of parametric equations on its own coordinate system. How are the two graphs related?

b. Find the x and y values for both sets of parametric equations when $t = 1$. Compare the results.

c. For each pair of equations, find a value of t that gives the position $(0, -3)$.

Matt Tuley/CPMP

CONNECTIONS

8 In Investigation 1, you represented a position vector, such as $\overrightarrow{OP}$, in terms of the rectangular coordinates (x, y) of its terminal point. In this case, x is the horizontal directed distance of the point from the y-axis and y is the vertical directed distance of the point from the x-axis. Sometimes you also used the form [*magnitude, direction*] or polar form $[r, \theta]$ to represent a position vector.

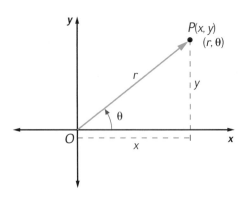

In the case of images on a radar screen, it is more useful to use the polar form to specify the location of a point. The labels r and θ are called the **polar coordinates** of point P.

a. Referring to the diagram above, explain why the coordinates (x, y) of a point P in a rectangular coordinate system are related to the coordinates $[r, \theta]$ of the point in a polar coordinate system by the following equations:

$$x = r \cos \theta \text{ and } y = r \sin \theta, \text{ where } r = \sqrt{x^2 + y^2}$$

b. Determine the rectangular coordinates of each point with the given polar coordinates:

    i. $A\left[10, \dfrac{\pi}{2}\right]$ **ii.** $B[5, 120°]$ **iii.** $C[4, 315°]$

c. Determine the polar coordinates of each point with the given rectangular coordinates:

    i. $Q(3, 3\sqrt{3})$ **ii.** $P(-5, 0)$ **iii.** $R(-6, 6)$

9 In the diagram at the right, $\overline{AB}$ is divided by point M in the ratio $AM{:}MB = 2{:}1$.

a. Copy and complete each statement below. All statements except the first should be completed using combinations of the vectors $\vec{a}$, $\vec{b}$, and $\vec{m}$.

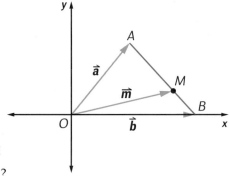

$\overrightarrow{AM} = \underline{\ ?\ }\ \overrightarrow{MB}$

    if and only if $\vec{m} - \vec{a} = 2(\underline{\ ?\ })$

    if and only if $\vec{m} - \vec{a} = \underline{\ ?\ } - 2\vec{m}$

    if and only if $\vec{m} + \underline{\ ?\ } = 2\vec{b} + \underline{\ ?\ }$

    if and only if $3\vec{m} = \underline{\ ?\ } + \underline{\ ?\ }$

    if and only if $\vec{m} = \underline{\ \ \ ?\ \ \ }$

b. Try to generalize the result in Part a. What if $\overline{AB}$ is divided by point M in the ratio 3:1? 5:1? k:1 for $k > 1$?

Yuri Arcurs/Alamy

10 In Course 2, you learned about matrices and their operations. In particular, you learned how to multiply a matrix by a scalar and how to add two matrices. Recall that a matrix with just one row, like $[6 \ \ -3]$ or $[-4 \ \ 0 \ \ 1 \ \ 3]$ is called a *row matrix*. A matrix with just one column is called a *column matrix*.

a. Perform the following matrix operations.

    i. $5[6 \ \ -3] + 2[-4 \ \ 5]$

    ii. $5\begin{bmatrix} 6 \\ -3 \end{bmatrix} + 2\begin{bmatrix} -4 \\ 5 \end{bmatrix}$

b. Perform the following operations on position vectors: $5(6, -3) + 2(-4, 5)$

c. How are the matrix operations in Part a similar to one another? Similar to the vector operations in Part b? How are they different? (Because of similarities like these, row matrices and column matrices are sometimes called *row vectors* and *column vectors*.)

11 A translation is completely determined by a vector, as was shown in Connections Task 8 (page 120) in Lesson 1. Reconsider the diagrams from that task, each showing a figure *F* and its image *G* under a translation.

Translation I	**Translation II**
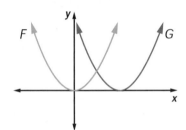	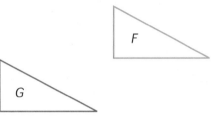

a. Suppose in Translation I, each image point of figure *G* is 5 units to the right of the corresponding preimage point (x, y) of figure *F*.

    i. Represent the translation vector in coordinate form.

    ii. If $(-1, 1)$ is a point on figure *F*, show how to use the translation vector to find the image of $(-1, 1)$.

b. Suppose in Translation II, each image point of figure *G* is 10 units to the left and 4 units below the corresponding preimage point (x, y) of figure *F*.

    i. Represent the translation vector in coordinate form.

    ii. If $(6, 4)$ is a point on figure *F*, show how to use the translation vector to find the image of $(6, 4)$.

c. The inverses of the translations in Parts a and b map figure *G* to figure *F*. What is the translation vector of the inverse translation in Part a? In Part b?

d. Of what geometric transformation does the scalar product of a position vector of the form $k(x, y)$ remind you? Discuss how scalar multiplication of a vector is like this kind of transformation and how it is different.

12 Solve each vector equation for $\vec{x}$.

a. $3(\vec{x} + \vec{u}) = 12\vec{v}$

b. $2\vec{x} + 5\vec{u} = 7\vec{v} + 5\vec{x}$

c. $6(\vec{x} - \vec{a}) = 4(\vec{x} + \vec{b})$

d. $2\vec{x} + (4, -1) = (5, 9) + \vec{x}$

13 Use the general coordinate forms of position vectors and properties of real numbers to prove these properties of dot products of vectors. Let $\vec{u} = (x_1, y_1)$, $\vec{v} = (x_2, y_2)$, and $\vec{w} = (x_3, y_3)$.

a. $\vec{u} \cdot \vec{v} = \vec{v} \cdot \vec{u}$

b. $\vec{u} \cdot (\vec{v} + \vec{w}) = \vec{u} \cdot \vec{v} + \vec{u} \cdot \vec{w}$

c. $\vec{v} \cdot \vec{v}$ is equal to the square of the magnitude of $\vec{v}$.

14 In Unit 1, *Families of Functions*, you used the "sliders" functionality of the CPMP-Tools CAS to investigate the effect of changing parameters of familiar functions on their graphs and tables of values. Use the parametric feature of the CPMP-Tools CAS or similar software to explore the connections among transformations, the symbolic form of parametric rules, and their graphs using the sliders as shown in the display at the right.

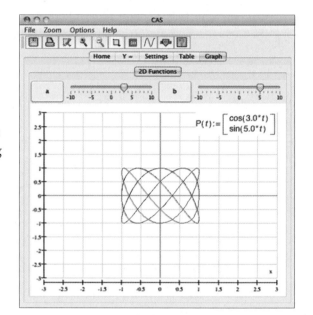

Explore the function

$$P(t) := \begin{bmatrix} a*\cos(t)+c \\ b*\sin(t)+d \end{bmatrix}.$$

a. What is the effect on the graph when one of either a or b is zero?

b. What is the effect on the graph for various values of a and b? Describe in terms of specific function transformations.

c. What is the effect on the graph for various values of c and d? Describe in terms of specific function transformations.

d. Pose and answer a question of your own about the nature of curves represented parametrically.

15 The graph of each pair of parametric equations below is a line.

Set 1	Set 2	Set 3
$x = 2t$	$x = -3t$	$x = 2t + 1$
$y = 3t$	$y = 2t$	$y = -t - 2$

Set 4	Set 5
$x = at, a \neq 0$	$x = at + c, a \neq 0$
$y = bt$	$y = bt + d$

a. For each pair of parametric equations, determine the slope of the line.

b. For each pair of parametric equations, combine the two equations into a single rule that expresses y as a function of x.

c. In the case of Set 5, what is the equation of the line if $a = 0$?

16 Geometric transformations such as rotations, line reflections, and size transformations (dilations) can be represented by matrices. Multiplying the transformation matrix by a column vector on the right (see Connections Task 10) transforms the position and/or size of the vector. For each of the following transformations represented by matrix multiplication, sketch the original vector and the image vector. Then identify the type of transformation.

a. $\begin{bmatrix} -1 & 0 \\ 0 & -1 \end{bmatrix} \begin{bmatrix} 6 \\ -3 \end{bmatrix}$
b. $\begin{bmatrix} 2 & 0 \\ 0 & 2 \end{bmatrix} \begin{bmatrix} 6 \\ -3 \end{bmatrix}$
c. $\begin{bmatrix} 1 & 0 \\ 0 & -1 \end{bmatrix} \begin{bmatrix} 6 \\ -3 \end{bmatrix}$

REFLECTIONS

17 Suppose the coordinates of the terminal point of a position vector are (a, b).

a. How could these numbers be used to calculate the length of the vector?

b. How do the signs ($+$ or $-$) on each coordinate help you sketch the vector? For example, if a and b are both negative, in which quadrant would you draw the vector?

c. Explain how the coordinates (a, b) can be used to determine the direction of the vector.

18 A basic structural property of the set of real numbers is that it is **closed under addition**, that is, the sum of any two real numbers is always a real number.

a. The set of real numbers is also **closed under multiplication**, but not under division. Why?

b. Is the set of vectors closed under vector addition? Why or why not?

c. Is the set of vectors closed with respect to dot products? Why or why not?

19 Suppose the angle between two vectors $\vec{u}$ and $\vec{v}$ is α.

a. $\cos \alpha$ can be written as a rational expression whose numerator is $\vec{u} \cdot \vec{v}$. In this form, what is the denominator of $\cos \alpha$?

b. Extending your reasoning in Part a, how could $\vec{u} \cdot \vec{v}$ be written in terms of $\vec{u}$, $\vec{v}$, and $\cos \alpha$?

c. If the magnitudes of $\vec{u}$ and $\vec{v}$ are 6 and 8 units, respectively, and $\alpha = 60°$, what is $\vec{u} \cdot \vec{v}$?

20 What can you say about the angle between two vectors whose dot product is positive? Negative? Zero?

21 Linear motion through the origin can be modeled with parametric equations of the form

$$x = At \cos \theta$$
$$y = At \sin \theta$$

where A and θ are constants.

a. What does A represent? What does θ represent?

b. How would these equations be modified if $(x, y) = (4, 0)$ when $t = 0$?

c. How would these equations be modified if $(x, y) = (0, -3)$ when $t = 0$?

EXTENSIONS

22 Equations for certain graphs are much simpler when expressed in polar coordinates (see Connections Task 8 on page 148) than in rectangular coordinates. Sometimes, graphs that have simple equations in rectangular coordinates have more complicated equations in polar coordinates.

a. Describe in words the graph of $y = 2$ (in rectangular coordinates) and $r = 2$ (in polar coordinates).

b. Write the equation $y = 2$ using polar coordinates. Write the equation $r = 2$ using rectangular coordinates.

c. Draw the graph of $r = \theta$ with θ in degrees. It might help to first complete a table of values like that below.

θ	0°	30°	60°	90°	120°	150°	180°	210°	240°	270°
r										

d. How would you describe the graph in Part c?

23 In the Course 3 *Similarity and Congruence* unit, you discovered that the medians of a triangle are concurrent at a point called the *centroid*. The distance from each vertex to the point of concurrency is $\frac{2}{3}$ the length of the median. That is, the centroid divides each median in the ratio 2:1. In this task, you will complete a vector proof of this important result.

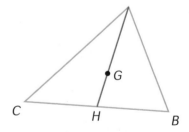

In the diagram at the right, G divides median $\overline{AH}$ in the ratio 2:1. You can use position vectors to prove that G is the point of concurrency of the three medians. Position $\triangle ABC$ so that C is at the origin and $\overline{CB}$ lies on the x-axis. Let, $\vec{g}, \vec{h}, \vec{a}, \vec{b}$, and $\vec{c}$ be the position vectors of points G, H, A, B, and C, respectively. Supply reasoning that supports each statement in the following proof.

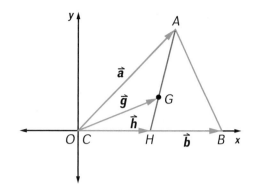

a. $\overrightarrow{OG} = \overrightarrow{OA} + \overrightarrow{AG}$

if and only if $\quad \overrightarrow{OG} = \overrightarrow{OA} + \frac{2}{3}\overrightarrow{AH}$ $\qquad$ (1)

if and only if $\quad \vec{g} = \vec{a} + \frac{2}{3}(\vec{h} - \vec{a})$ $\qquad$ (2)

if and only if $\quad = \vec{a} + \frac{2}{3}\left[\frac{1}{2}(\vec{b} + \vec{c}) - \vec{a}\right]$ $\qquad$ (3)

if and only if $\quad = \vec{a} + \frac{1}{3}\vec{b} + \frac{1}{3}\vec{c} - \frac{2}{3}\vec{a}$ $\qquad$ (4)

if and only if $\quad = \frac{1}{3}(\vec{a} + \vec{b} + \vec{c})$ $\qquad$ (5)

b. Use similar reasoning to show that if J divides the median $\overline{BM}$ in the ratio 2:1, then $\overrightarrow{OJ} = \vec{j} = \frac{1}{3}(\vec{a} + \vec{b} + \vec{c})$.

c. Prove that if K divides the median $\overline{CN}$ in the ratio 2:1, then $\overrightarrow{OK} = \vec{k} = \frac{1}{3}(\vec{a} + \vec{b} + \vec{c})$.

d. Why are the medians concurrent at point G?

e. Why does the centroid divide each median in the ratio 2:1?

24 The graph of this pair of parametric equations is a line.

$$x = 2t + 1$$
$$y = -t - 2$$

Write these equations in the general parametric form for linear motion with a constant velocity:

$$x = At \cos \theta + B$$
$$y = At \sin \theta + D$$

25 Delta Airlines® has hubs at Atlanta, Cincinnati, Detroit, Memphis, Minneapolis, New York, and Salt Lake City.

Flights from Memphis to Seattle and Detroit to Los Angeles leave at the same time and cruise at 32,000 ft. The direction of the flight out of Memphis is 143° and the direction of the flight out of Detroit is 190°. The still air speed of each airliner is 600 mph.

a. Memphis is approximately 400 miles west and 500 miles south of Detroit. Write parametric equations for the paths each plane will follow if they head directly toward their destinations.

b. On a particular day, a 70-mph upper-level northwest wind is blowing. At what direction should each plane fly to counteract the effects of the wind?

   i. What is the wind-affected speed of each plane?

   ii. Is there danger of a mid-air collision? If so, at what location? Explain your reasoning.

c. Seattle is about 1,800 miles from Memphis. Los Angeles is about 2,000 miles from Detroit. About how long will it take each plane to reach its destination at its wind-affected speed?

26 Two freighters leave port at the same time. On a coordinate system, one port is located at $A(20, 0)$ and the other at $B(-15, -4)$. The freighter *Mystic Star* leaves port A steaming west at 8 knots at a direction of 170°; the *Queensland* steams east from port B at a direction of 20° at 10 knots.

a. Write parametric equations for each of these routes.

b. Represent the path of each of these freighters on your graphing calculator or computer graphing software. Sketch the paths on your paper.

c. Is there a danger that the freighters will collide? If so, at what location? Explain.

d. What is the location of the *Mystic Star* when the *Queensland* crosses its path? What is the location of the *Queensland* when the *Mystic Star* crosses its path?

REVIEW

27 Determine the indicated side length and angle measures.

a. AC and m$\angle C$

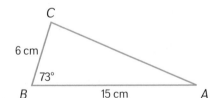

b. m$\angle E$ and m$\angle F$

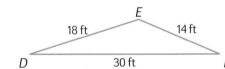

28 If possible, factor each expression as a product of linear factors.

a. $x^2 + 3x - 18$

b. $9 - 16x^2$

c. $x^2 + x + 5$

d. $25x^2 + 20x + 4$

29 The number of students enrolled in Ryder High School is currently 1,628. The enrollment is expected to increase by 1% each year.

   a. Use this information to determine the expected enrollment for each of the next five years. Is the sequence of enrollment numbers a geometric or arithmetic sequence? Explain your reasoning.

   b. Write a recursive formula beginning "$E_n = ...$" for the sequence of enrollment numbers.

   c. Write a function formula beginning "$E(n) = ...$" for the sequence of enrollment numbers.

   d. Ryder High School can accommodate a maximum of 2,000 students. If this growth rate continues, how long will it be before the school building is too small for the student population?

30 In the circle with center O at the right, $m\overset{\frown}{AC} = 130°$, $\overline{AD}$ is a diameter of the circle, and $AC = 10$ cm.

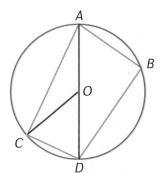

   a. Determine $m\angle AOC$.

   b. Determine $m\angle ABD$.

   c. Determine $m\angle ADC$.

   d. Determine the radius of the circle.

31 Let $f(x) = 3x^2 + 6x + 2$ and $g(x) = 5x - 3$. Find a rule in standard polynomial form for each new function.

   a. $h_1(x) = f(x) + g(x)$ **b.** $h_2(x) = f(x) \cdot g(x)$

   c. $h_3(x) = f(g(x))$ **d.** $h_4(x) = g(f(x))$

32 The height (in feet) of a basketball pass is a function of time (in seconds) since it was released and can be modeled by the function $h(t) = -16t^2 + 20t + 5$.

   a. Explain what the coefficients -16, 20, and 5 tell you about the situation.

   b. If the basketball is not caught or intercepted, how long will it be in the air?

   c. What is the maximum height of the basketball?

33 Angles of rotation can be measured in revolutions, degrees, and radians. Fill in each blank with the correct value.

   a. $\frac{1}{2}$ revolution = _____ degrees = _____ radians

   b. _____ revolutions = _____ degrees = $\frac{3\pi}{4}$ radians

   c. _____ revolutions = 30 degrees = _____ radians

34 Rewrite each product as a single algebraic fraction. Simplify the result as much as possible.

   a. $\left(\frac{4x^2}{9}\right)\left(\frac{15}{4x}\right)$ **b.** $\left(\frac{3x}{12}\right)\left(\frac{x+4}{x}\right)$ **c.** $\left(\frac{x^2-9}{x+1}\right)\left(\frac{x}{x+3}\right)$

LESSON 3

Modeling Nonlinear Motion

In Lesson 1, you modeled linear motions of boats, characters in a video game, and airplanes in flight. There are many other kinds of motions that are *nonlinear*. The wheels of an automobile or bicycle rotate around their centers. Satellites orbit Earth. A catcher for a softball or baseball team with a weak arm throws a "rainbow" to second base. A tennis player serves a ball so it clears the net and lands in the service box. A professional golfer drives a ball over 300 yards to the middle of the fairway. You and a group of friends may play volleyball or compete in a friendly game of darts. In each of these contexts, both direction and distance are important elements.

Carnivals and fairs often include games in which you throw a baseball at a pyramid of bottles. Imagine that the bottles are 6 meters away. Your goal is to throw a ball that knocks over all the bottles.

a What are important variables that may affect whether you are successful?

b What are some factors that affect the path of the ball?

c If you can throw hard, where should you aim?

d Suppose you threw the ball on a slight arc and it fell short of the table holding the bottles. How would you adjust your throw for the next try?

In this lesson, you will extend the methods you learned for simulating linear motion to include nonlinear motions. By building graphing calculator or computer-based simulation models, you will be able to see representations of the paths traveled by moving objects. You will even be able to simulate the motion of balls in flight and of seats on a rotating Ferris wheel.

INVESTIGATION 1

What Goes Up, Must Come Down

You have seen that linear motion can be described in terms of two components, a horizontal component and a vertical component. Objects such as boats, trains, or airplanes can move along a straight-line path because energy is applied to maintain the speed and direction. However, some moving objects, such as projectiles, are subject to forces such as friction or gravity, and these affect both the speed and direction of the object causing it to follow a nonlinear path.

As you work on the problems of this investigation, look for answers to the following question:

How can parametric equations be used to simulate nonlinear motion like that of a projectile?

Slow-Pitch Softball In the game of slow-pitch softball, the pitcher throws the ball underhand so that it goes high in the air and crosses the plate as it comes down. The batter tries to hit the ball on its way down.

1 Consider a pitcher on a slow-pitch softball team who throws the ball toward home plate at a speed of 12 meters per second at an angle of 55° with the horizontal. The pitcher tries to get the ball to drop nearly vertically across the plate.

   a. For the moment, assume that the speed of the ball is constant at 12 m/sec. On a coordinate system, sketch the vector showing the position of the ball after 1 second and the horizontal and vertical components.

   b. Write parametric equations describing the position, with respect to time, of a ball pitched at an angle of 55° to the ground with a constant velocity of 12 m/sec. Ignore the effect of gravity. What is the unit of the parameter t in the equations?

   c. Using technology, display the graph in **Dot** mode.

   d. What are some limitations of this model?

2 Refine your model in Problem 1 to take into account the effect of gravity on the softball pitch. Recall that the effect of gravity on the position of a falling object is represented by a vector $4.9t^2$ meters long and pointing straight down $(0, 4.9t(t \sin 270°)) = (0, -4.9t^2)$.

   a. Modify your parametric equations for the softball pitch to include the gravitational component.

   b. Sketch the vectors that show the position of the ball at 1 second, 1.7 seconds, and 3.3 seconds.

   c. Now investigate more closely the motion of the pitched ball.

      i. How long is the ball in the air?

      ii. What is its maximum height and when does it occur?

   d. In slow-pitch softball, the pitcher stands on a pitching rubber 13.7 meters from home plate. Will the pitched ball make it to the plate?

   e. What are some limitations of this refined parametric model?

3 Conduct the following experiments to help refine your model in Problem 2 of the motion of the pitched softball.

   a. Experiment 1: Physically simulate tossing a slow-pitch softball. Estimate the height at which you would release the ball. Use your result to modify the model of the slow-pitch toss.

   b. Experiment 2: Use a chalk mark or a piece of tape to simulate the front edge of the pitching rubber. Standing with both feet on the pitching rubber, step forward and simulate pitching a softball toward home plate. Estimate the distance in front of the pitching rubber that you release the ball. Use your result to further modify the model to account for the distance from the front of the pitching rubber to where the ball is released.

c. Use your modified parametric model to estimate the height of the pitch when it passes over home plate. For a player of average height, the strike zone is between 0.5 m and 1.5 m above the ground. Should this pitch be called a strike?

d. Maintaining the release angle at 55°, modify the initial velocity of the ball until the pitch crosses the plate inside this strike zone.

e. Now modify the angle of release so that the pitch crosses the plate inside the same strike zone when thrown with an initial velocity of 13 m/sec.

f. Explain why it is difficult to consistently pitch a ball with great accuracy.

4 In general, if there are several forces acting on a body, the resultant motion is the sum of the corresponding horizontal and vertical components. The parametric equations

$$x = At \cos \theta + Bt \cos \phi + C$$
$$y = At \sin \theta + Bt \sin \phi + D$$

model the location (x, y) of an object under two forces acting at angles θ and ϕ in a plane with initial velocities A and B, respectively.

a. What do the values of C and D represent?

b. When $Bt = \frac{g}{2}t^2$, where g is the gravitational constant, the terms $Bt \cos \phi$ and $Bt \sin \phi$ in the parametric model above represent the force of gravity. Explain why the parametric equations for the two forces would be either:

$$x = At \cos \theta + C \qquad \text{or} \qquad x = At \cos \theta + C$$
$$y = At \sin \theta - 4.9t^2 + D \qquad\qquad y = At \sin \theta - 16t^2 + D$$

Other Objects in Flight Problems 5–7 provide other contexts that can be modeled using parametric equations. Your class should scan the three problems and then, in consultation with your teacher, select one for each group to complete and report on to the entire class.

5 World-class horseshoe pitchers are very accurate. For example, Alan Francis, who won his 17th world championship in 2012, averages over 80% ringers. The men's horseshoe pitching court has metal stakes 40 feet apart. The stakes stand 18 inches out of the ground (**Source:** www.horseshoepitching.com/worldchamps/ WCRmens.html).

a. Alan pitches a horseshoe toward a stake at 45 feet per second, at a 14° angle to the ground. He releases the horseshoe at about 3 feet above the ground and 2 feet in front of the stake at one end. Write parametric equations modeling a typical throw.

b. How long is the thrown horseshoe in the air?

c. How close to 40 ft is the horizontal component when the horseshoe hits the ground?

d. If Alan releases a horseshoe at 13° or 15° instead of 14°, what happens to the length of his pitch? Will his pitch be a ringer?

6 In 2001, Ichiro Suzuki was named the American League's Rookie of the Year and Most Valuable Player. When he hits the ball well, it leaves the bat at about a 29° angle, 1 meter above the ground, with a velocity of 40 meters per second (ignoring wind).

a. If the outfield wall is 6 m high and 125 m from home plate, how high will Ichiro's hit likely be when the ball reaches the plane of the wall? Is it a home run?

b. Assuming nothing impedes the flight of the ball, how long is the ball in the air?

c. How far horizontally does the ball travel before it hits the ground?

d. What is the ball's maximum height?

e. How far horizontally would the ball travel (assuming it is not impeded) if it left the bat at a 31° angle?

7 The laminar water fountain at the Detroit Metro Airport is pictured below. The parabolic arcs are $\frac{1}{2}$-inch diameter streams of water in which all water molecules in the stream are moving at the same velocity, so there is almost no rotation in the water stream. Each water molecule acts as a projectile flying through the air, and when it lands there is

very little splash. Water streams are propelled into the air at angles between 45° and 70°. Each water stream follows a parabolic arc that is approximately the same shape as the path of the ball considered in Problem 3.

a. Suppose a molecule of water is propelled from a point 6 inches below fountain level at an angle of 60° with an initial velocity of 21.3 feet per second. Write modeling equations that describe the position of the molecule in the water stream for any time t.

b. Approximately how long is the water molecule in the air before hitting the water surface?

c. What is the maximum height above fountain level reached by the molecule?

d. The point at which the water is propelled is 12 feet away from the center of the receptacle where it needs to land. If the receptacle is circular with a diameter of 3 inches, will the water molecule land in the receptacle?

e. In a similarly designed fountain at McCormick Place in Chicago, an arc of water needs to land in a 3-inch circular receptacle in the floor whose center is 15 feet from the point at which the water is propelled. Suppose the water is propelled from a point 6 inches below fountain level at an angle of 55°. At approximately what initial velocity must the water be propelled in order to land in the receptacle?

SUMMARIZE THE MATHEMATICS

In this investigation, you learned how to use parametric equations to model projectile motion.

a Explain your solution to the problem your group completed from Problems 5–7.

b Name several factors that affect the path of a moving object. Which component (horizontal or vertical) does each factor affect?

c Explain how the horizontal and vertical components of a motion may be used to model projectile motion and display it graphically. What units are used for each variable?

d Explain how technology tools must be set up to graphically display projectile motion as a function of time.

Be prepared to share your ideas and reasoning with the class.

 CHECK YOUR UNDERSTANDING

Suppose Luisa begins her 10-meter platform dive with a velocity of about 2.3 meters per second. The angle at which Luisa leaves the platform is about 85° with the horizontal.

a. Write parametric equations modeling Luisa's position during the dive.

b. About how long is she in the air?

c. How high above the platform is she before she starts moving toward the water?

d. How far does she move horizontally before hitting the water?

e. Divers who push off nearly vertically could hit the platform on the way down. How many meters is Luisa from the platform when she passes it during the dive?

f. If the push-off angle were changed to 80°, how close to the platform would she come?

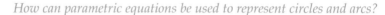

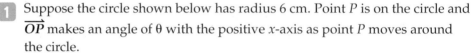

INVESTIGATION 2

Representing Circles and Arcs Parametrically

One of the most common nonlinear motions is the turning of objects around a point, that is, circular motion. The wheels on cars and bicycles, CDs, driveshafts for lawn mowers, carnival rides such as Ferris wheels, and many other toys, tools, and machines involve circular motion in some way.

The computer information industry now primarily uses circular data storage devices. For example, the DVD is a popular optical disc storage media format used to distribute software, games, and movies, and also for data backups. One important reason that circular discs like a DVD (or a higher-capacity Blu-Ray Disc) are still used for storage is that the entire recording surface of a circular DVD is almost instantaneously accessible for data retrieval by rotating the disc.

As you work on the problems in this investigation, look for answers to the following question:

How can parametric equations be used to represent circles and arcs?

1 Suppose the circle shown below has radius 6 cm. Point P is on the circle and $\overrightarrow{OP}$ makes an angle of θ with the positive x-axis as point P moves around the circle.

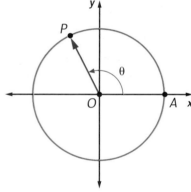

a. What is the magnitude of $\overrightarrow{OP}$?

b. What is the direction of $\overrightarrow{OP}$?

c. What are the components of $\overrightarrow{OP}$?

d. Write parametric equations describing the coordinates of point P.

e. Use your parametric equations and technology to produce a graph of the circle. Is the graph what you expected? Explain. If necessary, change your window settings to display the graph of the circle.

f. In this setting, what does the parameter **T** represent? If **Tmin = 0**, what is the smallest **Tmax** value needed to produce a complete circle? Explain.

2 Investigate and then explain how you could set your calculator or computer software so that it displays only the portion of the circle (in Problem 1) described in each case below.

a. Quarter-circle in the first quadrant

b. Half-circle to the left of the y-axis

c. Half-circle below the x-axis

d. Half-circle to the right of the y-axis

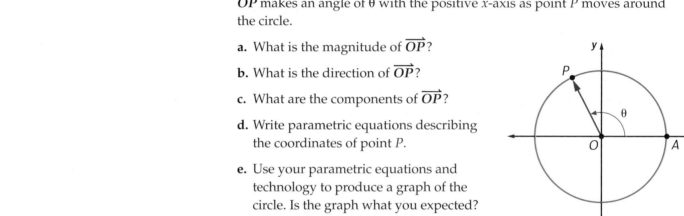

e. Quarter-circle above the *x*-axis and between the lines with equations $y = x$ and $y = -x$

f. An *arc* of a circle that a classmate describes to you

Degrees or Radians Recall that radians as well as degrees can be used to measure angles. While a degree is the measure of an angle determined by an arc that is $\frac{1}{360}$ of a complete circle, a radian is determined by an arc that is $\frac{1}{2\pi}$ of a complete circle. Since the circumference of a circle is $2\pi r$, the length of the arc corresponding to an angle of one radian is $\frac{1}{2\pi}(2\pi r) = r$ linear units. The diagrams below illustrate these ideas.

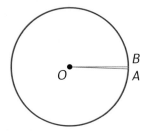
Measure of ∠*AOB* is 1 degree.

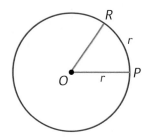
Measure of ∠*POR* is 1 radian.

When calculating or graphing trigonometric functions, it is *very important* to always check the mode. Before proceeding with the next problem, set your calculator or graphing software to **Parametric** and **Radian** modes.

3 Consider once again a circle with center at the origin and radius 6 cm.

   a. Explore how to produce the graph of this circle when θ is measured in radians. Write the parametric equations you would use.

   b. What **T** settings for the viewing window enable you to produce a complete circle without overlap? Record your **T** settings.

   c. Display the following graphs and record the **T** settings.

   i. Quarter-circle in Quadrant I

   ii. Half-circle below the *x*-axis

   iii. Half-circle to the left of the *y*-axis

   iv. Quarter-circle in Quadrant III

   v. Quarter-circle above the *x*-axis and between the lines with equations $y = x$ and $y = -x$

   d. In what direction, clockwise or counterclockwise, were the graphs in Part c drawn? Investigate how to draw them in the opposite direction. Make notes of your findings.

In this investigation, you learned how to use parametric equations to represent circles and arcs.

a Consider these pairs of parametric equations:

Set I	Set II
$x = 3 \cos t$	$x = 3t \cos 42°$
$y = 3 \sin t$	$y = 3t \sin 42°$

  i. Which pair of equations produces a line? A circle? Explain how you can tell by looking at the symbolic form of the rules.

  ii. What does "3" represent in each case?

  iii. What does "t" represent in each case?

b How could parametric equations be used to represent a circle with radius r centered at the origin? An arc on that circle?

c In what direction, clockwise or counterclockwise, would the circle in Part b be drawn? Explain. How would you modify the parametric equations to draw this circle in the opposite direction?

Be prepared to explain your ideas to the class.

 CHECK YOUR UNDERSTANDING

A circle has radius 8 and is centered at the origin.

a. Write parametric equations for this circle.

b. With your calculator or computer graphing software set in **Degree** and **Dot** modes and axes turned off, adjust the viewing window so that it shows 42 dots on the circle in one revolution. Record your settings.

c. With your technology set in **Radian** and **Dot** modes, adjust the viewing window so that it shows 25 dots on the circle in one revolution. Record your settings.

d. Modify your parametric equations in Part a so that the sequence of dots is drawn in the opposite direction starting at $t = 0$.

Simulating Orbits

You now know how to write parametric equations for a circle centered at the origin and how to use technology to display the circle. As you work on the problems in this investigation, look for answers to this question:

How can you use parametric equations to simulate circular and elliptical orbits?

In order to simulate rotating objects such as satellites and CDs, you need to draw on the idea of *angular velocity*. Recall that the **angular velocity** of a rotating object is described in terms of the degrees (or radians) through which the object turns in a unit of time. For example, 3,600 degrees per second and 20π radians per second each describe the angular velocity of an object making 10 complete revolutions per second. Thus, for a particular time t in seconds, the measure of the angle through which an object has turned is

$$\theta = 3{,}600°t \qquad \text{or} \qquad \theta = 20\pi t \text{ radians.}$$

Because the value of the parameter t determines the size of the angle θ, θ is a function of the time t.

1 Suppose a Blu-Ray Disc with 2.36-inch radius is making 8 counterclockwise revolutions per second. To track the position of a point P on the disc as a function of time in a coordinate model, assume it revolves about the origin O and that at $t = 0$, point P is at (2.25, 0).

a. What is the angular velocity of point P in degrees per second? In radians per second?

b. Through how many degrees and how many radians has point P turned when $t = \frac{1}{100}$ sec? When $t = \frac{1}{50}$ sec? When $t = \frac{1}{10}$ sec?

c. Write parametric equations that give the location of point P for any time t. Give both the degree and radian forms of the equations.

d. How is t used to determine the size of the angle of rotation?

e. Set your graphing calculator or graphing software window as follows: **Tmin = 0**, **Tmax = 1**, and **Tstep = 0.013**, with **Xmin = −5**, **Xmax = 5**, **Ymin = −5**, and **Ymax = 5**. Set the mode to **Dot** and **Radian**.

  i. Using the radian form of the parametric equations, display the graph. Adjust the window so that the graph appears circular.

  ii. How many times does point P rotate around the circle? Why does your answer make sense in terms of your window setting?

f. Suppose you wanted to simulate the rotation of point P around the circle exactly once. How would you change the **T** values? How could you simulate point P moving around the circle two times? Three times?

g. Repeat Part e but with the mode set to **Connected**. Describe your observations.

2 Sometimes when you expect to see a circle produced by a pair of parametric equations, you see something else. Enter the following parametric equations in your calculator or graphing software:

$$X_T = 4 \cos (20\pi T)$$
$$Y_T = 4 \sin (20\pi T)$$

Set your technology to **Connected** mode. Set **Tmin = 0** and **Tmax = 1**. Use the **Trace** capability to investigate each graph for differing values of **Tstep**.

a. Set **Tstep** $= \dfrac{1}{30}$. Describe the graph displayed.

b. Set **Tstep** $= \dfrac{1}{60}$. Describe the graph you see.

c. Set **Tstep** $= \dfrac{1}{16}$. Describe the graph displayed.

d. Set **Tstep** $= \dfrac{7}{120}$. Describe the graph you see.

e. Choose another value for **Tstep** and predict the resulting graph. Check your prediction.

3 Now consider a pulley that rotates counterclockwise at 5π radians per second. Represent the center of the pulley by the origin O of a coordinate system and let P be a point on the circumference of the pulley. The parametric equations for the terminal point of a rotation vector $\overrightarrow{OP}$ are:

$$x = 7 \cos\left(5\pi t + \frac{2\pi}{3}\right)$$
$$y = 7 \sin\left(5\pi t + \frac{2\pi}{3}\right)$$

a. Describe the location of point P at $t = 0$ by giving the components of $\overrightarrow{OP}$. What are the magnitude and direction of $\overrightarrow{OP}$?

b. Describe the location of point P at $t = 0.1$, $t = 0.2$, $t = 0.3$, and $t = 0.4$ seconds.

c. Simulate the motion of point P. Set **Tstep = 0.02**.

d. How should you choose the window setting so that point P makes only one revolution?

e. How do these parametric equations differ from those in Problems 1 and 2? How are they similar?

4 A Ferris wheel with a 20-foot radius and center 24 feet above the ground is turning at one revolution per minute. Suppose that on a coordinate system, the ground surface is represented by the x-axis and the center of the wheel is on the y-axis.

a. Explain why the path of a seat on the rotating Ferris wheel can be modeled by the following pair of parametric equations:

$$x = 20 \cos 2\pi t$$
$$y = 20 \sin 2\pi t + 24$$

b. Display the path of a rotating seat for one revolution of the Ferris wheel. Record the viewing window settings you used. Explain why each setting was selected.

c. At what position on the Ferris wheel is the seat located when $t = 0$?

d. How far above the ground is the seat that started at the "3 o'clock" position when $t = 0.1$ minutes? When $t = 0.5$ minutes? When $t = 0.75$ minutes?

e. How are the parametric equations modeling the position of a seat on the rotating Ferris wheel similar to and different from those modeling a point on the rotating disc in Problem 1? The pulley in Problem 3?

5 **Elliptical Orbits** Once fully launched, a satellite or space station does not move in a circular orbit, but in an *elliptical orbit*. Ellipses will be studied more completely in Unit 6, *Surfaces and Cross Sections*. For now, investigate how you can modify parametric equations for a circular path to produce an elliptical path.

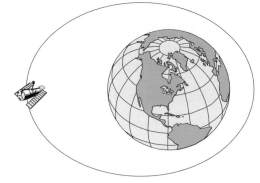

a. Modify these parametric equations of a circle so that they will produce an elliptical path that is stretched horizontally.

$$x = 8 \cos 2\pi t$$
$$y = 8 \sin 2\pi t$$

b. Modify the parametric equations in Part a so that they produce an elliptical path that is stretched vertically.

c. Write parametric equations for an elliptical path that crosses the x-axis at ± 10 and the y-axis at ± 5.

d. How would you modify the equations in Part c so that the "orbit" starts at $(-10, 0)$?

6 In completing Problems 4 and 5, you probably drew on your understanding of transformations of function graphs by translating and stretching. As you saw in Unit 1, *Families of Functions*, those kinds of transformations are closely connected to the symbolic form of function rules. You can extend the idea of customizing function graphs to graphs of parametric equations. Consider the path modeled by the following parametric equations:

$$x = 4 \cos t \text{ and } y = 4 \sin t, \text{ where } 0 \le t < 2\pi.$$

a. Describe the path including the starting point.

b. Modify the equations so that the path is centered at (2, 1).

c. Modify the original equations so that the path is traced in a clockwise direction.

d. Modify the original equations so that the path starts at (0, 4).

e. Modify the original equations so that the path is an ellipse crossing the x-axis at ±8 and the y-axis at ±2.

f. Modify your equations in Part e so that the path is traced in a clockwise direction starting at (−8, 0).

SUMMARIZE THE MATHEMATICS

In this investigation, you examined how the motion of a point moving in a counterclockwise circular path can be modeled with parametric equations. The fundamental equations are

$$x = A \cos Bt$$
$$y = A \sin Bt$$

where $A > 0$ and $B \neq 0$ is the angular velocity of the moving point.

a Describe the center and radius of this circular path.

b What is the location of a point $P(x, y)$ following this path when $t = 0$?

c How would you modify the equations so that the path is traced in a clockwise direction?

d How would you modify the equations so the center of the circular motion is at the point with coordinates (p, q)? Explain.

e How would you modify the equations to produce an elliptical path with center at the origin?

Be prepared to explain your ideas and reasoning to the class.

 CHECK YOUR UNDERSTANDING

A circular disc with center at $P(0, 2)$ has a radius of 3. The disc rotates counterclockwise at 6π radians per second. Point Q on the disc has coordinates (3, 2) when $t = 0$.

a. Sketch this situation.

b. Write parametric equations modeling the motion of point Q for any time t. Locate point Q when $t = 0.1$ and $t = 0.5$.

c. What setting of **Tmax** will ensure that point Q makes exactly one revolution? Explain.

d. Describe the settings for the parameter t that provide you with a good visual model of the path of point Q.

1 In archery, as in other target shooting, the archer sets her sights so that when an arrow is shot at a particular distance from the target, the arrow should hit the bull's-eye.

a. Suppose an arrow leaves a bow at about 150 feet per second. Taking into consideration the effect of gravity, estimate an angle at which you think the archer should shoot so that the arrow hits the bull's-eye located 100 feet away. (Assume the center of the bull's-eye is at the same height as the release point of the arrow.)

b. Find equations for the *x*- and *y*-components of the shot at any time *t* using your estimated angle.

c. Check the adequacy of your model. How long does it take the arrow to reach the target? Could it hit the correct spot?

d. If your model is not very accurate, modify the angle of aim until it gives better results. What angle seems to be the best?

2 Many people enjoy the game of darts. There are even national darts tournaments. In one dart game, you stand 8 feet from a dart board, aim at its center, and throw three darts per round, attempting to score points totaling 500.

a. Suppose you are most accurate when you throw a dart at an angle 20° to the horizontal. If the bull's-eye is at the same height as your release point, at about what initial velocity must you throw the dart to hit the bull's-eye?

b. How long is the thrown dart in the air?

c. Suppose your dart-throwing opponent is most accurate when he throws at an initial velocity of 30 feet per second. At what angle to the horizontal should he throw his dart to hit the bull's-eye?

d. About how long is his dart in the air?

3 Joan Embery has spent much of her life researching the behavior of gorillas. Before examining injured gorillas, she uses a tranquilizer dart gun to sedate them. Her tranquilizer dart gun shoots darts that travel at about 650 feet per second. Suppose Joan shot a dart (aimed horizontally at 5 ft above the ground) at a large injured gorilla 400 ft away.

  a. Would the dart reach the gorilla? Explain.

  b. How should Joan's aim be adjusted so that she can hit the gorilla at a point somewhere between 2 and 5 ft above the ground? Test your best model with a graphing calculator or graphing software and sketch the result.

  c. How much leeway does Joan have in choosing the angle at which to shoot?

4 Consider a circle with center at the origin and radius 5 cm. Suppose you want to display 20 dots of the half circle appearing in a counterclockwise direction from the line with equation $y = x$ in the first quadrant to the same line in the third quadrant. Explain how to set your calculator or computer graphing software to do this in **Degree** mode. In **Radian** mode. Test your ideas and adjust your answers as needed.

5 The diagram below shows the design of a pulley system. A pulley with 5-cm radius is centered at point $O(0, 0)$. A second pulley with 2-cm radius is centered at point $B(10, 0)$. The pulley at O is rotating counterclockwise at 2 revolutions per second.

  a. Find the angular velocity of the pulley with center at point B.

  b. Write parametric equations that model the position of point A as the pulley rotates around its center O starting at $(5, 0)$.

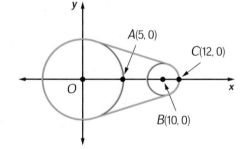

  c. Model the position of point C with parametric equations, if its starting position is $(12, 0)$.

  d. Graph both sets of parametric equations using **Simul** (simultaneous) mode.

  e. Set **Tmin** and **Tmax** so that one revolution around the circle with center O is graphed. What is the effect on the graph of the related motion around the circle with center B? Explain.

  f. Set **Tmin** and **Tmax** so that one revolution around the circle with center B is graphed. What is the effect on the graph of the related motion around the circle with center O? Explain.

6 Planets move in elliptical orbits around the Sun. Elliptical orbits can be modeled by stretching a circle in one direction to become an *ellipse* and viewing a point (representing a planet) moving around the ellipse.

a. A point begins at (3, 0) and moves counterclockwise in an elliptical orbit centered at the origin, making a complete revolution in 1 minute. Write parametric equations for this elliptical orbit if it crosses the x-axis at $x = \pm 3$ and the y-axis at $y = \pm 4$.

b. Modify the equations in Part a so that the orbit starts at (0, 4).

c. Modify the equations in Part a so that the ellipse is centered at (3, 0) and the path begins at the origin and moves in a clockwise direction.

CONNECTIONS

7 In Investigation 1 Problem 7 (page 160), you wrote parametric equations to model the position of a water molecule in a stream of water at any time t:

$$x = 21.3t \cos 60°$$
$$y = 21.3t \sin 60° - 16t^2 - 0.5$$

a. There are three variables in these two equations: x, y, and t. What aspect of the stream of water is represented by each variable?

b. Each parametric equation describes a separate function that provides information about the water stream. Graph each equation as a *separate function* of t on a coordinate grid. Describe the feature(s) of the water stream that is represented.

    I. $x = 21.3t \cos 60°$

    II. $y = 21.3t \sin 60° - 16t^2 - 0.5$

c. What information does the graph in the x-y coordinate system of the pair of parametric equations provide about the water stream?

d. The two parametric equations taken together can also be expressed as a single function of the form $y = f(x)$.

    i. Solve the first equation for t in terms of x. Then substitute for t in the second equation.

    ii. Graph your function from part i.

    iii. What aspect(s) of the water stream does this function and its graph represent?

8 Joan Elmore, five-time women's national horseshoe champion, made a phenomenal 84.62% ringers in the 2012 championship match. The women's horseshoe pitching court has metal stakes 30 feet apart that stand 18 inches out of the ground (**Source:** www.horseshoepitching.com/worldchamps/WCRwomens.html).

a. Suppose the typical throw for a woman professional like Elmore is geometrically similar (that is, all relevant linear dimensions including those related to the release point are in the ratio of 3 to 4) to that of the men's champion, Alan Francis. (See Investigation 1 Problem 5 page 159.) Aiming at the target stake, what are the following measures of Joan's throw?

    i. Height of release

    ii. Distance of release in front of stake

    iii. Maximum height of throw

    iv. Angle of release

b. The initial velocity of Alan Francis's throw is 45 feet per second. What would you expect to be a reasonable initial velocity for Joan Elmore's throw? Check your conjecture by examining the maximum height of the resulting path of the horseshoe and its height when it reaches the target stake. If your conjecture is not feasible, adjust the initial velocity until you find one that will work.

c. Compare the following aspects of the throws of the women's champion and the men's champion. Be as specific as you can.

    i. Initial velocity

    ii. Length of time that the horseshoe is in the air

9 In the figure at the right, there are four circles with the same center, O. The circles have radii of 1, 2, 3.5, and 5 units. $\overset{\frown}{AB}$ has length 1 unit.

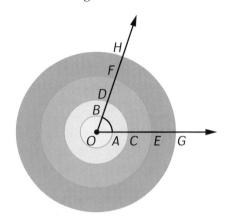

   a. What is the radian measure of $\angle O$?

   b. Size transformations with center O will map $\overset{\frown}{AB}$ to each of the other arcs. Find the magnitude of each size transformation.

   c. Find the lengths of $\overset{\frown}{CD}$, $\overset{\frown}{EF}$, and $\overset{\frown}{GH}$.

   d. How do these similar circles and their arcs contribute to your understanding of the definition of a radian?

10 A circular disc with center at the origin has a radius of 1 unit. The disc rotates counterclockwise at 1 radian per second. Point P on the disc has coordinates $(1, 0)$ when $t = 0$.

   a. Write parametric equations modeling the motion of point P for any time t. Locate point P when $t = 1$ and when $t = \pi$.

   b. Select a window and **Tmax** setting that will ensure point P makes exactly one revolution. Make your window selection so that it will also be useful in Part c. Then square up the viewing window and graph the pair of parametric equations representing the motion of point P.

   c. Using the same window, simultaneously graph the parametric equations $x = t$ and $y = \sin(x(t))$ with the equations representing the motion of point P.

   d. What is the connection between the graphs in Parts b and c?

   e. Explain how to use parametric equations in a similar way to show the connection between the circular motion of point P and the graph of $y = \cos\theta$.

REFLECTIONS

11 Would a thrown, kicked, or hit ball go further on the Moon, Earth, or Jupiter? Explain your reasoning.

12 The study of parametric equations is enhanced by the strategic use of graphing and table-building capabilities of technological tools. Identify the technology capability you find most helpful for each of the following tasks when modeling projectile motion such as that of a kicked soccer ball and explain your choice.

   a. Visualize the path.

   b. Determine the maximum height.

   c. Determine the horizontal distance traveled.

   d. Determine a point at which the vertical velocity is 0.

13 According to Morris Kline, a former Professor of Mathematics at New York University, "The advantage of radians over degrees is simply that it is a more convenient unit. Since an angle of 90° is of the same size as an angle of 1.57 radians, we now have to deal only with 1.57 instead of 90 units. The point involved here is no different from measuring a mile in yards instead of inches." Based on what you learned about circular motion in this lesson, why do you think Kline believes the radian is a "more convenient unit"?
(**Source:** Morris Kline. *Mathematics for Liberal Arts*. Addison Wesley, 1967. page 423)

14 Consider a circle with equation $x^2 + y^2 = 36$. Write parametric equations that model the motion of a point moving counterclockwise about this circle starting at the point (6, 0).

15 Adjust the equation in Reflections Task 14 so the graph is an ellipse centered at the origin and containing points (6, 0), (0, 3), (−6, 0), and (0, −3). Write parametric equations that model the motion of a point moving counterclockwise about this ellipse starting at the point (6, 0).

16 By now you may be adept with the important mathematical practices listed on page vii. Which of those practices were most helpful in completing Reflections Tasks 14 and 15? Explain your thinking.

EXTENSIONS

17 Parametric equations can be used to represent many familiar graphs. Use technology to display the graph of each pair of parametric equations below, and then make a sketch of each graph on your paper. Describe the shape of the graph. Then combine the equations into a single equation that relates x and y.

a. $x = t + 1, y = t + 2$ **b.** $x = 2 − 3t, y = t + 5$

c. $x = t, y = \dfrac{3}{t}$ **d.** $x = 4 \cos t, y = 4 \sin t$

e. $x = 4t − 2, y = 8t^2$

18 Suppose a pro golfer drives a ball about 315 yards, which includes 50 yards of roll after it hits the ground. He hits the ball so that its direction toward the pin makes about a 27° angle with the horizontal.

a. What is the initial velocity of the ball when it leaves the club?

b. What is the maximum height that the ball reaches?

c. How long is the ball in the air?

19 The double Ferris wheel is a popular carnival ride that multiplies the interest and thrills of a single Ferris wheel ride. In one double Ferris wheel, the center of a 50-foot bar is attached to the main support for the ride, and the bar revolves about its center point once every 60 seconds. Seats for riders are evenly spaced along two separate wheels, each revolving about the ends of the long bar. Each wheel has a radius of 10 feet and turns independently of the bar, making one complete revolution every 15 seconds. If available, study the "Double Ferris Wheel" custom app.

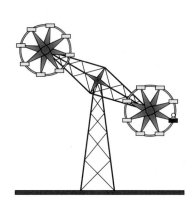
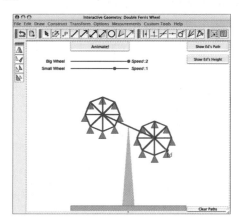

a. Imagine the Ferris wheel on a coordinate system with origin at the midpoint of the bar. In the simplest case, both the bar and the wheels rotate counterclockwise, and the rider begins at the rightmost point (35, 0).

  i. Write parametric equations to model the location at time t of one endpoint of the bar that is the center of one of the circular wheels.

  ii. Write parametric equations to model the location at time t of a rider relative to the center point of the wheel on which she is riding.

  iii. Combine these two models by adding them to create a model for the location of the rider on the wheel as the bar and the wheels rotate.

  iv. Sketch the path of the rider.

b. In **Function** mode, plot the vertical position of the rider over time, and in the same window, plot the horizontal position of the rider over time. Through how many revolutions does the ride turn each minute?

c. Adjust the parametric model so that the lower wheel's center is initially located at (0, −25) and the rider starts at the bottom of the ride, (0, −35).

d. Consider your model of the location of the rider in Part aii.

  i. Adjust the model so that the wheel containing the rider rotates clockwise, while the bar rotates counterclockwise.

  ii. Sketch the path of the rider.

  iii. In **Function** mode, plot the vertical position of the rider over time, and plot the horizontal position of the rider over time.

  iv. Through how many revolutions does the ride turn each minute?

20 In this lesson, you saw that in a coordinate system, a circular orbit with center at the origin can be described by parametric equations of the form:

$$x = r \cos At \quad \text{and} \quad y = r \sin At$$

Similarly, an elliptical orbit with center at the origin can be described by parametric equations of the form:

$$x = a \cos At \quad \text{and} \quad y = b \sin At, \text{ where } a \neq b$$

In the case of a satellite or space station orbiting Earth, the path is an elliptical orbit, but its center is not the center of Earth. Earth's center is at one of two foci of the orbit, as indicated in the diagram below. The *apogee* is the point on the orbit farthest from Earth. The *perigee* is the point on the orbit closest to Earth.

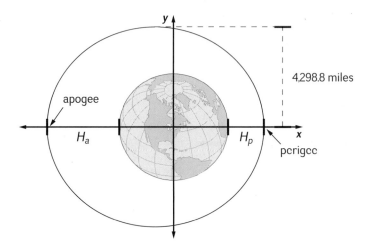

a. In the case of a space station, NASA expects the altitude of apogee H_a to be 400 miles and the altitude of perigee H_p to be 200 miles. Using the diagram above (not drawn to scale) and assuming the radius of Earth is about 4,000 miles, develop parametric equations to model (in two dimensions) Earth and the orbit of a space station around Earth.

b. Simulate the motion of the space station in orbit around Earth and verify the coordinates of the apogee and perigee.

REVIEW

21 In the diagram at the right, $\overline{AB} \parallel \overline{DE}$.

a. Prove that $\triangle ABC \sim \triangle DEC$.

b. Determine the length of $\overline{CD}$.

c. Determine the length of $\overline{DE}$.

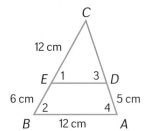

22 For each rational function below, describe the domain, sketch a graph, and describe any horizontal asymptotes.

a. $f(x) = \dfrac{1}{x}$

b. $g(x) = \dfrac{1}{x+3}$

c. $h(x) = \dfrac{1}{x} + 4$

23 The amplitude and period are important properties of the cosine and sine functions.

   a. Determine the amplitude and period (in radians) for each of the following trigonometric functions.

   I $y = \cos x$ **II** $y = 3 \cos (x + \pi)$ **III** $y = -\cos 4x$

   b. Draw a sketch of each function.

   c. How would you modify your sketches in Part a if the cosine function was replaced by the sine function?

24 Suppose that the wheels on a bike have a radius of 31 centimeters and that the rider pedals so that the rear sprocket has a velocity of 66 rpm.

   a. How many meters does the bike travel in one minute?

   b. What is the angular velocity of the rear sprocket in degrees per minute? In radians per minute?

   c. Through how many radians will the rear wheel turn in 12 seconds? In 30 seconds? In 100 seconds?

25 Rewrite each product in standard polynomial form.

   a. $(3x + 6)(x - 7)$

   b. $(x - 2)(x + 10)(x - 4)$

   c. $(x + 5)(x^2 + 2x + 4)$

26 Rewrite each radical expression with the smallest possible value under the radical.

   a. $\sqrt{45}$ **b.** $2\sqrt{90}$

   c. $\sqrt{27} + \sqrt{48} - 4$ **d.** $\sqrt{25 + 100}$

27 Consider the quadratic function $f(x) = x^2 - 4x + 12$.

   a. What is the y-intercept of the graph of $f(x)$?

   b. Determine the x-intercepts of the graph of $f(x)$.

   c. Determine the vertex of the graph of $f(x)$.

28 Xavier and Jamila both bought the same type of pens and notebooks for school. Xavier bought 8 notebooks and 5 pens and spent $14.25. Jamila bought 6 notebooks and 10 pens and spent $16. Suppose that tomorrow, you plan to shop for school supplies.

   a. Write a system of equations that can be used to determine the cost of each pen and each notebook.

   b. Write a matrix equation that represents your system of equations.

   c. Determine the cost of each pen and each notebook.

Looking Back

In this unit, you investigated vectors and their use in situations that involve magnitude and direction. The idea of the sum or resultant vector was found to be useful in determining courses of ships and airplanes that were moving with a constant velocity. You explored how vectors provide another method to prove geometric relationships. You also examined how vectors, their components, and the parametric equations derived from the components can be used to analyze linear and projectile motion. These ideas are also useful in representing circular motion of a point on a disc rotating at a constant velocity. Using ideas of transformations of graphs, you were able to develop models for elliptical orbits. The tasks in this final lesson give you the opportunity to review and apply these important ideas in new contexts.

1 The Gulf Stream is a warm ocean current flowing from the Gulf of Mexico along the east coast of the United States. It is about 50 nautical miles wide.

   a. Off the coast of New York City, the Gulf Stream flows at about 3 knots in a direction of about 55°. Sketch a vector representing the Gulf Stream current.

   b. Suppose the freighter *Morocco* is steaming out of New York City at 12 knots in a direction of 350° when it meets the Gulf Stream. Represent the *Morocco*'s course with a vector.

   c. Draw a vector model showing the effect of the Gulf Stream on the *Morocco* as it moves across the Gulf Stream.

   d. What course should the *Morocco* steer to cross the Gulf Stream and remain on its planned course?

   e. How long will it take the *Morocco* to cross the Gulf Stream?

2 Preliminary testing of robocarriers used in paper mills involves parallel tracks 85 meters long. Suppose a robocarrier programmed to travel 20 meters per minute (m/min) is placed on one track and a robocarrier programmed to travel at 30 m/min is placed on the second track.

   a. Use parametric equations to model the motion of the two robocarriers under the conditions that the testing of the second robocarrier begins 30 seconds after the first.

b. When one robocarrier first reaches the end of the test track, how far is the other from the end of the test track?

c. Does the second robocarrier overtake the first? If so, at what time?

d. How could the rates of the robocarriers be adjusted so that they arrive at the end of their tracks at about the same time?

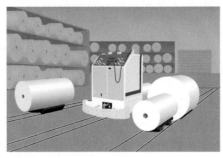

Robotic paper mover (robocarrier)

3 Suppose that position vector $\vec{a} = (x_1, y_1)$ and position vector $\vec{b} = (x_2, y_2)$ are parallel, and neither vector is horizontal or vertical. Prove that the dot product $\vec{a} \cdot \vec{b} = 0$ if and only if the slope of vector $\vec{a}$ is the negative reciprocal of the slope of vector $\vec{b}$.

4 Write parametric equations describing each of the following paths of an object moving at a constant velocity.

a. A vertical line through the point $(5, -2)$

b. A line through the points $(2, 4)$ and $(6, -2)$

c. A circle with radius 5 centered at the origin, traced counterclockwise, starting at the point $(0, 5)$ when $t = 0$

d. An ellipse centered at the origin, traced clockwise, crossing the x-axis at ± 8 and the y-axis at ± 12

5 Trina, a female professional golfer, can swing her driver with a velocity of about 132 feet per second. When she uses a driver with 10° loft, she propels the ball at about 150 ft/sec at an angle of about 30° (since the ball is met on the upswing and its compression adds to its velocity).

a. Write modeling equations that describe the position of the ball for any time t.

b. How long is the ball in the air?

c. If the ball bounces and rolls 30 to 50 yards after it first hits the ground, what is the total drive length in yards?

d. If Trina wanted to lengthen her drive, should she learn to hit the ball higher by 5° or swing the club faster by 5 ft/sec? Explain your answer.

6 The following problem was posed by Neal Koblitz in the March 1988 issue of the *American Mathematical Monthly* (page 256). The problem is one of several applied problems he gave to his calculus classes at the University of Washington. Solve this problem, as described on the next page, using methods developed in this unit.

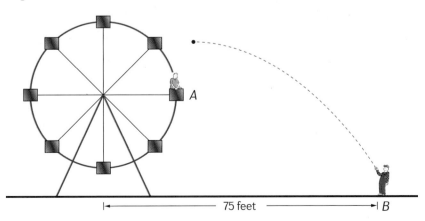

You are standing on the ground at point B (see diagram), a distance of 75 ft from the bottom of a Ferris wheel with radius 20 ft. Your arm is at the same level as the bottom of the Ferris wheel. Your friend is on the Ferris wheel, which makes one revolution (counterclockwise) every 12 seconds. At the instant when he is at point A, you throw a ball to him at 60 ft/sec at an angle of 60°. Let $g = -32$ ft/sec^2, and neglect air resistance. Find the closest distance the ball gets to your friend. (**Source:** "Problems that Teach the Obvious but Difficult," *American Mathematical Monthly*, March 1988, p. 256. Copyright 1988 Mathematical Association of America. All Rights Reserved.)

SUMMARIZE THE MATHEMATICS

Vectors and parametric equations are very useful for modeling and analyzing both linear and nonlinear motion.

a What is a vector? How can vectors be represented?

b Suppose vector $\vec{a}$ is expressed in polar form $[r, \theta]$. What are the magnitude, direction, and components of vector $\vec{a}$?

c If you know the magnitude and the direction of two vectors, describe how to find the magnitude and direction of the resultant vector.

d How are vector proofs of geometric relationships similar to, and different from, coordinate proofs and synthetic proofs such as those you prepared in previous courses?

e Suppose α is the angle between nonzero position vectors $\vec{a} = (x_1, y_1)$ and $\vec{b} = (x_2, y_2)$.

  i. Write $\vec{a} + \vec{b}$, $\vec{a} - \vec{b}$, and $\vec{a} \cdot \vec{b}$, in terms of the coordinates of the vectors.

  ii. Explain why the two vectors are perpendicular if and only if $\vec{a} \cdot \vec{b} = 0$.

f How many variables are involved in modeling motion with parametric equations? What is the significance of each variable?

g Describe parametrically and graphically the following paths of a point P.

  i. P is moving in the direction θ at a constant velocity of 7 m/sec.

  ii. P is propelled at an angle θ with an initial velocity of 100 m/sec and then flies freely in the air.

  iii. P travels on a circle with radius 2 m making 15 counterclockwise revolutions per minute.

h How are parametric equations for circular motion similar to, and different from, those for elliptical motion? For linear motion? For projectile trajectories?

Be prepared to explain your responses and reasoning to the entire class.

 CHECK YOUR UNDERSTANDING

Write, in outline form, a summary of the important mathematical concepts and methods developed in this unit. Organize your summary so that it can be used as a quick reference in future units.

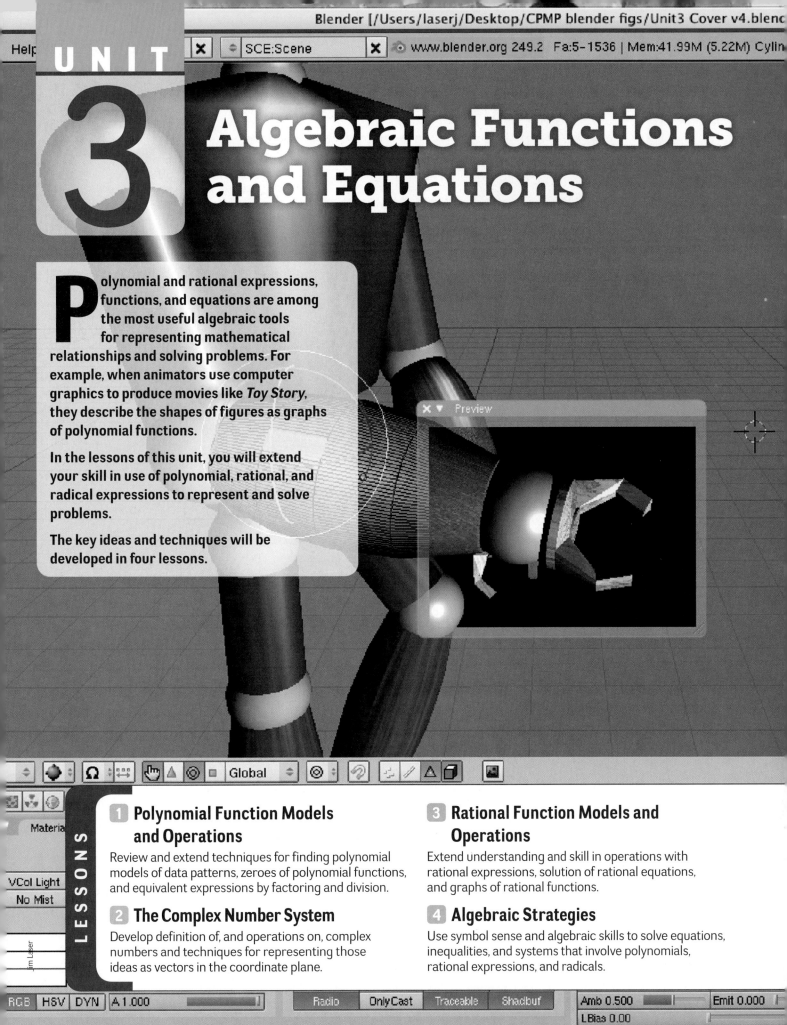

UNIT 3

Algebraic Functions and Equations

Polynomial and rational expressions, functions, and equations are among the most useful algebraic tools for representing mathematical relationships and solving problems. For example, when animators use computer graphics to produce movies like *Toy Story*, they describe the shapes of figures as graphs of polynomial functions.

In the lessons of this unit, you will extend your skill in use of polynomial, rational, and radical expressions to represent and solve problems.

The key ideas and techniques will be developed in four lessons.

LESSONS

1 Polynomial Function Models and Operations

Review and extend techniques for finding polynomial models of data patterns, zeroes of polynomial functions, and equivalent expressions by factoring and division.

2 The Complex Number System

Develop definition of, and operations on, complex numbers and techniques for representing those ideas as vectors in the coordinate plane.

3 Rational Function Models and Operations

Extend understanding and skill in operations with rational expressions, solution of rational equations, and graphs of rational functions.

4 Algebraic Strategies

Use symbol sense and algebraic skills to solve equations, inequalities, and systems that involve polynomials, rational expressions, and radicals.

Polynomial Function Models and Operations

When the first animated cartoons and movies were created, the apparent motion of every character was created by illustrators who made thousands of hand-drawn pictures, each slightly different than the one before.

Now most of that animation and many other visual effects in movies are created by computers. Creating computer images of the characters and action in such films often requires drawing and transforming smooth curves quickly.

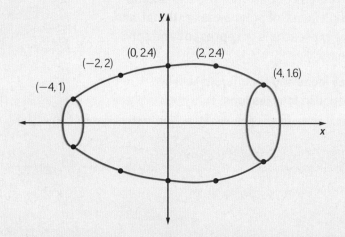

The curves are specified by giving coordinates of key *control points*, as shown in this sketch of part of the arm from the robot in the image above.

Jim Laser/CPMP

Examine the partial model of a computer-animated figure shown on the previous page.

a What strategies could you use to find an equation or function whose graph goes through the five control points with coordinates indicated on the graph?

b Will every curve that goes through those control points match the curve drawn on the diagram? Explain your thinking.

c How will the number of control points needed to find an equation representing a curve be related to the shape of the curve?

d How might the control points be useful in animating the movements of the figure?

Working on the problems of this lesson will review and extend algebraic skills and understanding needed to deal with modeling of geometric shapes and solving polynomial equations in other contexts.

INVESTIGATION 1

Constructing Polynomial Function Models

The partial model of a computer-animated figure was outlined by fairly simple curves, so you might try to model the shape with quadratic functions. You know from prior experience that coefficients for such a model can be determined by statistical curve-fitting software. You also know that finding functions whose graphs are more complex curves often requires polynomial expressions of degree three, four, or even higher.

As you work on the problems of this investigation, look for answers to this question:

What efficient strategies will find polynomial functions with graphs that have two or more local maximum and minimum points?

The picture at the right shows a section of rural road with a kind of S-shaped curve leading back to the hills. To plan construction of such curvy roads, highway engineers find it helpful to make scale drawings of the planned road section and to represent sections of the road centerline by mathematical functions.

The following sketch shows such a planned road section on a coordinate grid. The coordinate axes have been placed on the road drawing in a way that helps with the modeling task.

Digital Vision

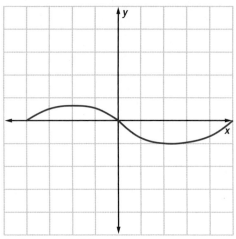

Scale 1 unit = 100 feet

1 One way to begin the search for a function whose graph matches the S-shaped curve of the road is to focus attention on the points where the road curve crosses the *x*-axis.

    a. Why can you be sure that the function $f(x) = x(x + 4)(x - 5)$ has a graph with the same *x*-intercepts as the drawing of the S-shaped road curve?

    b. Produce a graph of $f(x)$ and compare it to the shape of the road curve. See if you can explain why the graph is such a poor model for the road shape, even though it has the same *x*-intercepts.

    c. What simple change in the rule for $f(x)$ would produce a better fit? Test your idea.

2 Choose a set of control points on the road curve graph that you believe will lead to a model for the shape that is better than that obtained in Problem 1 Part c. Then use those points and a cubic or quartic curve-fitting tool to find coefficients for a polynomial function whose graph is determined by those points.

    Compare the graph of your new function model to the shape of the road curve and see if you can explain why the new graph is (or is not) a better match for the curve than the function in Problem 1 Part c.

In some situations, like drawing the shape of a character in a computer-animated movie, close is not good enough. The graph of a function must pass directly through the prescribed control points. There is a third technique for constructing polynomial functions whose graphs pass through specific sets of control points. It is called the **method of undetermined coefficients**. To see how this method works, consider again the design of the arm of the robot shown in the introduction to this lesson. A copy of a graphical model of the arm together with control points is shown at the top of the next page.

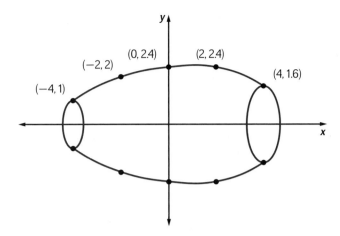

It looks as if the curve on the top of the arm might be modeled well by the graph of a quadratic function and three control points would be enough to determine the coefficients in a rule like $g(x) = ax^2 + bx + c$.

3 Suppose that the control points chosen are $(-4, 1)$, $(0, 2.4)$, and $(4, 1.6)$.

  a. Explain why $g(-4) = 1$, $g(0) = 2.4$, and $g(4) = 1.6$.

  b. Explain why the values of a, b, and c must satisfy all three of these equations:

$$16a - 4b + c = 1$$
$$0a + 0b + c = 2.4$$
$$16a + 4b + c = 1.6$$

  Adapt what you know about solving systems of linear equations with two unknowns to solve the system of three equations and three unknowns.

  c. Write the rule for $g(x)$ implied by your results in Part b and compare its graph to the upper curve in the drawing.

  d. Use what you know about symmetry and function transformations to write the rule for a function $h(x)$ whose graph will be a good match for the lower curve in the drawing.

  e. Explain in general why three control points are sufficient to determine the coefficients for a function like $f(x) = ax^2 + bx + c$. How many control points would be necessary to determine a general cubic function model? Explain your reasoning.

When you think about the fact that the modeling calculations in Problem 3 only deal with one part of a figure, you can easily imagine how animation work relies on powerful computers. The basic idea is illustrated fairly accurately by the example that you have worked on. With computers available, other mathematical strategies can be used to solve the resulting systems of linear equations.

4 Suppose that a highway engineer wanted to find a quartic function model for the S-shaped curve in Problems 1 and 2 and chose the control points $(-4, 0)$, $(-2, 0.5)$, $(0, 0)$, $(2, -1)$, and $(5, 0)$. Answer the following questions to explain how those points could be used in the method of undetermined coefficients to find the desired quartic model.

a. The quartic function will be in the form $j(x) = ax^4 + bx^3 + cx^2 + dx + e$.

    i. What system of linear equations with unknowns a, b, c, d, and e would be implied by the requirement that the graph of $j(x)$ pass through the five chosen control points?

    ii. Explain how you know that one equation is $256a - 64b + 16c - 4d + e = 0$.

b. How does the following matrix equation express the conditions on a, b, c, d, and e?

$$\begin{bmatrix} 256 & -64 & 16 & -4 & 1 \\ 16 & -8 & 4 & -2 & 1 \\ 0 & 0 & 0 & 0 & 1 \\ 16 & 8 & 4 & 2 & 1 \\ 625 & 125 & 25 & 5 & 1 \end{bmatrix} \begin{bmatrix} a \\ b \\ c \\ d \\ e \end{bmatrix} = \begin{bmatrix} 0 \\ 0.5 \\ 0 \\ -1 \\ 0 \end{bmatrix}$$

c. Why can the solution to the matrix equation in Part b be obtained by calculating the matrix product on the right side of the equation below?

$$\begin{bmatrix} a \\ b \\ c \\ d \\ e \end{bmatrix} = \begin{bmatrix} 256 & -64 & 16 & -4 & 1 \\ 16 & -8 & 4 & -2 & 1 \\ 0 & 0 & 0 & 0 & 1 \\ 16 & 8 & 4 & 2 & 1 \\ 625 & 125 & 25 & 5 & 1 \end{bmatrix}^{-1} \begin{bmatrix} 0 \\ 0.5 \\ 0 \\ -1 \\ 0 \end{bmatrix}$$

d. Using technology, complete work in Part c to find the quartic model for the S-shaped road curve and check to see how well it fits the desired shape.

e. Alternatively, find the quartic model for the road curve using the "Solving Linear Systems" feature of the *CPMP-Tools* CAS. This tool uses the familiar method of elimination of variables.

SUMMARIZE THE MATHEMATICS

The problems of this investigation illustrated three strategies for finding polynomial function models of graphical data patterns.

a What are the key steps in the modeling strategy that focuses on *x*-intercepts of a graph?

b What are the key steps in the modeling strategy that relies on statistical curve-fitting tools to find polynomial function models of graph patterns?

c What are the key steps in the method of undetermined coefficients for finding polynomial function models of graph patterns?

d What technology tool(s) did you find most helpful in finding polynomial models for curves with given control points? Explain.

Be prepared to discuss these three strategies and your ideas on technology use.

This pedestrian walkway in a neighborhood of Parque das Nações in Lisbon, Portugal, has a dramatically curved design.

Suppose that you were given the task of finding an algebraic model for the edge of a curved surface that is sketched in the diagram below.

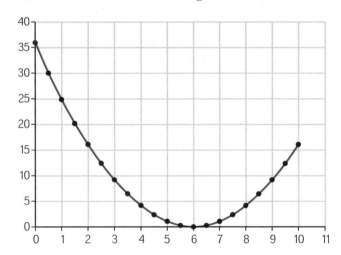

a. What kind of polynomial function seems likely to provide an accurate model for the curve?

b. What specific function is produced when a statistical curve-fitting program is applied to what you believe is a suitable set of control points?

c. What set of control points and system of linear equations could be used to find coefficients for a polynomial function model by the method of undetermined coefficients?

d. What matrix equation is equivalent to the system of equations in Part c and how could it be solved? What is the solution?

e. Solve the system of equations in Part c using the "Solving Linear Systems" CAS feature. Compare the graph of your polynomial function model with the curve sketched above.

Zeroes and Factors of Polynomials

The problems of Investigation 1 focused on strategies for finding polynomial functions whose graphs match given curved shapes. In many other problems, you will be given a polynomial function and asked to analyze it—to find zeroes, end behavior, local maximum and/or minimum points, and methods for efficiently calculating output values.

As you work on the problems of this investigation, look for answers to this question:

> *How can polynomials be written in equivalent forms that make the*
> *solution of problems simpler and computations faster?*

Equivalent Expressions are Useful Calculators and computers are capable of making many arithmetic calculations almost instantaneously. But, for computer-intensive problems, like producing animated movies, there are so many calculations involved that finding *efficient* algorithms is very important.

The Cray XK7 Titan supercomputer at the Oak Ridge National Laboratory fills an entire room.

1. The form of a function rule can greatly influence the time it takes to calculate function values or produce a graph. On some graphing calculators, the function

$$p(x) = x^5 - 15x^4 + 85x^3 - 225x^2 + 274x - 120$$

will appear in about 9.5 seconds, while the equivalent function

$$p(x) = (x - 1)(x - 2)(x - 3)(x - 4)(x - 5)$$

will appear in only 4.5 seconds.

a. Why do you think this result occurs?

b. The factored form of a polynomial can reduce the computation time, but not all polynomials can be written as a product of linear factors. To avoid exponents, polynomials can also be written in *nested multiplication* form. The nested multiplication form of the rule $p(x)$ is

$$p(x) = ((((x - 15)x + 85)x - 225)x + 274)x - 120.$$

Find the time required to graph this form of $p(x)$. Compare the time saved using the nested multiplication form over the standard polynomial form.

Courtesy Oak Ridge National Lab/U.S. Dept. of Energy

2 As a general rule, calculators and computers can perform addition and subtraction faster than multiplication or division. And all four basic arithmetic operations can be done faster than evaluating exponential and trigonometric functions.

Computation time is commonly measured in units called *cycles*. Suppose that calculating any power (like 1.3^4) takes two cycles of computation time, while multiplication, addition, and subtraction each take one cycle of computation time.

a. How many cycles of computation time are required to evaluate the polynomial

$$p(x) = x^5 - 15x^4 + 85x^3 - 225x^2 + 274x - 120$$

for any given value of x?

b. How many cycles of computation time are required to evaluate the same function with the rule written in equivalent factored form as

$$p(x) = (x - 1)(x - 2)(x - 3)(x - 4)(x - 5)?$$

c. How many cycles of computation time are required to evaluate the nested multiplication form

$$p(x) = ((((x - 15)x + 85)x - 225)x + 274)x - 120?$$

d. How are the cycle counts for the three forms related to the time saved using the factored form and the nested multiplication form rather than the standard form?

3 Before using a different expression to evaluate $p(x) = x^5 - 15x^4 + 85x^3 - 225x^2 + 274x - 120$, it makes sense to check that the proposed alternative is equivalent to the original. Use order of operations and repeated application of the distributive property or a CAS **expand** command to check that the two more efficient algorithms are equivalent to the polynomial expression for $p(x)$.

a. $(x - 1)(x - 2)(x - 3)(x - 4)(x - 5)$
   (*Hint:* Start with $(x - 1)(x - 2)$, then multiply that result by $(x - 3)$, and so on.)

b. $((((x - 15)x + 85)x - 225)x + 274)x - 120$
   (*Hint:* Start with $(x - 15)x$, then add 85, then multiply that result by x, and so on.)

Producing Useful Equivalent Expressions As you have seen in work on Problems 1 and 2 and in much earlier work with polynomials, equivalent expressions often reveal different properties of the corresponding functions and their graphs. The challenge is producing equivalent nested multiplication and factored expressions for given polynomials.

4 Compare the standard polynomial and nested multiplication expressions for $p(x)$ to find a pattern connecting the two forms.

$$x^5 - 15x^4 + 85x^3 - 225x^2 + 274x - 120$$
$$((((x - 15)x + 85)x - 225)x + 274)x - 120$$

Then use your ideas to write the following expressions in equivalent nested multiplication form Expand the resulting nested forms to check that they are equivalent to the original polynomials.

a. $x^3 + 9x^2 + 11x - 21$

b. $x^4 + 2x^3 - 13x^2 - 14x + 4$

c. $7w^4 - 3w^3 + 2w^2$

d. $t^4 + 2t^3 - 11t^2 - 12t + 36$

Expressing cubic, quartic, and higher-degree polynomials in factored form is a challenging problem. For example, in Problem 1 you were told that for all x

$$x^5 - 15x^4 + 85x^3 - 225x^2 + 274x - 120 = (x - 1)(x - 2)(x - 3)(x - 4)(x - 5).$$

Could you have found the linear factors on your own? This problem interested mathematicians for many centuries, and they developed a variety of special techniques for finding factors of polynomials with integer coefficients. With the development of modern computational tools, those formal techniques are now less important than they once were. However, both modern and ancient methods rely on the relationship between factors and zeroes that you know well for quadratic polynomial functions.

5 The next diagram shows a graph of the cubic polynomial

$$c(x) = x^3 - 2x^2 - 11x + 12.$$

 a. Use the graph to estimate the zeroes of $c(x)$.

 b. Use your estimates of zeroes for $c(x)$ to write a product of linear factors that you expect to be equivalent to the standard polynomial expression $x^3 - 2x^2 - 11x + 12$.

 c. Use several applications of the distributive property or a CAS to expand the product of linear factors to see if it is in fact equivalent to the original expression.

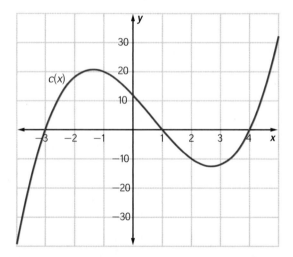

6 Adapt the strategy suggested in Problem 5 to write rules for each of the following polynomial functions as products of linear factors. Then check your work by expanding the resulting products using the distributive property or a CAS.

 a. $f(x) = x^3 + 9x^2 + 11x - 21$

 b. $g(x) = x^5 - 4x^4 - 4x^3 + 22x^2 + 3x - 18$

 c. $h(w) = w^5 - 3w^4 + 2w^3$

 d. $p(t) = 2t^4 + 4t^3 - 22t^2 - 24t + 72$

For the case of quadratic polynomials, you have developed skills in using algebraic reasoning to write the expressions as products of linear factors. The process was relatively easy for cases in which the lead coefficient was 1. For quadratics like $6x^2 + 23x + 20$, it is still possible (though not always easy) to write equivalent expressions that are products of linear factors.

7 Answer the following questions by using your personal symbol manipulation skills and understanding. Then check your work using appropriate CAS commands.

a. Write these products in expanded standard polynomial form.

    i. $(4x + 3)(x + 2)$ **ii.** $(2x - 3)(5x + 7)$

    iii. $(mx + n)(px + q)$ **iv.** $(-3x + 1)(4 + 5x)$

b. Compare the factored and expanded forms of quadratics produced in your work on Part a to see how the coefficients and constants in the expanded forms are related to those in the linear factors. Use patterns that you see to "work backward" from expanded to factored forms of the following quadratics.

    i. $3x^2 + 11x + 6$ **ii.** $2x^2 - 5x - 3$

    iii. $14x^2 + 17x - 6$ **iv.** $6x^2 + 23x + 20$

c. Use results of your work in Part b to solve these quadratic equations.

    i. $3x^2 + 11x + 6 = 0$ **ii.** $2x^2 - 5x - 3 = 0$

    iii. $14x^2 + 17x - 6 = 0$ **iv.** $6x^2 + 23x + 25 = 5$

8 Your work on the modeling task in Problem 1 of Investigation 1 showed that using the zeroes of a polynomial function helps in writing a rule for the function as a product of linear factors. But that rule might need some adjustment to make it match the function exactly. For each of the following functions:

- use the information about zeroes to write a preliminary factored form of the rule.

- compare tables and graphs produced by the factored and standard form rules.

- adjust the factored form in a way that makes it equivalent to the standard form.

a. $f(x) = 6x^2 + x - 2$ has zeroes $-\frac{2}{3}$ and $\frac{1}{2}$.

b. $g(x) = 5x^2 - 13x - 6$ has zeroes $-\frac{2}{5}$ and 3.

c. $h(x) = -8x^3 + 2x^2 + 3x$ has zeroes $\frac{3}{4}, -\frac{1}{2},$ and 0.

End Behavior of Polynomial Functions You probably noticed that graphs of polynomial functions often rise and fall and cross the x-axis several times. But as the absolute value of x gets large, the graphs stop changing directions and head off to $\pm\infty$.

9 Recall that the **degree** of a polynomial is the highest power of the independent variable in the expression. Use technology to study graphs of many different polynomials with degree ranging from 1 to 6 and look for an answer to this question:

> *How can inspection of the coefficients and degree of a polynomial reveal the end behavior of the graph of the corresponding polynomial function?*

It might help to see patterns in the results if you summarize your findings in a table that lists the polynomial degree and the *end behavior* of the graph of the corresponding function as $x \rightarrow \pm\infty$.

10 Study the following factored and nested multiplication forms of polynomials and predict the degree of the polynomial and end behavior of the graph of the corresponding polynomial function. Check your work by expanding the expressions by repeated application of the distributive property or by use of a CAS.

a. $x(x + 3)(x - 2)$

b. $(x^2 + 3)(5 - 2x^2)$

c. $x^3(x + 3)(2 - x)$

d. $(x^2 + 3)(2x^4 + 5) + 7x$

e. $(((x + 3)x - 5)x + 6)x$

f. $(((((x + 3)x - 5)x + 6)x - 4)x - 2)$

An important characteristic that is used to describe a polynomial function is the number of zeroes. For example, the polynomial function $y = (x - 1)^3(x + 2) = (x - 1)(x - 1)(x - 1)(x + 2)$ is said to have a **zero of multiplicity 3** at $x = 1$ and a zero of multiplicity 1 at $x = -2$.

11 The shape of the graph of a polynomial function near a zero provides information about the possible multiplicity of the zero.

a. Graph the polynomial function $f(x) = (x - 3)^n$ for different positive integer values of n. Examine the behavior of each graph near $x = 3$. What appears to be true about the behavior of the graph of a polynomial function near a zero of multiplicity n?

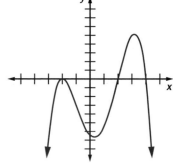

b. Test your conjecture in Part a by considering the graph of the polynomial function $g(x) = (x + 3)^2(x + 1)^3(x - 1)$.

c. Examine the graph of $y = h(x)$ at the right. The scale on both axes is 1. Find an equation for $h(x)$ if $h(-1) = -3$.

SUMMARIZE THE MATHEMATICS

In this investigation, you extended your understanding and skill in work with equivalent expressions for polynomials.

a What are the advantages of using standard polynomial form, factored form, or nested multiplication form in answering questions about polynomials?

b What strategies can be used to write standard-form polynomials as products of linear factors?

c How can you use the algebraic form of a polynomial to predict the end behavior of its graph?

d How does the shape of the graph of a polynomial function near a zero provide information about the possible multiplicity of the zero?

Be prepared to share your ideas and reasoning with the class.

✔ CHECK YOUR UNDERSTANDING

Consider the polynomial function $p(x) = x^3 - x^2 - 6x$.

a. Express the rule for $p(x)$ in equivalent form as:

   i. a product of linear factors.

   ii. a nested multiplication.

b. Find the number of computation cycles required to evaluate $p(2)$ using each of the three rule forms. Assume that evaluation of each power requires 2 cycles and each sum, difference, and product takes only 1 cycle.

c. Factor $5x^2 - 3x - 2$ and use the result to solve $5x^2 - 3x - 2 = 0$.

d. Without using technology, describe the end behavior of the graph of $y = -x^5 + 3x^2$. How do you know your answer is right?

INVESTIGATION ③

Division of Polynomials

The problems of Investigation 2 focused on strategies for expressing polynomials in useful equivalent forms. In particular, you developed methods for writing polynomials as products of linear factors to clearly show the zeroes of the corresponding functions.

As you work on the problems of this investigation, look for an answer to this question:

How can division be used to express a polynomial as a product of simpler factors?

1 The following diagram shows a graph of the cubic polynomial function $c(x) = x^3 - 2x + 4$.

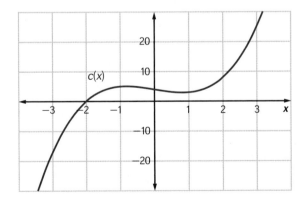

a. How do the shape and location of the graph suggest that $x = -2$ is the only zero of this function?

b. What is one linear factor of $c(x) = x^3 - 2x + 4$?

c. Why does the fact $c(-2) = 0$ suggest that $x^3 - 2x + 4 = (x + 2)q(x)$ for some polynomial $q(x)$?

d. Why does it make sense to find the function $q(x)$ in Part c, by calculating $(x^3 - 2x + 4) \div (x + 2)$?

2 To develop ideas about how to find the quotient of two polynomials, it helps to examine again one of the algorithms for division of whole numbers. Study the following long division calculations, record the results obtained, and explain how and why the algorithm works.

a. To find $489 \div 3$:

$$
\begin{array}{r}
3 \\
60 \\
100 \\
3\overline{)489} \\
-300 \\
\hline
189 \\
-180 \\
\hline
9 \\
-9 \\
\hline
\end{array}
$$

b. To find $1{,}145 \div 5$:

$$
\begin{array}{r}
9 \\
20 \\
200 \\
5\overline{)1{,}145} \\
-1{,}000 \\
\hline
145 \\
-100 \\
\hline
45 \\
-45 \\
\hline
\end{array}
$$

3 To find a polynomial $q(x)$ that satisfies the equation $x^3 - 2x + 4 = (x + 2)q(x)$, there is a polynomial division algorithm that is very similar to the procedure for long division of numbers.

a. Study the following example to see how the algorithm works—explain how terms in the quotient are found at each step.

Step 1.

$$
\begin{array}{r}
x^2 \\
x + 2\overline{)x^3 + 0x^2 - 2x + 4} \\
x^3 + 2x^2 \\
\hline
-2x^2 - 2x + 4 \\
\end{array}
$$

Multiply $(x + 2)$ by x^2
Subtract

Step 2.

$$
\begin{array}{r}
x^2 - 2x \\
x + 2\overline{)x^3 + 0x^2 - 2x + 4} \\
x^3 + 2x^2 \\
\hline
-2x^2 - 2x + 4 \\
-2x^2 - 4x \\
\hline
2x + 4 \\
\end{array}
$$

Multiply $(x + 2)$ by $-2x$
Subtract

Step 3.

$$
\begin{array}{r}
x^2 - 2x + 2 \\
x + 2\overline{)x^3 + 0x^2 - 2x + 4} \\
x^3 + 2x^2 \\
\hline
-2x^2 - 2x + 4 \\
-2x^2 - 4x \\
\hline
2x + 4 \\
2x + 4 \\
\hline
0 \\
\end{array}
$$

Multiply $(x + 2)$ by 2
Subtract

b. Expand the product $(x + 2)(x^2 - 2x + 2)$ to confirm that it equals $x^3 - 2x + 4$.

c. How do you know that $x^2 - 2x + 2$ cannot be factored further?

4 For each of the following polynomial functions:

- inspect a table or graph to find one zero $x = k$.

- divide the original polynomial by $(x - k)$ to get a quotient $q(x)$.

- express the original polynomial as a product $(x - k)q(x)$.

a. $p(x) = -x^3 - 2x^2 + 11x - 20$

b. $s(x) = x^3 + 2x^2 + 4x - 7$

c. $t(x) = x^4 - 2x^3 - 2x^2 - 17x + 42$

5 You might recall that some arithmetic division problems do not "come out even." Instead, they leave a remainder that is smaller than the divisor. For example, to calculate $2,398 \div 7$, you could use the work shown at the right.

$$
\begin{array}{r}
2 \\
40 \\
300 \\
7\overline{)2,398} \\
-2,100 \\
\hline
298 \\
-280 \\
\hline
18 \\
-14 \\
\hline
4
\end{array}
$$

a. Why does the equation $2,398 = 7(342) + 4$ show how one could check the division work?

b. Find the quotient and remainder resulting from $5,280 \div 13$. Then write an equation that shows how to check the results of the division work by multiplication and addition.

6 Some polynomial division problems also result in a quotient and nonzero remainder. Find the results of these divisions. Then write expressions involving multiplication and addition that could be used to check your work.

a. $(x^3 + 5x^2 - 4x - 20) \div (x - 1)$

b. $(x^3 + 2x^2 + 4x - 7) \div (x - 3)$

c. $(x^4 - 2x^3 - 17x + 42) \div (x - 1)$

d. $(x^3 - 1) \div (x - 2)$

7 In Problem 6, each polynomial division by a linear factor $(x - k)$ produced a quotient and a remainder. In each part of Problem 4, you chose the divisor $(x - k)$ so that k was a zero of the original polynomial and found that the remainders were all 0.

a. In Problem 6, the divisors were $(x - 1)$ and $(x - 3)$. Show that 1 is not a zero of the original polynomial functions in Parts a and c and that 3 is not a zero of the original polynomial function in Part b.

b. Show that in each part of Problem 6 the result of the division implies that if $p(x) = (x - k)q(x) + R$, then $R = p(k)$.

c. Check that in each part of Problem 4 where the result of the division is in the form $p(x) = (x - k)q(x) + 0$, it is the case that $p(k) = 0$.

8 The patterns you have observed in division of polynomials illustrate some general principles about that operation. Use algebraic reasoning to prove each of the following properties of all polynomials $p(x)$. If $p(x) = (x - k)q(x) + R$, then:

a. $R = p(k)$.

b. $(x - k)$ is a linear factor of $p(x)$ if and only if $p(k) = 0$.

9 The algorithm you have used to divide polynomials by linear terms in the form $(x - k)$ can be easily adapted to deal with other kinds of polynomial divisors as well. Analyze the procedure in Part a and then use similar reasoning to find the quotients in Parts b–e.

a. What reasoning leads to selection of the terms $4x^3$ and $-7x$ in the quotient at the right?

$$
3x^2 + 5 \overline{)\begin{array}{l} 4x^3 - 7x \\ 12x^5 - x^3 - 35x \\ \underline{12x^5 + 20x^3} \\ -21x^3 - 35x \\ \underline{-21x^3 - 35x} \end{array}}
$$

b. $(3x^3 + 17x^2 + 10x) \div (x^2 + 5x)$

c. $(28x^4 + 9x^3 - 9x^2 + 35x - 15) \div (7x - 3)$

d. $(3x^5 + 6x^4 + 2x^3 + 2x^2 + 3x - 1) \div (x^3 + x^2 + 1)$

e. $(2x^5 - 6x^4 + 4x^3 - 12x^2 - 3x + 7) \div (x^2 - 3x)$

10 Look back over the dividends, divisors, quotients, and remainders in the various division problems of this investigation to find answers and explanations for these questions.

a. Suppose that a numeric division problem leads to the result $a \div b = c$ with remainder d or $a = bc + d$ (for example, $27 \div 6 = 4$ with remainder 3 or $27 = 6(4) + 3$). What can you be sure of about the relationship between the divisor b and the remainder d?

b. Suppose that a polynomial division problem leads to an expression in the form

$$p(x) \div d(x) = q(x) \text{ with remainder } r(x) \text{ or}$$
$$p(x) = d(x)q(x) + r(x).$$

What can you be sure of about the relationship between the degree of the divisor polynomial $d(x)$ and the remainder $r(x)$?

11 **Solving Polynomial Equations and Inequalities** Factored forms of polynomial functions and their corresponding graphs provide helpful information for solving related polynomial equations and inequalities. Combine algebraic and graphic reasoning to solve the following equations and inequalities. Record the solutions to the inequalities in interval notation.

a. In Problem 4, you found that $t(x) = x^4 - 2x^3 - 2x^2 - 17x + 42 = (x - 2)(x - 3)(x^2 + 3x + 7)$. Find all exact real number solutions to $(x - 2)(x - 3)(x^2 + 3x + 7) = 0$.

b. Find all exact real number solutions to $x^3 - 3x^2 - 2x + 6 \leq 0$.

c. Find all exact real number solutions to $-x^4 + 9x^3 - 24x^2 + 20x > 0$.

12 Look back at your solutions to Problem 11. What does the degree of a polynomial tell you about the possible number of real zeroes for the corresponding function? Compare your solutions in Problem 11 and your related conjecture in this problem with those of your classmates. Resolve any differences.

SUMMARIZE THE MATHEMATICS

In this investigation, you learned how to use division to express given polynomials in equivalent forms.

a How can you find factors of the form $(x - k)$ for given polynomials?

b What are the possible results from division of a polynomial $p(x)$ by $(x - k)$?

   i. How can those results be checked by using multiplication and addition of polynomials?

   ii. What useful information is revealed in the expression used in the check?

c How is the solution to a polynomial equation or inequality, $p(x) = 0$, $p(x) < 0$, or $p(x) > 0$, related to the factored form of the polynomial $p(x)$?

Be prepared to share your methods and reasoning with the class.

 CHECK YOUR UNDERSTANDING

Consider the polynomial functions $p(x) = x^3 - x^2 - 6x$ and $s(x) = x^3 + 4x^2 + 7x + 6$ and the linear expression $x - 3$.

a. Use polynomial division to find $q_1(x)$, $q_2(x)$, R_1, and R_2 so that:

   i. $p(x) = (x - 3)q_1(x) + R_1$

   ii. $s(x) = (x - 3)q_2(x) + R_2$

b. Use the results of Part a to decide whether:

   i. $(x - 3)$ is a factor of $x^3 - x^2 - 6x$.

   ii. 3 is a zero of $p(x)$.

   iii. $(x - 3)$ is a factor of $x^3 + 4x^2 + 7x + 6$.

   iv. 3 is a zero of $s(x)$.

c. Combine algebraic and graphic reasoning to solve $x^3 - x^2 - 6x \geq 0$. Express the solution using a number line and interval notation.

APPLICATIONS

1 One of the most exciting events in the X-Games is the BMX Big Air competition. Participants ride their bikes down a very long and steep ramp, take off like ski jumpers, and do tricks in the air before landing. Long hours of practice are needed for such competitions.

Suppose that the design for a ramp was to satisfy the specifications in the following graph. It looks like the shape of the ramp could be modeled by the graph of a quadratic function.

BMX Big Air Ramp

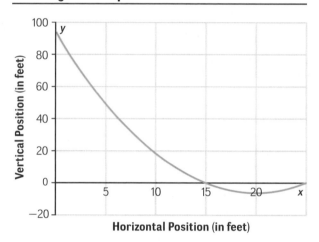

Horizontal Position (in feet)

a. Use information about the indicated x-intercepts of the ramp graph to find the rule for a quadratic function whose graph has the same x-intercepts. Write the rule using an expression that is the product of two linear factors.

b. Write the rule using an equivalent expression in standard polynomial form.

c. Compare a graph of the proposed model to the shape of the ramp and explain how and why (if at all) the two graphs differ.

2 Select a set of *control points* that you believe will be useful in finding a quadratic regression equation with graph that matches the given ramp shape better than the result of Applications Task 1.

a. Use the control points to find the desired quadratic regression model for the graph shape.

b. Compare the regression model to the given ramp shape and to the model in Applications Task 1.

3 One reasonable set of control points for the modeling process in the ramp modeling situation in Applications Task 1 is the y-intercept and the two x-intercepts.

a. Estimate coordinates for the intercept points.

b. Use the estimates of Part a and the method of undetermined coefficients to find a quadratic expression $ax^2 + bx + c$ for a function whose graph models the ramp shape.

4 NASCAR and Indy car racing generally occur on symmetric oval tracks with only left turns and no hills or dips. But sports car and Grand Prix racetracks have irregular shapes with a variety of left and right turns and changes in elevation. Design of tracks for sports car and Grand Prix racing must be done carefully to make them safe at the expected speeds of cars.

Suppose that in designing a Grand Prix track, the plan included an S-shaped curve like that shown next.

Grand Prix Track S-Curve

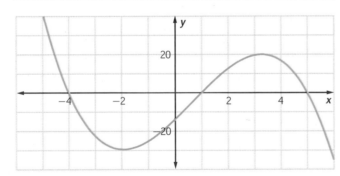

Scale: 1 unit = 50 feet

Select a set of control points that you believe will lead to a good regression equation model for the given race track graph pattern.

a. Use the chosen control points to produce a polynomial regression equation that matches the control points.

b. Produce a graph of the function developed in Part a and compare it to the given race track graph. Describe and explain reasons for any differences between the two graphs.

5 There are five natural control points to use in modeling the S-shaped curve of the race track graph in Applications Task 4—the three x-intercepts, the local minimum point, and the local maximum point.

   a. Estimate coordinates of the suggested control points.

   b. Use the coordinates of suggested control points to write a system of linear equations whose solution will provide coefficients in the expression for a quartic function whose graph passes through the points.

   c. Solve the system in Part b.

   d. Write the quartic polynomial determined by the coefficients in Part c. Then compare the graph of the corresponding quartic function to the given race track graph.

6 Use the method of undetermined coefficients to find quadratic or cubic functions whose graphs pass through these sets of points.

   a. $(-1, 0)$, $(2, 12)$, $(5, -12)$

   b. $(-1, 9)$, $(0, 3)$, $(5, 3)$

   c. $(-1, 9)$, $(0, 0)$, $(1, 1)$, $(4, 4)$

   d. $(-4, 0)$, $(2, -8)$, $(0, -8)$, $(3, -62)$

7 The x-intercepts of the race track graph in Applications Task 4 are $(-4, 0)$, $(1, 0)$, and $(5, 0)$.

   a. Use the information about x-intercepts to write the rule for a polynomial function whose graph has the same x-intercepts.

      i. Express the rule using a product of linear factors.

      ii. Express the rule using standard polynomial form.

   b. Produce a graph of the function developed in Part a and compare it to the given race track graph. Describe and explain reasons for any differences between the two graphs.

   c. Adjust the rule for the function developed in Part a so that its graph is a better model for the race track graph.

8 Produce graphs of the following polynomial functions and use estimates of their zeroes to write each polynomial as a product of linear factors (as far as that is possible).

   a. $y = x^2 + 3x - 10$

   b. $y = x^2 + 5.5x - 3$

   c. $y = x^3 - 3x^2 - 18x + 40$

   d. $y = x^3 + 2x^2 + 4x + 3$

   e. $y = -x^4 + 4x^3 - 5x^2 + 2x$

9 Write the given polynomial expressions in equivalent form as a product of linear factors. Then find the exact real number solutions.

a. $3x^2 + 5x + 2 = 0$

b. $5x^2 - 18x - 8 < 0$

c. $12x^2 + 8x - 15 = 0$

d. $4x^2 - x - 3 \geq 0$

e. $2x^3 - 6x = 0$

f. $x^3 + x^2 - 10x + 8 < 0$

10 Perform the indicated polynomial divisions. Then express the results as equations in the form $p(x) = f(x)q(x) + r(x)$.

a. $(x^2 - 3x - 10) \div (x - 5)$

b. $(3x^2 - x - 12) \div (x - 2)$

c. $(x^3 + 6x^2 + x - 4) \div (x + 1)$

d. $(x^3 + x^2 - x - 14) \div (x - 2)$

11 Perform the indicated polynomial divisions. Then express the results as equations in the form $p(x) = f(x)q(x) + r(x)$.

a. $(x^4 + 5x^3 + 7x^2 - 2x - 20) \div (x^2 + 3x + 5)$

b. $(3x^3 + 5x^2 + 17x + 10) \div (x^2 + x + 5)$

c. $(x^5 + 2x^4 + 11x^3 + 8x^2 + 28x) \div (x^3 + 4x)$

d. $(x^5 + 2x^4 + 11x^3 + 9x^2 + 2x + 3) \div (x^3 + 4x)$

CONNECTIONS

12 The following sketch shows how to make an open-top box from a flat sheet of cardboard with cuts (heavy lines) and folds (dashed lines) as indicated. The dimensions of the resulting box depend on the length x of the corner cuts.

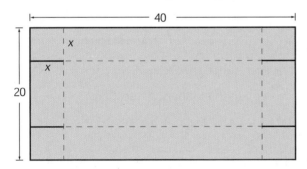 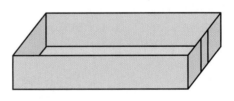

a. Show how the volume of the box depends on length x of the corner cuts.

   i. Express each dimension of the box—length, width, and height—as a function of x.

   ii. Use the formula for volume V of a box to write V as a function of x.

   iii. Write the volume expression in expanded standard polynomial form.

b. Analyze the volume function for the box by answering the following questions.

   i. Sketch a graph of $y = V(x)$ for values of x that make sense in this situation.

   ii. Estimate coordinates of the x-intercepts and local maximum and minimum points on the graph of $V(x)$.

   iii. Explain what the x-intercepts and local maximum and minimum points tell about the way the volume of such a box depends on the length x of the cuts.

13 The following graph of one period for the function $f(x) = \sin x$ looks very much like graphs of some cubic polynomials that you have studied in this lesson.

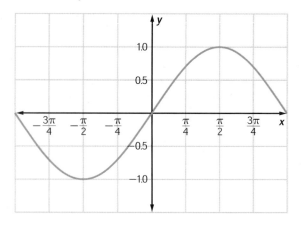

a. Use a method of your choice to develop the rule for a cubic polynomial function whose graph you suspect will match that of the sine function.

b. Compare tables and graphs of the sine function and the approximating cubic polynomial function and note any differences between the two functions.

14 An **even** function is one for which $f(-x) = f(x)$ for all x. An **odd** function is one for which $f(-x) = -f(x)$ for all x. For example, $f(x) = x^3$ is an odd function because $f(-x) = (-x)^3 = -x^3$ which is the opposite of $f(x) = x^3$.

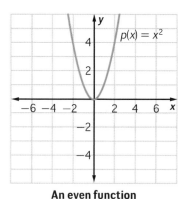

An even function

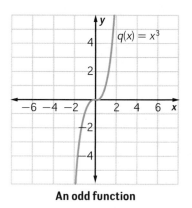

An odd function

a. Is $f(x) = x^4 + 3x^2$ an odd or even function? Explain.

b. Use algebraic reasoning to explain the following questions.

   i. Why is the polynomial $f(x) = a_0x^0 + a_2x^2 + a_4x^4 + a_6x^6 + a_8x^8$ an even function?

   ii. Why is any polynomial with only even-power terms an even function?

c. What symmetry is implied for the graph of an even function by the property that $f(-x) = f(x)$?

d. Use algebraic reasoning to explain the following questions.

   i. Why is the polynomial $g(x) = b_1x + b_3x^3 + b_5x^5 + b_7x^7 + b_9x^9$ an odd function?

   ii. Why is any polynomial with only odd-power terms an odd function?

e. What symmetry is implied for the graph of an odd function by the property that $f(-x) = -f(x)$?

15 Explain why the graph of any polynomial function, when viewed on a large enough interval, will look like the graph of a power function.

16 The graph in Connections Task 13 can be described as *concave up* in the interval $[-\pi, 0]$ and *concave down* in the interval $[0, \pi]$. In general, the graph of a function is **concave up** if it bends upward as you move from left to right. It is **concave down** if it bends downward.

  a. Look back at the graphs in Investigation 2 Problem 11 (page 192) and Investigation 3 Problem 1 (page 193). For each graph, describe the interval(s) for which it is concave up. Concave down.

  b. If possible, sketch the graph of a function on an interval for which it is:

  i. increasing and concave up   **ii.** decreasing and concave up

  iii. increasing and concave down   **iv.** decreasing and concave down

17 In your earlier study of recursion and iteration, you saw many examples of geometric sequences with terms in the general form $1, r, r^2, r^3, r^4, r^5, \ldots, r^n$. In many situations, it is useful to find the sum of the first n terms of such a sequence. In Course 3 Unit 7, *Recursion and Iteration*, you derived the formula

$$1 + r + r^2 + r^3 + r^4 + r^5 + \cdots + r^n = \frac{r^{n+1} - 1}{r - 1}.$$

Use polynomial division to confirm this formula for the three cases below by showing the steps in your calculations.

  a. $\dfrac{r^3 - 1}{r - 1}$ **b.** $\dfrac{r^4 - 1}{r - 1}$ **c.** $\dfrac{r^5 - 1}{r - 1}$

18 Find a cubic polynomial $p(x)$ so that the solution of $p(x) \leq 0$ is $[-2, 1]$ or $[4, \infty)$. Is your polynomial the only one that gives this solution? Explain.

REFLECTIONS

19 In using mathematical function graphs to model geometric shapes, polynomial functions are often preferred over other functions—like exponential, logarithmic, or trigonometric functions—whose graphs have similar shapes. What reasons can you think of for this emphasis on polynomials?

20 In the work of finding polynomial models for graph patterns, you explored three different methods: (1) using x-intercept points to determine zeroes and linear factors; (2) using control points and polynomial regression routines; (3) using control points and the method of undetermined coefficients involving systems of linear equations. What do you see as the advantages and disadvantages of each method?

21 If you were to use the method of undetermined coefficients to find a polynomial function whose graph passes through eight given points, what would be the largest possible degree of the polynomial? Explain your reasoning.

22 Most students find factoring of quadratics in the form $ax^2 + bx + c$ more difficult when $a \neq 1$. Based on your work in Investigation 2, what explanation can you give for this difficulty?

23 Often students find division of polynomials easier than division of numbers.

    a. Show the steps to find $938 \div 34$ using one of the standard long-division algorithms.

    b. Show the steps to find $(9x^2 + 3x + 8) \div (3x + 4)$ using division of polynomials.

    c. Study the work in calculation of Parts a and b and see if you can understand the preference for polynomial division over arithmetic division.

24 Division with remainders occurs in both whole number arithmetic and in the algebra of polynomials. The results can be expressed in quite similar forms. For example, $213 = 12(17) + 9$ shows the result of $213 \div 12$ and $2x^2 + x + 3 = (x + 2)(2x - 3) + 9$ shows the result of $(2x^2 + x + 3) \div (x + 2)$.

    How are the remainders related to the divisors in both arithmetic and polynomial division and why is that the case?

25 Andrea used the graph shown at the right to help her solve the equation $x^4 - x^3 - 27x^2 + 81x - 54 = 0$. She determined that the solutions of the equation were $x = 1$ and $x = 3$. Yolanda, who was working with Andrea, looked at the graph and said there had to be another solution. She then found that -6 was also a solution.

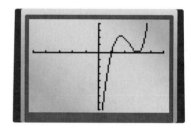

    • Verify that $1, 3$, and -6 are all solutions.

    • How did Yolanda know there had to be another solution?

26 Look back at what you were asked to prove in Investigation 3 Problem 8 Parts a and b (page 196).

    a. The result in Part a is often called the **Remainder Theorem**. State the Remainder Theorem as precisely as you can in words.

    b. The result in Part b is referred to as the **Factor Theorem**. State the Factor Theorem as precisely as you can in words.

EXTENSIONS

27 Finding polynomial functions whose graphs contain a given set of control points can be combined with parametric equations to produce more complicated curves. Examine the loop in the roller coaster track shown here in the picture and graph.

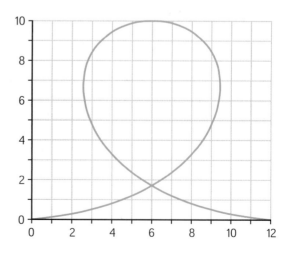

a. Identify a set of control points that you think would determine a curve similar to that shown. Then assign times starting at $t = 0$ for the coaster to be at each location and create a table of values for (t, x, y).

b. Use any method you know to find a polynomial function $P_1(t)$ whose values match the coordinate pairs (t, x) in your table of control point coordinates.

c. Find a second polynomial function $P_2(t)$ whose values match the coordinate pairs (t, y) in your table of control point coordinates.

d. Use technology to plot the curve given parametrically by the equations $x = P_1(t)$ and $y = P_2(t)$ over an appropriate interval of values for t.

28 Research the *CPMP-Tools* Spreadsheet Solver tool for solving systems of equations. Prepare a demonstration for your class on how this Spreadsheet capability can be used in solving problems like Applications Task 3. Be prepared to answer any questions your classmates may have.

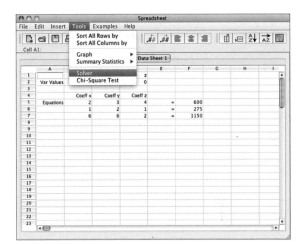

29 Find a third degree polynomial function whose graph contains the points $(-4, -80)$, $(0, 12)$, $(5, 82)$, and $(2, 10)$.

a. Find a *different* polynomial function whose graph contains the same four points.

b. Describe a general method for finding different polynomial functions whose graphs contain the four points.

c. Is it possible to find a different *third degree* polynomial function whose graph contains the same four points? Explain your reasoning.

30 One of the most useful patterns for factoring quadratic expressions is given by the equation $x^2 - a^2 = (x - a)(x + a)$. In advanced mathematics, it turns out to be useful to be able to factor similar expressions in the form $x^n - a^n$ for different integer values of n.

a. Explain why $(x - a)$ is always a factor of the expression $x^n - a^n$.

b. Use division of polynomials to find $q(x)$ so that $x^3 - 27 = (x - 3)q(x)$.

c. Use division of polynomials to find $q(x)$ so that $x^4 - 81 = (x - 3)q(x)$.

d. Use division of polynomials to find $q(x)$ so that $x^5 - a^5 = (x - a)q(x)$.

e. Use division of polynomials to find $q(x)$ so that $x^6 - a^6 = (x - a)q(x)$.

f. What pattern determines $q(x)$ so that $x^n - a^n = (x - a)q(x)$ for any positive integer n?

31 In earlier work with quadratic functions, you found that it was often informative to write a quadratic expression in equivalent vertex form. The examples you dealt with at that time had lead coefficient of 1, making the algebraic work somewhat simpler.

a. Give justification for each step in the following derivation of an equivalent vertex form of the quadratic expression $2x^2 - 12x + 13$.

$$
\begin{aligned}
2x^2 - 12x + 13 &= 2(x^2 - 6x) + 13 && (1) \\
&= 2(x^2 - 6x + 9) + 13 - 18 && (2) \\
&= 2(x - 3)^2 - 5 && (3)
\end{aligned}
$$

b. Apply similar reasoning to derive equivalent vertex forms of these quadratic expressions.

   i. $3x^2 - 30x + 63$

   ii. $5x^2 + 10x + 1$

   iii. $ax^2 + bx + c$

c. Use the vertex forms derived in Part b to solve each of the following quadratic equations for x.

   i. $3x^2 - 30x + 63 = 0$

   ii. $5x^2 + 10x + 1 = 0$

   iii. $ax^2 + bx + c = 0$

32 In work on the problems of Investigation 2, you compared the efficiency of several algorithms for evaluating polynomial functions. Consider the polynomial $p(x) = 4x^3 - 6x^2 - 8x + 12$ which can be written with a rule in nested form as $p(x) = ((4x - 6)x - 8)x + 12$.

a. List the steps involved in evaluating $p(2)$ using the order of operations indicated by the nested expression for $p(x)$.

b. The following display shows a shorthand method of evaluating polynomials like $p(x) = 4x^3 - 6x^2 - 8x + 12$.

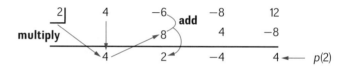

Explain how the procedure indicated in that diagram is related to the nested multiplication form in Part a.

c. The procedure shown schematically in Part b is called **synthetic substitution**. Use a similar procedure to evaluate $f(x) = 5x^4 - 2x^3 + 10x^2 - x - 5$ when $x = 2$. When $x = -3$.

33 The work on Investigation 3 showed that if $p(x) = (x - k)q(x) + R$, then $p(k) = R$. This suggests a connection between polynomial division and polynomial evaluation. To explore that connection, begin by comparing the polynomial division and synthetic substitution shown here.

$$
\begin{array}{r}
2x^2 - x - 3 \\
x + 1 \overline{\smash{\big)}\, 2x^3 + x^2 - 4x - 3} \\
\underline{2x^3 + 2x^2} \\
-x^2 - 4x - 3 \\
\underline{-x^2 - x} \\
-3x - 3 \\
\underline{-3x - 3} \\
0
\end{array}
$$

$$
\begin{array}{r|rrrr}
-1 & 2 & 1 & -4 & -3 \\
   &   & -2 & 1 & 3 \\
\hline
   & 2 & -1 & -3 & 0
\end{array}
$$

a. How is the fact that -1 is a zero of $p(x) = 2x^3 + x^2 - 4x - 3$ indicated in both the long division of polynomials and the synthetic substitution?

b. How do the coefficients of the quotient and the remainder in polynomial division show up in the synthetic substitution process?

c. Test your ideas from Parts a and b by finding the quotient and remainder for $(3x^3 - 7x - 4) \div (x + 1)$ in both ways.

d. Use synthetic substitution to rewrite the following expressions in equivalent forms without any indicated division.

  i. $(2x^3 - 5x^2 - 3x + 4) \div (x + 1)$

  ii. $\dfrac{2x^3 - 6x + 4}{x + 2}$

  iii. $(6x^3 + 5x^2 + 6x - 5) \div (2x - 1)$

e. How do the results of your work on Parts a–d help to explain why synthetic substitution is sometimes called *synthetic division*?

REVIEW

34 Show how to solve each of the following systems of linear equations in three ways: (1) using substitution; (2) using elimination of variables; (3) using matrix inverses.

a. $\begin{cases} 3x - y = 1 \\ -2x + y = 1 \end{cases}$ **b.** $\begin{cases} 3x + 2y = 5 \\ -2x + 3y = 14 \end{cases}$

35 Write these quadratic expressions in equivalent form as products of linear factors.

a. $x^2 - 3x - 40$ **b.** $x^2 - 9x + 20$

c. $6x^2 + 33x + 15$ **d.** $25x^6 - 49$

e. $x^4 - 16$

36 For a polynomial of degree $n \geq 1$, what is the maximum number of:

a. zeroes of the related function?

b. local maximum and minimum points on the graph of the related function?

37 Recall that in order to have an inverse, a function must be one-to-one.

a. Explain what it means for a function to be one-to-one.

b. Give a rule for a function that is not one-to-one.

c. Restrict the domain of your function in Part b so that the new function has an inverse.

38 A tugboat left its dock and traveled for 30 minutes at a speed of 25 knots with a direction of 70°. It then turned and traveled at the same speed for 15 minutes on a path with direction 320°.

a. Assume that the dock is at the origin of a coordinate system. Draw a sketch of the tugboat's path to this point. Then determine the coordinates of the location of the tugboat at this point in time.

b. The tugboat then traveled directly back to the dock. How far did it travel to get back? What was the direction of its path back to the dock?

39 Write each expression in standard polynomial form.

a. $(2x - 7) + 3(12 + 4x)$ **b.** $(5 - 4y) - (6y + 23)$

c. $(6x + 2)(3x + 5)$ **d.** $(2x + 1)(x^2 + 3x + 5) + 12$

40 Quadrilateral $ABCD$ is a parallelogram whose diagonals intersect at point E. If $m\angle EAD = 30°$, $m\angle EBA = 64°$, and $m\angle EBC = 50°$, find the measure of each indicated angle.

a. $m\angle ADC$ **b.** $m\angle EDA$

c. $m\angle AED$ **d.** $m\angle BAD$

41 Rosario wants all of the pizza she sells to cost the same per square inch. She sells a 10-inch diameter round pizza for $6.99.

a. How much should she charge for a 20-inch diameter round pizza?

b. Write a rule that relates the price p Rosario should charge for a round pizza and the diameter d of the pizza.

c. Complete this sentence: The cost of a pizza is _____ proportional to the _____ with constant of proportionality _____ .

42 Without using technology, sketch a graph of each of the following functions. Then check your work by using technology to graph the function.

a. $f(x) = (x - 5)^2$ **b.** $f(x) = x^2 - 5$

c. $f(x) = 5x^2$ **d.** $f(x) = -x^2 + 5$

43 Use the quadratic formula to solve each equation.

a. $x^2 - 6x + 2 = 0$ **b.** $3x^2 + 10x + 5 = 0$

c. $x(x + 8) = 5$

The Complex Number System

Complex numbers have a long and rich history starting with solving quadratic equations to contemporary computer-based work in fractal geometry pioneered by Benoit Mandelbrot, a French mathematician.

Solving equations is one of the most important tasks in mathematics. For thousands of years, mathematicians all over the world have worked to develop progressively more powerful techniques for solving equations. The quest for equation solutions has prompted development of many new ideas about numbers and algebra.

For example, over 2,500 years ago, philosophers and scientists were puzzled by the problem of finding a square with area 2 square units. They realized that this problem was equivalent to solving the equation $x^2 = 2$. But, no one could find a solution among the integers and rational numbers that they knew about.

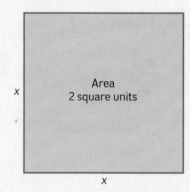

x

Area
2 square units

x

Pythagoras stunned the intellectual community when he proved that no rational number solution to the equation $x^2 = 2$ could ever be found. It was not until late in the 19th century that mathematicians solved the problem by providing a sound definition of *irrational* numbers.

In the middle of the 16th century, an Italian mathematician named Girolamo Cardano posed another simple but puzzling problem about equations and their solutions. He asked readers of his book *Ars Magna* to "Divide 10 into two parts whose product is 40."

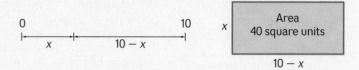

0   10 x Area
|—————|—————————| 40 square units
  x $10 - x$

$10 - x$

Solving this problem is equivalent to solving the equation $x(10 - x) = 40$ or $x^2 - 10x + 40 = 0$. If you apply the quadratic formula to this equation you will see that the solutions must be $x = 5 \pm \sqrt{-15}$, a result that troubled mathematicians deeply.

Work on the problems of this lesson will develop the complex numbers that solved Cardano's problem, techniques for operating on them, and ways of visualizing both the numbers and operations on a coordinate grid.

INVESTIGATION

A Complex Solution

The solution to Cardano's problem about dividing 10 into factors of 40 challenged mathematicians to make sense of square roots for negative numbers. It took several centuries for the mathematical community to become comfortable with this strange idea. But, as so often happens, new problems led to development of useful new mathematics; in this case, a very useful extension of the real number system.

As you work on the problems of this investigation, look for answers to these questions:

> *What kind of number system allows square roots for negative numbers?*
>
> *How can the familiar operations of arithmetic be generalized in useful ways to the new kinds of numbers?*

1 Consider again Cardano's quadratic equation $x(10 - x) = 40$.

    a. What algebraic reasoning shows that the given equation is equivalent to $x^2 - 10x + 40 = 0$?

    b. Use the quadratic formula to verify that the solutions of this equation are $x = 5 \pm \sqrt{-15}$.

    c. How does the graph of $y = x^2 - 10x + 40$ show that there are no real number solutions of the equation?

2 Consider another similar quadratic equation $x^2 - 6x + 13 = 0$.

    a. How does the quadratic formula imply that the solutions of this equation should be $x = 3 + 2\sqrt{-1}$ and $x = 3 - 2\sqrt{-1}$?

    b. How does the graph of $y = x^2 - 6x + 13$ show that there are no real number solutions of the given equation?

The problem in solving $x^2 - 10x + 40 = 0$ appears with the radical $\sqrt{-15}$; the problem in solving $x^2 - 6x + 13 = 0$ appears with the radical $\sqrt{-1}$. All of your prior experience with square roots tells you that *no* real number has -15 or -1 as its square. For hundreds of years, mathematicians seemed to accept as a fact that equations like $x^2 - 10x + 40 = 0$ and $x^2 - 6x + 13 = 0$ simply have no solutions.

Cardano suggested that all we needed was a new kind of number. But his idea was explored by mathematicians for several more centuries (often with strong doubts about using numbers with negative squares) until *complex numbers* became an accepted tool for both pure and applied mathematics. The resolution of the problem comes from reasoning that the solutions for $x^2 - 6x + 13 = 0$ should be numbers that can be expressed in the form

$$x = 3 + 2\sqrt{-1} \text{ and } x = 3 - 2\sqrt{-1}.$$

In a similar fashion, it makes sense that $\sqrt{-15} = \sqrt{15}\sqrt{-1}$. So, the solutions of $x^2 - 10x + 40 = 0$ should be numbers expressed in the form

$$5 + \sqrt{15}\sqrt{-1} \text{ and } 5 - \sqrt{15}\sqrt{-1}.$$

This kind of observation led mathematicians to develop what is now called the **complex number system** with elements in the form $a + bi$, where a and b are real numbers and $i = \sqrt{-1}$. So, $3 + 2\sqrt{-1}$ is written as $3 + 2i$ and $5 - \sqrt{15}\sqrt{-1}$ is written as $5 - \sqrt{15}\,i$. The new number system provides solutions not only for the problematic quadratic equations, but for *all other polynomial equations of the form* $p(x) = 0$.

Numbers of the form bi are called **imaginary numbers**. Complex numbers contain all the real numbers (complex numbers $a + bi$ for which $b = 0$) and all the imaginary numbers (complex numbers $a + bi$ where $a = 0$). The connection between i and $\sqrt{-1}$ suggests that $i^2 = -1$.

3 Use what you know about the quadratic formula and properties of radicals to solve the following equations. Write the results in standard complex number form $a + bi$.

    a. $x^2 + 6x + 34 = 0$

    b. $x^2 + 10 = 0$

    c. $2x^2 - 28x + 106 = 0$

    d. $x^2 + 7x + 6 = 0$

    e. $x^2 - 6x = -10$

Operations on Complex Numbers If mathematical expressions in the form $a + bi$ are to be considered numbers, it is natural to wonder whether there is a sensible way to define arithmetic operations like addition, subtraction, multiplication, and division for these new objects. When this question was first raised, there were no real-life applications of complex numbers to suggest operational definitions. So, the most natural guide was to think of complex numbers like $3 + 2i$ and $5 - 7i$ as algebraic expressions similar to $3 + 2x$ and $5 - 7x$. Problems 4 to 10 ask you to devise rules for arithmetic with complex numbers. You can check the ideas you come up with using a computer algebra system, because almost every CAS is programmed to do complex number arithmetic.

4 Think about rules that would seem the most natural definitions for addition and subtraction of two complex numbers. Compare your ideas with those of others and resolve differences.

For any two complex numbers $a + bi$ and $c + di$:

a. $(a + bi) + (c + di) =$ _____ $+$ _____

b. $(a + bi) - (c + di) =$ _____ $+$ _____

5 According to the operational definitions that you agreed on, what complex numbers in the form $a + bi$ represent the following sums and differences?

a. $(4 + 6i) + (3 + 2i)$ b. $(4 + 6i) - (3 + 2i)$

c. $(4.5 + 7.2i) + (9 + 3.1i)$ d. $(4.5 + 7.2i) - (9 + 3.1i)$

e. $(12 + 9i) + (-3 - 9i)$ f. $(12 + 9i) - (-3 - 9i)$

6 If you compare the complex number $a + bi$ to the algebraic expression $a + bx$ and remember that $i^2 = -1$, what complex numbers in the form $a + bi$ would represent these products?

a. $(4 + 6i)(3 + 2i)$ b. $(5 - 7i)(9 + 3i)$

c. $(2 + 9i)(-3 - 2i)$ d. $(4 + 7i)(4 - 7i)$

7 What general rule for multiplication of complex numbers is suggested by the examples in Problem 6?

$$(a + bi)(c + di) = \text{\_\_\_\_\_} + \text{\_\_\_\_\_}$$

Compare your ideas with those of your classmates and resolve differences.

8 Pairs of complex numbers in the form $a + bi$ and $a - bi$ are called **conjugates** of each other. The product of any conjugate pair has a very useful special property. Calculate the results in Parts a–c and then describe what seems to be the special property of products for conjugate pairs.

a. $(4 + 6i)(4 - 6i)$ b. $(5 - 7i)(5 + 7i)$

c. $(-2 + 9i)(-2 - 9i)$ d. $(-3 - 4i)(-3 + 4i)$

e. Describe the pattern of results in Parts a–d by completing the sentence that begins:

"The product of two complex numbers that are conjugates of each other is always a _____ number equal to … ."

f. How are your results in Parts a–e explained by the algebraic principle that $(x + y)(x - y) = x^2 - y^2$?

9 Division of complex numbers is a bit more involved than the other three arithmetic operations. Study the following argument that suggests a definition for division of complex numbers. Be prepared to explain the reasoning behind each step and how the argument shows a way of expressing the quotient of any two complex numbers (divisor not zero) in the standard complex number form.

$$(a + bi) \div (c + di) = \frac{a + bi}{c + di} \qquad (1)$$

$$= \frac{a + bi}{c + di} \cdot \frac{c - di}{c - di} \qquad (2)$$

$$= \frac{(ac + bd) + (bc - ad)i}{c^2 + d^2} \qquad (3)$$

$$= \frac{(ac + bd)}{c^2 + d^2} + \frac{(bc - ad)}{c^2 + d^2}i \qquad (4)$$

10 Use the rule that you developed in Problem 9 to find these quotients of complex numbers. Write the answers in $a + bi$ form.

a. $(4 + 6i) \div (4 - 6i)$

b. $(3 - 2i) \div (5 + 7i)$

c. $(-2 + 9i) \div (-3 + 2i)$

SUMMARIZE THE MATHEMATICS

In this investigation, you explored the historical and mathematical origins of the complex number system and the reasoning that leads to definitions for operations on these new numbers.

a How does solving an equation like $x^2 - 10x + 40 = 0$ demonstrate the need for complex numbers?

b How are all elements of the system of complex numbers expressed in terms of real numbers and the special imaginary number $i = \sqrt{-1}$?

c What are the standard rules for calculating:

- $(a + bi) + (c + di)$?
- $(a + bi) - (c + di)$?
- $(a + bi)(c + di)$?
- $(a + bi) \div (c + di)$?

Be prepared to share your thinking with the class.

✔ **CHECK YOUR UNDERSTANDING**

Consider the quadratic equation $x^2 + 6x + 25 = 0$.

a. Find the complex number solutions for the equation and express them in the form $a + bi$.

b. Calculate the sum, difference, product, and quotient of the two complex numbers that are solutions to the given equation. Express the results in the standard form $a + bi$.

INVESTIGATION **2**

Properties of Complex Numbers

The problems of Investigation 1 focused on reasons why complex numbers were created and the ways that it makes sense to add, subtract, multiply, and divide them. To use complex numbers in further mathematical work, it is helpful to know how to represent complex numbers geometrically and to establish the algebraic rules that govern manipulation of complex number expressions into equivalent forms.

As you work on the problems of this investigation, look for answers to these questions:

How can complex numbers be represented geometrically on a coordinate grid?

How do the algebraic properties of complex numbers compare to those of real numbers?

Graphical Representation of Complex Numbers Each complex number can be represented in the form $a + bi$, where a and b are real numbers and i represents the imaginary number $\sqrt{-1}$. Around the turn of the 19th century, three different mathematicians—Caspar Wessel (1797), Jean Robert Argand (1806), and Carl Friedrich Gauss (1811)—suggested that the complex numbers could also be represented by points on a coordinate grid. The complex number $a + bi$ would correspond to the point with coordinates (a, b).

You know that points of the coordinate grid also determine vectors in standard position. So, complex numbers and vectors have a close and useful connection as well. Examples are shown on the following graph. Scales on both axes are 1 unit.

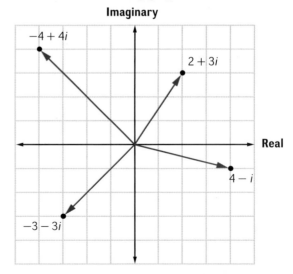

1 On a single coordinate grid, locate and label points and vectors corresponding to these complex numbers.

a. $2 + 5i$ **b.** $-3 + i$ **c.** $1 - 3i$ (*Note:* It is customary to write simply $2i$ and 3 instead of $0 + 2i$ and $3 + 0i$.)

d. $-5 - 3i$ **e.** $0 + 2i$ **f.** $3 + 0i$

2 The **absolute value** (or **modulus**) of a complex number $a + bi$, denoted $|a + bi|$, is defined to be the length of the vector from the origin $(0, 0)$ to the point (a, b).

a. What formula can be used to calculate the absolute value of any complex number $a + bi$?

b. Find the absolute values of the complex numbers plotted in Problem 1.

3 Consider next the geometric relationship between complex numbers and their conjugates.

a. On the diagram you drew in Problem 1, locate and label the points representing conjugates of the given complex numbers. Draw segments connecting each complex number to its conjugate.

b. Describe the geometric pattern relating each complex number to its conjugate.

4 Consider next the geometric representation of complex number addition and subtraction.

a. Draw vectors on a coordinate grid corresponding to the complex numbers $4 + 3i$ and $-3 + 2i$. Then draw the vectors that represent:

   i. $(4 + 3i) + (-3 + 2i)$ **ii.** $(4 + 3i) - (-3 + 2i)$

b. What connection between complex number and vector operations is suggested by the results of your work in Part a?

c. Test your conjecture about the connection between complex number and vector operations by drawing vectors to represent $3 - 2i$ and $-2 - i$ and then those that represent:

   i. $(3 - 2i) + (-2 - i)$ **ii.** $(3 - 2i) - (-2 - i)$

d. Look back at your drawings for the subtraction problems in Parts a and c. Use them to help determine the distance between the two numbers when graphed on the complex number plane.

e. How is the distance between two complex numbers plotted on the complex number plane related to the vector representing the difference between the two complex numbers?

Algebraic Properties of Complex Numbers When you manipulate algebraic expressions involving real numbers and variables, you often rearrange the order and grouping of numbers and expressions and expand or factor products. Because complex numbers have some properties that are different from those of real numbers, it is important to see whether rules for manipulating complex number expressions are different too.

Problems 5 to 9 ask you to compare the properties of real and complex number operations and ordering. In each problem:

- use some specific numeric examples and geometric reasoning to see if you think the property of real numbers generalizes to the complex number system.

- if possible, use the definitions of complex numbers and operations to prove that the property holds in all cases.

5 The numbers 0 and 1 are additive and multiplicative identity elements in the real number system, because for any real number r we know that $0 + r = r$ and $1 \cdot r = r$. This leads one to ask about the corresponding identity elements in the system of complex numbers.

a. Is there an *additive identity element* in the system of complex numbers? That is, is there a number in the form $a + bi$ with the property that $(a + bi) + (x + yi) = x + yi$ for any complex number $x + yi$?

b. Is there a *multiplicative identity element* in the system of complex numbers? That is, is there a number in the form $a + bi$ with the property that $(a + bi) \cdot (x + yi) = x + yi$ for any complex number $x + yi$?

c. Discuss the following proof for the additive identity element. Then use the reasoning pattern in this example to write a proof for a multiplicative identity element.

> **The number $0 + 0i$ seems likely to be the additive identity element in the complex number system. For example,**
>
> $$(0 + 0i) + (2 + 3i) = (0 + 2) + (0 + 3)i = 2 + 3i.$$
>
> **In general, $(0 + 0i) + (a + bi) = (0 + a) + (0 + b)i = a + bi.$**

Because of the results in this problem, the additive and multiplicative identities in the complex number system are typically written simply as 0 and 1, rather than $0 + 0i$ and $1 + 0i$.

6 Every real number has an additive inverse, because for any real number r we know that $r + (-r) = 0$. Every nonzero real number has a multiplicative inverse, because for any nonzero number r we know that $\frac{1}{r} \cdot r = 1$.

a. Does every complex number have an *additive inverse*? That is, for any complex number $a + bi$, can we always find another complex number $x + yi$ so that $(a + bi) + (x + yi) = 0$?

b. Does every nonzero complex number have a *multiplicative inverse*? That is, for any nonzero complex number $a + bi$, can we always find another complex number $x + yi$ so that $(a + bi) \cdot (x + yi) = 1$?

(*Hint:* How is a number like $\frac{1}{5 + 7i}$ expressed in equivalent $x + yi$ form?)

7 Addition and multiplication of real numbers are *commutative operations*, but subtraction and division are not commutative. This means that for any real numbers, $a + b = b + a$ and $ab = ba$. But, in general, $a - b \neq b - a$ and $a \div b \neq b \div a$.

a. Is addition of complex numbers commutative? That is, for complex numbers $a + bi$ and $c + di$, is it always true that $(a + bi) + (c + di) = (c + di) + (a + bi)$?

b. Is multiplication of complex numbers commutative? That is, for complex numbers $a + bi$ and $c + di$, is it always true that $(a + bi) \cdot (c + di) = (c + di) \cdot (a + bi)$?

c. Is subtraction of complex numbers commutative? That is, for complex numbers $a + bi$ and $c + di$, is it always true that $(a + bi) - (c + di) = (c + di) - (a + bi)$?

d. Is division of complex numbers commutative? That is, for complex numbers $a + bi$ and $c + di$, is it always true that $(a + bi) \div (c + di) = (c + di) \div (a + bi)$?

8 Addition and multiplication of real numbers are *associative operations*, but subtraction and division are not associative. This means that for any real numbers, $(a + b) + c = a + (b + c)$ and $(ab)c = a(bc)$. But, in general, $(a - b) - c \neq a - (b - c)$ and $(a \div b) \div c \neq a \div (b \div c)$.

a. Is addition of complex numbers associative? That is, for complex numbers $a + bi$, $c + di$, and $e + fi$, is it always true that

$$[(a + bi) + (c + di)] + (e + fi) = (a + bi) + [(c + di) + (e + fi)]?$$

b. Is multiplication of complex numbers associative? That is, for complex numbers $a + bi$, $c + di$, and $e + fi$, is it always true that

$$[(a + bi)(c + di)](e + fi) = (a + bi)[(c + di)(e + fi)]?$$

c. Is subtraction of complex numbers associative? That is, for complex numbers $a + bi$, $c + di$, and $e + fi$, is it always true that

$$[(a + bi) - (c + di)] - (e + fi) = (a + bi) - [(c + di) - (e + fi)]?$$

d. Is division of complex numbers associative? That is, for complex numbers $a + bi$, $c + di$, and $e + fi$, is it always true that

$$[(a + bi) \div (c + di)] \div (e + fi) = (a + bi) \div [(c + di) \div (e + fi)]?$$

9 In algebra, using real numbers, many of the key operations like expanding, factoring, and combining "like terms" depend on application of the *distributive property* of multiplication over addition and subtraction. For any real numbers a, b, and c:

$$a(b + c) = ab + ac$$
$$a(b - c) = ab - ac$$

a. Does multiplication distribute over addition in the system of complex numbers? That is, for complex numbers $a + bi$, $c + di$, and $e + fi$, is it always true that

$$(a + bi)[(c + di) + (e + fi)] = (a + bi)(c + di) + (a + bi)(e + fi)?$$

b. Does multiplication distribute over subtraction in the system of complex numbers? That is, for complex numbers $a + bi$, $c + di$, and $e + fi$, is it always true that

$$(a + bi)[(c + di) - (e + fi)] = (a + bi)(c + di) - (a + bi)(e + fi)?$$

SUMMARIZE THE MATHEMATICS

In this investigation, you explored ways of representing complex numbers and operations on coordinate grids. You also checked to see which of the algebraic properties of operations on real numbers hold in the complex number system as well.

a How can the location and addition of complex numbers and their conjugates be represented by points and vectors on a coordinate grid?

b In what ways, if any, do properties of operations in the complex number system differ from those of the real number system?

Be prepared to share your ideas and reasoning with the class.

 **CHECK YOUR UNDERSTANDING**

Consider the complex numbers $z = 4 + 2i$, $w = -1 + 3i$, and $v = -3 - i$.

a. Plot points and vectors representing z, w, and v on a coordinate grid. Then calculate $z + w$ and $z - w$ and plot the points and vectors representing those results.

b. Find the absolute value, conjugate, and additive inverse of z.

c. Find the multiplicative inverse of w.

d. Use two of the given numbers to disprove commutativity of complex number division.

e. Use the given numbers to disprove associativity of complex number subtraction.

f. Calculate $w + v$, $z(w + v)$, zw, zv, and $zw + zv$ to illustrate distributivity of multiplication over addition for complex numbers.

APPLICATIONS

1 Solve the following equations and write the solutions in standard complex number form $a + bi$.

a. $x^2 + 2x + 2 = 0$

b. $5x^2 + 4x + 1 = 0$

c. $5x^2 + 2x + 6 = 4$

d. $x^2 + 3x + 1 = 0$

e. $x^2 + 2x + 4 = -1$

f. $x^2 + 1 = 0$

g. $3x^2 + 5x + 3 = 0$

h. $x^3 - 1 = 0$

2 Calculate $p + q$, $p - q$, pq, and $p \div q$ for each pair of complex numbers.

a. $p = 3 - 2i$ and $q = 1 + 4i$

b. $p = -3 + i$ and $q = 4 + 2i$

c. $p = 1 + 2i$ and $q = 5i - 3$

d. $p = 2i$ and $q = 4 - 3i$

3 Solve the following equations for $z = a + bi$. Consider each equation with the properties of complex numbers in mind to find shortcuts for the solving process.

a. $z + (3 - 2i) = (0 + 0i)$

b. $z(3 - 2i) = (1 + 0i)$

c. $z(3 - 2i) = (3 - 2i)$

d. $z + (3 - 2i) = (3 - 2i)$

4 Solve the following equations for $z = a + bi$.

a. $(3 + 2i) + z = (5 - 4i)$

b. $(3 + 2i)z = (5 - 4i)$

c. $5z + (3 + 2i) = (2 - i)$

d. $(5 + 2i)z + (3 + 2i) = (2 - i) - (1 + i)z$

5 Although 400 years ago mathematicians resisted the idea of square roots of negative numbers, complex numbers are used today in many scientific fields. For example, in electronics the *impedance* (opposition to flow of current) Z is a combination of *resistance R* and *reactance X*. These quantities are measured in ohms, and they are related by the formula $Z = R + Xi$.

a. If a circuit with impedance $65 + 30i$ has two sources of impedance and one of those sources measures $40 + 19i$ ohms, what is the contribution to impedance of the other source?

b. The voltage V, current I, and impedance Z at any location and time in an alternating current (AC) electrical circuit are related by a formula $V = IZ$ that is similar to Ohm's Law.

  i. Find the voltage when current is $2 - 5i$ amperes and impedance is $3 + 4i$ ohms.

  ii. Find the current when the voltage is $12 - 8i$ volts and the impedance is $7 + 5i$ ohms.

  iii. Find the resistance when the current is $6 + 4i$ amperes and the voltage is $8 - 11i$ volts.

C. Borland/PhotoLink/Getty Images

6 Consider the complex numbers represented by points on the coordinate grid at the right. Scales on both axes are 1 unit.

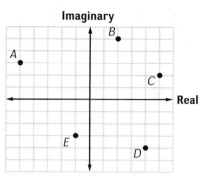

a. Find the absolute value of each complex number.

b. Write the conjugate of each number in $a + bi$ form.

c. Compare the absolute values of complex numbers and their conjugates and explain the pattern you observe.

7 For the following pairs of complex numbers:

- draw the vectors representing each number in the pair.

- draw the vectors representing the sum and difference of the pair of numbers.

a. $3 + i$ and $1 - 3i$ b. $-2 + 2i$ and $-4 - 3i$

8 Write the additive and multiplicative inverses of these complex numbers in $a + bi$ form.

a. $3 + i$ b. $1 - 3i$

c. $-2 + 2i$ d. $-4 - 3i$

9 Use calculations with the complex numbers $4 - 2i$ and $2 + 3i$ to illustrate or disprove the following statements about complex number operations.

a. Addition of complex numbers is commutative.

b. Subtraction of complex numbers is commutative.

c. Multiplication of complex numbers is commutative.

d. Division of complex numbers is commutative.

10 Use calculations with the complex numbers $4 - 2i$, $-5 + i$, and $2 + 3i$ to illustrate or disprove the following statements about complex number operations.

a. Addition of complex numbers is associative.

b. Subtraction of complex numbers is associative.

c. Multiplication of complex numbers is associative.

d. Division of complex numbers is associative.

11 Use calculations with the complex numbers $4 - 2i$, $-5 + i$, and $2 + 3i$ to illustrate or disprove the following statements about complex number operations.

a. Multiplication distributes over addition from the left; that is:

$$(a + bi)[(c + di) + (e + fi)] = (a + bi)(c + di) + (a + bi)(e + fi)$$

b. Multiplication distributes over subtraction from the left; that is:

$$(a + bi)[(c + di) - (e + fi)] = (a + bi)(c + di) - (a + bi)(e + fi)$$

c. Division distributes over addition from the left; that is:

$$(a + bi) \div [(c + di) + (e + fi)] = [(a + bi) \div (c + di)] + [(a + bi) \div (e + fi)]$$

d. Division distributes over addition from the right; that is:

$$[(a + bi) + (c + di)] \div (e + fi) = [(a + bi) \div (e + fi)] + [(c + di) \div (e + fi)]$$

CONNECTIONS

12 When you use the quadratic formula to solve an equation in the form $ax^2 + bx + c = 0$ for x (assuming a, b, and c are integers), at what point in the process and how can you be sure that:

a. the solutions will be rational numbers?

b. the solutions will be irrational real numbers?

c. the solutions will not be real numbers?

13 Explain why every cubic polynomial function has 0 or 2 nonreal complex number zeroes.

14 In solving equations, it is often useful to apply the *addition and multiplication properties of equality*. For any real numbers a, b, and c:

$$\text{If } a = b, \text{ then } a + c = b + c.$$
$$\text{If } a = b, \text{ then } ac = bc.$$

a. Under what conditions will $m + ni = r + si$?

b. Do the two given properties of equality hold in general for complex numbers? Give a counterexample or an algebraic proof of each conjecture below.

  i. If $m + ni = r + si$, then $(m + ni) + (a + bi) = (r + si) + (a + bi)$.

  ii. If $m + ni = r + si$, then $(m + ni)(a + bi) = (r + si)(a + bi)$.

15 Consider equations in the form $az + b = cz + d$ with $a \neq c$.

a. Solve the equation for z in terms of a, b, c, and d.

b. Would the same formula provide the solution if the coefficients are complex numbers? Why or why not?

16 Describe the results if you were to plot points corresponding to all complex numbers z that meet these conditions.

a. $|z| = 3$

b. The imaginary part of z is equal to the square of the real part of z.

17 In Investigation 1, you multiplied complex numbers of the form $(a + bi)(c + di)$.

a. Extend this procedure to multiply the following algebraic expressions.

    i. $(x + 3i)(x - 3i)$ **ii.** $(2x - i)(2x + i)$

    iii. $(3x + 5i)(3x - 5i)$ **iv.** $(ax + bi)(ax - bi)$

b. Use the pattern you found in Part a to factor the following over the complex numbers.

    i. $x^2 + 25$ **ii.** $x^2 + k^2$

    iii. $16x^2 + 1$ **iv.** $32x^2 + 2$

18 Name and give the coordinate rules $(x, y) \rightarrow (\ \ , \ \ )$ of the geometric transformations that map every complex number $a + bi$ onto:

a. the conjugate of $a + bi$.

b. the additive inverse of $a + bi$.

c. k times $a + bi$, where k is a real number.

19 Consider next the geometric effect of adding one fixed complex number $1 + 2i$ to every other complex number.

a. Plot the points corresponding to these complex numbers on a coordinate grid.

    i. $z_1 = 3 + i$ **ii.** $z_2 = 1 - 3i$

    iii. $z_3 = -2 + 2i$ **iv.** $z_4 = -4 - 3i$

b. Then draw the following vectors.

    • from the origin to each point z_n

    • from each point z_n to the sum $z_n + (1 + 2i)$

    • from the origin to each sum $z_n + (1 + 2i)$

c. What geometric transformation maps each complex number point $x + yi$ onto the point $(x + yi) + (1 + 2i)$?

d. Give the coordinate rule $(x, y) \rightarrow (\ \ , \ \ )$ for the transformation in Part c.

REFLECTIONS

20 Powers of the imaginary number i have an interesting property.

a. Using the fact that $i^2 = -1$, calculate i^3, i^4, i^5, i^6, i^7. Summarize your findings.

b. Use your discovery in Part a to find a simpler expression for i^{15}. For i^{74}.

21 Why do you suppose that mathematicians found it so difficult to accept complex numbers as a reasonable extension of the integers and rational numbers?

22 When asked to propose a definition for multiplication of complex numbers, one's first reaction might be that $(a + bi)(c + di) = ac + bdi$. How would you convince someone that this rule might not be the best choice?

23 The conjugate of a complex number $z = a + bi$ can be written as $\bar{z} = \overline{a + bi} = a - bi$. How is the conjugate of the sum of two complex numbers related to the sum of their conjugates?

24 The complex number system was constructed in stages that began with the set W of whole numbers $\{0, 1, 2, 3, 4, \ldots\}$ and gradually introduced other important sets of numbers as practical problems and mathematical problems required them.

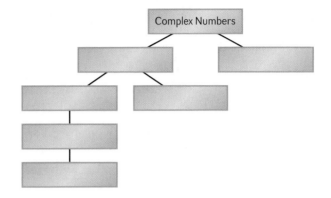

    Complete a copy of this diagram to show the relationships among the whole numbers, integers, rational numbers, irrational numbers, real numbers, imaginary numbers, and complex numbers.

25 The **closure property of addition** and the **closure property of multiplication** are important properties of the real number system. That is, the sum (product) of any two real numbers is another real number.

  a. Does the complex number system have a closure property for addition? Explain your reasoning.

  b. Does the complex number system have a closure property for multiplication? Explain your reasoning.

EXTENSIONS

26 You can use an indirect argument to prove that $\sqrt{2}$ is irrational. Begin by assuming the contrary. Give reasoning that supports the following statements.

Assume that $\sqrt{2}$ is rational.   (1)

Then $\sqrt{2} = \frac{a}{b}$ where a and b are integers, $b \neq 0$
    and $\frac{a}{b}$ is expressed in simplest form.   (2)

So, $2 = \frac{a^2}{b^2}$.   (3)

Since $2b^2 = a^2$, a^2 is even.   (4)

So, a is an even integer and can be expressed as $a = 2k$
    where k is an integer.   (5)

Then $2b^2 = (2k)^2$ or $2b^2 = 4k^2$.   (6)

Therefore, $b^2 = 2k^2$, which implies that b^2 is even and so b is even.   (7)

Since a and b are both even, each has a factor of 2.   (8)

But this is a contradiction.   (9)

So, our assumption is false and $\sqrt{2}$ is irrational.

27 Consider quadratic equations in the form $ax^2 + bx + c = 0$ with solutions z_1 and z_2. Find formulas in simplest possible form (in terms of a, b, and c) for these combinations of the solutions.

a. $z_1 + z_2$ **b.** $z_1 - z_2$ **c.** $z_1 \cdot z_2$

28 The complex number $3 + 2i$ and its conjugate $3 - 2i$ are roots of the same quadratic equation with lead coefficient 1.

a. Use those roots to write that equation in the form $(x - m)(x - n) = 0$.

b. Expand the expression on the left side of the equation in Part a to standard polynomial form.

c. Explain why the roots of any quadratic equation with real number coefficients will always be conjugate complex numbers when $b^2 - 4ac < 0$.

d. Conversely to Part c, use algebraic reasoning to show that any pair of complex conjugates $m + ni$ and $m - ni$ are roots of a single quadratic equation with real number coefficients.

29 In solving a new mathematical problem, it often helps to reduce it to a simpler problem by temporarily ignoring some details. Consider the task of solving the quartic equation $x^4 + 6x^2 - 16 = 0$.

a. Suppose that $u = x^2$. Rewrite the given equation in equivalent form with the letter u.

b. Solve the resulting equation for u.

c. Now use the relationship $u = x^2$ to solve for x.

d. Adapt the *substitution strategy* suggested in Parts a–c to solve these higher-degree equations and check the solutions you come up with.

    i. $4x^4 - 81 = 0$ **ii.** $2x^4 + 3x^2 - 2 = 0$ **iii.** $x^4 + 6x^2 + 9 = 0$

30 Plot the following complex numbers on a coordinate grid: $5 + 0i$, $4 + 3i$, $3 + 4i$, $0 + 5i$, $-3 + 4i$, $-4 + 3i$, $-5 + 0i$, $-4 - 3i$, $-3 - 4i$, $0 - 5i$.

a. Calculate iz for each given complex number z.

b. Compare the locations of each original number and its image after multiplication by i and describe the geometric transformation that seems to be mapping each number onto its image.

c. Provide an algebraic proof, showing that your conjecture is correct.

31 Consider the relationship between $|z + w|$ and the sum $|z| + |w|$.

a. How are those two quantities related in the case that z and w are integers or rational numbers?

b. Use calculations with some specific complex number values for z and w to see how the two quantities might be related in the case that z and w are complex numbers.

c. Use ideas from your vector representation of complex numbers and sums in Investigation 2 to give a geometric argument justifying your conjecture in Part b.

32 The picture at the right highlights a very interesting and important subset of the coordinate plane called the *Mandelbrot set*. It has a simple definition, but a very subtle shape.

Consider the quadratic function $f(z) = z^2 + (0.1 + 0.1i)$, that maps every complex number z (point of the plane) to another complex number (point). If you set your calculator in complex mode, you can explore the effect of the mapping and see how the Mandelbrot set is defined.

a. Enter the number **0**. Then enter **(Ans)²+0.1+0.1i** and press **ENTER** several times. This has the effect of evaluating $f(0), f(f(0)), f(f(f(0))), \ldots$. That is, it iterates the function $f(z)$ starting from $z = 0$. What pattern do you notice in the results of successive iterations?

b. Now repeat the iteration process of Part a, always starting with $z = 0$, but using these rules.

i. $f(z) = z^2 + (0.2 + 0.2i)$

ii. $f(z) = z^2 + (0.3 + 0.3i)$

iii. $f(z) = z^2 + (0.4 + 0.4i)$

iv. $f(z) = z^2 + (-0.2 + 0.2i)$

v. $f(z) = z^2 + (-0.5 + 0.5i)$

vi. $f(z) = z^2 + (0.8 + 0.8i)$

c. The functions you explored in Parts a and b all had rules in the form $f(z) = z^2 + c$, where c was a complex number parameter that determined the path of results from iteration of the function.

i. What do the results of your experimentation in Part b suggest about the effect of the parameter c on the "long-run" path of results from iteration of $f(z)$?

ii. Use your observations to complete a definition for the Mandelbrot set:

The complex numbers c for which repeated iteration of $f(z) = z^2 + c \ldots$.

33 In your work on Problems 5 to 9 of Investigation 2, you probably discovered that operations with complex numbers have most of the same properties as operations with real numbers. The story about ordering of numbers is quite different. The real numbers can be split into three well-defined sets—positive, negative, and zero. Then it is possible to define the "less than" relationship by saying that $a < b$ if and only if $b - a$ is positive.

You also know some useful properties of operations on inequalities:

Property 1: If $a < b$, then $a \pm c < b \pm c$.
Property 2: If $a < b$ and $c > 0$, then $ac < bc$.

a. Can you devise a way of ordering the complex numbers so that for any two distinct numbers $a + bi$ and $c + di$, either $a + bi < c + di$ or $c + di < a + bi$?

b. If you come up with such an ordering of the complex numbers, does it have the same additive and multiplicative properties as order of real numbers?

Jim Laser/CPMP

REVIEW

34 Suppose that you were using a calculator that had no square root key. Explain how you could find estimates for these numbers, check your results, and improve your estimates.

    a. $\sqrt{3}$ **b.** $\sqrt{12}$ **c.** $\sqrt{15}$

35 $\square WXYZ$ is a parallelogram and $WX = 28$ cm, $XY = 15$ cm, and $m\angle WXY = 125°$.

    a. Draw a sketch of this parallelogram.

    b. Find $m\angle XYZ$ and $m\angle YZW$.

    c. Find the area of $\square WXYZ$.

36 Solve each of the following equations by algebraic reasoning. If possible, check your solutions by inspecting tables and/or graphs of related functions.

    a. $3x + 5 = \dfrac{2}{x}$ **b.** $3x + 5 = \dfrac{2}{x - 4}$ **c.** $3x + 5 = \dfrac{2x - 3}{x + 2}$

37 Gloria is at the top of a watchtower and spots an elephant in the wild. The angle of depression of her line of sight to the elephant is 30°. If the tower is 94 feet tall and Gloria is a little taller than 6 feet, about how far is the elephant from the base of the tower?

38 For each of the following functions, sketch a graph and state the domain and range of the function. Then identify the horizontal and/or vertical asymptotes.

    a. $y = \dfrac{-1}{x}$ **b.** $y = \dfrac{1}{2x + 10}$ **c.** $y = \dfrac{1}{x - 3} + 5$

39 A cylindrical can of green beans has a diameter of 7 cm and is 11 cm tall.

    a. Find the volume of the can.

    b. Find the surface area of the can including the top and bottom of the can.

    c. A paper label is used to go around the can and identify its contents. It does not cover the top or bottom of the can. There is a 1-cm overlap where the label is glued to itself. What is the area of the paper used to make the label?

40 Express results of the following fraction computations in the form $\dfrac{a}{b}$ where a and b have no common factors.

    a. $\dfrac{30}{42}$ **b.** $\dfrac{7}{15} + \dfrac{5}{15}$ **c.** $\dfrac{7}{15} + \dfrac{3}{5}$

    d. $\dfrac{7}{5} + \dfrac{1}{4}$ **e.** $\dfrac{2}{15} \cdot \dfrac{5}{6}$ **f.** $\dfrac{7}{12} \div \dfrac{5}{7}$

(l)IT Stock Free/Alamy, (r)FLPA/Alamy

LESSON 3

Rational Function Models and Operations

In the modern world, nearly everyone carries (and sometimes drops) a variety of small and fragile electronic devices like cell phones, calculators, tablets, and music players. Engineers who design those tools work hard to come up with devices that can withstand repeated drops before breaking. On the other hand, there are engineering problems in which the goal is to drop an object so that it (or a target on the ground) *will* break.

One of the most intriguing natural examples of dropping with intent to break is exhibited by sea gulls and crows who feed on mollusks that have shells, like snails and clams. Biologists have observed a species of crows that pick up *whelks,* lift them into the air, and drop them on rocks to break open the shells.

What has especially intrigued biologists who observe the whelk-dropping behavior of northwestern crows is the uncanny way that they seem to rise consistently to a height of about 5 meters before dropping the shells onto the ground.

Work on the problems of this lesson will extend your skill in using rational functions to model data patterns, in performing algebraic operations on rational expressions, and in solving equations and inequalities that involve rational expressions. That combination of skills will help in explaining the amazing whelk-dropping strategy of crows.

INVESTIGATION **1**

Rational Function Models

The explanation of why crows choose 5 meters as the optimal height for dropping whelks requires extending ideas from the *Polynomial and Rational Functions* unit of *Core-Plus Mathematics* Course 3. As you work on the problems of this investigation, look for answers to these questions:

> *What patterns of change are modeled well by rational functions?*
>
> *How can rational functions be expressed in different equivalent forms?*

To investigate the curious behavior of whelk-dropping crows, Canadian biologist Reto Zach performed an experiment in which he dropped whelks from many different heights and recorded the number of drops it took to break the whelk shell in each case. His data gave a pattern like that in the following table and graph.

Drop Height (in meters)	Average Number of Drops
2	20
3	10.5
4	7.5
5	6
6	5
7	4.3
8	3.8
10	3.2
15	2.4

Number of Drops to Break a Whelk Shell

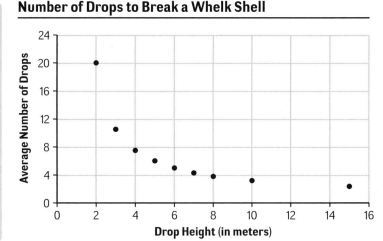

1 Study the pattern of data, the shape of the graph, and the problem situation being modeled. Then answer the following questions about the kind of function that is apt to provide a good model for the relationship between number of drops and drop height.

   a. Why is the relationship unlikely to be quadratic?

   b. Why is the relationship unlikely to be exponential decay?

   c. Why is some type of inverse variation a promising type of model for the relationship?

2 One proposed model for the relationship between number of drops and drop height was the function $N(h) = 1 + \dfrac{200}{10h - 9}$. Test the plausibility of this model by analyzing the rule itself, without using a graph of the rule.

   a. How will the values of $N(h)$ change as the value of h increases?

   b. How will the values of $N(h)$ change as the value of h decreases toward 1?

   c. What is the lower bound for the value of $N(h)$ when $h > \dfrac{9}{10}$?

3 The rule for $N(h)$ can also be written in the form of a single *rational expression*. Provide justifications for each step in the following derivation of that rule.

$$N(h) = 1 + \frac{200}{10h - 9}$$
$$= \frac{10h - 9}{10h - 9} + \frac{200}{10h - 9} \qquad (1)$$
$$= \frac{10h + 191}{10h - 9} \qquad (2)$$

4 Data from the whelk-dropping experiment suggest that crows ought to simply drop the shelled mollusks from great heights in order to minimize the number of drops required. But they do not seem to do that.

   Why might a crow choose multiple drops from a lower height rather than a single drop from a great height?

5 One reason that the crows might not choose to lift the whelks very high is the work required by that task. In physics, work done in moving an object is calculated by combining the force required and the distance the object moves. Data in the following table show the amount of work (in joules) that might be required to lift a typical whelk to various heights.

Height (in meters)	2	4	6	8	10	12	14	16
Work (in joules per drop)	6	12	18	24	30	36	42	48

   a. What function $W(h)$ shows how work required for lifting a whelk depends on height?

   b. Continue using the proposed model for the relationships between number of drops and drop height as in Problem 2. Suppose that a crow chose to consistently drop whelks from a height of 8 meters.

   i. How many drops from that height would be required to break a whelk?

   ii. How much total work would be required to drop a whelk until it breaks?

c. Suppose that the crow consistently dropped whelks from a height of 4 meters.

    i. How many drops from that height would be required to break a whelk?

    ii. How much total work would be required to drop a whelk until it breaks?

d. For crows that choose some consistent height h for their whelk drops:

    i. what function $WB(h)$ shows how to calculate the average *work* required to break a whelk shell as a function of drop height?

    ii. write the rule for $WB(h)$ as a single rational expression.

e. Produce a graph of $WB(h)$ and use it to estimate the drop height which, if used consistently, will break whelk shells with the least total work by the crows.

SUMMARIZE THE MATHEMATICS

In this investigation, you explored two examples of *rational functions*—functions with rules in the form of fractions with polynomials in the numerator and denominator.

a What patterns in the (*drop height, number of drops*) in the whelk-dropping data suggested that a model based on inverse variation would be best?

b What rational expression occurs in the rule for:

    i. $N(h)$?

    ii. $WB(h)$?

c Describe the domains and ranges for the two functions. Then explain what those properties of the functions tell about the problem situation involving crows and whelk dropping.

    i. $N(h)$ giving average number of drops to break as a function of drop height

    ii. $WB(h)$ giving average work to break as a function of drop height

d Identify local maximum and minimum points of the two functions. Then explain what those points tell about the problem situation involving crows and whelk dropping.

e Describe asymptotes of the two functions. Then explain what those limiting lines tell about the problem situation involving crows and whelk dropping.

Be prepared to share your ideas and reasoning with the class.

Many common food items are packaged in cylindrical cans. Designing the size and shape of those cans requires consideration of visual appeal, ease of handling, durability, and manufacturing cost. Suppose that a new product is to be sold in cans that hold 355 milliliters (ml), an amount that is equal to 355 cubic centimeters or about 12 fluid ounces.

a. What formula shows the relationship among the radius r in centimeters, height h in centimeters, and volume 355 ml of the can?

b. The cost of manufacturing the can depends on its surface area. What formula shows how the surface area SA of a cylinder depends on the radius r and height h? (*Hint:* Imagine the can with the circular top and bottom cut out and the side unfolded to form a rectangle.)

c. Use the information from Part a to express the height of the can in terms of the radius.

d. Use the information from Parts b and c to express the surface area of the can as a function of the radius alone.

e. Use the result of Part d to estimate the radius that will produce minimum surface area and the height corresponding to that radius.

INVESTIGATION 2

Properties of Rational Functions

The problems of Investigation 1 showed how rational expressions and functions can arise in analysis of certain relationships among variables. You were reminded of some important things to think about in analyzing rational functions—domain and range, local max/min points of graphs, and asymptotes—and you saw how operations of addition and multiplication can be used to produce new and informative rational expressions and functions.

To use rational functions effectively in solving problems, it helps to be skillful in adding, subtracting, multiplying, dividing, and simplifying rational expressions and in recognizing important graph features. As you work on this investigation, look for answers to these questions:

How can rational expressions be analyzed to reveal important features of the corresponding function graphs?

What rules govern operations with rational expressions?

Asymptotes of Rational Function Graphs Consider the rational function $WB(h) = \dfrac{30h^2 + 573h}{10h - 9}$ that gives work (in joules) required to break a whelk shell if it is consistently dropped from a height of h meters. The following graph of $WB(h)$ shows how work changes as drop height increases. For values of $h > 5$, the graph appears to be almost a straight line.

Work to Break Shell

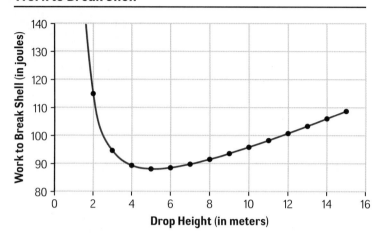

The graph of $WB(h)$ has the line $y = 3h + 60$ as what is called an *oblique* or *slant asymptote*. As h increases, the graph of $WB(h)$ approaches but never touches the graph of $y = 3h + 60$. It also has the line $h = 0.9$ as a vertical asymptote.

You can use algebraic reasoning to determine the exact locations of those two asymptotes. As you work on Problems 1–3, look for strategies that can solve that problem for any rational function.

1. Graph the following simpler rational functions and identify what appear to be the horizontal, vertical, and oblique asymptotes of each. Be sure to look at the graphs for both positive and negative values of x. In each case, the window $-10 < x < 10$ and $-30 < y < 30$ should give you a good start toward finding a suitable view of the graph.

   a. $g(x) = \dfrac{2x - 1}{x - 3}$

   b. $h(x) = \dfrac{2x^2 - x + 4}{x - 1}$

   c. $f(x) = \dfrac{-x^2 + 4x + 8}{x + 1}$

   d. $k(x) = \dfrac{x^3 - 9x}{x^2 - 6x + 9}$

2. Now consider how locations of asymptotes for a rational function graph can be found by algebraic reasoning. For example, consider again the function $f(x) = \dfrac{2x^2 - x + 4}{x - 1}$.

   a. How can you tell by studying the expression of that function rule that the graph of $f(x)$ will have the vertical line $x = 1$ as an asymptote?

   b. You know from arithmetic that a fraction like $\dfrac{214}{13}$ also indicates the division $214 \div 13$ and that the result can be expressed as 16 R 6 or as $16 + \dfrac{6}{13}$. Use division of polynomials to show that $\dfrac{2x^2 - x + 4}{x - 1}$ can be written as an equivalent expression in the form $A(x) + \dfrac{B}{x - 1}$.

c. Explain how and why the algebraic form $A(x) + \dfrac{B}{x-1}$ indicates the vertical and oblique asymptotes for the graph of $f(x) = \dfrac{2x^2 - x + 4}{x - 1}$.

3 Adapt the strategy suggested in Problem 2 to find asymptotes for the graphs of these functions by algebraic reasoning. Produce graphs of the given functions to check your reasoning.

a. $g(x) = \dfrac{2x - 1}{x - 3}$

b. $h(x) = \dfrac{-2x^2 + 9x - 7}{x - 2}$

c. $f(x) = \dfrac{-x^2 + 4x + 8}{x + 1}$

d. $m(x) = \dfrac{x^3 + 7x}{x^2 - x - 2}$

e. $k(x) = \dfrac{x^3 - 9x}{x^2 - 6x + 9}$

f. $k(x) = \dfrac{x^4 + 3x}{x^2 - 4}$

4 Consider again the rational function $WB(h) = \dfrac{30h^2 + 573h}{10h - 9}$ that gives the work (in joules) required to break a whelk shell if it is consistently dropped from a height of h meters.

a. Use division of polynomials to find an expression for the rule of $WB(h)$ in the form $C(h) + \dfrac{D}{10h - 9}$.

b. Use either of the two expressions for the rule of $WB(h)$ to explain why $h = 0.9$ is a vertical asymptote for the graph.

c. What does the asymptote $h = 0.9$ tell about the way the *work* required to break the shell changes as the drop height is reduced toward only 0.9 meters.

d. Use the expression for the rule of $WB(h)$ in the form $C(h) + \dfrac{D}{10h - 9}$ to explain why the graph has an oblique or slant asymptote $y = 3h + 60$, meaning that the graph approaches that line as h increases. Explain what that asymptote tells about the way that *work* required to break the shell changes as drop height increases.

Equivalent Expressions In your work with the algebraic model for work done by crows trying to break whelk shells, you found that different, but equivalent, expressions for the work function were helpful in answering different questions. You might remember a similar experience in the *Polynomial and Rational Functions* unit of Course 3, when you simplified, added, subtracted, and multiplied rational expressions involved in analysis of profit for a concert business.

For example, the profit per customer in operating the concert came from two sources—ticket sales and concession sales—and both depended on the price charged for admission to the concert. To find total profit per customer, you had to add these rational expressions:

$$\dfrac{-25x^2 + 875x - 4{,}750}{-25x + 750} \quad \text{and} \quad \dfrac{-175x + 5{,}250}{-25x + 750}$$

Work on Problems 5–7 will review and extend your skill in combining rational expressions and simplifying them to equivalent useful forms.

5 Consider the arithmetic fractions $\frac{4}{6}$ and $\frac{6}{15}$.

    a. How can those fractions be "simplified" so that numerator and denominator involve smaller numbers? What reasoning justifies each simplification?

    b. What are the sum, difference, product, and quotient of the two fractions? In each case, show steps in your work so that someone else could understand your reasoning.

6 Generalize your knowledge of operations with numerical fractions to simplify each of the following rational expressions as much as possible. In each case, show steps in your work so that someone else could understand your reasoning.

    a. $\dfrac{x^2 + 2x}{3x + 6}$ **b.** $\dfrac{x^2 - 8x + 12}{x - 2}$

    c. $\dfrac{2x^2 - 7x - 15}{x^2 - 3x - 10}$ **d.** $\dfrac{2x^3 - 8x}{x^2 - 2x}$

7 In earlier work with rational expressions and functions, you discovered that zeroes of the denominator often (but not always) indicated vertical asymptotes for graphs of the corresponding functions. Analyze the following rational functions by:

    • graphing the functions in the window $-10 \le x \le 10$ and $-30 \le y \le 30$.

    • identifying the cases where restrictions on the domain are related to vertical asymptotes and those that are not.

    • explaining why some domain restrictions determine vertical asymptotes and others do not.

    a. $f(x) = \dfrac{3x^2 + 15x}{x + 5}$ **b.** $g(x) = \dfrac{x^2 + 3x}{x + 2}$

    c. $h(x) = \dfrac{9x + 6}{x^2 - 4}$ **d.** $j(x) = \dfrac{9x + 18}{x^2 - 4}$

8 Generalize your knowledge of operations with numerical fractions to find $p(x) + q(x)$, $p(x) - q(x)$, $p(x)q(x)$, and $p(x) \div q(x)$ for each pair of rational functions given below. Simplify the resulting expressions as far as possible. In each case, show steps in your work so that someone else could understand your reasoning.

    a. $p(x) = \dfrac{3}{x}$ and $q(x) = \dfrac{-2}{7x}$

    b. $p(x) = \dfrac{4}{x - 1}$ and $q(x) = \dfrac{5}{x + 1}$

    c. $p(x) = \dfrac{3}{x + 1}$ and $q(x) = \dfrac{5x}{x - 2}$

    d. $p(x) = \dfrac{x - 2}{x + 2}$ and $q(x) = \dfrac{5x + 10}{x^2 - 4}$

Solving Rational Equations and Inequalities You know that many of the most important questions in algebra require solving equations and inequalities. For example, suppose that the rational expressions

$$\frac{-25x^2 + 875x - 4{,}750}{-25x + 800} \text{ and } \frac{-175x + 5{,}250}{-25x + 800}$$

show how profit per customer from ticket sales and concession sales, respectively, are related to the average price x for admission to a concert. To find the break-even points for those parts of the concert business, you need to solve these equations:

$$\frac{-25x^2 + 875x - 4{,}750}{-25x + 800} = 0 \text{ and } \frac{-175x + 5{,}250}{-25x + 800} = 0.$$

All of this work is made much easier if you realize that the expressions for profit from ticket and concession sales can be simplified to

$$\frac{x^2 - 35x + 190}{x - 32} \text{ and } \frac{7x - 210}{x - 32}.$$

9 Consider first the simpler of the two equations $\frac{7x - 210}{x - 32} = 0$ which represents no profit per customer from concession sales.

a. Why can this rational equation be solved by simply solving $7x - 210 = 0$?

b. What is the solution?

c. Use algebraic operations on expressions and equations to transform the equation $\frac{7x - 210}{x - 32} = 5$ to an equation in the form $\frac{A(x)}{B(x)} = 0$. Then solve that equation and check that the solution satisfies the original equation.

d. Show how to combine your algebraic reasoning in Parts a–c with graphical analysis of the function $p(x) = \frac{7x - 210}{x - 32}$ to solve the inequality $\frac{7x - 210}{x - 32} > 5$. Record your solution on a number line. Explain what the result tells about profit from concession sales.

10 Consider next the equation $\frac{x^2 - 35x + 190}{x - 32} = 0$.

a. Why can this rational equation be solved by simply solving $x^2 - 35x + 190 = 0$?

b. What is the solution?

c. Combine your reasoning in Parts a and b with graphical analysis of the function $c(x) = \frac{x^2 - 35x + 190}{x - 32}$ to solve the inequality $\frac{x^2 - 35x + 190}{x - 32} > 0$ which asks about profit per person from concert ticket sales. Explain what the solution tells about profit from concert ticket sales.

11 Use algebraic and graphic reasoning to solve the following rational equations and inequalities. Show the steps in your work so that someone else can understand your reasoning. Check your work by using a CAS.

a. $\dfrac{x^2 + 5x - 6}{x^2 + 5} = 0$

b. $\dfrac{x^2 + 5x - 6}{x^2 + 5} < 0$

c. $\dfrac{5}{x} = 2 - \dfrac{1}{x}$

d. $\dfrac{5}{x} < 2 - \dfrac{1}{x}$

SUMMARIZE THE MATHEMATICS

In this investigation, you extended your understanding and skill in working with rational expressions, functions, and equations.

a What strategies can be used to locate vertical, horizontal, and oblique asymptotes for graphs of rational functions?

b What strategies can be used to simplify rational expressions? What cautions do you need to apply when using the simplified expressions to solve problems from which they arise?

c What strategies can be used to find the sum, difference, product, and quotient of two rational expressions?

d What strategies can be used to solve equations and inequalities involving rational expressions?

Be prepared to compare your strategies with your classmates.

✓ CHECK YOUR UNDERSTANDING

Consider the rational function $f(x) = \dfrac{-x^2 + 3x + 5}{x - 3}$.

a. Write the rule for $f(x)$ with an equivalent expression in the form $A(x) + \dfrac{B}{x - 3}$.

b. Graph $y = f(x)$ and label all asymptotes with their equations.

c. Solve algebraically the equation $f(x) = 0$.

d. Solve the inequality $f(x) < 0$.

e. Write the algebraic expression  $\dfrac{x + 2}{x - 3} + \dfrac{2x + 1}{x + 1}$ in equivalent form as a single fraction. Use the result to solve the equation $\dfrac{x + 2}{x - 3} + \dfrac{2x + 1}{x + 1} = 0$.

APPLICATIONS

1 The *Demon Drop* is a ride at Dorney Park & Wildwater Kingdom in Ohio that simulates freefall from the top of a 10-story building. Riders sit in four-passenger cars that are raised to the top of a 131-foot tower. The cars are then dropped without warning 99 feet almost straight down, before entering a pullout curve where a braking system stops the car.

Suppose that the track of the *Demon Drop* is the graph of a function $h(x)$, where x is the horizontal distance from the tower in feet and $h(x)$ is the height of the track at a point x feet from the tower. A sample of $(x, h(x))$ values are given in the table below.

x	1	10	20	30	40	50
h(x)	130	40	35	33	32.5	32

a. What rule in the form $h(x) = A + \dfrac{B}{x}$ models the pattern of values in the table reasonably well?

b. What does the value of A tell about the track?

2 Suppose that to help with braking of the speeding *Demon Drop* cars, the function defining the track is modified so that it has the rule $j(x) = \dfrac{0.5x^2 + 20x + 100}{x}$.

a. Graph this new modeling function on $(0, 60]$ and label the local minimum point with its coordinates.

b. Write an equivalent rule for $j(x)$ in the form $j(x) = A(x) + \dfrac{B}{x}$. Then use that form to explain the fact that the graph of $j(x)$ is nearly linear for $x > 20$.

3 The surface area of a cylindrical can with radius r and height h is given by the formula $A = 2\pi r^2 + 2\pi rh$. The volume is given by the formula $V = \pi r^2 h$. Suppose that a soup can is to have surface area of 750 cm².

  a. Use the surface area formula and the constraint that area must equal 750 cm² to express the height of the can in terms of the radius.

  b. Use the information from Part a to express the volume of the can as a function of the radius alone.

  c. Estimate the radius that will produce maximum volume for the fixed surface area and the height corresponding to that radius. Use entries from an appropriate table or graph to show how you arrived at your answer.

4 Find equations for all asymptotes of graphs for the following functions.

  a. $f(x) = \dfrac{4x}{x^2 - 2x + 1}$

  b. $g(x) = \dfrac{5x^2 + 3x + 7}{x^2 - 9}$

  c. $h(x) = \dfrac{-3x^2 + x - 12}{x + 2}$

  d. $j(x) = \dfrac{x^2 + 6.5x + 5}{2x + 1}$

5 Simplify the rules for the following rational functions. Then compare the domains of the original and simplified expressions.

  a. $f(x) = \dfrac{x^2 - 4x - 12}{x^2 + 2x}$

  b. $g(x) = \dfrac{2x^2 - x - 15}{x^2 - 9}$

  c. $h(x) = \dfrac{x^3 - 7x^2 - 8x}{x^2 + x}$

  d. $j(x) = \dfrac{x - 1}{x^2 + 4x - 5}$

6 Write each of the following expressions in equivalent form as single rational expression.

  a. $\dfrac{6x}{x + 2} + \dfrac{x - 3}{x - 5}$

  b. $\dfrac{4}{x} - \dfrac{3x + 7}{2x - 1}$

  c. $\dfrac{x - 4}{2x + 3} \cdot \dfrac{3x - 1}{x^2}$

  d. $\dfrac{5x}{-2x^2 + 1} \div \dfrac{7x}{3}$

7 Write each of the following rational expressions in equivalent form $A(x) + \dfrac{B(x)}{C(x)}$, where $C(x)$ is the original denominator and the degree of $B(x)$ is less than the degree of $C(x)$.

  a. $\dfrac{-2x^2 + 3x - 1}{x + 2}$

  b. $\dfrac{16x^2 - 10x + 5}{x - 3}$

  c. $\dfrac{x^3 - 3x^2 + 2}{x^2 + x + 2}$

  d. $\dfrac{3x^3 - 10x^2 - 28x + 11}{x - 5}$

8 Solve the following equations for x. In each case, show work in a way that allows another person to understand your reasoning.

a. $\dfrac{3x^2 - 5x - 2}{x^2 + 1} = 0$

b. $\dfrac{x^2 - 8x + 15}{x - 4} = 0$

c. $\dfrac{x}{x - 1} = \dfrac{2x + 1}{x - 4}$

d. $\dfrac{x^2 - x - 12}{x + 4} = 0$

e. $\dfrac{-x^2 - x - 2}{x^2 - 1} = 0$

9 Solve the following inequalities for x and record your solutions on number line graphs. Using interval notation. In each case, show work that allows another person to understand the reasoning that led to your solution.

a. $\dfrac{3x^2 - 5x - 2}{x^2 + 1} > 0$

b. $\dfrac{x^2 - 8x + 15}{x - 4} < 0$

c. $\dfrac{x^2 - x - 12}{x + 4} > 0$

d. $\dfrac{-x^2 - x - 2}{x^2 - 1} > 0$

10 When a person starts work at a new job or when a company begins making a new product, there is a learning process during which the rate (and thus cost) of producing each new item decreases as the worker(s) get "smarter" and "quicker."

Suppose that a factory making emergency housing trailers for FEMA has the following experience in building a new model.

Trailer Number	1	2	3	4	5	10	15	20	25
Time to Build (in hours)	18	12	10	9	8.4	7.2	6.8	6.6	6.5

a. Use what you have learned to find a rational function model for the relationship between construction time and trailer number. Test the match of your model to the data pattern.

b. Find another model in the exponential decay family that also fits the data pattern.

c. Compare the two proposed models and explain any ideas you have about which function family probably provides the more accurate model.

d. Why is there no linear model that will provide a good fit to the data pattern? What is it about the problem situation that supports that conclusion?

CONNECTIONS

11 When you ride the *Demon Drop* at Dorney Park & Wildwater Kingdom, the first part of your ride is a free-fall straight down. You might recall from earlier work with quadratic functions that the distance fallen (in feet) is a function of time (in seconds) with rule $d(t) = -16t^2$.

a. What function gives the height of the falling car as a function of time, assuming that the trip starts from a height of 130 feet?

b. The Dorney Park Web site claims that the first 60 feet of the trip are pure free-fall. How long does that part of the trip take?

c. How far does the car fall in the last 0.1 seconds of the free-fall?

d. What speed of descent (in feet per second) is implied by your answer to Part c?

e. If 1 mile per hour corresponds to about 1.47 feet per second, what speed (in miles per hour) is implied by your answer to Part d?

12 To find models for the *Demon Drop* ride, you might consider a variety of familiar functions. Consider again the data relating height of the track $h(x)$ to distance x from the drop tower.

x	0	10	20	30	40	50
h(x)	130	40	35	33	32.5	32

a. Use regression to find the best-fitting quadratic model for the data. Then compare the graph of that quadratic function to the data and to the graph of the model you developed in Applications Task 1.

b. Use regression to find the best-fitting exponential model for the data. Then compare the graph of that exponential function to the data and to the graph of the model you developed in Applications Task 1.

13 The spinner shown at the right was used in a probability experiment to see how long it would take until the first even number would appear.

After many repetitions of the experiment, the proportions of various possible outcomes were as in the following table.

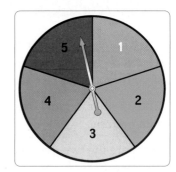

Trials Until First Even	1	2	3	4	5	6	7	8
Proportion of Outcomes	0.38	0.25	0.15	0.09	0.05	0.03	0.02	0.01

a. Bianca studied the data and reasoned that it would take "an infinite number" of trials of the experiment to get an even number on 0 spins, so she thought that the function with rule in the form $P(n) = \frac{0.38}{n}$ would probably fit the given data pattern. Is that function a good model for the relationship between proportion of outcomes and number of trials until the first even number appears? Explain.

b. Katia remembered a way of analyzing such probability experiments. She explained:

> *The probability of getting the first even number on the first spin should be $\frac{2}{5}$; the probability of getting the first even number on the second spin should be $\left(\frac{3}{5}\right)\left(\frac{2}{5}\right)$; the probability of getting the first even number on the third spin should be $\left(\frac{3}{5}\right)^2\left(\frac{2}{5}\right)$; and so on.*

  i. Is Katia's analysis of the experiment correct?

  ii. Does Katia's idea produce *proportions of outcomes* data that are closer to the experimental results than Bianca's model?

  iii. What function would match Katia's idea of predicting the proportion of trials for which the first even-number outcome would occur on spin number n?

c. What do the results of your work in Parts a and b suggest about the relative merits of data analysis and logical analysis in searching for good models of phenomena?

14 It is common to see data patterns that have close to an L-shape as in the plot below. In such situations, you might consider modeling the data with a rational function $g(x) = \frac{ax + b}{x - c}$.

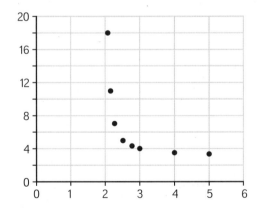

a. How is the value of a related to the graph of the function?

b. How is the value of b related to the graph of the function?

c. How is the value of c related to the graph of the function?

15 Explain why the equation $\frac{x + 2}{x - 3} + \frac{2x + 1}{x + 1} = 0$ is equivalent to $\frac{x + 2}{x - 3} = \frac{-2x - 1}{x + 1}$. Then use the fact that $\frac{a}{b} = \frac{c}{d}$ is equivalent to $ad = bc$ to solve that equation in a different way than used in the Check Your Understanding on page 236.

16 Rational functions are used extensively in optics and thus in cameras. In the diagram below, F is the focal length of the lens, u is the distance between the object and the lens, and v is the distance between the lens and the image surface inside the camera. In this case, $\frac{1}{F} = \frac{1}{u} + \frac{1}{v}$.

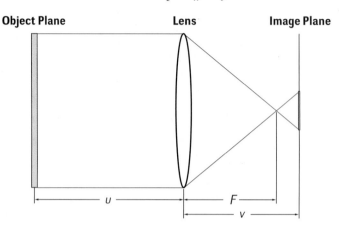

Object Plane   **Lens**   **Image Plane**

a. Rewrite the equation $\frac{1}{F} = \frac{1}{u} + \frac{1}{v}$ to express v as a function of u and F.

b. Suppose that the lens of a particular camera has focal length of 20 millimeters. What should be the distance between the lens and the image surface if the camera is to be focused on:

   i. an object that is 6 meters (6,000 millimeters) away from the lens?

   ii. an object that is 0.5 meters (500 millimeters) away from the lens?

c. Why can you be sure that the two triangles between the lens and the image plane are similar?

d. What rational expression shows the ratio of the width of the lens w_L to the width of the image w_I in terms of F and v?

REFLECTIONS

17 Rational function graphs often look like variations on exponential decay curves. In what ways are the two families of graphs similar and in what ways are they different?

18 What function families that you have encountered in other mathematical study have asymptotes for their graphs and which do not?

19 Why must you be careful in viewing and interpreting a calculator- or computer-produced graph of a rational function?

20 In what ways is work with rational expressions similar to work with arithmetic fractions, and in what ways is that work different?

21 To solve the equation $\dfrac{x^2 + 5x + 6}{x + 3} = 0$, you might reason that any fraction is 0 when the numerator is 0. The solutions of $x^2 + 5x + 6 = 0$ are -3 and -2. Substitute each of those values for x in the given equation and explain the results.

EXTENSIONS

22 In previous work, you saw that the weight in pounds of an object above Earth is a function of its weight on the surface of the Earth w_0 and its distance h above the surface in miles with rule:

$$w(h) = \left(\frac{3{,}950}{3{,}950 + h}\right)^2 w_0$$

a. How does this model imply that for crows lifting whelks above the surface of an ocean shoreline, the force (weight) can be considered constant throughout the lift, and thus the work is simply the product of the weight and the lift height?

b. At what height above Earth's surface would a 150-pound "object" weigh only 50 pounds?

c. How high would a whelk-carrying crow have to fly for the weight of that whelk to change by as little as 0.1%?

23 For each of the following functions:

- find all zeroes.

- find equations of all asymptotes.

- describe the end behavior (as $|x| \rightarrow \infty$).

- estimate coordinates of all local max/min points.

- sketch the graph.

Show work that would allow someone else to follow your solution procedures and justification of answers.

a. $f(x) = \dfrac{x^3 - x^2 + 1}{x^2 - 2x + 1}$

b. $g(x) = \dfrac{2x^3 - 5x^2 + 7x - 2}{x^2 + 2}$

24 Estimate intercepts and asymptotes of the following graphs and use the information to find possible rules for functions that have those graphs.

a.

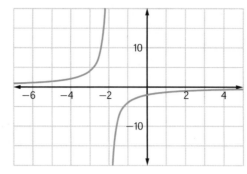

b.

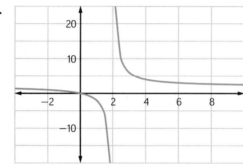

c.

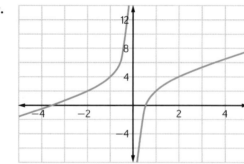

d.

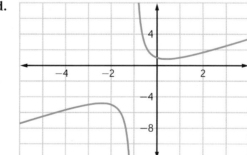

e.

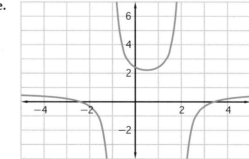

25 The following graph shows a function $h(t) = 32 + \dfrac{80 \cos (2t - 0.7)}{t + 1}$ that gives the height in feet of a bungee jumper as a function of time in seconds. Although this is not a rational function, it has traits similar to those of a rational function.

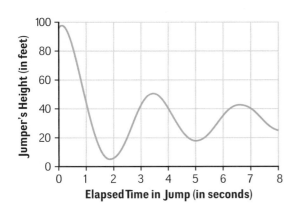

a. What are the horizontal and vertical asymptotes of this function, and how can each be determined by reasoning about the form of the function rule?

b. What does the horizontal asymptote tell about the bungee apparatus?

c. Another common model for the *damped oscillation* that occurs in processes like a bungee jump is given by $j(t) = 32 + 75(0.6)^t \cos (2t - 0.7)$. Compare the graph of this function to that given above.

d. Find all horizontal and vertical asymptotes for the graph of $j(t)$. Explain how those properties of the function can be determined by analyzing the rule.

26 Consider all rational functions with rules in the form $f(x) = \dfrac{ax + b}{cx + d}$ $(cd \neq 0)$.

a. What is the domain for any particular $f(x)$?

b. What is the rule (in terms of a, b, c, and d) for the inverse function f^{-1}? What is the domain of f^{-1}?

c. Suppose that $g(x) = \dfrac{mx + n}{px + q}$. Use algebraic reasoning to show that the rule for $f(g(x))$ has the same general form as the rules for $f(x)$ and $g(x)$.

27 The simplest form of a function that combines direct and inverse variation by addition has the rule $f(x) = x + \frac{1}{x}$. Use this function to complete the following tasks.

a. Solve each of these equations first for real number solutions and then for complex number solutions.

  i. $x + \frac{1}{x} = 0$ ii. $x + \frac{1}{x} = 1$ iii. $x + \frac{1}{x} = 2$

  iv. $x + \frac{1}{x} = 3$ v. $x + \frac{1}{x} = 4$

b. For which values of a does the equation $x + \frac{1}{x} = a$ have no real number solutions? Explain.

REVIEW

28 What are the equations for these lines?

a. A horizontal line with y-intercept $(0, k)$

b. A vertical line with x-intercept $(m, 0)$

29 Factor each of the following expressions into linear factors.

a. $x^2 + 5x$ b. $x^2 + 10x + 16$

c. $3x^2 - 7x - 20$ d. $4x^3 - 36x$

30 Each year at the annual Holton County Fair, the Holton Community College has a booth to help raise money for scholarships. The spinner with 5 equal sectors shown below is used. To play, you pay $5 and then spin the wheel twice. If you get the same prize on both spins, you win that prize. Latisha decides to play.

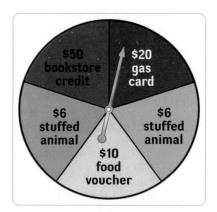

a. What is the probability that Latisha will win the $50 bookstore credit?

b. What is the probability that she will win a stuffed animal?

c. What is the probability that she will not win a prize?

d. On average, how much money can Holton Community College expect to make each time someone plays the game?

31 Determine if each mathematical statement is true or false. If false, change the right-hand side of the statement to make it true.

 a. $(ax + b)(bx + a) = abx^2 + (a^2 + b^2)x + ab$

 b. $(yz - 5)^2 = y^2z^2 - 25$

 c. $b^2y^2 + 5b^2y + 4b^2 = b^2(y + 4)(y + 1)$

 d. $\sqrt{6}(x + 2y) + \sqrt{6}(2x - y) = \sqrt{6}(3x + y)$

 e. If $i^2 = -1$, $(x + 3i)^2 = x^2 + 6ix + 9$.

32 Consider a coordinate plane with a circle centered at the origin O and radius 5 inches.

 a. Draw a sketch of this circle.

 b. Is the point with coordinates $(3, -4)$ on the circle? Justify your reasoning.

 c. Use the symmetry of the circle to identify the coordinates of three other points that are on the circle.

 d. Find the coordinates of a second point in the fourth quadrant that is on the circle. Explain how you know your point is on the circle.

 e. Suppose that point $P(x, y)$ is on the circle and that point A has coordinates $(5, 0)$. If $\angle AOP$ is an angle in standard position and $m\angle AOP = \theta$, express the coordinates of P in terms of trigonometric functions of θ.

33 Use reasoning about square roots to solve each equation without using technology.

 a. $\sqrt{x} = 9$

 b. $2\sqrt{x} = 6$

 c. $\sqrt{x + 20} = 10$

 d. $\sqrt{\dfrac{4}{x}} = 8$

34 Solve each equation for y.

 a. $y^2 - x = 9$

 b. $4y^2 = x^2 + 9$

 c. $\dfrac{y^2}{16} + x^2 = 25$

Algebraic Strategies

At this point in your study of algebra and functions, there are surely many kinds of problem situations that you can model and analyze almost without thinking. That kind of highly developed skill is very useful. But it is also important to become skillful in adapting and combining well-known methods to deal with new kinds of problems and unexpected algebraic forms.

For example, you have studied and used quadratic functions and their parabolic graphs in many different situations. So, you recognize parabolas when you see them as function graphs, and you know a variety of algebraic methods for finding zeroes, maximum or minimum points, and intercepts of the graphs. It turns out that parabolas have important properties as *geometric shapes*. There are problems where analysis most naturally begins with the geometric definition:

A **parabola** is the set of points that are equidistant from a fixed point (called the **focus**) and a fixed line (called the **directrix**).

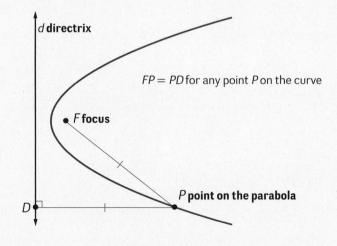

$FP = PD$ for any point P on the curve

F **focus**

P **point on the parabola**

d **directrix**

D

Work on the problems of this lesson will extend your skill in using algebraic concepts and skills to model and analyze problem situations. The aim is to develop the kind of big-picture understanding and strategic thinking that is useful, regardless of the specific function or expression encountered.

INVESTIGATION

Dealing with Radicals

Algebraic representation of the conditions that define any parabola depends on use of the distance formula and thus an expression involving a radical. As you work on the problems of this investigation, look for answers to this question:

How can functions and equations involving radicals be expressed in different equivalent ways?

1 One way of thinking about the geometric shape of a parabola is shown in the following diagram, where the directrix has been placed on the *y*-axis of a coordinate grid.

  a. What point on the diagram seems to represent the focus?

  b. What expression represents the distance from any point on the parabola $P(x, y)$ to the focus?

  c. What expression represents the distance from any point on the parabola $P(x, y)$ to the directrix?

  d. What equation represents the condition that the distance from any point on the parabola $P(x, y)$ to the focus is equal to the distance from that point to the directrix?

  e. The equation derived in Part d involves a radical expression. Rewrite that equation in an equivalent form that does not involve a radical.

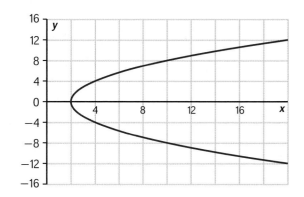

2 Re-examine the drawing of the parabola given in Problem 1.

  a. Explain why the parabola is not the graph of a function of *x*.

  b. How could you reproduce the drawing with a function-graphing tool?

3 Recall that a circle with center at the origin and radius r is the set of all points with coordinates satisfying the equation $x^2 + y^2 = r^2$.

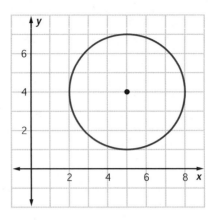

a. How could you write the circle equation in a related form and use that form to produce the desired drawing with a function-graphing tool?

b. The drawing at the right shows a circle of radius 3 and center (5, 4).

   i. What equation describes the coordinates of all points (x, y) on the circle?

   ii. How could you write the circle equation from part i in related forms that can be used to produce a drawing of the circle with a function-graphing tool?

4 American news media and political parties conduct regular surveys of public opinion on the performance of elected officials and important policy questions.

For example, polls regularly ask samples of citizens whether they approve of the way the President is doing his/her job.

If a poll of n people reports that a proportion p of the sample approve of the President's recent performance, then the estimated standard deviation of the population proportion σ_p is given by the formula $\sigma_p = \sqrt{\dfrac{p(1 - p)}{n}}$.

a. What are the estimated standard deviations of poll results in these cases?

   i. $n = 500$ and $p = 0.6$

   ii. $n = 1,000$ and $p = 0.6$

   iii. $n = 500$ and $p = 0.2$

b. If a poll taker expects that the proportion approving of a particular government policy will be about 0.3, what sample size n would be required to have an estimated standard deviation of 0.01 for the survey proportion?

c. If the sample size in Part b were doubled, what effect would the extra effort and expense of the larger survey have on the estimated standard deviation?

d. What general formula shows the required sample size n for a survey as a function of the proportion p and standard deviation σ_p?

Your work on Problems 1–4 led to algebraic expressions and equations involving radicals. You found methods of solving those problems by transforming the expressions and equations into convenient equivalent forms.

5 Use algebraic reasoning to solve these equations. Record the steps in your work so that someone else could follow your reasoning. Then check the solutions you derive by substituting the values back in the original equations and by using a CAS for the same solving tasks.

a. $\sqrt{x^2 - 8x} = 3$

b. $\sqrt{\dfrac{6x^2 + 95}{5}} = x + 2$

c. $\sqrt{x + 1} = 3 + \sqrt{x - 2}$

d. $\sqrt{\dfrac{x + 2}{x - 3}} = \sqrt{\dfrac{x + 4}{x - 5}}$

SUMMARIZE THE MATHEMATICS

In this investigation, you formulated and solved equations that involved radical expressions.

a What strategies are helpful in solving equations that involve radicals?

b What cautions must be observed when preparing to use solutions of radical equations that are derived by the strategies described in Part a?

Be prepared to explain your ideas to the class.

 CHECK YOUR UNDERSTANDING

Developing skill in working with radicals is important for future study of science, technology, engineering, and mathematics.

a. Sketch a graph of a parabola with focus (2, 3) and directrix $y = 1$. Write a function rule for the parabola.

b. Show that any point $(x, x^2 + 3)$ is equidistant from the point $\left(0, 3\frac{1}{4}\right)$ and the line $y = 2\frac{3}{4}$. What can you conclude?

c. Solve $\dfrac{\sqrt{x - 3}}{\sqrt{x + 2}} = \dfrac{2}{\sqrt{x}}$ for x. Show steps used in deriving the solution(s) and how the solution(s) can be checked.

Seeing the Big Picture

In work on the problems of Investigation 1, you developed ways of adapting familiar algebraic methods to solve equations involving radical expressions. The first step was almost always some sort of squaring to remove the radicals, so that solution strategies you know for solving polynomial and rational equations could be applied.

When you face an algebraic problem, even one that looks fairly familiar, it is often helpful to analyze the structure of the problem before plunging ahead into symbolic calculations. Many problems can be solved with insight rather than calculation. In many other situations, some thought about the nature of expected solutions will help you to avoid major errors that can result from minor mistakes in calculation.

As you work on the problems of this investigation, look for answers to this question:

How do the forms of algebraic problems suggest
overall solution strategies and possibilities?

Algebraic Structure and Strategies Problems 1–4 ask you to review the algebraic structure of functions, expressions, and equations, and strategies for manipulating expressions and equations into equivalent forms. Then Problems 5–10 give you a chance to put that strategic thinking into practice.

1 Suppose that $A(x)$ is a polynomial function and k is a fixed number.

   a. How are the following properties of $A(x)$ related to the degree of the polynomial?

      i. The domain and range of $A(x)$

      ii. The number of zeroes of $A(x)$

      iii. The number of local maximum and minimum points

      iv. The end behavior of the graph of $A(x)$

   b. Describe strategies you know for finding approximate or exact answers for the following questions. Be prepared to explain your reasoning in each case.

      i. What are the zeroes of $A(x)$?

      ii. What are the local maximum and minimum points on the graph of $A(x)$?

      iii. What are the solutions for the equation $A(x) = k$?

      iv. What are the solutions for the inequalities $A(x) \geq k$ and $A(x) \leq k$?

      v. What are the solutions for the equation $\sqrt{A(x)} = k$?

2 Suppose that $A(x)$ and $B(x)$ are polynomials and k is a fixed number. Describe strategies you know for finding approximate or exact answers for these questions. Be prepared to explain why the reasoning in each strategy is justified.

a. What are the zeroes of the function $\dfrac{A(x)}{B(x)}$?

b. What are the local maximum and minimum points on the graph of $\dfrac{A(x)}{B(x)}$?

c. What are the solutions for the equation $\dfrac{A(x)}{B(x)} = k$?

d. What are the solutions for the inequalities $\dfrac{A(x)}{B(x)} \geq k$ and $\dfrac{A(x)}{B(x)} \leq k$?

e. What are the solutions for the equation $\sqrt{\dfrac{A(x)}{B(x)}} = k$?

f. What are the asymptotes of the graph of $y = \dfrac{A(x)}{B(x)}$?

3 In problems with two or more independent variables, important questions often can be expressed as systems of linear equations to be solved. It is common to write systems of two linear equations in two variables in the form:

$$\begin{cases} ax + by = c \\ dx + ey = f \end{cases}$$

More complex systems involve more variables and more equations in a similar form.

a. What does it mean to solve a 2×2 linear system like the one above?

b. What are the possible numbers of solutions for a 2×2 system?

c. How can you determine the number of solutions for a 2×2 system by analyzing the relationship of the coefficients a, b, d, and e and the constants c and f?

d. What strategies can be applied to find solutions of a 2×2 system?

e. What strategies can be applied to find solutions of a system with more variables and equations?

4 To answer questions about exponential growth and decay, you need to analyze and manipulate expressions with basic elements in the form $a(b^x)$.

a. What do the parameters a and b tell about the shape and location of the graph of $y = a(b^x)$?

b. Write the following expressions and equations in different equivalent forms.

   i. $(b^x)(b^y)$ ii. $(b^x)^y$

  iii. b^{-x} iv. $\log y = x$

    v. $\log ab$ vi. $\log a^b$

c. What strategies do you have for solving equations in these forms?

   i. $a(b^x) = k$

  ii. $a(b^x) + k = m$

Putting Algebraic Principles and Strategies to Work The next several problems ask you to use your understanding about properties of algebraic functions, graphs, and expressions to answer mathematical questions without the toil of extensive symbol manipulation.

5 Without actually graphing the given functions, decide which of the following could have the graph shown at the right. Explain how you know your choices are correct and how you eliminated the other choices.

a. $a(x) = x(x + 1)(x - 2)^2$

b. $b(x) = x^4 - 3x^3 + 4x + 2$

c. $c(x) = x(x + 1)(x - 2)$

d. $d(x) = x(x + 1)$

e. $e(x) = -x^4 + 3x^3 - 4x$

f. $f(x) = x^4 - 3x^3 + 4x$

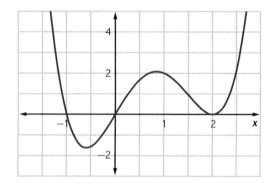

6 If the following expressions were to be expanded and written in standard polynomial form, what would you expect to be the degree of each result? Check your conjecture and make note of any errors in your reasoning.

a. $(3 + x)(x^2 - 5)$

b. $(3 + x)(x^2 - 5)^2$

c. $(x^2 + 3)^2 - x^4$

7 Rational expressions I–IV are all equivalent. Use the expressions to determine answers for the questions that follow about the function that corresponds to these expressions. In each case, explain which expression you used to find the requested property of the function and why you chose that form.

Expression I: $\dfrac{x^3 - 4x^2 - 10x - 10}{x^2 + x - 6}$

Expression II: $(x - 5) + \dfrac{x - 40}{x^2 + x - 6}$

Expression III: $\dfrac{x^3 - 4x^2 - 10x - 10}{(x + 3)(x - 2)}$

Expression IV: $(x - 5) + \dfrac{x - 40}{(x + 3)(x - 2)}$

a. What are the vertical asymptotes of the graph of the function?

b. What are the zeroes of the function?

c. What are the oblique asymptotes of the graph of the function?

d. What is the value of the function when $x = 6$?

e. What are the domain and range of the function?

8 Expressions I–III are equivalent rules for a function $f(x)$. Use the expressions to determine answers for the questions that follow. In each case, explain which expression you used to find the requested property of the function, why you chose that form, and how you could have determined the result with one of the other forms as well.

> **Expression I:** $15(1 - 0.9^x) + 5$
>
> **Expression II:** $20 - 15(0.9^x)$
>
> **Expression III:** $20 - 15(10^{-0.046x})$

a. What is the horizontal asymptote for the graph of $f(x)$?

b. What is the y-intercept of the graph of $f(x)$?

c. For what value of x is $f(x) = 10$?

9 Without performing any calculations, Anita said that the equation $\frac{x+5}{x+3} + \frac{x}{x+4} = 0$ would have at most 2 solutions.

a. Do you agree with her conjecture? How do you suppose Anita came up with that conjecture?

b. What are the solutions of the equation?

c. Would the conjecture about number of possible solutions be different for $\frac{x+5}{x+3} + \frac{x}{x+4} = 1$? Why or why not?

d. What are the solutions of the new equation? Compare your solutions with those of your classmates and resolve any differences.

10 Alesandra made some similar conjectures about the solutions of systems of linear equations.

a. Without making any calculations toward a solution, she said that the system $\begin{cases} y = 2x + 5 \\ y = 3x + 2 \end{cases}$ will have a single solution with both x and y positive numbers.

   i. Is she correct?

  ii. How do you suppose Alesandra came up with that conjecture?

b. Without making any calculations toward a solution, she said that the system $\begin{cases} -x + 3y = 3 \\ 2x - 6y = -5 \end{cases}$ will have no solution.

   i. Is she correct?

  ii. How do you suppose Alesandra came up with that conjecture?

c. When presented with the slightly different system $\begin{cases} -x + 3y = 3 \\ 3x - 6y = -5 \end{cases}$,

Alesandra claimed that this system would have a unique solution (x, y) and, furthermore, that both x and y would be positive numbers.

    i. Is Alesandra's conjecture about this system correct?

    ii. How do you suppose she came up with her ideas about the solution possibilities?

SUMMARIZE THE MATHEMATICS

In this investigation, you extended your understanding and skill in work with algebraic expressions, functions, and equations.

a   What strategies can be used to find zeroes and local maximum and minimum points of polynomial functions? To find solutions for equations involving polynomial expressions?

b   What strategies can be used to find zeroes, asymptotes, and local maximum and minimum points of rational functions? To find solutions for equations and inequalities involving rational expressions?

c   What strategies can be used to predict and/or check the solutions for systems of linear equations?

Be prepared to explain your ideas to the class.

 CHECK YOUR UNDERSTANDING

Consider this difference of rational expressions: $\dfrac{2x}{x + 1} - \dfrac{x + 2}{x + 3}$

a. Write a single rational expression that is equivalent to the given expression but uses polynomials in standard form in the numerator and denominator.

b. Write a rational expression that is equivalent to that in Part a but has the numerator and the denominator expressed as products of linear factors (if possible).

c. Choose from among the equivalent expressions you wrote to help answer the following questions. Be prepared to explain your choice of algebraic form as a starting point for each question.

    i. What are the vertical asymptotes for the graph of $y = \dfrac{2x}{x + 1} - \dfrac{x + 2}{x + 3}$?

    ii. What are the solutions for the equation $\dfrac{2x}{x + 1} - \dfrac{x + 2}{x + 3} = 0$?

    iii. What are the solutions for the equation $\dfrac{2x}{x + 1} - \dfrac{x + 2}{x + 3} = 4$?

APPLICATIONS

1. Suppose that a parabola is drawn on a coordinate grid so that the focus is at $(2, 0)$ and the directrix has equation $x = -2$.

   a. Write expressions that give the distance from any point (x, y) on the parabola to:

      i. the focus.

      ii. the directrix.

   b. Write an equation that expresses the condition that any point on the parabola is equidistant from the focus and the directrix.

   c. Use the equation in Part b to express y as a function of x in a way that can be used to help draw the parabola.

2. Repeat the analysis of Task 1 to find a general form of the equation for a parabola with focus $(a, 0)$ and directrix $x = -a$. Then use that general form to explore the effect of increasing or decreasing the parameter a on the shape of the resulting parabola. See if you can explain why changing the value of a has the effect that it does.

3. Write the equation for a circle of radius 2 and center at $(3, 4)$. Then solve that equation to express y as a function of x in a way that can be used to draw the circle using a graphing tool.

4. The standard deviation for a binomial random variable is given by the formula $\sigma = \sqrt{np(1 - p)}$, where n represents the number of trials of an experiment and p represents the probability of success on each individual trial.

   a. Find σ in case:

      i. $n = 10$ and $p = 0.4$

      ii. $n = 20$ and $p = 0.4$

      iii. $n = 10$ and $p = 0.2$

      iv. $n = 10$ and $p = 0.8$

   b. If $p = 0.7$, find n so that $\sigma < 3$.

   c. If n is a fixed number, what value of p will lead to the maximum value of σ?

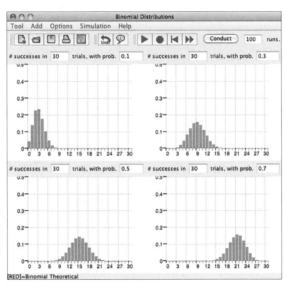

5 Solve each of the following equations showing your calculations and reasoning. Then check your answers by substitution in the original equation.

a. $\sqrt{x^2 + 3x} = 2$

b. $\sqrt{\dfrac{13x^2 - 9}{3}} = 2x$

c. $3 + \sqrt{x + 1} = \sqrt{x + 22}$

d. $\sqrt{\dfrac{2x + 1}{x - 7}} = \sqrt{\dfrac{x}{x - 3}}$

6 Without actually graphing the functions in Parts a–f, decide which of those could have the graph shown below. Explain how you know your choices are correct and how you eliminated the other choices.

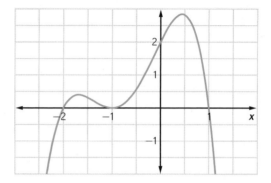

a. $a(x) = -(x - 1)(x + 1)(x + 2)$

b. $b(x) = -x^4 - 3x^3 + x^2 - 3x - 2$

c. $c(x) = -x^4 - 3x^3 - x^2 + 3x + 2$

d. $d(x) = (x - 1)(x + 1)^2(x + 2)$

e. $e(x) = (x - 1)(x + 1)(x + 2)$

f. $f(x) = -(x + 2)(x + 1)^2(x - 1)$

7 Consider this polynomial in factored form: $(x - 1)(x + 1)^2(x + 2)$

a. What is the degree of the polynomial? How can that degree be determined without actually expanding the product of all factors?

b. What is the constant term in the expanded form of this polynomial? How can that value be determined without expanding the product of all factors?

8 Write systems of linear equations in the form $\begin{cases} ax + by = c \\ dx + ey = f \end{cases}$ that meet the following conditions and explain how you can tell, without actually finding the solutions, that your examples do what is requested.

a. Exactly one solution

b. Infinitely many solutions

c. No solution

9. Complete the rational expressions begun below so that the resulting functions meet the given conditions. Then explain how you know, without actually producing the graphs, that your examples are correct.

   a. $y = \dfrac{1}{a(x)}$ with graph having vertical asymptotes $x = -2$ and $x = 3$.

   b. $y = b(x) + \dfrac{1}{x}$ with graph having oblique asymptote $y = 5x + 3$.

   c. $y = c(x) + \dfrac{1}{d(x)}$ with graph having oblique asymptote $y = 5x + 3$ and vertical asymptote $x = 1$.

10. Write rules for functions whose graphs have the shape of the family of exponential growth or decay patterns and the following special properties.

    a. Decreasing function with horizontal asymptote $y = 3$ and y-intercept $(0, 8)$

    b. Increasing function with horizontal asymptote $y = 3$ and y-intercept $(0, 8)$

    c. Increasing function with horizontal asymptote $y = 8$ and y-intercept $(0, 3)$

CONNECTIONS

11. Solving a linear equation in the form $ax + b = cx + d$ should be almost automatic by this point in your mathematical studies.

    a. What formula shows how to calculate x from the values of a, b, c, and d?

    b. How does the formula reveal the cases in which every value of x satisfies the equation?

    c. How does the formula reveal the cases where there is no solution to the equation?

12. Notice that the matrix equation

$$\begin{bmatrix} 7 & 5 \\ 4 & 3 \end{bmatrix} \begin{bmatrix} x \\ y \end{bmatrix} + \begin{bmatrix} 3 \\ -4 \end{bmatrix} = \begin{bmatrix} 4 & 0 \\ 3 & 1 \end{bmatrix} \begin{bmatrix} x \\ y \end{bmatrix} + \begin{bmatrix} 5 \\ 3 \end{bmatrix}$$

    is of the linear form $AX + B = CX + D$, where A, B, C, D, and X are matrices.

    a. What values of x and y satisfy the equation?

    b. Under what conditions will the matrix equation $M_1X + A = M_2X + C$ have a unique solution?

13. Describe the solution to this system of inequalities.

$$\begin{cases} y \leq \sqrt{16 - x^2} \\ y \geq -\sqrt{16 - x^2} \end{cases}$$

14 The graph of the function $f(x) = 3^{-x^2}$ is very similar in shape to the normal probability distribution.

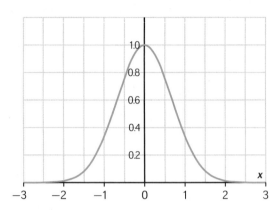

a. Explain how the apparent symmetry of the graph about the y-axis can be proven algebraically using the function rule.

b. Explain how the function rule indicates that the x-axis is an asymptote for the graph.

REFLECTIONS

15 The problems of Investigation 1 led to algebraic expressions involving radicals in two main ways—using the distance formula and working with the standard deviation of a probability distribution.

What geometric principle is the basis of the distance formula? How does the algebraic expression of that principle explain the appearance of a radical in the standard expression of the formula?

16 What do you look for in the algebraic rule for a polynomial or rational function or a function involving radicals when your task is to find:

a. the domain of the function?

b. the zeroes of the function?

c. vertical asymptotes for the graph of the function?

d. oblique asymptotes for the graph of the function?

e. horizontal asymptotes for the graph of the function?

17 What do you look for in a system of linear equations in two or more variables to see whether the system might have infinitely many solutions, no solution, or a unique solution?

18 What strategies do you apply to solve equations that involve radicals?

19 Many algebra students operate with the informal strategy guideline, "You can do anything to an equation as long as you do the same thing to both sides." Irene used this strategy to "solve" the equation below that involves radicals. See if you can explain why the result is problematic and why it suggests limitations to the "do anything as long as you do it to both sides" guideline.

$$\text{If } \sqrt{x+1} = 2 + \sqrt{x},$$
$$\text{Then } x + 1 = 4 + 4\sqrt{x} + x.$$
$$\frac{-3}{4} = \sqrt{x}$$
$$\frac{9}{16} = x$$
$$\text{But } \sqrt{\frac{9}{16} + 1} \neq 2 + \sqrt{\frac{9}{16}}.$$

EXTENSIONS

20 Consider parabolas that occur as graphs of functions $y = ax^2$. What are the focus and directrix of the parabolic graph when:

a. $a = 1$?

b. $a = 2$?

c. $a = 3$?

d. $a = k$?

21 Consider this system of three linear equations in three variables.

$$\begin{cases} 3x + 2y - 5z = 2 \\ -6x - 4y + 10z = -4 \\ 9x + 6y - 15z = 6 \end{cases}$$

a. Show that $(1, 2, 1)$, $(0, 3.5, 1)$, and $(4, 0, 2)$ are all solutions of the system.

b. Explain how you can determine that this system of equations will in fact have an infinite number of solutions by comparing coefficients of the variables and the constants in the three equations.

c. Explain how you can prove that the following slightly different system has no solution, again by comparing the coefficients and constants in the three equations.

$$\begin{cases} 3x + 2y - 5z = 2 \\ -6x - 4y + 10z = -4 \\ 9x + 6y - 15z = 5 \end{cases}$$

22 Suppose that a mathematics course you are taking has 10 short quizzes in every marking period. Your score on quiz number 1 is represented by x_1, your score on quiz number 2 by x_2, and so on.

a. What measure of your performance is represented by $\dfrac{\sum_{i=1}^{10} x_i}{10}$?

b. Solve $\dfrac{\sum\limits_{i=1}^{10} x_i}{10} = 7$ for x_{10} and explain what question the solution to that equation answers.

c. If your average score for the 10 quizzes in a marking period is 7.5, what symbolic expression represents the standard deviation of the set of scores?

23 In analyzing data from an experiment comparing effects of two medical treatments, one often encounters an equation in the form $\sqrt{\dfrac{s_1^{\,2}}{n_1} + \dfrac{s_2^{\,2}}{n_2}} = E.$

Solve this equation for n_1 in terms of the other variables.

24 In Extensions Task 27 of Lesson 3, you explored the solution possibilities for equations in the general form $x + \dfrac{1}{x} = a$.

a. What equivalent form of such an equation makes it possible to apply familiar equation-solving strategies?

b. How can your answer to Part a be used to answer this question: Is it possible to find three distinct numbers p, q, and r so that $p + \dfrac{1}{p} = q + \dfrac{1}{q} = r + \dfrac{1}{r} = a$?

REVIEW

25 Solve each equation.

a. $3^x + 6 = 370$

b. $4(2.5^x) = 100$

c. $-5(6^x) + 204 = 39$

26 Determine the solutions for these systems of equations.

a. $\begin{cases} 0.5x + 7y = 35 \\ x + 14y = 20 \end{cases}$

b. $\begin{cases} 4x - y = -12 \\ x + 3y = 10 \end{cases}$

c. $\begin{cases} 12x + 3y = 60 \\ 8x + 2y = 40 \end{cases}$

27 Rewrite each expression as a single rational expression. Then simplify each expression as far as possible.

a. $1 + \dfrac{3}{x}$

b. $\dfrac{x}{4} \div \dfrac{x+1}{8}$

c. $\dfrac{\frac{x+2}{x}}{x^2 + 2x}$

d. $3x - \dfrac{5x}{x+1}$

e. $\dfrac{\frac{a}{b}}{a}$

f. $\dfrac{\frac{m}{3}}{\frac{n}{6}}$

28 On the grid below, the line contains the origin and forms a 30° angle with the positive x-axis.

a. The x-coordinate of point A is 2. Determine the y-coordinate of A.

b. The y-coordinate of point B is -1. Determine the x-coordinate of B.

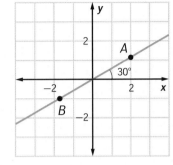

29 The coordinates of point X are $(2, 6)$. Give the coordinates of the image of X if you apply each indicated transformation.

a. Reflection across the x-axis

b. Reflection across the y-axis

c. Rotation of 180° centered at the origin

d. Counterclockwise rotation of 90° centered at the origin

30 Consider the graph of $f(x)$ shown below.

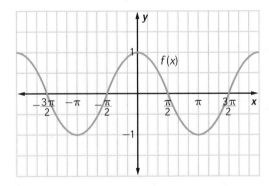

a. Write a function rule that matches the graph.

b. Sketch a graph of $y = 2f(x)$.

c. Sketch a graph of $y = f(2x)$.

d. How is the graph of $g(x) = \sin x$ related to the graph shown above?

LESSON

5

Looking Back

The lessons, investigations, and specific problems of this unit challenged you to recall what you had learned about polynomial and rational functions and to develop strategies for combining those functions to build models for more complex relationships. You also developed understanding and skill in work with complex numbers and algebraic expressions that involve radicals. Finally, you were asked to formulate some general algebraic strategies that can help in solving equations with new forms.

The tasks in this final lesson of the unit give you a chance to review your skill and understanding of functions, equations, and complex numbers.

1 For most of us in the United States, it is hard to imagine a place we would like to visit that cannot be reached by public transportation or a drive on good roads. But in the wilderness of North America, Europe, and Asia there are very few roads and little public transportation.

To haul valuable minerals from mines that are far from civilization, many countries "build" *ice roads* across the frozen surfaces of rivers and lakes. For more information about ice road trucking, visit www.history.com.

The critical question in using ice roads is the relationship between ice thickness and weight of trucks driving on the road. The following table shows some (*thickness, weight*) guidelines.

Ice Thickness (in inches)	2	3	5	10
Vehicle Weight (in pounds)	400	900	2,500	10,000

Of course, the risks of misjudging the carrying weight of ice are great and there are factors other than ice thickness to consider.

a. Study the pattern of (*thickness, weight*) values to see if you can find a rule relating the variables.

b. Use the first three data pairs to write a system of linear equations that determine coefficients in a quadratic model for the data pattern. Use matrix methods or generalization of other algebraic strategies to solve the system. Then compare the predictions of that model to the given data and evaluate the fit of the model.

National Geographic RF/Getty Images

c. What factors other than ice thickness can you imagine that will affect the "carrying weight" of an ice road?

2 Many amusement parks have "grand prix" racetracks for go-carts. The speeds are slower than those of real racecars on real grand prix racecourses. But the thrill is still attractive to many customers.

Suppose that designers of a "grand prix" go-cart track wanted to include a segment that has the shape shown in the diagram below.

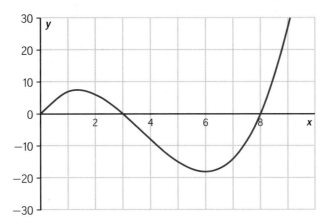

a. Write a rule for a polynomial function that has the same x-intercepts as the graph shown. Then compare the graph of that function to that above and evaluate the fit of the model.

b. Identify several control points that you believe can be used to develop a better model for the graphed racetrack segment, and use statistical regression to find a polynomial model based on those points. Compare the resulting graph to the graph above and evaluate the fit of the model.

c. Use key control points from Part b to set up a system of linear equations that will help in finding coefficients of a cubic polynomial model for the given diagram. Solve the system, compare the resulting graph to the graph above, and evaluate the fit of the model.

3 Consider the quadratic equation $x^2 - 6x + 10 = -3$.

a. Find all solutions of that equation and express the results in simplest possible $a + bi$ form.

b. Plot the two solutions of the equation on a coordinate grid and draw the position vectors representing each solution. Describe a geometric transformation that would map each point representing a solution onto the other.

c. Calculate the sum, difference, product, and quotient of the two solutions and express the results of each calculation in $a + bi$ form. Where possible, describe how those results are represented on the plot that shows the two original numbers as points on a coordinate grid.

4 It is hard to imagine life without the electricity that enables so many modern conveniences. But the lighting that allows us to walk and drive safely or watch outdoor events at night has been around for less than 200 years.

You know from prior studies that the intensity of light on any surface is directly proportional to the emitted power of the light source and inversely proportional to the square of the distance of the lighted surface from the source.

a. If light intensity is represented by I, emitted light power by P, and distance from source by d, what formula models the proportional relationship of those variables?

Suppose that you are walking along a sidewalk illuminated by two overhead lights positioned as in the following diagram with the light towers 20 feet high, 35 feet apart, and emitting 300 watts of light power.

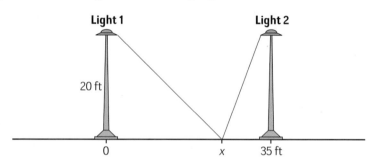

b. What algebraic expression represents the distance from Light 1 to the point with horizontal position x feet from the base of that light?

c. What algebraic expression represents the distance from Light 2 to the point x feet from the base of Light 1?

d. What algebraic expression represents the light intensity (in watts per square feet) from Light 1 on the spot with horizontal position x feet from the base of Light 1?

e. What algebraic expression represents the light intensity (in watts per square feet) from Light 2 on the spot with horizontal position x feet from the base of Light 1?

f. The spot that is x feet from the base of Light 1 actually receives illumination from both Light 1 and Light 2. What algebraic expression seems likely to represent the total light intensity at that spot?

g. What is the total light intensity at a spot on the sidewalk that is 15 feet from the base of Light 1?

5 For each of the rational functions that follow, use algebraic methods where possible to:

- find all zeroes.

- find equations for all asymptotes.

- describe the end behavior.

- find coordinates of local maximum and/or minimum points.

- sketch a graph.

a. $y = \dfrac{2x + 1}{x - 3}$

b. $y = \dfrac{x - 4}{x^2 + x - 2}$

c. $y = \dfrac{4x^2 - 8x - 12}{x^2 + 3x + 2}$

d. $y = \dfrac{x^2 + 10}{x^2 - 2x - 8}$

e. $y = \dfrac{x^3 - x^2 + 1}{x^2 - 2x + 1}$

f. $y = \dfrac{-3x - 8}{x^2 + 3x - 10}$

6 Analyze the graph of the polynomial function $f(x)$ given below to help answer the following questions.

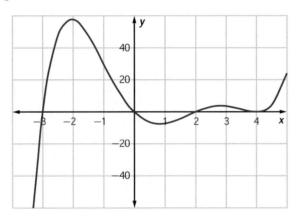

a. If the polynomial is expressed as a product of linear factors, what factors should be included?

b. What is the probable degree of the polynomial? Explain your reasoning.

c. What is the constant term of the polynomial?

7 Solve each equation showing your calculations and reasoning. Check your answers.

   a. $\sqrt{x^2 + 6x} = 4$

   b. $2x - 1 = \sqrt{x}$

SUMMARIZE THE MATHEMATICS

In this unit, you investigated a variety of situations that involved polynomial and rational function models, you extended your understanding and skill in working with those functions and their graphs. Similarly, you extended your understanding of complex numbers as a mathematical system and you extended your skill in work with algebraic equations—including those that involve radicals.

a What properties of a situation help you to decide that a polynomial or rational function might be useful as a model for relationships among variables?

b What different equivalent forms can be used to express polynomial functions? What are the advantages of each form? How can each form be transformed into the others?

c What mathematical problems led to development of the complex number system? What are the rules for arithmetic operations with complex numbers?

d What different equivalent forms can be used to express rational functions? What are the advantages of each form and how can each form be transformed into the others?

e What strategies are useful in solving equations that involve radical expressions?

Be prepared to share your responses and reasoning with the class.

 CHECK YOUR UNDERSTANDING

Write, in outline form, a summary of the important mathematical concepts and methods developed in this unit. Organize your summary so that it can be used as a quick reference in future units.

Trigonometric Functions and Equations

Previously in *Core-Plus Mathematics*, you learned to use the trigonometric functions sine, cosine, and tangent to determine unknown measurements of sides and angles in triangles. You also studied how the sine and cosine functions can be used to model periodic phenomena and linear and nonlinear motion.

In this unit, you will add three more functions to the family of trigonometric functions. Your focus will be on reasoning to produce and verify equivalent trigonometric expressions and to solve equations involving those functions. You will also use trigonometric relations to deepen your understanding of the geometry of complex numbers.

The key ideas and methods will be developed in three lessons.

LESSONS

1 Reasoning with Trigonometric Functions

Extend the family of trigonometric functions, derive fundamental identities, and use those identities together with geometric and algebraic reasoning to prove or disprove other potential identities.

2 Solving Trigonometric Equations

Extend proficiency in solving applied problems using trigonometric methods and develop skill in using symbol sense and algebraic reasoning to solve trigonometric equations.

3 The Geometry of Complex Numbers

Represent complex numbers in trigonometric form. Connect this representation to geometric transformations and to determination of nth roots of complex numbers.

Reasoning with Trigonometric Functions

In Unit 2, *Vectors and Motion*, you learned that projectile motion such as the path of a baseball or golf ball is the sum of a horizontal and a vertical component. When the initial velocity is V and the angle of elevation is θ, the initial magnitude of the horizontal component is $V\cos\theta$ and the initial magnitude of the vertical component is $V\sin\theta$.

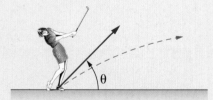

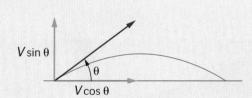

Ignoring air resistance, the height (in feet) of the projectile after t seconds is given by

$$h(t) = (V\sin\theta)t + 16t(t\sin 270°) \text{ or } h(t) = (V\sin\theta)t - 16t^2.$$

While in the air, the horizontal distance (in feet) the projectile travels in t seconds is given by

$$d(t) = (V\cos\theta)t.$$

Golfers, baseball players, and football kickers are often interested in the nonzero value of $d(t)$ when $h(t) = 0$. That is, they are interested in how far the ball will travel before it returns to the ground. This horizontal distance (in feet), called the *range*, is given by

$$R(\theta) = \frac{V^2\cos\theta\sin\theta}{16},$$

where V is in feet per second.

The range function $R(\theta)$ involves the product of two trigonometric functions. This product can be expressed in an equivalent form as a single trigonometric function with which you are familiar. In this lesson, you will develop key *trigonometric identities* that are useful in rewriting expressions involving trigonometric functions in equivalent and often easier to use forms.

 INVESTIGATION **1**

Trigonometric Identities

An **identity** is a statement that is true for all replacements of a variable for which the statement is defined. You have seen identities in your previous work in algebra. For example, $a\left(\dfrac{1}{a}\right) = 1$ is an identity for all nonzero real numbers and $\sqrt{ab} = \sqrt{a}\,\sqrt{b}$ is an identity for all non-negative real numbers. Examples of identities involving all real numbers are $a(b + c) = ab + ac$ and $(2x + 1)^2 = 4x^2 + 4x + 1$. A statement such as $(x + 3)(x - 1) = 0$ is *not* an identity for all real numbers. Why not?

Identities are useful because they permit you to rewrite expressions so that they are more recognizable, easier to interpret, or easier to work with in solving problems. Consider again the rule given at the beginning of this lesson for the horizontal distance (in feet) that a projectile will travel.

$$R(\theta) = \frac{V^2 \cos \theta \sin \theta}{16}$$

For a fixed value of V, R is a function of θ, the angle at which the projectile is launched.

The fact that $R(\theta)$ involves the product "$\cos \theta \sin \theta$" makes it difficult to visualize the shape of its graph. In Investigation 2, you will establish the identity $2 \cos \theta \sin \theta = \sin 2\theta$ or, equivalently, $\cos \theta \sin \theta = \frac{\sin 2\theta}{2}$. Using this identity, you can rewrite $R(\theta)$ as $\frac{V^2 \sin 2\theta}{32}$, which is an easier form to interpret. What do θ, V, and $R(\theta)$ represent in the context shown below?

There are many additional useful identities involving trigonometric functions. As you work on the problems of this investigation, look for answers to these questions:

What strategies are useful in proving trigonometric identities?

How are the sine, cosine, and tangent values of an angle and its opposite related?

1 In Unit 2, *Vectors and Motion*, you learned that for any given position vector with length r, the terminal point of the vector is always on a circle of radius r. Thus, the coordinates of the terminal point can be expressed in terms of the cosine and sine of the direction angle of the vector:

$$(x, y) = (r \cos \theta, r \sin \theta)$$

In the case of the unit circle displayed at the right, $r = 1$. Use the definition of $\tan \theta$ and the diagram to prove that

$$\tan \theta = \frac{\sin \theta}{\cos \theta}.$$

What restrictions must be placed on θ?

2 A key identity you have encountered in your previous study is the statement $(\sin \theta)^2 + (\cos \theta)^2 = 1$. For simplicity, $(\sin \theta)^2$ is often written $\sin^2 \theta$. When using technology, the expression $\sin^2 \theta$ may need to include the parentheses, for example $(\sin(\theta))^2$.

a. Without using technology, describe the graph of $f(\theta) = \sin^2 \theta + \cos^2 \theta$.

b. For what values of θ is the statement $\sin^2 \theta + \cos^2 \theta = 1$ defined?

c. Use the diagram above to prove $\sin^2 \theta + \cos^2 \theta = 1$ is an identity for all values of θ.

d. $\sin^2 \theta + \cos^2 \theta = 1$ is called a **Pythagorean identity**. Discuss why this name is appropriate.

3 **Deriving Opposite-Angle Identities** Now suppose point A is on the terminal side of an angle θ in standard position, r is the length of $\overline{OA}$, and:

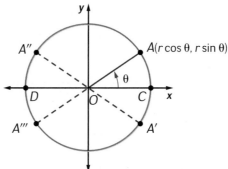

- point A' is the image of point A reflected across the x-axis.

- point A'' is the image of point A reflected across the y-axis.

- point A''' is the image of point A rotated 180° or π radians about the origin.

a. On a copy of the diagram above, label the coordinates of the points A', A'', and A''' in terms of θ.

b. Use the coordinates that you found in Part a to justify each step in the following derivation of the **opposite-angle identities** for the cosine and sine functions.

$$\cos(-\theta) = \cos\theta \text{ and}$$
$$\sin(-\theta) = -\sin\theta$$

(1) Since the coordinates of point A are $(r\cos\theta, r\sin\theta)$, the coordinates of point A' are $(r\cos\theta, -r\sin\theta)$.

(2) The measure of $\angle COA'$ is $-\theta$. Thus, A' has coordinates $(r\cos(-\theta), r\sin(-\theta))$.

(3) It follows that $r\cos(-\theta) = r\cos\theta$ and $\cos(-\theta) = \cos\theta$. Similarly, $\sin(-\theta) = -\sin\theta$.

c. Write an argument showing that $\tan(-\theta) = -\tan\theta$.

4 The symmetry of a circle centered at the origin of a coordinate system can help you discover and derive other trigonometric identities. In the diagram above, the symmetry of the circle ensures $m\angle COA'' = \pi - \theta$ and similar other angle relationships. Use the coordinates found in Problem 3 Part a to complete identities involving each expression below. Then, discuss your proposed identities with the class and choose one sine, one cosine, and one tangent identity to prove.

a. $\cos(\pi - \theta) =$ b. $\sin(\pi - \theta) =$

c. $\tan(\pi - \theta) =$ d. $\cos(\pi + \theta) =$

e. $\sin(\pi + \theta) =$ f. $\tan(\pi + \theta) =$

Reasoning Strategies Proving an identity involves writing a sequence of equivalent expressions, using definitions, previously proven identities, and algebraic procedures to show that two given expressions are equivalent.

5 Suppose that in solving a problem, you encounter two expressions, for example, $\sin^2\theta$ and $(1 - \sin^2\theta)\tan^2\theta$, and you want to determine whether or not they are equivalent for all values θ for which both expressions are defined.

a. How could you use tables of values to support or contradict the claim that $(1 - \sin^2\theta)\tan^2\theta = \sin^2\theta$ is an identity?

b. How could you use graphs to support or contradict the claim that the statement is an identity?

c. Find a value of θ for which the statement $(1 - \sin^2 \theta) \tan^2 \theta = \sin^2 \theta$ is *not* true. Does this mean that the statement is not an identity? Explain your reasoning.

d. Definitions and the two fundamental identities you proved in Problems 1 and 2 can be used to prove that $(1 - \sin^2 \theta) \tan^2 \theta = \sin^2 \theta$ is an identity. Justify each step in the following reasoning chain.

$$(1 - \sin^2 \theta) \tan^2 \theta = (1 - \sin^2 \theta) \frac{\sin^2 \theta}{\cos^2 \theta} \tag{1}$$

$$= \cos^2 \theta \frac{\sin^2 \theta}{\cos^2 \theta} \tag{2}$$

$$= \sin^2 \theta \tag{3}$$

e. Explain why the symbolic reasoning used above *proves* that $(1 - \sin^2 \theta) \tan^2 \theta = \sin^2 \theta$ is an identity, while the use of tables or graphs only makes it plausible that the statement is an identity.

The reasoning strategy used in Problem 5 was to rewrite the left (and more complicated) side of $(1 - \sin^2 \theta) \tan^2 \theta = \sin^2 \theta$ into a simpler form with the expression $\sin^2 \theta$ as the goal. Sometimes it is helpful to rewrite each side of a proposed identity *independently* into forms that are equal, as is shown in Problem 6.

6 Study the following proof of the identity $(1 - \sin \theta)(1 + \sin \theta) = \dfrac{\cos \theta \sin \theta}{\tan \theta}$, in which the expression on each side is rewritten *independently* of the other. The vertical line is used to emphasize this fact.

$$(1 - \sin \theta)(1 + \sin \theta) \qquad\qquad \frac{\cos \theta \sin \theta}{\tan \theta}$$

$$= 1 - \sin^2 \theta \qquad\qquad = \frac{\cos \theta \sin \theta}{\frac{\sin \theta}{\cos \theta}}$$

$$= \cos^2 \theta \qquad\qquad = \cos \theta \sin \theta \frac{\cos \theta}{\sin \theta}$$

$$= \frac{\cos^2 \theta \sin \theta}{\sin \theta}$$

$$= \cos^2 \theta$$

The reasoning on the left side of the vertical line shows $(1 - \sin \theta)(1 + \sin \theta) = \cos^2 \theta$. Using the fact that each step on the right side can be reversed, it follows that $\cos^2 \theta = \dfrac{\cos \theta \sin \theta}{\tan \theta}$. Thus,

$$(1 - \sin \theta)(1 + \sin \theta) = \frac{\cos \theta \sin \theta}{\tan \theta}.$$

a. Justify the reasoning steps on the left side of the line. On the right side of the line.

b. How does the fact that each step on the right side can be reversed complete the proof of the identity?

c. Would reasoning first on the right side and then reversing each step on the left side also complete a proof of the identity? Explain your thinking.

d. How does this strategy for proving identities illustrate the idea of "working forward and working backward?"

7 To show that a statement involving trigonometric functions of θ is *not* an identity, it is sufficient to provide a counterexample, that is, one value of θ for which both sides of the statement are defined but not equal. Show that $(2 \sin \theta)(\cos \theta) = \sin \theta$ is not an identity.

8 Decide whether each of the following statements is or is not an identity. If you think a statement is an identity, use reasoning to prove it. If it is not, provide a counterexample.

a. $2 \sin^2 \theta - 1 = 1 - 2 \cos^2 \theta$

b. $\tan^2 \theta = \dfrac{1 - \cos^2 \theta}{\cos^2 \theta}$

c. $\dfrac{\sin \theta + \cos \theta}{\sin \theta} = 1 + \cos \theta$

d. $\dfrac{\sin \theta}{\cos \theta} + \dfrac{\cos \theta}{\sin \theta} = \dfrac{1}{\sin \theta \cos \theta}$

e. $\sin^2 \theta - \cos^2 \theta = 1$

f. $(1 + \sin^2 \theta)(1 + \cos^2 \theta) = 2 + \sin^2 \theta \cos^2 \theta$

SUMMARIZE THE MATHEMATICS

Trigonometric identities are special types of statements involving trigonometric functions.

a How is a trigonometric identity similar to, and different from, an algebraic property like the Commutative Property of Addition?

b How is a trigonometric identity different from a trigonometric equation?

c Explain why examining tables of values and/or graphs of functions is not sufficient to prove a statement involving those functions is an identity. Explain how you would prove a statement is an identity.

d How can you use the graphs of the sine, cosine, and tangent functions to help you recall the identities for sin (−θ), cos (−θ), and tan (−θ)? To recall special sums and differences of the sine and cosine such as sin (π + θ) = −sin θ and cos (π − θ) = −cos θ?

Be prepared to share your ideas and reasoning with the class.

 CHECK YOUR UNDERSTANDING

For each statement, decide whether or not it is an identity. If a statement is an identity, use reasoning to prove it. If it is not, provide a counterexample.

a. $(1 - \cos \theta)(1 + \cos \theta) = \sin^2 \theta$

b. $\dfrac{1}{\cos \theta} - \dfrac{\sin^2 \theta}{\cos \theta} = \cos \theta$

c. $(\sin \theta + \cos \theta)^2 = \sin^2 \theta + \cos^2 \theta$

d. $\dfrac{\tan \theta \cos \theta}{\sin \theta} = 1$

Sum and Difference Identities

In the *Trigonometric Methods* unit of Course 2, you saw that in the case of a right triangle *DEF* with ∠*F* a right angle, sin *D* = cos *E* = cos (90° − *D*). That is, the sine of an acute angle in a right triangle is equal to the cosine of its complement. How could you write cos *D* in terms of the complement of ∠*D*?

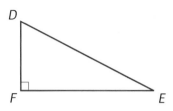

By now you should be asking yourself: "Is it possible to generalize these results to any angle θ?"

As you work on the problems in this investigation, look for answers to these questions:

> *How can you express the sine or cosine of an angle θ in terms of $\frac{\pi}{2} - \theta$, the complement of θ?*

> *What are trigonometric identities for the sum and difference of two angles?*

> *How can these identities be used to verify other identities?*

1 **Deriving Cofunction Identities**
In Investigation 1, you saw how rotational symmetry of a circle with center at the origin of a coordinate system or line symmetry involving the coordinate axes were useful tools in deriving trigonometric identities. In this problem, you will see how reflection symmetry of a circle across the line *y* = *x* can be used in deriving other identities.

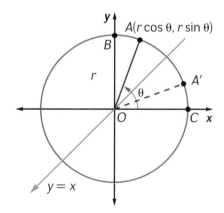

a. In the diagram at the right, point *A′* is the image of point *A* reflected across the line *y* = *x*.

  i. What are the coordinates of point *A′* in terms of θ?

  ii. Explain why m∠*A′OC* = $\frac{\pi}{2} - \theta$.

b. Reasoning from your results in Part a, write identities relating $\sin\left(\frac{\pi}{2} - \theta\right)$ and $\cos\left(\frac{\pi}{2} - \theta\right)$ to those of cos θ and sin θ.

c. How do these **cofunction identities** differ from the opposite-angle identities you proved in Investigation 1, Problem 3 (page 273)?

Deriving Sum and Difference Identities

In your previous work, you used vectors to model navigation problems. For example, suppose two ships leave a harbor at approximately the same time and travel in different directions as shown in the diagram at the right. (The angle measure symbols "α" and "β" are the Greek letters "alpha" and "beta," respectively). How could you use the Law of Cosines to determine how far apart the ships are at any point in time?

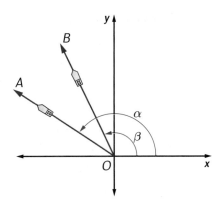

This use of the Law of Cosines would require that you find the value of $\cos (\alpha - \beta)$ (or $\cos \angle BOA$). In the next problem, you will examine a derivation of a useful identity involving $\cos (\alpha - \beta)$.

2 In the diagram at the right, points A, B, C, and D are on the circle with center O and radius r. Suppose $\overline{OA}$ and $\overline{OB}$ make angles of α and β with the positive x-axis, respectively.

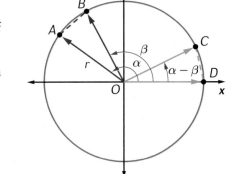

a. Suppose $\overline{OC}$ makes an angle of $\alpha - \beta$ with the positive x-axis and point D is on the x-axis. Find the rectangular coordinates of points A, B, C, and D in terms of the sine and cosine functions.

b. Analyze how the distance formula is used below to derive an identity involving $\cos (\alpha - \beta)$. Supply reasons for each step in the derivation.

$$AB = CD \tag{1}$$

$$\sqrt{(r \cos \alpha - r \cos \beta)^2 + (r \sin \alpha - r \sin \beta)^2} = \sqrt{[r \cos (\alpha - \beta) - r]^2 + [r \sin (\alpha - \beta)]^2} \tag{2}$$

$$(r \cos \alpha - r \cos \beta)^2 + (r \sin \alpha - r \sin \beta)^2 = [r \cos (\alpha - \beta) - r]^2 + [r \sin (\alpha - \beta)]^2 \tag{3}$$

$$(\cos \alpha - \cos \beta)^2 + (\sin \alpha - \sin \beta)^2 = [\cos (\alpha - \beta) - 1]^2 + \sin^2 (\alpha - \beta) \tag{4}$$

$$\cos^2 \alpha + \sin^2 \alpha + \cos^2 \beta + \sin^2 \beta - 2(\cos \alpha \cos \beta + \sin \alpha \sin \beta) = \cos^2 (\alpha - \beta) + \sin^2 (\alpha - \beta) - 2 \cos (\alpha - \beta) + 1 \tag{5}$$

$$1 + 1 - 2(\cos \alpha \cos \beta + \sin \alpha \sin \beta) = 1 - 2 \cos (\alpha - \beta) + 1 \tag{6}$$

$$\cos \alpha \cos \beta + \sin \alpha \sin \beta = \cos (\alpha - \beta) \tag{7}$$

c. How could you use function graphs to illustrate the *difference identity* $\cos (\alpha - \beta) = \cos \alpha \cos \beta + \sin \alpha \sin \beta$?

d. Using the difference identity $\cos (\alpha - \beta)$, show that $\cos \left(\dfrac{3\pi}{2} - \theta \right) = -\sin \theta$.

e. Without using technology, evaluate $\cos 130° \cos 40° + \sin 130° \sin 40°$.

f. Write the **difference identity for the cosine function** in words, beginning:

The cosine of the difference of two angles α and β is _____.

3 The difference identity $\cos(\alpha - \beta) = \cos\alpha\cos\beta + \sin\alpha\sin\beta$ can be used to derive other useful identities.

a. Use the fact that $\alpha + \beta = \alpha - (-\beta)$ to derive a *sum identity for cos $(\alpha + \beta)$*. Begin as follows: $\cos(\alpha + \beta) = \cos[\alpha - (-\beta)]$. Write the **sum identity for the cosine function** in words.

b. In Problem 1, you showed that $\cos\left(\frac{\pi}{2} - \theta\right) = \sin\theta$. You can use this identity to derive a *sum identity for sin $(\alpha + \beta)$*. One possible way to begin is as follows:

$$\sin(\alpha + \beta) = \cos\left[\frac{\pi}{2} - (\alpha + \beta)\right]$$

$$= \cos\left[\left(\frac{\pi}{2} - \alpha\right) - \beta\right]$$

$$= \ldots$$

c. Write the **sum identity for the sine function** in words.

d. Use the technique suggested in Part a to derive a *difference identity for sin $(\alpha - \beta)$*. Then write the **difference identity for the sine function** in words.

e. Using an identity, show that $\sin\left(\frac{3\pi}{2} + \theta\right) = -\cos\theta$.

f. Without using technology, find the exact values for

$\sin\frac{5\pi}{12}\cos\frac{\pi}{12} - \cos\frac{5\pi}{12}\sin\frac{\pi}{12}$.

4 Use identities and algebraic reasoning to rewrite each expression as an equivalent expression involving a single trigonometric function as indicated.

a. Rewrite $\sin 2x \cos 3x + \cos 2x \sin 3x$ as an equivalent expression involving only the sine function.

b. Rewrite $\sin 3x \cos 2x - \cos 3x \sin 2x$ as an equivalent expression involving only the sine function.

c. Rewrite $\cos 2x \cos x - \sin 2x \sin x$ as an equivalent expression involving only the cosine function.

5 **Deriving Double-Angle Identities** At the beginning of this lesson, you reasoned that the horizontal distance, in feet, that a projectile such as a hit golf ball or kicked soccer ball will travel in the air is given by

$R(\theta) = \dfrac{V^2 \cos\theta \sin\theta}{16}$, where
V is given in feet per second, and θ is the angle of elevation at which the ball is hit. In this problem, you will derive an easier form of $R(\theta)$ to interpret and to calculate with.

a. How could you express $\sin 2\theta$ as the sine of the sum of two angles?

b. Using your answer from Part a, prove the identity $\sin 2\theta = 2 \sin\theta \cos\theta$.

c. Prove that the range formula above can be re-expressed in simpler form as $R(\theta) = \dfrac{V^2 \sin 2\theta}{32}$.

d. Suppose a golfer can consistently hit the ball so that it leaves the ground with an initial velocity V. Explain why the maximum range is attained when $\theta = 45°$.

e. Imagine kicking a soccer ball with an initial velocity V. Prove that the horizontal distance the ball will travel in the air will be the same for two different values of θ, namely, $\theta = 45° + \alpha$ and $\theta = 45° - \alpha$, where $0 \le \alpha < 45$.

6 The identity $\sin 2\theta = 2 \cos \theta \sin \theta$ that you proved in Problem 5 Part b is called the **double-angle identity for the sine function**.

a. Use similar reasoning to derive a **double-angle identity for the cosine function**.

b. Write, in words, the two double-angle identities you derived.

7 Recall from previous courses that using relationships from geometry, you can find exact values of the trigonometric functions for angles whose measures are multiples of 30° or 45°, that is, $\frac{\pi}{6}$ or $\frac{\pi}{4}$. By using sums and differences of these angle measures, you can find exact values of the trigonometric functions for many other angles.

a. In the diagrams below, $\triangle ABC$ is an equilateral triangle with side length 2 and $\overline{BD}$ is an altitude. Also, $\triangle PQR$ is an isosceles right triangle with legs of length 1. In each case, explain how to find the other angle and side measurements in the diagrams.

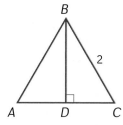 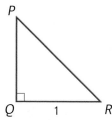

b. Use the measurements you found for the diagrams above to help you complete a copy of the following table giving exact values of $\sin \theta$ and $\cos \theta$ for selected values of θ (in degrees and radians). *NOTE:* $\sin 45°$ is expressed as $\frac{\sqrt{2}}{2}$ since $\frac{1}{\sqrt{2}} = \frac{1}{\sqrt{2}} \frac{\sqrt{2}}{\sqrt{2}} = \frac{\sqrt{2}}{2}$ and the latter form is easier to interpret.

θ	0° or 0	30° or $\frac{\pi}{6}$	45° or $\frac{\pi}{4}$	60° or $\frac{\pi}{3}$	90° or $\frac{\pi}{2}$	180° or π	270° or $\frac{3\pi}{2}$
$\sin \theta$	?	?	$\frac{\sqrt{2}}{2}$	?	?	?	?
$\cos \theta$	?	?	?	?	?	?	?

c. Use your completed table and the identities from Problems 5 and 6 to find exact values for each expression below.

i. $\cos 15°$ **ii.** $\sin \frac{\pi}{12}$ **iii.** $\cos \frac{11\pi}{12}$

iv. $\sin 195°$ **v.** $\sin 285°$

In this investigation, you proved identities relating the sine and cosine functions. You also derived sum and difference identities and double-angle identities for the sine and cosine functions. You used these identities to transform trigonometric expressions into equivalent forms and to find exact values of special angles.

a Write identities for $\sin(\alpha + \beta)$, $\cos(\alpha + \beta)$, $\sin(\alpha - \beta)$, and $\cos(\alpha - \beta)$.

b What strategies were used to prove these identities?

c Explain how the double-angle identities for the sine and cosine functions can be thought of as direct applications of the corresponding sum identities.

d Explain how exact values of the sine and cosine functions for angles that are multiples of $\frac{\pi}{12}$ can be found using the sum and difference identities developed in this investigation.

Be prepared to compare your identities and methods to those of your classmates.

 CHECK YOUR UNDERSTANDING

Reflect on the mathematical methods used in this investigation as you complete the following tasks.

a. Using the diagram and information at the right, what is the radius of the circle? Find a simplified expression for the length PQ.

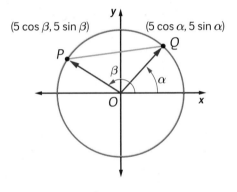

b. Use algebraic reasoning to derive an identity expressing $\sin\left(\frac{3\pi}{2} - \alpha\right)$ in terms of $\sin(\alpha)$ or $\cos(\alpha)$. Repeat for $\cos\left(\frac{3\pi}{2} - \alpha\right)$.

c. Derive an identity expressing $\sin\left(\alpha + \frac{\pi}{2}\right)$ in terms of $\sin(\alpha)$ or $\cos(\alpha)$. Repeat for $\cos\left(\alpha + \frac{\pi}{2}\right)$.

d. Find the exact value for each expression below.

  i. $\cos\frac{9\pi}{12}\cos\frac{5\pi}{12} - \sin\frac{9\pi}{12}\sin\frac{5\pi}{12}$

  ii. $\sin\frac{7\pi}{12}$

Extending the Family of Trigonometric Functions

Up to this point in *Core-Plus Mathematics*, you have studied the three basic trigonometric functions, sine, cosine, and tangent. In this investigation, three additional trigonometric functions are defined, and some of their properties are examined.

As you work on the problems in this investigation, look for answers to the following questions:

What are the three reciprocal trigonometric functions?

What are the characteristics of the graphs of these functions?

What are the fundamental trigonometric identities involving these functions?

In Course 2, the trigonometric functions cosine, sine, and tangent were defined in terms of the coordinates of a point $A(x, y)$ on the terminal side of an angle in standard position.

$$\cos \theta = \frac{x}{r} \qquad \sin \theta = \frac{y}{r} \qquad \tan \theta = \frac{y}{x}, x \neq 0$$

where $r = \sqrt{x^2 + y^2}$.

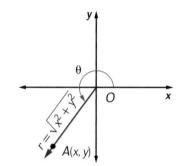

1. What other ratios can be formed using the coordinates of point A above? For each ratio, indicate any restrictions on the coordinates x and y.

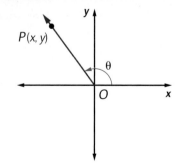

Secant, Cosecant, and Cotangent The three ratios you found in Problem 1 can be used to define three additional trigonometric functions. If $P(x, y)$ is any point (other than the origin) on the terminal side of an angle in standard position and $r = \sqrt{x^2 + y^2}$, then:

$$\text{secant of } \theta = \sec \theta = \frac{r}{x}, x \neq 0$$

$$\text{cosecant of } \theta = \csc \theta = \frac{r}{y}, y \neq 0$$

$$\text{cotangent of } \theta = \cot \theta = \frac{x}{y}, y \neq 0$$

2 Assume that θ is an angle in standard position in a coordinate plane and that $P(x, y)$ is a point on the terminal side of the angle.

a. Sketch θ and find $\csc \theta$, $\sec \theta$, and $\cot \theta$ when the coordinates of point P are as indicated.

    i. $(3, -2)$ **ii.** $(-4, -3)$

    iii. $(2, 5)$ **iv.** $(-3, 4)$

b. Recall that the value of a trigonometric function is positive or negative depending on the quadrant in which the terminal side of the angle lies. For each quadrant in which the terminal side of an angle θ may lie, determine:

- the corresponding range of angle measures for θ.

- whether $\sin \theta$, $\cos \theta$, $\tan \theta$, $\csc \theta$, $\sec \theta$, and $\cot \theta$ are positive or negative.

Summarize your findings in a chart similar to the one below in which answers for Quadrant I have been filled in.

Quadrant II

$90° < \theta < \underline{}°$

$\underline{} < \theta < \underline{}$ radians

$\sin \theta > 0$
$\cos \theta \underline{} 0$
$\tan \theta \underline{} 0$
$\csc \theta \underline{} 0$
$\sec \theta \underline{} 0$
$\cot \theta \underline{} 0$

Quadrant I

$0° < \theta < 90°$

$0 < \theta < \dfrac{\pi}{2}$ radians

$\sin \theta > 0$
$\cos \theta > 0$
$\tan \theta > 0$
$\csc \theta > 0$
$\sec \theta > 0$
$\cot \theta > 0$

Quadrant III

$\underline{}° < \theta < \underline{}°$

$\underline{} < \theta < \underline{}$ radians

$\sin \theta \underline{} 0$
$\cos \theta \underline{} 0$
$\tan \theta \underline{} 0$
$\csc \theta \underline{} 0$
$\sec \theta \underline{} 0$
$\cot \theta \underline{} 0$

Quadrant IV

$\underline{}° < \theta < 360°$

$\underline{} < \theta < 2\pi$ radians

$\sin \theta \underline{} 0$
$\cos \theta \underline{} 0$
$\tan \theta \underline{} 0$
$\csc \theta \underline{} 0$
$\sec \theta \underline{} 0$
$\cot \theta \underline{} 0$

c. The chart in Part b does not include values of the six trigonometric functions when the terminal side of θ is on the *x*- or *y*-axis.

    i. For what values of θ, $0 \leq \theta \leq 360°$, is its terminal side on the *x*-axis? Express these values in degrees and in radians. What are the values of each of the six functions for these angle measures?

    ii. For what values of θ, $0 \leq \theta \leq 360°$, is its terminal side on the *y*-axis? Express these values in degrees and in radians. What are the values of the six functions for these angle measures?

3 Now investigate how knowing the value of one of the trigonometric functions of an angle θ in standard position permits you to find values of the other five trigonometric functions. Assume the terminal side of θ is in the first quadrant. You may find it helpful to draw sketches to aid in your reasoning.

    a. If $\sin \theta = \frac{3}{5}$, find $\cos \theta$, $\tan \theta$, $\sec \theta$, $\csc \theta$, and $\cot \theta$.

    b. If $\cot \theta = \frac{3}{4}$, find the values of the other five trigonometric functions of θ.

    c. If $\sec \theta = \sqrt{10}$, find the values of the other five trigonometric functions of θ.

4 In Problem 3, the terminal side of θ was in the first quadrant.

    a. Find another quadrant in which $\sin \theta = \frac{3}{5}$. Then find the values of the other five trigonometric functions when the terminal side of θ is in that quadrant.

    b. Find another quadrant in which $\cot \theta = \frac{3}{4}$. Then find the values of the other five trigonometric functions when the terminal side of θ is in that quadrant.

    c. Find another quadrant in which $\sec \theta = \sqrt{10}$. Then find the values of the other five trigonometric functions when the terminal side of θ is in that quadrant.

5 The ratios used to define the secant, cosecant, and cotangent functions are the *reciprocals* of the ratios used to define the cosine, sine, and tangent functions, respectively.

    a. Explain why $\cos \theta \sec \theta = 1$. Why $\sin \theta \csc \theta = 1$. Why $\tan \theta \cot \theta = 1$.

    b. Use the equations in Part a to derive the following *reciprocal identities*.

        i. $\sec \theta = \frac{1}{\cos \theta}$, $\cos \theta \neq 0$

        ii. $\csc \theta = \frac{1}{\sin \theta}$, $\sin \theta \neq 0$

        iii. $\cot \theta = \frac{1}{\tan \theta}$, $\tan \theta \neq 0$

    c. State the **reciprocal identities** in words.

    d. Explain why $\cot \theta = \frac{\cos \theta}{\sin \theta}$, $\sin \theta \neq 0$.

6 Now explore the graph of $y = \sec\theta$ for $-2\pi \le \theta \le 2\pi$.

a. The graph of $y = \cos\theta$ is shown below for $-2\pi \le \theta \le 2\pi$. As the values of the cosine function increase (decrease), what happens to the values of the secant function?

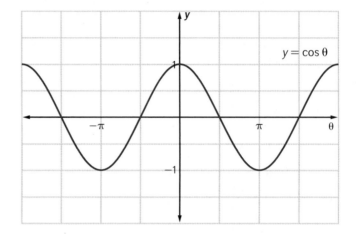

b. On a copy of the graph of $y = \cos\theta$, sketch the graph of its reciprocal function $y = \sec\theta$. Identify common points on the graphs.

c. Identify the values of θ, $-2\pi \le \theta \le 2\pi$, for which $y = \sec\theta$ is not defined. How is this revealed in the definition of the function? How is it seen in the graph of the function?

d. Describe the range of $y = \sec\theta$.

e. Is the function $y = \sec\theta$ periodic? If so, give the period. Why does this make sense in terms of the definition of the function?

f. Explain how you can quickly sketch the graph of $y = \sec\theta$ by remembering the shape of, and key points on, the graph of $y = \cos\theta$.

7 Repeat Problem 6 for the function $y = \sin\theta$ and its reciprocal function $y = \csc\theta$.

8 Repeat Problem 6 for the function $y = \tan\theta$ and its reciprocal function $y = \cot\theta$.

9 Reasoning with the reciprocals of the sine, cosine, and tangent functions is similar to reasoning with these trigonometric functions themselves.

a. Sketch a right triangle with acute angle θ and $\sec\theta = 2$. Evaluate the remaining five trigonometric functions of θ.

b. Sketch two different position vectors $\overrightarrow{OA}$ and $\overrightarrow{OB}$ with direction angles α and β, respectively, and $\cot\alpha = \cot\beta = -1$. Give the coordinates of points A and B. Find the remaining trigonometric function values for α and for β.

c. Sketch two different position vectors $\overrightarrow{OA}$ and $\overrightarrow{OB}$ with direction angles α and β, respectively, and $\csc \alpha = \csc \beta = -2$. Give the coordinates of points A and B. Then find the remaining trigonometric function values for α and for β.

d. Find all the solutions to each equation below on the interval $0 \leq \theta \leq 2\pi$.

    i. $\sec \theta = -2$

    ii. $\csc \theta = 4$

10 In Investigation 1, you proved the identity $\sin^2 \theta + \cos^2 \theta = 1$. Other similar Pythagorean identities involving the reciprocal functions can also be proved.

a. Prove in two different ways that $\sec^2 \theta - \tan^2 \theta = 1$ is an identity.

    i. Begin by writing secant and tangent in terms of the sine and cosine functions.

    ii. Begin with the identity $\sin^2 \theta + \cos^2 \theta = 1$.

b. Is $\tan^2 \theta + 1 = \sec^2 \theta$ an identity? Explain how you know.

There is one more Pythagorean identity relating the cotangent and cosecant functions: $\cot^2 \theta + 1 = \csc^2 \theta$. You will be asked to prove this identity in the Check Your Understanding.

    You now have at your disposal eight fundamental identities that you can use to rewrite trigonometric expressions in equivalent forms. They are summarized below. In addition, you have an extensive toolkit of other identities, including opposite angle identities, cofunction identities, angle sum and difference identities, and double-angle identities, to use in rewriting trigonometric expressions in equivalent forms and in proving other identities.

Fundamental Identities

Reciprocal Identities	**Quotient Identities**	**Pythagorean Identities**
$\csc \theta = \dfrac{1}{\sin \theta}$, $\sin \theta \neq 0$	$\tan \theta = \dfrac{\sin \theta}{\cos \theta}$, $\cos \theta \neq 0$	$\sin^2 \theta + \cos^2 \theta = 1$
$\sec \theta = \dfrac{1}{\cos \theta}$, $\cos \theta \neq 0$	$\cot \theta = \dfrac{\cos \theta}{\sin \theta}$, $\sin \theta \neq 0$	$1 + \tan^2 \theta = \sec^2 \theta$, $\cos \theta \neq 0$
$\cot \theta = \dfrac{1}{\tan \theta}$, $\tan \theta \neq 0$		$1 + \cot^2 \theta = \csc^2 \theta$, $\sin \theta \neq 0$

11 In Unit 7 of *Core-Plus Mathematics* Course 2, you were introduced to the trigonometric functions sine, cosine, and tangent in terms of angles in standard position and for acute angles in a right triangle. In Unit 6 of Course 3, the circular function interpretations of the cosine and sine were developed. The identities you have been deriving in this lesson are referred to as "trigonometric identities." Do these identities also hold for the circular functions cosine and sine and for the derived functions tangent, secant, and cosecant? Explain your reasoning.

12   For each statement below, decide whether or not it is an identity. If it is an identity, use reasoning to prove it. If it is not, provide a counterexample.

   a. $\csc \theta - \cos \theta \cot \theta = \sin \theta$

   b. $\dfrac{\tan^2 \theta + 1}{\tan^2 \theta} = \csc^2 \theta$

SUMMARIZE THE MATHEMATICS

In this investigation, you extended the family of trigonometric functions to include the secant, cosecant, and cotangent functions.

a How are the secant, cosecant, and cotangent functions defined?

b How would you define the secant, cosecant, and cotangent functions of an acute angle in a right triangle?

c For what values of θ, $-360° \le \theta \le 360°$ $(-2\pi \le \theta \le 2\pi)$, are each of the six trigonometric functions defined?

d Explain how sketches of the graphs of $y = \sec \theta$, $y = \csc \theta$, and $y = \cot \theta$ can be obtained from examination of the graphs of $y = \cos \theta$, $y = \sin \theta$, and $y = \tan \theta$, respectively.

Be prepared to share your ideas and reasoning with your classmates.

 CHECK YOUR UNDERSTANDING

Use the definitions of the six trigonometric functions and previously proven identities to help complete the following tasks.

a. Suppose $\sin \theta = \dfrac{12}{13}$ and $\sec \theta = -\dfrac{13}{5}$. How many values of θ between 0 and 2π satisfy both equations? Find all such θ.

b. Prove the Pythagorean identity: $\cot^2 \theta + 1 = \csc^2 \theta$

c. Prove the identity: $(1 - \sin \theta)(1 + \sin \theta) = \dfrac{1}{1 + \tan^2 \theta}$

d. Express $\dfrac{\sin^2 \theta \sec \theta}{2 \tan \theta}$ in terms of $\sin \theta$.

APPLICATIONS

1 In Investigation 1, you used the symmetry of a circle centered at the origin of a coordinate system to prove opposite-angle identities and identities such as $\sin(\pi - \theta) = \sin\theta$. Use reasoning based on symmetry and the values $\sin\frac{\pi}{12} \approx 0.26$, $\cos 70° \approx 0.34$, and $\tan\frac{5\pi}{12} \approx 3.73$ to find values of each of the following.

a. $\sin\frac{11\pi}{12}$

b. $\cos 250°$

c. $\tan\frac{7\pi}{12}$

d. $\sin\left(-\frac{\pi}{12}\right)$

e. $\cos 430°$

f. $\tan\left(-\frac{5\pi}{12}\right)$

2 For each statement, decide whether or not it is an identity. If a statement is an identity, use reasoning to prove it. If it is not, provide a counterexample.

a. $(1 + \sin\theta)(1 - \sin\theta) = \cos^2\theta$

b. $\tan\theta \sin\theta = 1 - \cos\theta$

c. $\sin\theta\left(\frac{1}{\sin\theta} - \sin\theta\right) = \cos^2\theta$

d. $\cos^2\theta - \sin^2\theta = 1 - 2\sin^2\theta$

e. $\dfrac{\sin^2\theta}{\tan\theta\cos\theta} = \tan\theta\cos\theta$

3 Identities can often be used to rewrite trigonometric expressions in an equivalent form involving only one trigonometric function.

a. Rewrite $2\cos^2 x - \sin^2 x$ in an equivalent form involving only the sine function. In an equivalent form involving only the cosine function.

b. Rewrite $\dfrac{\cos\theta - \sin\theta}{\cos\theta}$ in an equivalent form involving only the tangent function.

4 A frequently used formula in the electromagnetic wave theory of light is

$$E'' = -E\left(\frac{k\cos r - \cos i}{k\cos r + \cos i}\right),$$

where E'' is the amplitude of reflected light, E is the amplitude of incident light, k is the index of refraction, i is the angle of incidence, r is the angle of refraction, and $k = \frac{\sin i}{\sin r}$. Prove that the above formula is equivalent to

$$E'' = -E\left(\frac{\sin(i - r)}{\sin(i + r)}\right).$$

5 Prove each identity, and indicate conditions on α and β, if any, for which the identity is undefined.

a. $\cos(\alpha + \beta) - \cos(\alpha - \beta) = -2\sin\alpha\sin\beta$

b. $\cos(\alpha + \beta)\cos(\alpha - \beta) = \cos^2\alpha - \sin^2\beta$

c. $\sin(\alpha + \beta) + \sin(\alpha - \beta) = 2\sin\alpha\cos\beta$

d. $\sin(\alpha + \beta)\sin(\alpha - \beta) = \sin^2\alpha - \sin^2\beta$

6 The sum identities for the sine and cosine functions can be used to derive a similar identity for the tangent function:

$$\tan(\alpha + \beta) = \frac{\tan\alpha + \tan\beta}{1 - \tan\alpha\tan\beta}$$

a. Study the following derivation of the **sum identity for the tangent function**. Provide additional steps if needed to convince yourself that the symbolic reasoning is sound.

$$\tan(\alpha + \beta) = \frac{\sin(\alpha + \beta)}{\cos(\alpha + \beta)} \qquad (1)$$

$$= \frac{\sin\alpha\cos\beta + \cos\alpha\sin\beta}{\cos\alpha\cos\beta - \sin\alpha\sin\beta} \qquad (2)$$

$$= \frac{\dfrac{\sin\alpha\cos\beta + \cos\alpha\sin\beta}{\cos\alpha\cos\beta}}{\dfrac{\cos\alpha\cos\beta - \sin\alpha\sin\beta}{\cos\alpha\cos\beta}} \qquad (3)$$

$$= \frac{\tan\alpha + \tan\beta}{1 - \tan\alpha\tan\beta} \qquad (4)$$

b. Look back at Step 3 in the derivation above. Why was the choice of dividing the numerator and the denominator by $\cos\alpha\cos\beta$ a wise choice?

c. What restrictions are there on α, β, and $\alpha + \beta$?

d. How could you use the tangent sum identity to prove that
$$\tan(\alpha - \beta) = \frac{\tan\alpha - \tan\beta}{1 + \tan\alpha\tan\beta}?$$

7 Screw jacks used for jacking up houses or heavy machinery must be designed so that they do not turn down under the load. The effort F necessary to obtain equilibrium on a screw jack with a pitch angle α and load W is given by the formula:

$$F = \frac{Wr}{a}\tan(\alpha - \theta).$$

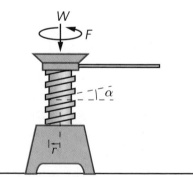

a. For what value of θ, the angle of friction, will no effort be required to keep the jack from turning down under the weight of the load?

b. If $\alpha = 45°$, express F as a function of θ. W, r, and a are constants in a particular context.

8 In the diagram at the right, the effort F necessary to hold a heavy object like a safe in position on a ramp is given by

$$F = \frac{W(\sin \alpha + \mu \cos \alpha)}{\cos \alpha - \mu \sin \alpha} \text{ where}$$

W is the weight of the safe and μ is the coefficient of friction. If θ is the angle of friction and $\mu = \tan \theta$, show that $F = W \tan (\alpha + \theta)$.

9 Use exact values of trigonometric functions of special angles and appropriate identities to determine exact values of the following.

a. $\sin \dfrac{5\pi}{12}$

b. $\cos 255°$

c. $\sin 160° \cos 70° - \cos 160° \sin 70°$

10 In Investigation 2 (page 279), you proved the *double-angle identity* for the sine function, $\sin 2\theta = 2 \sin \theta \cos \theta$.

a. Prove that $\cos 2\theta$ is equivalent to each of the following expressions.

   i. $\cos^2 \theta - \sin^2 \theta$

   ii. $2 \cos^2 \theta - 1$

   iii. $1 - 2 \sin^2 \theta$

b. Derive the double-angle identity for the tangent function.

11 For each statement below, prove or disprove it is an identity.

a. $\tan \theta = \dfrac{\sin 2\theta}{1 + \cos 2\theta}$

b. $\cos 2\theta = \dfrac{1 - \tan^2 \theta}{1 + \tan^2 \theta}$

c. $\tan \alpha = \cot 2\alpha - \csc 2\alpha$

d. $\sin 2\alpha = \dfrac{2 \tan \alpha}{1 + \tan^2 \alpha}$

12 Prove each of these identities.

a. $\tan \theta = \dfrac{\sec \theta}{\csc \theta}$. For what values of θ is this identity valid?

b. $\sec \theta \cot \theta = \csc \theta$. For what values of θ is this identity valid?

13 Use reasoning and the definitions of the trigonometric functions to complete the following tasks.

a. Suppose $\cos \theta = \dfrac{12}{13}$. How many solutions are there to this equation where $-\dfrac{\pi}{2} \le \theta < \dfrac{\pi}{2}$? Find the exact solutions.

b. For each solution in Part a, evaluate the five remaining trigonometric functions of θ.

c. If $\csc \theta = -\dfrac{5}{4}$ and $-\dfrac{\pi}{2} \le \theta < \dfrac{\pi}{2}$, evaluate the five remaining trigonometric functions of θ.

14 In applications involving trigonometric functions, it is often useful to rewrite trigonometric expressions in equivalent forms—optimally, forms using only one trigonometric function. This task will provide additional practice with reasoning strategies that are helpful in such situations.

a. Rewrite $\dfrac{\sec\theta + 1}{\sec\theta}$ in terms of $\cos\theta$.

b. Rewrite $\dfrac{1}{1 + \cot^2\theta}$ in terms of $\cos\theta$.

c. Rewrite $\dfrac{\sec\theta\sin\theta}{\tan\theta + \cot\theta}$ in terms of $\sin\theta$.

d. Rewrite $\dfrac{2\tan\theta}{1 + \tan^2\theta}$ in terms of $\sin 2\theta$.

CONNECTIONS

15 Recall that a function f is called an *even function* provided $f(-x) = f(x)$ for all values of x in the domain of f. A function g is called an *odd function* provided $g(-x) = -g(x)$ for all values of x in the domain of g.

a. Explain using the unit circle why the cosine function is an even function.

b. Explain using the unit circle why the sine function is an odd function.

c. Is the tangent function an even function, an odd function, or neither? Justify your answer.

d. Describe the symmetry of the graph of any even trigonometric function. Of any odd trigonometric function. Why does this make sense?

16 Professional golfer Paula Creamer drives a golf ball at about 147 ft/sec at an angle of about 29°.

a. Write parametric equations that describe the horizontal and vertical components of the position of the ball for any time t.

b. Determine the horizontal distance the ball travels before it hits the ground using the three methods outlined below.

Method I Use the parametric equations to analyze the path of the ball.

Method II Use the range formula that is given at the beginning of this lesson, namely $R(\theta) = \dfrac{V^2\cos\theta\sin\theta}{16}$.

Method III Use the equivalent formula, $R(\theta) = \dfrac{V^2\sin 2\theta}{32}$, derived in Problem 5 on page 278.

Professional golfer Paula Creamer

17 In this task, you will need to connect two ideas: finding values of all trigonometric functions of an angle when one value is given and applying sum or difference identities. Suppose you know that $\cos \alpha = \frac{3}{5}$, $\cos \beta = -\frac{5}{13}$, and α and β are first- and second quadrant angles, respectively. Without using technology, determine fractional values for the following. It may helpful to make a sketch on a coordinate grid to aid your thinking.

  a. $\cos (\alpha - \beta)$

  b. $\sin (\alpha + \beta)$

18 In Course 2, you used the Law of Sines and the Law of Cosines to determine unknown measures of sides and angles of triangles. Use these laws and appropriate identities from this lesson to prove the following statements.

  Assume $\triangle ABC$ is any triangle with side lengths a, b, and c opposite $\angle A$, $\angle B$, and $\angle C$, respectively.

  a. If $m\angle B = 2(m\angle A)$, then $b = 2a \cos A$.

  b. If $m\angle B = 2(m\angle A)$, then $b^2 = (a + c)^2 - 4ac \cos^2 A$.

19 In Investigation 1, you proved that $\cos (\pi + \theta) = -\cos \theta$. But you might have asked yourself what clues may have suggested that this was an identity?

  a. Explain why this identity is suggested by the unit circle interpretation of the cosine function.

  b. Now think about the relationship in terms of function graphs.

   i. How are the graphs of $y = \cos (\pi + \theta)$ and $y = \cos \theta$ related?

   ii. Why is it reasonable that the graph of $y = \cos (\pi + \theta)$ is the same as the graph of $y = -\cos \theta$?

  c. Use graphical reasoning to help you write an identity statement for $\csc (\pi + \theta)$.

   i. Does your identity make sense in terms of the unit circle interpretation of the trigonometric function?

   ii. Prove your identity.

REFLECTIONS

20 What do you need to do to prove that a statement is *not* an identity? To prove that a statement is an identity?

21 Why must proof of an identity be based on reasoning with definitions and other identities rather than reasoning from graphs or tables? What role might tables and graphs play in exploring identities?

22 In solving an equation, you frequently add (subtract) or multiply (divide) both sides of the equation by the same number. Why can you not use this algebraic procedures in proving an identity?

23 In applied fields where exact measurement of small angles is not possible, linear approximations are often used. Assume that angles can be measured accurately in radians to two decimal places.

    a. The approximation $\sin \theta \approx \theta$ is used in the field of optics. If approximations must be correct to two decimal places, find the largest acute angle (in radians) for which the relationship holds.

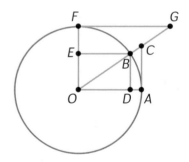

    b. The approximation $\tan \theta \approx \theta$ is used in the field of mechanics. If approximations must be correct to two decimal places, find the largest acute angle for which this relationship holds.

    c. Why are the relationships in Parts a and b not identities?

24 Look back at Problem 2 Part d of Investigation 2 (page 277). How could you show that $\cos\left(\dfrac{3\pi}{2} - \theta\right) = -\sin \theta$ using a line reflection and the symmetry of a circle centered at the origin of a coordinate system?

25 Sometimes it is suggested to students that a good strategy for proving a trigonometric identity is to first rewrite all functions using only sines and cosines. Why might this seem like a useful idea? A poor idea?

26 The circle shown on the right is a unit circle, that is, its radius is $OA = OB = OF = 1$. Thus, $OD = EB = \cos \angle AOB$ and $OE = DB = \sin \angle AOB$. Identify a labeled segment whose length corresponds to each of the other four trigonometric functions of $\angle AOB$. Explain your reasoning.

27 Without graphing, describe the graph of the function $f(x) = \sin x \csc x + \cos x \sec x + \tan x \cot x$. What is the domain of $f(x)$?

EXTENSIONS

28 A two-phase alternating current circuit is supplied power by two generators. The voltages produced by the two generators are given by

$$v_1 = V_p \cos\left(\theta + \frac{\pi}{p}\right) \text{ and } v_2 = V_p \cos\left(\theta - \frac{\pi}{p}\right),$$

where V_p is the peak voltage and $p > 0$ is a constant. The average voltage, $\dfrac{v_1 + v_2}{2}$, is reported as $\dfrac{v_1 + v_2}{2} = V_p \cos \theta \cos \dfrac{\pi}{p}$. Prove that the reported average is correct.

29 When a point $A(x, y)$ is rotated through an angle θ about the origin to the point $A'(x', y')$, the new coordinates are $x' = x \cos \theta - y \sin \theta$ and $y' = x \sin \theta + y \cos \theta$. This can be represented in matrix form as follows:

$$R_\theta \begin{bmatrix} x \\ y \end{bmatrix} = \begin{bmatrix} \cos \theta & -\sin \theta \\ \sin \theta & \cos \theta \end{bmatrix} \begin{bmatrix} x \\ y \end{bmatrix} = \begin{bmatrix} x' \\ y' \end{bmatrix}$$

The matrix R_θ is a *rotation matrix*. A rotation of θ followed by a rotation of ϕ can be represented in matrix form by the product $R_\phi R_\theta$.

a. What is the full angle of rotation that is the composite of a rotation of θ followed by a rotation of ϕ?

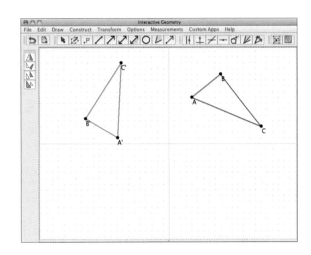

b. According to the matrix definition above, what is the matrix $R_{\phi + \theta}$?

c. $R_{\phi + \theta} = R_\phi R_\theta$. Use matrix multiplication on the right to carry out the composition.

d. How can Parts b and c be used to derive identities for $\sin (\phi + \theta)$ and for $\cos (\phi + \theta)$?

e. How could matrices be used to derive identities for $\sin (\phi - \theta)$ and for $\cos (\phi - \theta)$?

30 In mathematics, there is often more than one way to prove a statement is true. Yihnan David Gau proposed the following visual proof of the identity $\sin 2t = 2 \sin t \cos t$. (**Source:** "Proof Without Words: Double Angle Formulas," *Mathematics Magazine*, Vol. 71, no. 5 (December 1998), p. 385.)

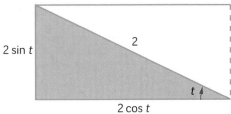

$$2 \sin t \cos t \ = \ \sin 2t$$

a. Explain how the diagrams prove the identity.

b. For what values of t is this visual proof valid?

31 Use the double-angle formulas for the sine and cosine functions to derive the following **half angle identities**. The $\pm$ sign indicates that the values are sometimes positive and sometimes negative, depending on the size of α. In each case, start by considering $\cos \alpha = \cos \dfrac{2\alpha}{2}$.

a. $\cos \dfrac{\alpha}{2} = \pm\sqrt{\dfrac{1 + \cos \alpha}{2}}$

b. $\sin \dfrac{\alpha}{2} = \pm\sqrt{\dfrac{1 - \cos \alpha}{2}}$

c. How would you decide whether to use the positive or negative value of the expression?

32 For each statement, prove or disprove it is an identity.

a. $\csc 2\alpha = \dfrac{\sec \alpha \csc \alpha}{2}$

b. $\sec 2\alpha = \dfrac{\sec^2 \alpha}{2 - \sec^2 \alpha}$

c. $\cot 2\alpha = \dfrac{1 + \cos 4\alpha}{\sin 4\alpha}$

d. $\sin 3\alpha = 3 \sin \alpha - 4 \sin^3 \alpha$

e. $\dfrac{\sin 2\alpha}{\sin \alpha} - \dfrac{\cos 2\alpha}{\cos \alpha} = \sec \alpha$

f. $\sec^4 \theta - \tan^4 \theta = \sec^2 \theta + \tan^2 \theta$

g. $\cos 4\alpha = 8 \cos^4 \alpha - 8 \cos^2 \alpha + 1$

h. $\sin 4\alpha = 4 \sin \alpha \cos \alpha (2 \cos^2 \alpha - 1)$

REVIEW

33 Consider the point P with coordinates (a, b). Determine the coordinates of the image point P' under each transformation.

a. Reflection across the x-axis

b. Reflection across the y-axis

c. Rotation of $180°$ centered at the origin

d. Reflection across the line $y = x$

e. Reflection across the line $y = 5$

34 Copy and complete each statement.

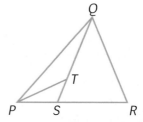

a. $\overrightarrow{QR} + \overrightarrow{RS} = \underline{\quad ? \quad}$

b. $\overrightarrow{QR} + \overrightarrow{RP} + \overrightarrow{PT} = \underline{\quad ? \quad}$

c. $\overrightarrow{PT} + \overrightarrow{QP} + \underline{\quad ? \quad} = \overrightarrow{QS}$

35 Rewrite each expression in equivalent form as a single algebraic fraction. Then simplify the result as much as possible.

a. $\dfrac{1}{b} - b$

b. $1 - \dfrac{x}{y}$

c. $\dfrac{1}{a} + \dfrac{a}{b}$

d. $y\left(\dfrac{1}{y} - y\right)$

e. $\dfrac{1}{ab} - \dfrac{a}{b}$

f. $\dfrac{1}{a^2} - \dfrac{1}{b^2}$

36 Consider the position vectors $\vec{u} = (-2, 7)$ and $\vec{v} = (3, 4)$.

a. Find the magnitude and direction of each vector.

b. Find the magnitude and direction of $\vec{u} + \vec{v}$ and $\vec{u} - \vec{v}$.

c. Give the coordinate representations of $3\vec{u}$ and $-2\vec{v}$.

37 In $\triangle ABC$ shown here, $\overline{BD}$ is an altitude.

a. Give reasons justifying the following statements.

i. $h = a \sin C$

ii. *area* $\triangle ABC = \frac{1}{2} ab \sin C$

iii. *area* $\triangle ABC = \frac{1}{2} bc \sin A$

iv. *area* $\triangle ABC = \frac{1}{2} ac \sin B$

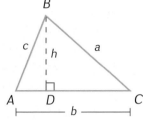

b. Write in words a description of how to calculate the area of a triangle using the sine function.

c. Find the area of $\triangle PQR$ if $p = 10$, $q = 14$, and $m\angle R = 73°$.

38 The expression $\frac{a}{b} \div \frac{c}{d}$ can also be written $\dfrac{\frac{a}{b}}{\frac{c}{d}}$. For example, $\dfrac{\frac{3}{2}}{\frac{6}{5}} = \frac{3}{2} \div \frac{6}{5} = \frac{3}{2} \cdot \frac{5}{6} = \frac{5}{4}$. Use this relationship to rewrite each expression in equivalent form as a single algebraic fraction. Then simplify the result as far as possible.

a. $\dfrac{\frac{1}{2}}{\frac{5}{k}}$

b. $\dfrac{\frac{1}{3}}{x}$

c. $\dfrac{\frac{1}{a^2}}{\frac{1}{a}}$

d. $\dfrac{\frac{x+y}{y}}{\frac{x-y}{y}}$

e. $\dfrac{1 - \frac{a}{b}}{\frac{1}{b}}$

f. $\dfrac{\frac{1}{x^2}}{2 - \frac{1}{x^2}}$

39 Rewrite each expression in standard polynomial form.

a. $(x - 4)(x + 4)$

b. $(6 - 3y)^2$

c. $(a - 2b)(a + 2b)$

d. $(x + 5)(x^2 - 5x + 25)$

e. $6x + 3(x - 1)(2x + 5)$

f. $(3x + 5)^2 - (x + 1)(x + 3)$

40 Without using technology, sketch each function on a coordinate grid and identify the range of each function. Then use your calculator or computer to check your answers.

a. $y = 3 \sin x$

b. $y = \sin x + 3$

c. $y = 2 \cos x - 1$

d. $y = \cos (x - \pi)$

41 Solve each equation for x.

a. $x^2 - x - 2 = 0$

b. $4x^2 - 12 = 69$

c. $9x^3 - x = 0$

d. $6x^2 + 5x = -1$

Solving Trigonometric Equations

In your prior studies, you saw that variations of the sine (or cosine) function could be used to model cyclical patterns such as water depth in a harbor, height of a rider on a Ferris wheel, or number of hours between sunrise and sunset at various locations on Earth.

Using the Spring Equinox (March 21) as a starting point, the number of hours of sunlight in Chicago, Illinois, can be modeled by the function $S(t) = 3 \sin \frac{2\pi}{365} t + 12$, where t is time in days after March 21. (This assumes a year of 365 days.)

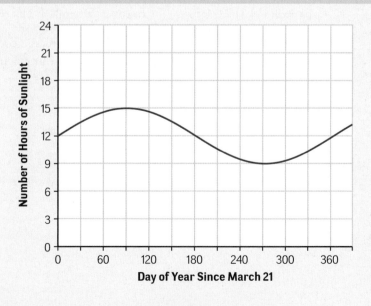

Number of Hours of Sunlight

Day of Year Since March 21

As is the case for algebraic equations, you can estimate the solutions of trigonometric equations by using tables and graphs, or you can often use symbolic reasoning strategies or a computer algebra system to find more exact solutions. In this lesson, you will learn how to solve trigonometric equations using strategies similar to those used to solve linear and quadratic equations. However, you will also draw upon your knowledge of the periodicity of trigonometric functions, and, in some cases, your knowledge of basic trigonometric identities.

INVESTIGATION **1**

Solving Linear Trigonometric Equations

In the Course 3 unit *Inverse Functions*, you solved equations like $5 \sin x = 3$, or, more generally, equations of the form $aT(x) = b$, where $T(x)$ is the sine, cosine, or tangent function and $a \neq 0$. In this investigation, you will build on that knowledge in two ways. You will learn strategies for solving the same sorts of equations and more complex equations like $2 \sin \left(2x + \frac{\pi}{6}\right) = \sqrt{3}$.

As you complete this investigation, look for answers to these two related questions:

> *Given an equation $aT(bx + c) = d$, where $T(x)$ is a trigonometric function and $a \neq 0$, $b \neq 0$, how can you tell if solutions exist?*
>
> *If solutions exist, how can you find them?*

The question of whether a trigonometric equation has a solution and, if so, how many, can often be answered by examining the equation and using your understanding of the trigonometric functions involved.

1 By examining the symbolic form of each equation below, indicate whether or not a solution exists and explain your reasoning. You do not need to solve these equations.

a. $3 \sin x = -1$ b. $3 \cos x = -4$

c. $3 \tan x = -1$ d. $3 \tan x = -4$

e. $3 \sec x = -1$ f. $3 \csc x = -4$

2 Suppose you are given a trigonometric equation in the form $aT(x) = b$ with $a \neq 0$. What conditions on a and b ensure that $aT(x) = b$ has a solution in each of the following cases?

a. $a \sin x = b$ b. $a \sec x = b$

c. $a \cot x = b$

3 Now consider how to solve an equation that is linear in $\sin x$, such as $2 \sin x + 5 = 6$.

a. Does this equation have a solution? Explain your reasoning.

b. Use algebraic reasoning and the appropriate inverse trigonometric function to find one exact solution of $2 \sin x + 5 = 6$ in the interval $0 \leq x < 2\pi$.

c. Does the equation have additional solutions on the interval $0 \leq x < 2\pi$? If so, find all such solutions and record them in exact radian form. Be prepared to explain your solution method.

d. Suppose you wished to solve $2 \sin x + 5 = 6$ over the entire domain of $f(x) = \sin x$.

    i. Explain why $\frac{\pi}{6} + 2\pi n$, for any integer n, describes some of the solutions.

    ii. Write an expression that represents the remaining solutions.

4 Consider similar equations containing other trigonometric functions. Find one solution (expressed in degrees) of each equation. Then write one or two expressions giving all possible solutions in degrees.

a. $2 \tan x + 4 = 3$

b. $2 \sec x = 3$

c. How do the periods of the tangent and secant functions help in writing expressions that give all possible solutions?

Using Inverse Trigonometric Functions In the previous two problems, you found all the solutions to three trigonometric equations. The equation $2 \sin x + 3 = 4$ has two solutions that occur in the interval $0 \le x < 2\pi$, namely $\frac{\pi}{6}$ and $\frac{5\pi}{6}$. You then used each of these and the fact that the period of the sine function is 2π to write two expressions that together represent *all* possible solutions.

For the equation $2 \tan x + 4 = 3$, you found one solution in the interval $-90° < x < 90°$, for example $-26.6°$. You then wrote an expression giving all solutions by adding and subtracting whole number multiples of the period of the tangent, $180°$.

The key to solving a trigonometric equation of the form $aT(x) + b = c$, $a \ne 0$, is to try to find one or two solutions in an interval of length one period. If $T(x)$ is a reciprocal trigonometric function, the equation can always be transformed to an equivalent equation $aS(x) + b = c$, where $S(x)$ is the sine, cosine, or tangent function. If a solution exists, one can always be found by using the appropriate inverse function on your calculator. You can then use that solution to generate expressions for all solutions. Graphs can help guide your reasoning.

As you saw in the *Inverse Functions* unit of Course 3, the inverse sine, inverse cosine, and inverse tangent functions are defined only if their ranges are restricted. The three inverse trigonometric functions on your calculator will produce solutions in the ranges shown below.

Ranges for Inverse Trigonometric Functions

$$-90° \le \sin^{-1} x \le 90° \text{ or } -\frac{\pi}{2} \le \sin^{-1} x \le \frac{\pi}{2}$$

$$0° \le \cos^{-1} x \le 180° \text{ or } 0 \le \cos^{-1} x \le \pi$$

$$-90° < \tan^{-1} x < 90° \text{ or } -\frac{\pi}{2} < \tan^{-1} x < \frac{\pi}{2}$$

5 Next, consider the equation $2 \cos x + \sqrt{3} = 0$.

   a. Determine if the equation has a solution for x.

   b. Find two exact solutions in degrees that are in the interval $0° \le x < 360°$.

   c. Use your two solutions to write expressions giving all solutions.

6 Find all solutions to each trigonometric equation. When possible, give exact solutions. Express solutions in both degrees and in radians.

   a. $3 \sin x - 2 = 1$

   b. $5 \cos x + 1 = 3$

   c. $2 \cot x - 3 = 2$

   d. $3 \sec x = 2$

NOTE For the remainder of this unit, express solutions to trigonometric equations in radians, unless the context or given equation suggests otherwise.

7 You now have symbolic reasoning strategies to solve equations of the form $aT(x) + b = c, a \neq 0$, where $T(x)$ is a trigonometric function. Next, consider strategies for solving equations of the form $aT(bx + d) = c$, where $a \neq 0$, $b \neq 0$, and $T(x)$ is a trigonometric function. In particular, consider the equation $\sin 2x = \frac{1}{2}$.

a. Explain why this equation is of the form $aT(bx + d) = c$ by giving the value of each parameter.

b. Working with classmates, brainstorm a symbolic reasoning strategy to solve $\sin 2x = \frac{1}{2}$ for x. Compare your solution strategy with that of other students.

c. Now compare your solution strategy to the following clever method used by a group of students at Nekoosa High School. Their method uses the idea of a *substitution of variable*.

> We know how to solve $\sin \theta = \frac{1}{2}$.
>
> To solve $\sin 2x = \frac{1}{2}$, let $2x = \theta$.
>
> Then, $\sin 2x = \sin \theta = \frac{1}{2}$.
>
> From Problem 3, we know the solutions of $\sin \theta = \frac{1}{2}$ are
>
> $\theta = \frac{\pi}{6} + 2\pi n$ or $\theta = \frac{5\pi}{6} + 2\pi n$, for any integer n.
>
> So, $2x = \frac{\pi}{6} + 2\pi n$, and thus $x = \frac{\pi}{12} + \pi n$
>
> or $2x = \frac{5\pi}{6} + 2\pi n$, and thus $x = \frac{5\pi}{12} + \pi n$, for any integer n.

8 Use the substitution of variable technique and the fact that $\tan^{-1}(1) = \frac{\pi}{4}$ to solve $\tan\left(2x + \frac{\pi}{2}\right) = 1$. Check your solutions graphically and numerically.

9 Use symbolic reasoning strategies of your choice to solve the following trigonometric equations.

a. $2 \cos (x + 30°) = 1$ b. $2 \cos (2x + 30°) = 1$

c. $\sqrt{3} \tan 3x = 1$ d. $2 \sin\left(2x + \frac{\pi}{6}\right) = \sqrt{3}$

10 At the beginning of this lesson, the number of hours of sunlight to be expected in Chicago, Illinois, was modeled by the function

$$S(t) = 3 \sin \frac{2\pi}{365}t + 12,$$

where t is the number of days after March 21.

a. Explain why the function
$$S(d) = 3 \sin\left[\frac{2\pi}{365}(d - 80)\right] + 12,$$
where d represents *day of the year* beginning with January 1 as day 1, predicts the number of sunlight hours in Chicago.

b. In Chicago, the "longest day" has about 15 hours of sunlight. The "shortest day" has about 9 hours of sunlight. Use the function in Part a to write and solve an equation whose solution is the day of the year when Chicago is predicted to have maximum sunlight hours. Minimum sunlight hours.

c. Reason in a similar manner to predict on what days of the year Chicago will have 12 hours of sunlight.

d. On what days of the year would you predict 12 or more hours of sunlight for Chicago? Explain your reasoning

SUMMARIZE THE MATHEMATICS

In this investigation, you examined strategies for solving linear trigonometric equations.

a How is solving the equation $a \tan x + b = c$, $a \neq 0$, which is linear in $\tan x$, similar to, and different from, solving $ax + b = c$, which is linear in x?

b Describe the procedures you would use to solve linear trigonometric equations such as the following:

   i. $a \cos (bx - d) = c$, $a \neq 0$, $b \neq 0$
   ii. $a \cot (bx + d) = c$, $a \neq 0$, $b \neq 0$

c Explain why equations linear in a trigonometric function will have either no solution or infinitely many solutions.

d How might your answer to Part c be modified if the equation were derived from a trigonometric function modeling a real-world situation?

Be prepared to share your thinking and procedures with the class.

✔ CHECK YOUR UNDERSTANDING

Because of ocean tides, the depth of the River Thames in London varies as a function of time that involves the sine. Suppose the depth d in meters as a function of t, the hour of the day, is modeled by

$$d(t) = 3 \sin \left(\frac{\pi}{6}(t - 4) \right) + 8,$$

where $t = 0$ corresponds to midnight, or 12:00 A.M.

The River Thames in London

Use $d(t) = 3 \sin \left(\frac{\pi}{6}(t - 4)\right) + 8$ to write and solve equations or inequalities for the following tasks. Check your solutions graphically and numerically.

a. Predict the minimum depth of the river and the times at which it occurs.

b. Predict the depth of the river at 2:00 P.M.

c. At approximately what times is the depth 10 m?

d. During what time periods will the depth of the river exceed 8 m?

INVESTIGATION **2**

Using Identities to Solve Trigonometric Equations

In the first investigation, you developed strategies for solving linear trigonometric equations involving a single trigonometric function. As you work on the problems of this investigation, make notes of answers to this question:

How can you solve equations that involve more than one trigonometric function?

1 Begin by examining the solution of $\sin 2x \cos x + \cos 2x \sin x = 1$ by a group of students at Rapid City High School in South Dakota.

a. Give reasons for each step or correct any missteps.

$$\sin 2x \cos x + \cos 2x \sin x = 1 \qquad (1)$$
$$\sin (2x + x) = 1 \qquad (2)$$
$$\sin 3x = 1 \qquad (3)$$
$$\text{If we let } 3x = \theta; \text{ then } \sin \theta = 1. \qquad (4)$$
$$\text{So, } \theta = \frac{\pi}{2} + 2\pi n, \text{ for any integer } n. \qquad (5)$$
$$\text{Hence, } 3x = \frac{\pi}{2} + 2\pi n, \text{ for any integer } n. \qquad (6)$$
$$\text{Thus, } x = \frac{\pi}{6} + \frac{2\pi n}{3}, \text{ for any integer } n. \qquad (7)$$

b. Discuss with your classmates what you think were the key strategies in solving the equation.

c. Check the solution in the original equation for $n = 1, 2, -1,$ and -2.

2 Next consider the solution of a trigonometric equation that is *quadratic in form*.

   a. Give reasons for each step in the following solution of $\cos^2 x + \sin x + 1 = 0$.

$$\cos^2 x + \sin x + 1 = 0 \quad\quad\quad\quad\quad (1)$$
$$1 - \sin^2 x + \sin x + 1 = 0 \quad\quad\quad\quad\quad (2)$$
$$\sin^2 x - \sin x - 2 = 0 \quad\quad\quad\quad\quad (3)$$
$$(\sin x - 2)(\sin x + 1) = 0 \quad\quad\quad\quad\quad (4)$$
$$\text{So, } \sin x - 2 = 0 \text{ or } \sin x + 1 = 0. \quad\quad\quad\quad\quad (5)$$
$$\sin x = 2 \text{ has no real number solutions.} \quad\quad\quad\quad\quad (6)$$
$$\text{If } \sin x + 1 = 0, \text{ then } \sin x = -1. \quad\quad\quad\quad\quad (7)$$
$$\text{Therefore, } x = \frac{3\pi}{2} + 2\pi n, \text{ for any integer } n. \quad\quad\quad\quad\quad (8)$$

   b. Discuss with your classmates what you think were the key strategies in solving this equation.

3 Examine this CAS solution of the equation $4 \sin^2 x = 1$. The marker ▶ on the right of the display in the second line indicates that more follows.

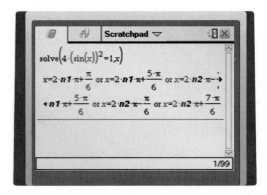

   a. Do the solutions seem reasonable? Why or why not?

   b. Sophie produced graphs for the functions $y = 4 \sin^2 x$ and $y = 1$ to provide a visual check for the reasonableness of the CAS solution given above. The graphical display led Sophie to conclude that the four expressions giving all solutions could be written as two expressions. Suggest one way to do this.

4 Solve each of the following equations using symbolic reasoning. Check your solutions using technology. (In **AUTO** mode, the CAS solves for exact solutions when possible. If not, then the CAS displays approximate solutions.)

   a. $2 \sin (\theta + 37°) = 1$

   b. $4 \cos^2 x = 1$

   c. $2 \sin x \cos x = \sqrt{2} \cos x$

   d. $2 \cos^2 x - 5 \cos x + 2 = 0$

   e. $\tan^2 x - \sec x - 1 = 0$

5 Solve each of the following equations using a technology tool. How could you check the reasonableness of your solutions?

  a. $4 \sin t \cos t = \sqrt{3}$

  b. $\tan x \sec x = \tan x$

6 Combine what you have learned about solving trigonometric equations by reasoning symbolically without technology, by using a CAS, and by using graphs of the related functions to solve the two equations below. Then discuss with your classmates the advantages and disadvantages of each method.

  a. $\cos 2t + \cos t = 0$

  b. $\sin 2x \cos x - \cos 2x \sin x = -\dfrac{\sqrt{3}}{2}$

SUMMARIZE THE MATHEMATICS

In this investigation, you examined how solving trigonometric equations can be guided by your previous experiences in solving familiar linear and quadratic equations.

a How is solving trigonometric equations different than solving linear and quadratic equations?

b Describe the steps you would take, in the order you would do them, when solving the trigonometric equation $\sin 2x = \sin x$, for $0 \le x < 2\pi$.

c How could you check your solutions in Part b?

d How would you modify your solutions in Part b if there were no restrictions on x?

Be prepared to explain your equation-solving ideas and reasoning to the class.

 CHECK YOUR UNDERSTANDING

Solve each of the following trigonometric equations for t. Check your solutions using technology.

  a. $\sin 3t \cos t - \cos 3t \sin t = 1$

  b. $2 \cos^2 t = 3 \sin t + 3$

  c. $2 \sin^2 3t - 5 \sin 3t = 3$

APPLICATIONS

1 After a pizza oven is heated to a specific temperature, for example, 500°F, the actual temperature in the oven oscillates from slightly below 500°F to slightly above 500°F. The temperature in one pizza oven is a function of time x in minutes with rule $t(x) = 10 \sin x + 500$.

a. Why is the sine function, rather than the cosine function, a better choice of model for this situation?

b. Interpret the meaning of the parameters 10 and 500 in the rule for $t(x)$.

c. When does the temperature in the oven first reach 505°F?

d. At what other times is the temperature in the oven 505°F?

e. How many times during the first hour will the temperature of the oven be 505°F?

f. What is the maximum temperature the oven reaches at a setting of 500°F?

2 Find exact solutions for each trigonometric equation.

a. $\sin x = -\dfrac{\sqrt{2}}{2}$

b. $\cos 2x = -\dfrac{\sqrt{2}}{2}$

c. $\sin\left(x + \dfrac{\pi}{4}\right) = -\dfrac{\sqrt{2}}{2}$

d. $\tan 0.5x = \sqrt{3}$

e. $\tan(x - 30°) = \sqrt{3}$

f. $2 \sin x + \sqrt{3} = 0$

g. $4 \tan x - 2 = 2$

h. $\sqrt{3} \sec x + 2 = 0$

3 Suppose a weight is suspended from the ceiling by a spring. When the weight is pulled down and released, it will oscillate up and down, bouncing above and below its at-rest position. If the effects of friction and air resistance were ignored, the oscillations would repeat indefinitely. Suppose you pull the weight down 8 cm from its at-rest position and let it go. The displacement d of the weight from its at-rest position is a function of the number of seconds since it was released, and is given by the rule $d(t) = -8 \cos 2\pi t$.

At-rest Position

a. How long will it take for the weight to make one complete oscillation?

b. When will the weight first be back at the same height from which it was released? Show two different ways to answer this question.

c. When will the weight be 3 cm below its at-rest position?

d. Will the weight ever be 10 cm below its at-rest position? If so, when? If not, why not?

4 The electrical current used in most households of the United States is AC (alternating current). The voltage alternates smoothly between 110 and −110 volts (V), approximately 60 times per second. The modeling function for this pattern of change in voltage is $V(t) = 110 \cos 120\pi t$, where t is measured in seconds.

a. Explain why the coefficient of t is 120π.

b. For what values of t will the voltage be −50 V?

c. For what values of t will the voltage be 100 V?

5 As you have previously seen, the cyclic pattern of the water depth in a tidal harbor can be modeled by a trigonometric function. Suppose the depth D, in feet, in one harbor is given by $D(t) = 9 \cos\left(\frac{\pi}{6}t\right) + 15$, where t is measured in hours after the first high tide on a given day.

a. Why is the cosine function, rather than the sine function, a better choice of model for this situation?

b. Explain the meaning of the parameters 9, $\frac{\pi}{6}$, and 15 in the given function rule.

c. At what times is the depth of the water 15 ft?

d. At what times is the depth of the water 8 ft?

e. At what times is the depth of the water 4 ft?

f. For how many hours each day will the water in the harbor be deeper than 10 ft?

6 Solve each trigonometric equation using a method of your choice. Give exact solutions where possible.

a. $4 \sin^2 x - 3 = 0$

b. $\cos x + 2 = 3 \cos x$

c. $2 \sin^2 2x + \sin 2x = 0$

d. $\tan^2 3x + \tan 3x = 0$

e. $\sin^2 2x + 5 \sin 2x + 6 = 0$

f. $\cos 2x - \sin x = 0$

g. $\sin 2x \sin x + \cos 2x \cos x = 1$

h. $\cos 2x + 3 \cos x = 1$

7 The populations of some animals vary with the time of year and can be modeled by trigonometric functions. Suppose that the rabbit population P in Sleeping Bear Dunes National Forest, in thousands, is modeled by the function
$P(t) = 4.25 \sin\left(\frac{\pi}{6}t - \frac{\pi}{2}\right) + 8.25,$
where t is measured in months since January 2014.

   a. At what times will the rabbit population be 6,000?

   b. For what time periods during the year 2014 was the rabbit population less than 10,000?

   c. What was the maximum population of rabbits in Sleeping Bear Dunes National Forest in 2014? The minimum population?

CONNECTIONS

8 If possible, find an exact zero for each of the following functions. Then look back at your work and describe how the zeroes for each function are related to the zeroes of $s(x) = \sin x$.

   a. $f(x) = 2 \sin x$

   b. $g(x) = \sin x + 2$

   c. $h(x) = \sin (x + 2)$

   d. $j(x) = \sin 2x$

9 The solutions of $\sin x = \frac{1}{2}$ in degrees are $30° + 360°n$ or $150° + 360°n$, for any integer n.

   a. Find the solutions of $\sin 2x = \frac{1}{2}$.

   b. Find the solutions of $\sin 3x = \frac{1}{2}$.

   c. Find the solutions of $\sin 0.5x = \frac{1}{2}$.

   d. Describe how the solutions of $\sin ax = \frac{1}{2}$ are related to the solutions of $\sin x = \frac{1}{2}$.

10 The solutions in radians of $\tan x = 1$ are $\frac{\pi}{4} + \pi n$, for any integer n.

   a. Find the solutions of $\tan \left(x - \frac{\pi}{3}\right) = 1$.

   b. Find the solutions of $\tan \left(x + \frac{\pi}{2}\right) = 1$.

   c. Find the solutions of $\tan (x + 2) = 1$.

   d. Describe how the solutions of $\tan (x + a) = 1$ are related to the solutions of $\tan x = 1$.

©Creatas/PunchStock

11 Illustrate two methods of solving the equation $2 \cos^2 x - \cos x - 1 = 0$, one using factoring directly and the other using substitution of variable ($u = \cos x$).

12 The equations you have solved in this lesson have involved trigonometric functions and constant terms. If an equation also contains an algebraic term, the solution becomes more difficult to find.

a. Solve $\cos x = x$ for the smallest positive solution to six decimal points. *Hint:* Try function iteration. Check your answer using a graphing tool or a spreadsheet.

b. How does the form of the equation in Part a suggest that function iteration would be an appropriate solution method?

c. Repeat Part a for the equation $\sin x = x^2$. Check your answer using a graphing tool or a spreadsheet.

REFLECTIONS

13 What aspect of solving a linear trigonometric equation is the most difficult for you? How can or did you overcome that difficulty?

14 Solving equations by symbolic manipulation is an important mathematical skill, but modern technology often provides alternative methods for solving equations. Look back at Applications Task 3 Part c, Task 4 Part b, and Task 5 Part c. Choose one that you previously completed.

a. Solve it using the table or graphing capabilities afforded by technology.

b. Solve it using the symbolic manipulation capabilities of a CAS.

c. Discuss the advantages and disadvantages of these solution methods compared to the method you first used.

15 Esperanza solved the equation $\sin x = \cos x$ as follows:

$$\sin x = \cos x$$
$$\sin^2 x = \cos^2 x$$
$$\sin^2 x = 1 - \sin^2 x$$
$$2 \sin^2 x = 1$$
$$\sin^2 x = \frac{1}{2} \text{ and } \sin x = \pm\frac{1}{\sqrt{2}}$$

Thus, $x = 45° + 180°n$ and $x = 135° + 180°n$, for any integer n.

a. Is Esperanza's reasoning correct? Explain.

b. Is Esperanza's solution correct? If not, how can you guard against making similar errors?

16 Consider the ranges of the inverse trigonometric functions (page 299). Suppose you are trying to determine the measure of an angle θ of a triangle. Can you uniquely determine θ if you know cos θ? If you know sin θ? If you know tan θ? Explain your reasoning.

17 Describe how you would solve the general trigonometric equation $a \sin x = b$, $a \neq 0$, for x.

a. How does your thinking allow you to assess the reasonableness of the CAS-produced result below?

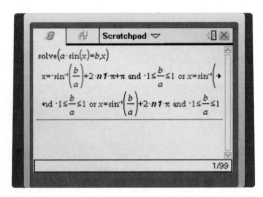

b. Why is there the restriction $-1 \le \dfrac{b}{a} \le 1$?

EXTENSIONS

18 The mean daily Fahrenheit temperature in Fairbanks, Alaska, can be modeled by the function $T(x) = 37 \sin\left(\dfrac{2\pi}{365}(x - 101)\right) + 25$, where x is the day of the year, with $x = 0$ representing January 1.

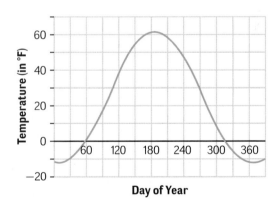

a. On what days would you predict the average temperature in Fairbanks to be 0°F? Explain how you determined your answer.

b. Write an equation whose solution provides an answer to the question in Part a.

   i. What strategies might you use to solve the equation?

   ii. Choose one of those strategies and solve the equation.

19 Equations similar to the following can occur during the process of working with mathematical models and structures. Examine each equation to determine if it has a solution. Solve the equations that do have solutions.

  a. $3 \sec 2x + 2 = 5$

  b. $1 - \cot 3x = -5$

  c. $4 \csc^2 x = 3$

20 On April 17, 1953, Mickey Mantle, Hall of Fame centerfielder of the New York Yankees, hit the longest home run ever hit in a regular-season major league baseball game. According to the *Guinness Book of Sports Record*, it was calculated to have traveled a distance of 565 feet from home plate in the air.

  a. The horizontal distance that a baseball travels before it returns to the height at which it was hit is given by the formula for horizontal distance from Lesson 1 (Problem 5 Part c, page 278), $R(\theta) = \dfrac{V^2 \sin 2\theta}{32}$. Here V is the initial velocity in feet per second and θ is the angle of elevation. If the ball was 3 feet above the ground when Mantle hit it with an initial velocity of 145 feet per second, what was the angle of elevation of the ball when it left the bat? Assume the ball traveled 560 feet horizontally before returning to a height of 3 feet (and went about 5 more feet before hitting the ground).

  b. Again assuming Mantle hit the ball when it was 3 feet above the ground and the ball traveled 560 feet horizontally before returning to a height of 3 feet, what is the possible minimum initial velocity V of Mantle's record home run?

21 Solve each equation below using symbolic reasoning. Check your solutions by solving the equation using a different method.

  a. $\cos^2 0.5x - 0.5 \cos x = 0.5$

  b. $2 \sin^2 x + \sin x = 1$

  c. $4 \sin^4 x + \sin^2 x = 3$

  d. $\sin 2x + \cos 3x = 0$

22 In certain applications of calculus, it is necessary to express a product of two trigonometric functions as a sum or difference. This can often be accomplished by using one of the following **product-sum identities**. Use sum and difference identities for sine and cosine to prove each identity.

a. $\cos \alpha \cos \beta = \frac{1}{2}[\cos (\alpha + \beta) + \cos (\alpha - \beta)]$

b. $\sin \alpha \sin \beta = \frac{1}{2}[\cos (\alpha - \beta) - \cos (\alpha + \beta)]$

c. $\sin \alpha \cos \beta = \frac{1}{2}[\sin (\alpha + \beta) + \sin (\alpha - \beta)]$

d. $\cos \alpha \sin \beta = \frac{1}{2}[\sin (\alpha + \beta) - \sin (\alpha - \beta)]$

23 Some applications of calculus require that the sum or difference of two trigonometric functions be rewritten as a product. This can be done by using alternate forms of the product-sum identities. Derive these forms using the identities in Extensions Task 22 and the substitution $\alpha = \frac{1}{2}(u + v)$ and $\beta = \frac{1}{2}(u - v)$. The four new identities relate the sum (difference) of sines and cosines of u and v to the product of sines and cosines of half sums and half differences. The identities are called **sum-product identities**.

24 Use the identities in Extensions Tasks 22 and 23 as needed to solve each equation below.

a. $\sin x + \sin 2x + \sin 3x = 0$

b. $\cos x - \cos 3x - \cos 5x = 0$

c. $\sin x + \sin 2x - \sin 4x = 0$

REVIEW

25 Rewrite each expression as a product of linear factors.

a. $a^2 + 7a + 12$

b. $3x^2 - 13x - 10$

c. $4s^2 + 12s + 9$

d. $t^3 - 8t^2 - 20t$

e. $30x^2 + 3x - 9$

f. $x^4 - 6x^2 - 27$

26 Solve each quadratic equation.

a. $x^2 = x + 2$

b. $2x^2 + x - 1 = 0$

c. $2x^2 - 5x = 3$

d. $2x^2 + 2 = 5x$

e. $6x^2 + 5x + 1 = 0$

27 Examine the sequence: 96, 48, 24, 12, …

    a. Assuming the pattern continues, is this sequence arithmetic, geometric, or neither? Explain your reasoning.

    b. Find the eighth term of the sequence.

    c. Find a recursive formula for the sequence.

    d. Find a function formula for the sequence.

28 As a fundraiser, the senior class at Matanuska Valley High School is considering selling class sweatshirts. They first need to analyze their costs. The printing company will charge $150 to set up the artwork and $8 per sweatshirt.

    a. Write a formula for the average cost of a sweatshirt if the class purchases n sweatshirts.

    b. How many sweatshirts do they need to purchase so that the average cost per sweatshirt is less than $10?

    c. Identify any horizontal or vertical asymptotes of the graph of the average cost function. Then explain their meaning in terms of this situation.

29 Let $\vec{u}$ be a vector with magnitude 5 and direction 210° and $\vec{v}$ be a vector with magnitude 3 and direction 315°.

    a. Draw a sketch of each vector as a position vector on a coordinate plane.

    b. Represent each vector with exact coordinates. Do not use approximations for the coordinates.

    c. Find the coordinate representation of $\vec{u} + \vec{v}$.

    d. Find the magnitude and direction of $-3\vec{v}$.

30 Rewrite each expression in standard form $a + bi$.

    a. $(3 + 2i) + (9 - 5i)$ **b.** $(1.7 - i) - (0.35 - 3.5i)$

    c. $(3 + 5i)(2 + 2i)$ **d.** $(4 + 7i)(4 - 7i)$

    e. $(9 + 2i)^2$ **f.** $\dfrac{8 + i}{i}$

31 In the following, $i^2 = -1$. Rewrite each product as a polynomial with complex coefficients.

    a. $(x - 5i)(x + 5i)$

    b. $(3x + i)(3x - i)$

    c. $(x - (2 + 3i))(x + (2 + 3i))$

The Geometry of Complex Numbers

In Unit 3, *Algebraic Functions and Equations*, the imaginary number $i = \sqrt{-1}$ and complex numbers $a + bi$ were revisited. There you discovered and proved properties of the complex number system that were remarkably similar to those for real numbers. You applied your understanding of complex numbers to solve problems involving electrical circuits and to create fractal images.

In Unit 3, you also learned how to represent complex numbers as ordered pairs and as position vectors in the complex number plane. The diagrams below illustrate some of the arithmetic of complex numbers represented as vectors.

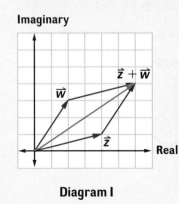

Diagram I

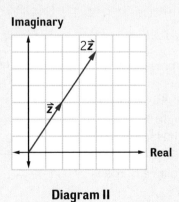

Diagram II

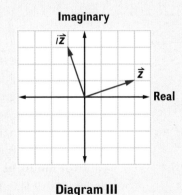

Diagram III

In this lesson, you will continue to investigate the geometry of complex numbers. In particular, you will explore how the *trigonometric form* of complex numbers eases the multiplication and division of complex numbers and allows you to determine powers and roots of complex numbers.

INVESTIGATION 1

Trigonometric Form of Complex Numbers

Addition and subtraction of complex numbers are relatively easy when the numbers are expressed in standard form, $a + bi$. However, multiplication and division of complex numbers are most easily done when the numbers are expressed in *trigonometric form*.

As you work on the problems in this investigation, look for answers to the following questions:

Given a complex number in standard form, how can you express it in trigonometric form and vice versa?

How can you find the product and quotient of two complex numbers expressed in trigonometric form?

How can you geometrically interpret complex number multiplication? Division?

1 Using the vector representation of $a + bi$ shown at the right, express $a + bi$ in terms of the magnitude of the vector, r, and its direction angle θ.

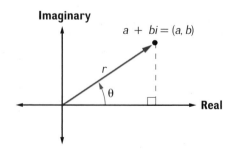

In the diagram, the length r of the position vector is called the **absolute value** or **modulus** of $a + bi$ and is denoted $|a + bi|$.

2 Explain why the complex number $a + bi$ can also be expressed in the form $r(\cos \theta + i \sin \theta)$, called the **trigonometric form** of the complex number.

3 Express each of the following complex numbers in trigonometric form. Use exact values where possible.

a. $4 + 4i$

b. $-3i$

c. $-2 + 3i$

d. $3 - 4i$

4 Express each complex number in standard form $a + bi$. Use exact values where possible.

a. $2(\cos 60° + i \sin 60°)$

b. $4\left(\cos \frac{7\pi}{6} + i \sin \frac{7\pi}{6}\right)$

c. $3\left(\cos \frac{\pi}{2} + i \sin \frac{\pi}{2}\right)$

d. $5(\cos \pi + i \sin \pi)$

e. $8(\cos 15° + i \sin 15°)$

5 Now examine multiplication of complex numbers in trigonometric form. Consider the complex numbers

$$z = 5\left(\cos \frac{\pi}{4} + i \sin \frac{\pi}{4}\right) \text{ and } w = 3\left(\cos \frac{\pi}{2} + i \sin \frac{\pi}{2}\right).$$

a. Give reasons for each step in the following calculation of the product of these two numbers.

$$zw = \left[5\left(\cos \frac{\pi}{4} + i \sin \frac{\pi}{4}\right)\right]\left[3\left(\cos \frac{\pi}{2} + i \sin \frac{\pi}{2}\right)\right] \tag{1}$$

$$= 15\left[\left(\cos \frac{\pi}{4} + i \sin \frac{\pi}{4}\right)\left(\cos \frac{\pi}{2} + i \sin \frac{\pi}{2}\right)\right] \tag{2}$$

$$= 15\left[\left(\cos \frac{\pi}{4} \cos \frac{\pi}{2} - \sin \frac{\pi}{4} \sin \frac{\pi}{2}\right) + i\left(\cos \frac{\pi}{4} \sin \frac{\pi}{2} + \sin \frac{\pi}{4} \cos \frac{\pi}{2}\right)\right] \tag{3}$$

$$= 15\left[\left(\cos \left(\frac{\pi}{4} + \frac{\pi}{2}\right) + i \sin \left(\frac{\pi}{4} + \frac{\pi}{2}\right)\right)\right] \tag{4}$$

$$= 15\left(\cos \frac{3\pi}{4} + i \sin \frac{3\pi}{4}\right) \tag{5}$$

b. Let $v = 2\left(\cos \frac{3\pi}{4} + i \sin \frac{3\pi}{4}\right)$. Use similar reasoning to find the product vw in trigonometric form.

c. Look for a pattern that relates the trigonometric form of two complex numbers to the trigonometric form of their product. Describe the pattern in words.

6 Next consider the general case of two complex numbers

$$z = r(\cos \alpha + i \sin \alpha) \text{ and } w = s(\cos \beta + i \sin \beta).$$

Supply reasons for each step in the following calculation of the product of these two numbers.

$$zw = [r(\cos \alpha + i \sin \alpha)][s(\cos \beta + i \sin \beta)] \tag{1}$$
$$= rs[(\cos \alpha + i \sin \alpha)(\cos \beta + i \sin \beta)] \tag{2}$$
$$= rs[(\cos \alpha \cos \beta - \sin \alpha \sin \beta) + i(\cos \alpha \sin \beta + \sin \alpha \cos \beta)] \tag{3}$$
$$= rs[\cos (\alpha + \beta) + i \sin (\alpha + \beta)] \tag{4}$$

7 The trigonometric form of complex numbers reveals an interesting connection between multiplication of complex numbers and geometric transformations in a plane such as rotations and size transformations.

a. On a coordinate grid, sketch the vector representations for the two complex numbers in Problem 5 Part a. On the same grid, sketch the vector for the product zw.

i. Compare the magnitude of the product vector to the magnitudes of the factors.

ii. Compare the directed angle of the product vector to the directed angles of the factors.

b. Make the same comparisons for the product vw in Problem 5 Part b as you made for the product in Part a above.

c. Explain what Parts a and b and Problem 6 suggest about the multiplication of complex numbers in terms of geometric transformations.

d. How would you modify your answer if either of the two complex number factors had absolute value 1?

8 Now consider the connection between complex number arithmetic and rotation of a point about the origin. What complex number operation can be used to rotate the point corresponding to $z = a + bi = r(\cos \theta + i \sin \theta)$ through an angle α about the origin? Explain your reasoning.

9 If $z = r(\cos \theta + i \sin \theta)$, $z \neq 0$, what is the trigonometric form of $z^{-1} = \frac{1}{z}$? Compare your trigonometric form of the multiplicative inverse of z with that of your classmates. Resolve any differences.

10 Consider once again the two general complex numbers in Problem 6,

$$z = r(\cos \alpha + i \sin \alpha) \text{ and } w = s(\cos \beta + i \sin \beta), \text{ with } w \neq 0.$$

a. Use the fact that $\frac{z}{w} = z \cdot w^{-1}$ to prove $\frac{z}{w} = \frac{r}{s}[\cos (\alpha - \beta) + i \sin (\alpha - \beta)]$.

b. Write in words how to divide two nonzero complex numbers expressed in trigonometric form.

c. On a coordinate grid, sketch the vectors for

$$w = 8(\cos 150° + i \sin 150°) \text{ and } z = 2(\cos 30° + i \sin 30°).$$

On the same grid, sketch the vector for the quotient $\frac{w}{z}$.

d. How can you interpret division of complex numbers in terms of geometric transformations?

SUMMARIZE THE MATHEMATICS

In this investigation, you explored multiplication and division of complex numbers in trigonometric form.

a Explain how you can determine the trigonometric form of a complex number expressed in standard form $a + bi$.

b Explain how you can determine the standard form $a + bi$ of a complex number expressed in trigonometric form $r(\cos \theta + i \sin \theta)$.

c Write a summarizing statement about how to multiply two complex numbers that are expressed in trigonometric form.

d Describe the connection between multiplication of complex numbers in trigonometric form and geometric transformations.

e Write a summarizing statement about how to divide one complex number by a nonzero complex number, each expressed in trigonometric form.

f Describe the connection between division of complex numbers in trigonometric form and geometric transformations.

Be prepared to explain your ideas and methods to the entire class.

 CHECK YOUR UNDERSTANDING

Use connections between various representations of complex numbers to complete the following tasks.

a. Write each complex number in trigonometric form: $z_1 = 2 + 5i$ and $z_2 = -6 + 3i$

b. Write each complex number in standard form: $w = 6\left(\cos \frac{\pi}{2} + i \sin \frac{\pi}{2}\right)$ and $z = 3\left(\cos \frac{\pi}{6} + i \sin \frac{\pi}{6}\right)$

c. Calculate wz and write your answer in standard $a + bi$ form.

d. Calculate $\frac{w}{z}$ and write your answer in standard $a + bi$ form.

e. Use complex number arithmetic to rotate the vector corresponding to the complex number $w = 6\left(\cos \frac{\pi}{2} + i \sin \frac{\pi}{2}\right)$ counterclockwise $\frac{2\pi}{3}$ radians about the origin. What complex number in standard form corresponds to the rotation vector?

De Moivre's Theorem

In Investigation 1, you learned how to multiply complex numbers expressed in trigonometric form. In particular, if $z = r(\cos \alpha + i \sin \alpha)$ and $w = s(\cos \beta + i \sin \beta)$, then

$$zw = rs[\cos(\alpha + \beta) + i \sin(\alpha + \beta)].$$

In terms of transformations, multiplying w by z is equivalent to a **spiral similarity transformation**; that is, the composite of a size transformation with center at the origin and magnitude $r\,(r \neq 1)$ and a rotation with center at the origin and directed angle α.

This connection between multiplication of complex numbers expressed in trigonometric form and transformations of the coordinate plane leads to a beautiful and useful principle known as De Moivre's Theorem. That theorem, in turn, leads to solution of equations that go far beyond the simple quadratic equations that generated the need for complex numbers. The theorem is attributed to Abraham De Moivre (1667–1754), a French mathematician who is credited with many other important discoveries, especially in probability theory.

As you complete the problems in this investigation, look for answers to the following questions:

What is De Moivre's Theorem?

How can De Moivre's Theorem be used to calculate powers and roots of complex numbers?

1. **Powers of Complex Numbers** Consider the complex number $z = r(\cos \theta + i \sin \theta)$.

   a. Write $z^2 = z \cdot z$ in trigonometric form. Interpret z^2 geometrically.

   b. Using your result from Part a, write $z^3 = z^2 \cdot z$ in trigonometric form. Then write z^4 in trigonometric form.

   c. Extend the reasoning in Part b to complete the following statement, known as **De Moivre's Theorem**:

      If $z = r(\cos \theta + i \sin \theta)$ and n is any positive integer, then $z^n = $ _____.

      (You will be asked to complete a proof of De Moivre's Theorem in Extensions Task 22.)

   d. If $z = 3\left(\cos \frac{\pi}{5} + i \sin \frac{\pi}{5}\right)$, find z^{10}.

      i. Explain why z^{10} is a real number.

      ii. What other power(s) of z less than 10 are real numbers?

Roots of Complex Numbers De Moivre's Theorem provides an efficient way to calculate powers of any complex number when written in trigonometric form $z = r(\cos \theta + i \sin \theta)$. Reversing the reasoning suggested by that result provides a way of finding all complex number roots of polynomial equations in the form $z^n - a = 0$, or equivalently $z^n = a$, for any positive integer n. Those solutions, in turn, provide insight into some important geometric problems.

2 Consider first the cubic equation $z^3 - 1 = 0$, or $z^3 = 1$. To find the roots, you need numbers which when raised to the third power equal 1. One obvious solution is $z = 1$. But there are two other cube roots! Supply explanations for the steps in the following mathematical argument.

(1) The number 1 can be represented in trigonometric form as
$1 = (\cos 0 + i \sin 0)$, $1 = (\cos 2\pi + i \sin 2\pi)$, $1 = (\cos 4\pi + i \sin 4\pi)$, $1 = (\cos 6\pi + i \sin 6\pi)$, and so on.

(2) If $w = r(\cos \theta + i \sin \theta)$ is a solution of $z^3 = 1$, then $w^3 = r^3(\cos 3\theta + i \sin 3\theta)$ must be one of the trigonometric forms of 1 in Step 1.

(3) These conditions imply that $r = 1$.

(4) These conditions also imply that $\theta = 0$, $\theta = \frac{2\pi}{3}$, and $\theta = \frac{4\pi}{3}$ determine solutions. (There are other values of θ that meet the given conditions, but they are all related to the basic three solutions by addition of multiples of the period 2π.)

(5) The three values of θ obtained in Step 4 and $r = 1$ determine the *complex roots* of $z^3 = 1$. In standard form, the three solutions are

$$\cos 0 + i \sin 0 = 1 + 0i$$
$$\cos \frac{2\pi}{3} + i \sin \frac{2\pi}{3} = -\frac{1}{2} + \frac{\sqrt{3}}{2}i$$
$$\cos \frac{4\pi}{3} + i \sin \frac{4\pi}{3} = -\frac{1}{2} - \frac{\sqrt{3}}{2}i$$

The complex number roots of the equation $z^3 = 1$ are called the cube roots of unity.

3 The diagram below shows a circle of radius 1 in the complex number plane.

a. On a copy of the diagram, draw point representations of the three cube roots of unity. Describe how the points are distributed on the circle.

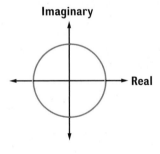

b. Now use reasoning similar to that in Problem 2 to find the four *quartic roots of unity*; that is, find the four complex number solutions of the equation $z^4 = 1$. Plot these solutions on a unit circle and describe their placement.

c. Describe a general procedure to find and represent the nth roots of unity. Compare your methods with those of others and resolve any differences.

4 Use reasoning similar to that in Problem 2 to find and represent the fifth roots of 243 in standard form $a + bi$.

5 How would you find solutions for equations of the form $z^n = a$ for any positive integer power n and any nonzero real number a? In general, how would the point representations of these solutions in the complex number plane be related? Compare your answers with your classmates and resolve any differences.

6 In Problem 5, you investigated methods for finding all roots of a given nonzero real number. In this problem, you will apply similar reasoning to find all roots of an imaginary number.

   a. Consider the equation $z^3 = i$.

   If $z = r(\cos \theta + i \sin \theta)$, then $z^3 = r^3(\cos 3\theta + i \sin 3\theta)$.

   Since $i = \cos \left(\frac{\pi}{2} + 2k\pi\right) + i \sin \left(\frac{\pi}{2} + 2k\pi\right)$, where k is any integer, what can you conclude about r? About 3θ? About θ?

   b. Use your conclusions in Part a to write the cube roots of i in trigonometric form.

   c. Display the cube roots of i on a unit circle in the complex number plane. Compare their placement with the cube roots of 1.

   d. Predict the locations of the cube roots of $-i$ on the unit circle. Check your conjecture.

7 You can extend your reasoning in Problem 6 to find the nth root of any nonzero complex number. Find the fourth roots of $-8 + 8\sqrt{3}i$ and display them on a circle in the complex number plane.

SUMMARIZE THE MATHEMATICS

In this investigation, you explored the general problem of calculating the nth power of any complex number and the related problem of solving equations like $z^n = a$ for any positive integer n and any nonzero value of a.

   a If $z = r(\cos \theta + i \sin \theta)$, what is the trigonometric form of z^n?

   b How does the connection between complex number multiplication and size transformations and rotations explain the pattern in Part a?

   c What are nth roots of unity, and how can their trigonometric and standard complex number forms be constructed using De Moivre's Theorem? How can they be displayed on the complex number plane?

   d Describe how to find the nth roots of $a + bi$ if $|a + bi| = 1$.

Be prepared to explain your ideas and reasoning to the entire class.

Use De Moivre's Theorem to help complete the following tasks involving powers and roots of complex numbers.

a. Plot $z = i$ on the complex number plane.

   i. Calculate z^3 and plot the result.

   ii. Explain how the plot could be predicted from the geometry of rotations and size transformations.

b. Plot $z = 1 + i$ on the complex number plane.

   i. Calculate z^6 and plot the result.

   ii. Explain how the plot could be predicted from the geometry of rotations and size transformations.

   iii. Express your answer for part i in standard complex number form.

c. On a copy of the unit circle at the right, draw the point corresponding to $(a + bi)^3$.

d. Find the angles of the position vectors corresponding to the seventh roots of unity. Then locate points corresponding to the roots on a unit circle in the complex number plane.

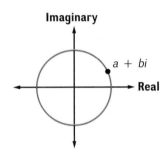

APPLICATIONS

1 Find the absolute value (modulus) of each complex number and then express the complex number in trigonometric form.

a. $-2 + 2\sqrt{3}i$

b. $4\sqrt{3} - 4i$

2 Calculate each product zw and sketch a vector diagram on the complex number plane illustrating the operation. Write your answer in standard $a + bi$ form.

a. $z = 2(\cos 35° + i \sin 35°)$
$w = 5(\cos 55° + i \sin 55°)$

b. $z = 3(\cos 35° + i \sin 35°)$
$w = 6(\cos 100° + i \sin 100°)$

c. $z = 5(\cos 0 + i \sin 0)$
$w = 3\left(\cos \dfrac{3\pi}{2} + i \sin \dfrac{3\pi}{2}\right)$

d. $z = 6\left(\cos \dfrac{3\pi}{4} + i \sin \dfrac{3\pi}{4}\right)$
$w = 0.5\left(\cos \dfrac{\pi}{2} + i \sin \dfrac{\pi}{2}\right)$

3 Calculate each quotient $\dfrac{z}{w}$ and write your answers in standard complex number form.

a. $z = 8(\cos 150° + i \sin 150°)$
$w = 2(\cos 30° + i \sin 30°)$

b. $z = 9(\cos 190° + i \sin 190°)$
$w = 3(\cos 40° + i \sin 40°)$

c. $z = 2\left(\cos \dfrac{4\pi}{3} + i \sin \dfrac{4\pi}{3}\right)$
$w = 6\left(\cos \dfrac{\pi}{6} + i \sin \dfrac{\pi}{6}\right)$

d. $z = 4(\cos \pi + i \sin \pi)$
$w = \cos \dfrac{\pi}{6} + i \sin \dfrac{\pi}{6}$

4 Complex numbers have important applications in electronics and electrical engineering. The trigonometric form of complex numbers is particularly useful when dealing with applied problems involving both multiplication and division. Consider the *voltage divider formula*

$$V_1 = \frac{V_S Z_1}{Z_T}$$

that is used to calculate the voltage V_1 across any element in an AC (alternating current) circuit. See the diagram at the top of the next page. Here V_S is the applied voltage, Z_1 is the impedance of the element, and Z_T is the total impedance.

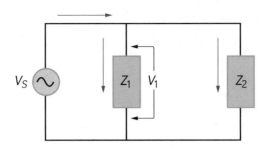

a. Calculate V_1 if $V_S = 20(\cos 0 + i \sin 0)$ volts, $Z_1 = 75\left(\cos \frac{\pi}{4} + i \sin \frac{\pi}{4}\right)$ ohms, and $Z_T = 180\left(\cos \frac{\pi}{3} + i \sin \frac{\pi}{3}\right)$ ohms.

b. Express your voltage answer in Part a in $a + bi$ form.

c. Determine the value of Z_1 in the voltage divider under conditions in which $V_1 = 12(\cos 10° + i \sin 10°)$ volts, $V_S = 28(\cos 0° + i \sin 0°)$ volts, and $Z_T = 200(\cos 75° + i \sin 75°)$ ohms.

5 One of the standard techniques of computer graphic design is to identify coordinates for key points of a figure and then produce the figure by connecting those points with line segments or curves. For example, the arrowhead shown at the right can be drawn by connecting four points with four segments. The scales on both axes are 1 unit.

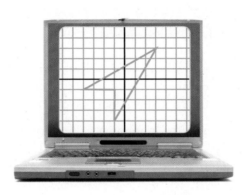

a. List the coordinates of key points that can be connected to form the figure. Write complex numbers corresponding to the key points.

b. Suppose the arrowhead is transformed by multiplying each of the key points by $3\left(\cos \frac{3\pi}{4} + i \sin \frac{3\pi}{4}\right)$. Describe geometrically the transformation that maps the original arrowhead onto its image. Then find the coordinates of the transformed key points and sketch the image figure to check your description.

c. Find a complex number multiplier that transforms the arrowhead into an image with side lengths multiplied by 4 and leaves the tilt of the arrowhead and its image the same, that is, each image side is parallel to its preimage. Write the multiplier in both trigonometric and standard complex number forms. How would you change the multiplier if side lengths were to be half the original?

6 Calculate the following powers of complex numbers. In each case, determine the absolute value and the quadrant or axis location for the point corresponding to the calculated power.

a. $[3(\cos \pi + i \sin \pi)]^5$

b. $\left[5\left(\cos \frac{\pi}{2} + i \sin \frac{\pi}{2}\right)\right]^7$

c. $\left[2\left(\cos \frac{\pi}{3} + i \sin \frac{\pi}{3}\right)\right]^4$

d. $(-1 + i)^8$

7 Find all specified roots of the given complex numbers and express all roots in standard $a + bi$ form. In each case, sketch the roots on a circle in the complex number plane and identify the radius of the circle.

a. Sixth roots of 1

b. Sixth roots of -1

c. Fourth roots of -81

d. Fifth roots of $32i$

e. Cube roots of $2 - 2i$

CONNECTIONS

8 Recall from your earlier work that, for any complex number $z = a + bi$, the number $a - bi$ is called the *conjugate* of z. The notation $\bar{z}$ (read "z bar") is used to denote the conjugate of z.

a. If the trigonometric form of the complex number $a + bi$ is $r(\cos \theta + i \sin \theta)$, what is the trigonometric form of its conjugate?

b. Describe geometrically the relationship between a complex number and its conjugate.

c. If $r + si$ is one solution of a quadratic equation $ax^2 + bx + c = 0$ with real number coefficients, what is the other solution?

d. Summarize in words your finding in Part c.

9 Find all complex number solutions of the equation $z\bar{z} + 2(z - \bar{z}) = 10 + 6i$. See Connections Task 8 for the definition of $\bar{z}$.

10 Suppose a figure is defined by connecting in order a set of n points corresponding to the n complex numbers $a_k + b_k i = w_k = s_k(\cos \beta_k + i \sin \beta_k)$, for $k = 1, 2, \dots, n$.

a. What is the effect on the figure if $z = -6 + 10i$ is added to each complex number in the figure?

b. What is the effect on the figure if each complex number is multiplied by 3? By i? By $-i$?

c. What is the effect on the figure if each complex number in the figure is multiplied by $z = 2.5\left(\cos \frac{\pi}{4} + i \sin \frac{\pi}{4}\right)$?

11 Look back at the geometric representations of your answers to Applications Task 7 Parts a, b, and c.

a. Try to generalize your results. In particular, describe the distribution on a circle of the points corresponding to the nth roots of any nonzero real number.

b. If you connect consecutive roots in order with line segments, what type of geometric figure is formed?

c. For a given real number, how many real number nth roots are possible?

d. For which integers n is there only one real number root?

12 The **Fundamental Theorem of Algebra** states:

A polynomial of degree n has exactly n complex number roots, counting multiplicity.

a. Find the complex number roots of each of the following polynomial equations.

  i. $x^2 + x + 11 = 0$

  ii. $x^3 - 27 = 0$

  iii. $x^4 - 16 = 0$

b. How is the Fundamental Theorem of Algebra related to your work in Investigation 2?

13 Consider the functions with rules of the form $f(z) = Az + B$, where z is a complex variable and A and B are constant complex numbers. Describe the geometric transformation that is equivalent to $f(z)$ in each of the following cases.

a. $A = 1$

b. $B = 0$ and $|A| = 1$

c. $B = 0$ and A is a positive real number different from 1

14 Calculating rational powers of negative real numbers is sometimes problematic. For example, $(-8)^{\frac{1}{3}} = \sqrt[3]{-8} = -2$. But $(-8)^{\frac{1}{3}} = (-8)^{\frac{2}{6}} = \left((-8)^{\frac{1}{6}}\right)^2$ yet $(-8)^{\frac{1}{6}}$ is undefined in the set of real numbers since a negative real number does not have a sixth root that is a real number.

a. Use De Moivre's Theorem to determine the complex cube roots of -8. The complex sixth roots of 64.

b. Explain how the cube roots of -8 and the sixth roots of 64 are related to one another.

c. Show that $\sqrt{2}i$ is a sixth root of -8, and that $(\sqrt{2}i)^2 = -2$.

15 Complex numbers can be expressed in standard form, as points in the complex number plane, as position vectors in the complex number plane, and in trigonometric form. Under what circumstances would you choose to use each of these representations of complex numbers?

16 What trigonometric identities were critical in establishing the algorithm for multiplying two complex numbers expressed in trigonometric form?

17 Recall that for any nonzero real number r, $r^0 = 1$. Reasoning analogously, define $z^0 = 1$ for any nonzero complex number. Verify that De Moivre's Theorem also holds for $n = 0$.

18 If the nth roots of a complex number $z = r(\cos \theta + i \sin \theta)$ are graphed on a circle in the complex number plane, what can you say about the radius of the circle? About the measure of the angle formed by the position vector representations of two consecutive roots?

19 Look back at your answer for Applications Task 7 Part c. Verify that each solution you found is a fourth root of -81.

20 Do an Internet search of "Abraham De Moivre." Write a brief (one page maximum) report of some of your most interesting findings about the man, not about his mathematics.

EXTENSIONS

21 Use De Moivre's Theorem to derive identities for $\cos 3\theta$ and for $\sin 3\theta$ in terms of $\cos \theta$ and $\sin \theta$.

22 De Moivre's Theorem: If $z = r(\cos \theta + i \sin \theta)$, then $z^n = r^n(\cos n\theta + i \sin n\theta)$, where n is any positive integer can be proved in two ways—using the *Principle of Mathematical Induction* that you will study in Unit 8, *Counting Methods and Induction*, or using the *Least Number Principle*.

a. The **Least Number Principle** states that every nonempty set of positive integers has a least element. We will assume the statement as a postulate. Why does this assumption make sense?

b. The Least Number Principle can be used to provide a proof by contradiction of De Moivre's Theorem. Give reasons for each of the following steps in the proof.

De Moivre's Theorem is true for $n = 1, 2, 3$, and 4.	(1)
Let k be the smallest positive integer for which De Moivre's Theorem is *not* true. It follows that the theorem is true for $n = \underline{k} - 1$.	(2)
So, $z \cdot z^{k-1} = r(\cos \theta + i \sin \theta) \cdot r^{k-1}(\cos (k-1)\theta + i \sin (k-1)\theta)$.	(3)
$z^k = r^k(\cos k\theta + i \sin k\theta)$	(4)
Hence, the theorem is true for $n = k$.	(5)
But this is a contradiction.	(6)
Therefore, De Moivre's Theorem is true for all positive integers n.	(7)

23 The nth roots of unity all have absolute value equal to 1. That means that they correspond to points on a unit circle in the complex number plane.

    a. Construct a table showing the angles (in radians) that determine roots of unity for $n = 3, 4, 5, 6, 7$, and 8. Inspect the results and find a formula for the angles that give nth roots of unity for any n.

    b. What sort of polygon is formed when the nth roots of unity are connected in order of increasing angle measure (starting at $\alpha = 0$)? How is your answer related to the measure of the central angles of the polygon?

24 Look back at Connections Task 8. Prove that if z is a complex number, then $(\overline{z})^n = \overline{z^n}$, for any positive integer n.

25 Consider all functions $f(z) = Az + B$, where z is a complex number variable and A and B are specific complex number constants.

    a. Show that the composite of any two such functions is another function of the same type.

    b. Show that whenever $|A| = 1$, the resulting function preserves distances between pairs of complex numbers; that is, show that $|f(z_1) - f(z_2)| = |z_1 - z_2|$.

    c. Show that if $A \neq 1$, the function has exactly one *fixed point*; that is, there is exactly one value of z for which $f(z) = z$. Explain how that allows you to conclude that the function corresponds to a spiral similarity of the plane about the fixed point.

    d. Show that if $A = 1$ and $B \neq 0$, the function has no fixed points. Explain how that allows you to conclude that the transformation is a translation of the plane.

    e. Explore similar cases for functions of the form $g(z) = A\overline{z} + B$, and see what you can conclude about the transformation of the plane described by such functions for various combinations of A and B.

26 Prove that if $z^n = r(\cos \theta + i \sin \theta)$ where n is a positive integer, then the nth roots of z^n are given by $z = \sqrt[n]{r}\left[\cos\left(\dfrac{\theta}{n} + \dfrac{2\pi k}{n}\right) + i \sin\left(\dfrac{\theta}{n} + \dfrac{2\pi k}{n}\right)\right]$, where k is any integer.

REVIEW

27 Determine all solutions to each equation. Report your solutions in radians.

    a. $4 \cos 3x = 2$

    b. $\tan (x - 5) = -3$

    c. $-2 \sin 2x = \sqrt{2}$

28 Describe how the graph of each function below is related to the graph of $f(x) = x^2$.

    a. $g(x) = x^2 - 2$

    b. $g(x) = (x - 2)^2$

    c. $g(x) = 2x^2$

    d. $g(x) = -x^2 + 3$

    e. $g(x) = -(x + 3)^2$

29 Point D is the common midpoint of distinct line segments $\overline{AC}$ and $\overline{EB}$.

    a. Draw a diagram representing this situation.

    b. Prove that $ABCE$ is a parallelogram.

30 Factor each trinomial into the product of two binomials. The coefficients and constant terms of the binomials can be real or imaginary numbers.

    a. $16x^2 - 1$ **b.** $16x^2 + 1$

    c. $x^2 - 6x + 9$ **d.** $x^2 - 6ix - 9$

31 Juana deposits $500 in a special saving account that earns 6% interest compounded annually. She does not withdraw any money.

    a. Write an exponential function rule that can be used to calculate the amount of money in the account after t years.

    b. How much money will be in the account after 15 years?

    c. How long will it take for Juana's money to triple in value? Explain your reasoning.

32 Given $f(x) = \log 2x$ and $g(x) = (x + 2)^3$, evaluate each of the following.

    a. $f(50) - g(-4)$ **b.** $f\left(\dfrac{1}{2}\right)$

    c. $f(g(5))$ **d.** $g(f(5))$

    e. $g^{-1}(-1)$ **f.** $f^{-1}(3)$

33 Solve each equation.

    a. $10^x = 375$ **b.** $4(10^{2x}) = 400$

    c. $6(10^{x-3}) = 300$ **d.** $8(3^x) = 158$

Looking Back

In this unit, you revisited and broadened your knowledge of trigonometric functions and their applications. By using reasoning and symbolic manipulation strategies, you were able to derive trigonometric identities that are helpful in solving equations that involve trigonometric functions used to model projectile motion and periodic phenomena. You extended the trigonometric function family to include the secant, cosecant, and cotangent functions and explored some of their properties. Finally, you used trigonometric functions to further your study of complex numbers and their applications and geometric representations.

As you complete the tasks that follow, think about how reasoning with and manipulating trigonometric expressions enhances your understanding of trigonometric functions and their uses. Consider also the precision and economy of thought (and work) that is often gained through this kind of reasoning. Reflect on how you could check your solutions with technology tools.

1 As you have previously seen, the number of minutes S between sunrise and sunset in any one location for a particular day of the year can be modeled by a variation of the basic sine or cosine function. One approximate model for the number of minutes between sunrise and sunset in Washington, D.C., is $S(d) = 180 \sin (0.0172d - 1.376) + 720$, where d is the day of the year with January 1 as day 1.

    a. On what day is the amount of time between sunrise and sunset closest to 900 minutes?

    b. On which days during a year will there be more than 12 hours between sunrise and sunset in Washington, D.C.?

    c. Will there ever be a day where there are less than 8 hours between sunrise and sunset in Washington, D.C.? Explain.

    d. Estimate the day of the year when the amount of time between sunrise and sunset in Washington, D.C. is greatest. Least.

2 The amount of power that a city uses varies with the time of day. Suppose that the power requirements in megawatts (MW) for Dos Rios are modeled by the function $P(t) = 40 - 20 \cos \left(\frac{\pi}{12}t - \frac{\pi}{4}\right)$, where t is the number of hours after midnight on Sunday night.

    a. Explain why this is a reasonable mathematical model. Over what domain do you think this model is valid?

b. At what times will the city consume 40 MW of power?

c. Under present conditions on power demands, will the city ever need 80 MW or more of power? Explain your thinking.

d. During what time periods will the city need less than 30 MW of power?

3 Suppose you are observing the motion of a clock pendulum. As you saw in Unit 1, *Families of Functions*, the directed distance the pendulum is to the left or right of its vertical position can be modeled by a cosine function. For one clock, the modeling rule is $d(t) = 5 \cos \pi t$, where t is the number of seconds since the pendulum was released at the right endpoint of its swing.

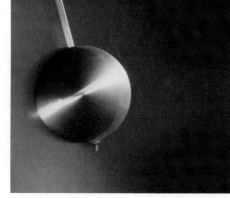

a. How long does it take for one complete swing of the pendulum?

b. At what times since its release will the pendulum be exactly vertical?

c. During what time periods since its release will the pendulum be to the left of vertical?

4 Determine if each of the following is an identity. If so, prove it; if not, give a counterexample.

a. $(1 - \tan \theta)^2 = \sec^2 \theta - 2 \tan \theta$

b. $\sin \theta + \cot \theta \cos \theta = \csc^2 \theta$

c. $(\sin \theta + \cos \theta)(\tan \theta + \cot \theta) = \sec \theta + \csc \theta$

d. $\cos (\alpha - \beta) \cos (\alpha + \beta) = \cos^2 \alpha - \sin^2 \beta$

e. $\tan^2 \theta = \sec^2 \theta - \sin^2 \theta - \cos^2 \theta$

f. $\tan \alpha = \dfrac{1 - \cos 2\alpha}{\sin 2\alpha}$

5 Find the exact values of each of the following.

a. $\cos 75°$

b. $\tan \dfrac{\pi}{12}$

6 Solve each equation on the interval $-2\pi \le x < 2\pi$.

a. $\sin^2 x - \cos^2 x - \cos x - 1 = 0$

b. $6 \tan^2 x + 5 \tan x + 1 = 0$

c. $\cos 2x + \cos x = 0$

7 Consider the general trigonometric equation $a \sin x + b = c$.

a. Describe how you would solve this general equation.

b. How does your thinking allow you to assess the reasonableness of the CAS-produced result below?

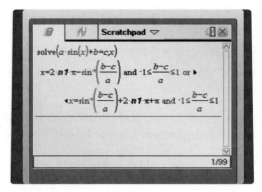

c. Why do the solutions have the restriction $-1 \leq \dfrac{b - c}{a} \leq 1$?

8 In the complex number plane, the vertices of a square are $(2, 2i)$, $(-2, 2i)$, $(-2, -2i)$, and $(2, -2i)$.

a. Sketch the square and write the complex numbers that correspond to the vertices:

  i. in standard form.

  ii. in trigonometric form

b. Express $1 - i$ in trigonometric form. Then find the image of the square in Part a under the transformation accomplished by multiplying each vertex by $1 - i$.

  i. Does it make any difference if the multiplication is done with $1 - i$ on the left or on the right? Explain.

  ii. Describe the effects of this transformation on the square.

9 The four fourth roots of a complex number z lie on a circle centered at the origin of the complex number plane. One root is $1 + 2\sqrt{2}\, i$.

a. What is the radius of the circle?

b. Find z.

c. Find the three other fourth roots of z expressed in standard form $a + bi$.

d. Represent each of the four fourth roots of z on a circle centered at the origin of the complex number plane with the radius you found in Part a.

SUMMARIZE THE MATHEMATICS

In this unit, you extended your understanding and skill in working with trigonometric functions and with complex numbers and discovered additional connections between complex numbers and geometry.

a What is a trigonometric identity?

   i. What strategies are useful in proving trigonometric identities?

   ii. What are the fundamental trigonometric relationships that are most useful in proving trigonometric identities?

b How are the sine and cosine of the sum of two angles related to the sines and cosines of the individual angles? What about the sine and cosine of the difference of two angles?

c Describe a reasoning strategy for solving $a \sin (cx + d) = k$ for x. Note any restrictions that must be made so that a solution exists. How could you check your answer using technology?

d How can you quickly sketch the graph of $y = \sec x$? Of $y = \csc x$? Of $y = \cot x$?

e Given a complex number in standard form $a + bi$, how can you derive the trigonometric form of the number?

f Explain with illustrations how geometric transformations are connected to complex numbers and the operations of multiplication and division.

g What is De Moivre's Theorem, and how is it useful in finding powers and roots of a complex number?

Be prepared to explain your ideas to the class.

 CHECK YOUR UNDERSTANDING

Write, in outline form, a summary of the important mathematical concepts and methods developed in this unit. Organize your summary so that it can be used as a quick reference in future units.

UNIT 5
Exponential Functions, Logarithms, and Data Modeling

Many important scientific problems require finding mathematical models for relationships that are expressed only in observational data. Furthermore, those problems often involve numbers of very different magnitudes, making it hard to visualize the relationship in a data plot. These two problems of data analysis and modeling can often be solved by transforming the given data with inverse functions, especially square roots and logarithms.

In this unit, you will extend your skill in use of exponents and logarithms to represent and solve problems.

The key ideas and techniques will be developed in two lessons.

Exponents and Natural Logarithms

In many earlier units and problems of *Core-Plus Mathematics*, you have used exponential functions to model relationships between variables—especially variables whose values change over time like populations, financial investments, and medications that metabolize in a patient's body. For example,

- In the 2010 Census, the United States population was over 308 million and growing at a rate of about 0.8% per year. Thoughtful planning by government agencies requires ability to predict the population many years into the future.

- Many financial planners assume that investments will increase at an annual rate of at least 5%. If a person has $50,000 of retirement savings, it is important to be able to predict the value of that investment many years in the future.

- When a person with a bacterial infection receives a 250-mg dose of antibiotic, the amount of medicine that is active in the patient's body might decline over time at a rate of 10% per hour. To assure that the infection is treated effectively, it is important to know when the active medication reaches the level where a new dose should be given.

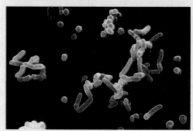

(t)Hisham Ibrahim/Getty Images, (c)©Royalty-Free/Corbis, (b)Image Source/Getty Images

Work on the problems of this lesson will review and extend algebraic skills and understanding needed to deal with modeling of exponential growth and decay situations. You will learn about the irrational number *e* that mathematicians and scientists consider the *natural* base for exponential and logarithmic functions. Then you will develop skill in using base-*e* exponential and logarithmic functions to solve problems.

INVESTIGATION 1

What is e^x?

Every problem that involves exponential growth or decay seems to have its own natural rate of growth or decay. Those different rates lead to exponential expressions with different bases in each case.

As you work on the problems of this investigation, look for answers to this question:

> *What is the number e and how can it be used as a kind of universal base for expressing every type of exponential growth or decay function?*

Credit Card Cautions Credit cards are almost more common than cash in today's consumer world. People around the world use billions of credit cards every day. Unfortunately, some people also find that their unpaid credit balance is a serious personal finance problem. Credit card companies charge interest at *annual percentage rates* (APR) from 7% to as much as 36%, and they usually compound those interest charges when payments are not made in a timely fashion.

 Suppose that a fiscally careless person runs up a credit card bill of $1,000, but then loses his/her job and cannot make any monthly payments.

a. If the credit card company charges interest at an annual rate of 18% compounded monthly, what will be the outstanding balance in the person's account after:

   i. 1 month?

   ii. 2 months?

   iii. 3 months?

b. What recursive formula shows how to calculate the credit card balance from one month to the next?

c. What formula shows how to calculate the credit card balance after x months?

d. What will be the balance on the credit card account if no payments are made for an entire year?

e. Compare the account balance after addition of interest charges for 12 months to the starting balance of $1,000 to find the actual percent increase in the amount due by completing the following tasks.

  i. Calculate the quotient (*balance due after 12 months*) ÷ 1,000.

  ii. Express the result as a percent.

  iii. Use that result to find the actual 12-month percent increase in the balance due.

2 The actual 12-month percent increase in balance due (that you found in Problem 1 Part eiii) is often called the **effective annual interest rate**. It is the simple interest rate that yields the same amount after one year as the compound annual percentage rate of interest.

a. Suppose that another credit card company tries to attract customers by proposing the same 18% annual percentage rate, but compounding interest charges quarterly (every three months). Would this offer be more attractive than the compounding offer in Problem 1? Why or why not?

b. Check your understanding of this new credit card offer to deal with the $1,000 balance. What will be the outstanding balance in the account after:

  i. 3 months?

  ii. 6 months?

  iii. 9 months?

  iv. 12 months?

c. What formula shows how to calculate the credit card balance after x quarters?

d. What is the *effective annual interest rate* of this credit plan?

3 Suppose that a third credit card company offers to take on customers with prior credit problems at the same 18% annual percentage rate, but with interest charges compounded every week. See how much more costly it would be to use this credit card in dealing with the $1,000 balance.

a. What will be the outstanding balance on the account after 1 week? After 2 weeks? After 52 weeks?

b. What formula shows how to calculate the credit card balance after x weeks?

c. What is the *effective annual interest rate* of this credit plan?

4 Next consider similarities and differences in credit card plans with different annual interest rates and different compounding schemes. For each of the following conditions:

- write a formula for calculating the balance after x compounding periods.

- compare the annual percentage rate and the effective annual interest rate.

a. Initial balance $1,000, annual interest rate 19%, and monthly compounding

b. Initial balance $5,000, annual interest rate 22%, and quarterly compounding

c. Initial balance $1,500, annual interest rate 13%, and weekly compounding

5 Suppose that a credit card account balance starts at B_0, the company charges interest at an annual interest rate r (expressed as a decimal), and interest is added with compounding n times in each year. Assuming that no payments are made on the account:

a. What formula shows how to calculate the outstanding balance $B(x)$ after x compounding periods?

b. How many compounding periods are there in t years?

c. What formula shows how to calculate the outstanding balance $C(t)$ after t years with initial balance C_0?

6 When you deposit money in an investment like a bank savings account, that deposit earns money for you in interest paid by the bank. What formula would give the value of such an investment after t years where the initial deposit is I_0 and interest is paid at an annual interest rate r (expressed as a decimal) with compounding n times in each year?

What is the Number e? In your answers to Problems 1–6, you produced a variety of functions with rules in the familiar form $f(x) = a(b^x)$, where the value of b depends on the annual interest rate and the number of times in a year that interest is compounded. You probably discovered that the actual effect of frequent compounding is less impressive than you expected.

It turns out that there is a way to explain the effects of frequent compounding and to deal with the different exponential growth conditions by using one special number as the base. The key to that strategy lies in understanding behavior of the expression $\left(1 + \frac{1}{n}\right)^n$ for large values of n.

7 Explain why it makes sense to think of the expression $\left(1 + \frac{1}{n}\right)^n$ as representing the value of a $1 credit card balance being charged a 100% annual interest rate that is compounded at n equal intervals in each year. Then evaluate $\left(1 + \frac{1}{n}\right)^n$ for values of n in the following table.

n	1	3	5	10	100	500	1,000	10,000	100,000
$\left(1+\frac{1}{n}\right)^n$									

The number to which values in the table in Problem 7 are converging is a special mathematical constant that is labeled throughout the scientific world with the letter e, in honor of the mathematician Leonhard Euler who discovered some of the most important properties of the number. **The value of e is an irrational number, approximately 2.71828.** Like the more familiar geometric constant π, it is not the solution of any simple polynomial equation.

8 For increasingly large values of n, the expression $\left(1 + \frac{1}{n}\right)^n \approx e$.

a. Use the relationship between $\left(1 + \frac{1}{n}\right)^n$ and e and algebraic properties to explain steps in the following connection of e to frequent compounding of interest.

Step 1. $\left(1 + \frac{1}{\frac{n}{r}}\right)^{\frac{n}{r}} \approx e$

Step 2. $\left(1 + \frac{r}{n}\right)^{\frac{n}{r}} \approx e$

Step 3. $\left(1 + \frac{r}{n}\right)^n \approx e^r$

Step 4. $\left(1 + \frac{r}{n}\right)^{nt} \approx e^{rt}$

b. Complete the following sentence in a way that explains the relationship of the number e and frequent compounding of interest.

The value of a debt (or investment) that starts at A and is charged (or paid) interest frequently at an annual rate of r (expressed as a decimal) can be approximated by _____.

9 In Problem 3, you considered the balance of a credit card with initial balance of \$1,000 owed and an 18% annual interest rate compounded weekly. The balance is $B(t) = 1{,}000\left(1 + \frac{0.18}{52}\right)^{52t}$ after t years with no payoff amounts.

a. Write a rule for an approximation of $B(t)$ in terms of e.

b. How does the approximate balance produced using $B(t)$ in terms of e compare to the balance computed from the form showing weekly compounding? Why does this make sense?

c. How would the two forms compare if the annual interest rate was computed daily? Check your conjecture.

10 Now examine more closely the function $f(x) = e^x$.

a. To which function family does $f(x)$ belong?

b. Sketch a graph of $f(x) = e^x$ and state the domain and range of the function.

c. How does the graph of $f(x) = e^x$ compare to the graph of $g(x) = 2^x$? To the graph of $h(x) = 3^x$? To the graph of $j(x) = 10^x$? Explain.

d. How would the graph of $j(t) = e^{0.18t}$ compare to the graph of $f(t) = e^t$? Check your conjecture and explain your findings.

e. Does $f(x) = e^x$ have an inverse that is a function? Explain your reasoning.

✓ CHECK YOUR UNDERSTANDING

When banks compete for your savings account, they usually advertise their annual interest rate quite prominently. It is common to pay interest with monthly or quarterly compounding. But some aggressive banks might offer *daily* or even *continuous* compounding.

Suppose that you invest $500 in a bank that advertises daily compounding at an annual percentage rate of 2%.

a. What formula gives the value of that investment $V(x)$ after x days? (Banks usually use 360 days as equivalent to a year.)

b. What function $V(t)$ gives the value of the investment after t years?

c. How can the value of the function in Part b be approximated by an exponential function with base e?

d. Use all three expressions for investment value to estimate the *effective annual interest rate* of the savings account plan.

Getty Images/Photodisc

Applications of e^x and ln x

The problems of Investigation 1 focused on compound growth situations that are modeled well by exponential functions. In particular, you learned about the special number $e \approx 2.71828$ and how it can be used as a base in expressions for exponential functions.

If a quantity is increasing or decreasing at a constant percent rate, with frequent compounding in each unit of time, the size of that quantity at any time t can be approximated well by an exponential function with rule in the general form $f(t) = Ae^{rt}$, where r is the percent rate expressed as a decimal. Questions about those exponential functions often require finding the values of A, r, t, or $f(t)$ that are determined by specific problem conditions.

As you work on the problems of this investigation, look for answers to this question:

How can problems involving exponential growth and decay be expressed and solved using base-e exponential and logarithmic functions?

1 Madagascar is an island nation in the Indian Ocean off the southeastern coast of Africa. The main island is the fourth largest island in the world. Census trends suggest that the population of Madagascar (in millions) at a time t years from 2012 can be predicted reasonably well by the function $M(t) = 21.9e^{0.03t}$.

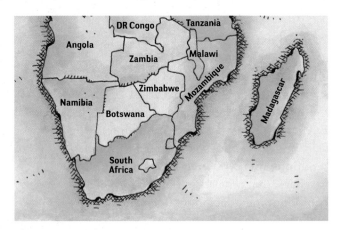

a. What information about the Madagascar population and its growth is suggested by the parameters in the rule for $M(t)$?

b. What is the predicted population of Madagascar in 2017?

c. When is the population of Madagascar predicted to reach 27 million?

d. When was the population of Madagascar about 17 million?

2 In work on Problem 1, you probably used tables and graphs to estimate answers for Parts c and d. These population questions can also be solved using what you know about base-10 logarithms. Solve the following equations using common logarithms. Then compare the solutions to your answers from Problem 1.

a. $27 = 21.9e^{0.03t}$

b. $17 = 21.9e^{0.03t}$

Solution of equations involving exponential expressions with base e is made somewhat simpler by introduction of logarithms with base e. The function that provides those logarithms is called the **natural logarithm** function. While it could be written $f(x) = \log_e x$, tradition in mathematical and scientific practice uses the notation $f(x) = \ln x$ or $f(x) = \ln (x)$. By definition,

$$\ln a = b \text{ if and only if } e^b = a.$$

3 On most technology, there is a ▣ **LN** button or software feature that returns base-e logarithms of entered numbers. You can also enter the function $f(x) = \ln x$ to investigate tables and graphs of the natural logarithmic function.

a. What are the domain and range of $f(x) = \ln x$?

b. For both the domain and range of $f(x) = \ln x$, explain:

- how your answer is shown by patterns in tables of values for $f(x) = \ln x$.

- how your answer is shown by patterns in the graph of $f(x) = \ln x$.

- how your answer can be explained logically, using the relationship of the natural logarithmic function $f(x) = \ln x$ and the exponential function $g(x) = e^x$.

c. Sketch graphs of the natural logarithmic function $f(x) = \ln x$ and the exponential function $g(x) = e^x$. Explain how the shape and relationship of those graphs illustrates the fact that the two functions are inverses of each other.

4 Use the definition of the base-e logarithm function and the natural logarithm capability of technology to solve these equations for x. If there is no real solution, explain why that makes sense.

a. $e^x = 20$ b. $e^x = 25$

c. $e^x = 1$ d. $e^x = -10$

e. $e^{2x + 3} = 15$ f. $15e^x = 30$

5 Hamilton County, Indiana, had a population of about 183 thousand in 2000 and about 275 thousand in 2010. Suppose that the population of this county is growing in a pattern that can be modeled by an exponential function with rule $P(t) = P_0 e^{rt}$, where t stands for years since 2000.

a. What is $P(0)$ and what does that value tell about the formula for $P(t)$?

b. What is $P(10)$? How can that value, the result from Part a, and the formula for $P(t)$ be combined to give an equation in which the only unknown is r?

Indiana

Hamilton County

c. Use the base e logarithm function and strategies developed in work on Problem 4 to solve the equation in Part b for r. Then explain what the solution tells about the growth rate of the Hamilton County population.

d. If Hamilton County continues to grow at the same rate, what population is predicted by this model in 2020?

6 Suppose that early in a flu epidemic, the number of cases doubles every 5 days, and that the number of cases at any time t days after the epidemic begins can be modeled by a function $C(t) = C_0 e^{rt}$.

a. Explain why the given information implies that $C(5) = 2C(0)$ and $C_0 e^{5r} = 2C_0 e^{0r}$.

b. Use natural logarithms and algebraic reasoning to solve $C_0 e^{5r} = 2C_0 e^{0r}$ for r.

c. If counting of flu cases begins when 250 have been reported, what number of cases is predicted for a time 14 days later? For 28 days later?

7 The Nobel Prize in Chemistry for 1911 was awarded to Madame Marie Curie for her discovery of the radioactive elements radium and polonium. Amounts of those elements decay exponentially in a pattern that can be modeled by $R(t) = R_0 e^{rt}$.

a. Suppose that Madame Curie produced a 100-gram sample of radium-226 (^{226}Ra) in 1911 and that 99 grams remained at her death in 1934. What values of R_0 and r define a model for the decay of that sample?

b. How much of the original sample remained in 2014?

c. What is the *half-life* of ^{226}Ra?

8 All living matter contains both stable and radioactive forms of the element carbon. In all living matter, the ratio of radioactive carbon (^{14}C) to stable carbon (^{12}C) is the same. However, when any living matter dies, the radioactive isotope begins to decay and the ratio of radioactive to stable carbon declines exponentially. This idea is used in dating of archaeological discoveries, because (^{14}C) has a known half-life of about 5,730 years.

a. If the function $R(t) = R_0 e^{rt}$ gives the ratio of radioactive to stable carbon in an organic object that has been dead for t years:

i. what does R_0 represent?

ii. what value of r is implied by the half-life of 5,730 years for radioactive carbon (^{14}C)?

b. Suppose that an archaeologist discovers an ancient relic and finds that its ratio of radioactive to stable carbon is $0.2R_0$. About how long ago was that object last "alive"?

SUMMARIZE THE MATHEMATICS

In this investigation, you extended your understanding and skill in work with exponential and logarithmic functions.

a If a quantity changing by exponential growth or decay is modeled by $A(t) = A_0 e^{rt}$, what do the values of A_0 and r tell about the situation?

b In what ways can the values of A_0 and r be determined for any particular situation?

c How is the natural or base-e logarithmic function $\ln x$ defined? Sketch a graph of $y = \ln x$ and indicate the domain, range, and any asymptotes of the function.

d How can natural logarithms be used to solve equations in the form $c = A_0 e^{rt}$ for r or t?

Be prepared to explain your ideas and reasoning to the class.

 CHECK YOUR UNDERSTANDING

According to the United States Census Bureau, Wyoming is the least populous state. In 1980, the population of Wyoming was 469,557; in 1990, the population was 453,588; and in 2000, the population was 493,782.

a. Assuming that the population of Wyoming is changing exponentially, what function $P(t) = P_0 e^{kt}$ is suggested by the given data for 1980 and 1990? Test the fit of that model by evaluating $P(20)$.

b. The actual 2010 population of Wyoming, according to the U.S. Census, was 563,626. Find another model $Q(t)$ for future projection of Wyoming's population using the given data for 1990 and 2000. Use that model to predict Wyoming's population in the year 2010. In the year 2015.

c. Use natural logarithms to solve the equation $Q(t) = 550,000$ and explain what the solution tells about the population of Wyoming.

d. Assuming $t > 0$, write each of the following expressions in a different equivalent form.

   i. $\ln (e^{4t})$ **ii.** $e^{\ln 4t}$ **iii.** $\ln 1$

INVESTIGATION 3

Properties of e^x and $\ln x$

Some students who have been working with exponential functions using a variety of bases and with base-10 logarithms are often puzzled when the irrational number e is introduced and used exclusively thereafter. The special useful character of e will become evident as you learn more mathematics, especially calculus. At this point in the story, it is best to explore how all exponential and logarithmic functions are related to each other.

As you work on the problems of this investigation, look for answers to these questions:

How can any exponential function $y = b^x$ be expressed in equivalent form using e as the base?

How are logarithms with different bases related to each other?

How can properties of logarithms and exponents be used to solve equations and rewrite algebraic expressions in useful equivalent forms?

The Families of Exponential and Logarithmic Functions In much the same way that all polynomial and rational functions share common symbolic forms, there are strong and useful connections among the varieties of exponential and logarithmic functions.

1 The following graphs should match what you would expect for the domain, range, and patterns of change for the exponential functions $y = e^x$ and $y = e^{-x}$.

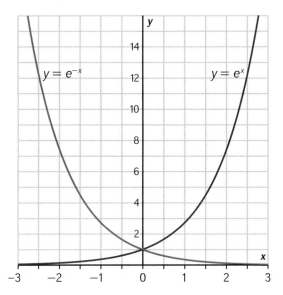

a. What are the domain and range of each function?

b. How do your answers to Part a explain the fact that for every positive number b, there is a number k satisfying the equation $b = e^k$?

c. Solve each of these equations for k. Check your answers by substituting the solution for k into the original equation.

    i. $6.5 = e^k$ **ii.** $0.8 = e^k$ **iii.** $b = e^k$

d. Use results of your work in Part c to show how each of the following exponential functions can be expressed in equivalent form using exponential base e.

    i. $y = 15(6.5^x)$ **ii.** $y = 75(0.8^x)$ **iii.** $y = b^x$

e. The family of functions with rules in the form $f(x) = Ae^{kx}$ includes models for all possible exponential growth and decay situations. So, it is helpful to know how the parameters A and k determine the pattern of change in the values and the graph of each specific function.

i. What does the value of A tell about the function $f(x) = Ae^{kx}$ and its graph?

ii. Use properties of exponents to prove your answer to part i.

iii. Which values of k determine exponential growth models and which give decay models?

iv. How does the fact that $e^{kx} \approx (2.7^k)^x$ help to explain your answer to part iii?

2 You now know two important logarithmic functions, $y = \log x$ and $y = \ln x$, that are defined as:

$$y = \log x \text{ if and only if } 10^y = x \quad \text{and} \quad y = \ln x \text{ if and only if } e^y = x.$$

a. What are the domain and range of the functions $y = \log x$ and $y = \ln x$?

b. Use the definitions of the two logarithmic functions and the related exponential functions to explain why:

i. $\log 1 = 0$ and $\ln 1 = 0$.

ii. $\log x > 0$ and $\ln x > 0$ for $x > 1$.

iii. $\log x < 0$ and $\ln x < 0$ for $0 < x < 1$.

c. Produce graphs of the two functions in the window $0 \le x \le 10$ and $-3 \le y \le 3$. Then explain how those graphs illustrate your answers about the domain and range of each.

3 The following identities are often useful in rewriting expressions involving logarithms.

a. Explain why if $a > 0$, $a = 10^{\log a}$. Why $a = e^{\ln a}$.

b. Express the two identities in words.

Properties of Natural Logarithms In earlier work with base-10 logarithms, you learned that it is often helpful to write given logarithmic expressions in different equivalent forms. It should be reasonable to expect that some rules for rewriting base-10 logarithms have analogs that work with base-e logarithms.

4 Write the following expressions in different forms that you suspect will be equivalent to the originals. Then complete the given numerical examples to check your conjectures.

a. $\ln ab = $ _____ , for a and b both positive numbers. For example, $\ln (5 \cdot 4) = $ _____ .

b. $\ln \left(\frac{a}{b}\right) = $ _____ , for a and b both positive numbers. For example, $\ln (20 \div 2) = $ _____ .

c. $\ln a + \ln b = $ _____ , for a and b both positive numbers. For example, $\ln 5 + \ln 4 = $ _____ .

d. $\ln (a^b) = $ _____ , for positive number a and any number b. For example, $\ln (3^4) = $ _____ .

5 Use the fact that $a = e^{\ln a}$, $b = e^{\ln b}$ and properties of exponents to complete and give reasons for each step in the following proofs of your conjectures in Problem 4 about properties of the natural logarithm function.

a. For any positive numbers a and b:

$$ab = e^{\ln a}e^{\ln b} \qquad (1)$$
$$= e^{\underline{\hspace{1.5cm}}} \qquad (2)$$
So, $\qquad \ln ab = \underline{\hspace{1.5cm}} \qquad (3)$

b. For any positive numbers a and b:

$$\frac{a}{b} = \frac{e^{\ln a}}{e^{\ln b}} \qquad (1)$$
$$= e^{\underline{\hspace{1.5cm}}} \qquad (2)$$
So, $\qquad \ln\left(\frac{a}{b}\right) = \underline{\hspace{1.5cm}} \qquad (3)$

c. For any positive numbers a and b:

$$e^{\ln a + \ln b} = e^{\ln a}e^{\ln b} \qquad (1)$$
$$= e^{\underline{\hspace{1.5cm}}} \qquad (2)$$
So, $\ln a + \ln b = \underline{\hspace{1.5cm}} \qquad (3)$

d. For any positive number a and any number b:

$$a^b = \left(e^{\ln a}\right)^b \qquad (1)$$
$$= e^{\underline{\hspace{1.5cm}}} \qquad (2)$$
So, $\qquad \ln\left(a^b\right) = \underline{\hspace{1.5cm}} \qquad (3)$

Connecting Logarithmic Functions Logarithms are defined for any base $a > 0$ as $\log_a b = c$ if and only if $a^c = b$. If you are working on a problem that requires solving the equation $3^x = 75$, the obvious solution is $x = \log_3 75$. But few, if any, calculators have buttons giving $\log_3$ values. Fortunately, in the same way that all exponential functions are closely related and can be expressed in terms of any one particular base (especially 10 or e), all possible logarithmic functions are closely related.

6 There is a *change of base* procedure for finding $\log_a b$ for any positive numbers $a \neq 1$ and $b \neq 1$.

a. Explain how properties of logarithms justify each step in the following algebraic derivation of the **change of base formula $\log_a b = \dfrac{\log_{10} b}{\log_{10} a}$**.

Suppose $\qquad\qquad b = a^n$.

Then $\qquad\qquad \log_{10} b = \log_{10} a^n. \qquad (1)$
$$\log_{10} b = n \log_{10} a \qquad (2)$$
$$\log_{10} b = \log_a b \, \log_{10} a \qquad (3)$$
So, $\qquad\qquad \log_a b = \dfrac{\log_{10} b}{\log_{10} a}. \qquad (4)$

b. Express the change of base formula in words.

c. Use the change of base formula to solve $3^x = 75$.

d. Modify the argument in Part a to prove that $\log_a b = \dfrac{\ln b}{\ln a}$ for any positive numbers a and b.

e. Use the change of base formula involving $\ln x$ to solve $3^x = 75$.

7 Use algebraic reasoning to solve each of the following equations for x.

a. $7^x = 5$ **b.** $8^x = 20$

c. $4.1^{2x} + 5 = 3$ **d.** $e^{3x + 5} = 18$

Equations, Exponents, and Logs The definitions and properties of exponents and logarithms that you have developed in this investigation and in earlier units can be used in a variety of ways to transform algebraic expressions and to solve algebraic equations.

8 Data in the following table show the relationship between distance from the Sun d (in millions of miles) and orbit time t (in Earth days) for the major planets of our solar system.

Planet	Mercury	Venus	Earth	Mars	Jupiter	Saturn	Uranus	Neptune
Distance	36	67	93	142	484	887	1,784	2,795
Orbit Time	88	224	365	686	4,344	10,768	30,660	60,225

Source: NASA Planetary Fact Sheets; nssdc.gsfc.nasa.gov/planetary/planetfact.html

Analysis of the pattern in these data led scientists to the equation $\log t = 1.5 \log d - 0.39$ relating orbit time to distance from the Sun.

a. Explain how properties of exponents and logarithms justify each step in the following algebraic reasoning that produces an equation showing t as a function of d.

$$\text{If} \qquad \log t = 1.5 \log d - 0.39,$$
$$\text{Then} \qquad 10^{\log t} = 10^{1.5 \log d - 0.39}. \qquad (1)$$
$$t = (10^{1.5 \log d})(10^{-0.39}) \qquad (2)$$
$$t = 10^{-0.39}(10^{\log d})^{1.5} \qquad (3)$$
$$t \approx 0.407 d^{1.5} \qquad (4)$$

b. Compare the predicted orbit times given by the model in Part a to the original data and explain why differences might be as expected.

c. Use algebraic reasoning—starting from the result in Part a or from the original equation $\log t = 1.5 \log d - 0.39$ itself—to find an equation that expresses distance d as a function of orbit time t. Check your work by comparing the predictions of this model to the original data.

d. How are the results of Parts a and c more useful than the original equation $\log t = 1.5 \log d - 0.39$?

9 Adapt the reasoning used in solving Problem 8 to transform each equation to an equivalent form that gives y as a function of x. Then identify the function family to which it belongs.

a. $\log y = \log (x - 1) + 1$ **b.** $\log y = -\log (x + 1) + 2$

c. $\log y = 1 - 2 \log (x + 3)$

d. $\ln y = 2 - 2 \ln (x - 1)$

e. $\ln y = k \ln (3x - 1) + 5$

f. $\log y = 3x - 1$

g. $\ln y = 2x - 7$

h. $\ln y = 0.5 \ln x + 3$

 Use algebraic reasoning to solve each of these equations for x. Check your solutions by substitution in the original equations. If no solution exists, explain why that is the case.

a. $\log (x + 1) - \log (x - 1) = 2$

b. $\ln (x + 1) + \ln (x - 1) = 0$

c. $\ln (3x - 1) - \ln (x + 1) = 0$

d. $\log (2x + 1) - 2 \log (x - 1) = 1$

SUMMARIZE THE MATHEMATICS

In this investigation, you extended your understanding and skill in work with exponential and logarithmic functions and in working with expressions and equations involving those functions.

a How can any exponential function $y = a(b^x)$ be expressed with an equivalent rule in the form $y = c(e^{kx})$?

b How can any logarithm $\log_a b$ be calculated by using $\log_{10}$? By using $\ln$?

c If an equation relating two variables x and y involves $\log x$ and $\log y$, what algebraic methods are likely to be useful in transforming the equation to an equivalent form in which y is shown as a function of x directly?

Be prepared to explain your ideas and reasoning to the class.

✔ CHECK YOUR UNDERSTANDING

The in-state cost for tuition, room, and board at 4-year public colleges and universities has been increasing recently in a pattern that can be modeled reasonably well by the function $c(t) = 14{,}333(1.06^t)$, where t is years since 2008 and $c(t)$ is in dollars per year.
(**Source:** College Board)

a. What equivalent rule for $c(t)$ uses e as the exponential base?

b. Solve the equation $14{,}333(1.06^t) = 20{,}000$ and explain what the solution tells about costs for higher education.

c. Show how to evaluate $\log_{1.06} (1.395)$ and explain why this provides the solution to the equation in Part b.

d. Transform the equation $\ln y = 4 \ln x - 1$ to a form showing that y is a power function of x.

APPLICATIONS

Many American young people pay for at least part of their college education by taking out loans. Typical student loans require no repayment until after graduation. Most loans, except those with special government sponsorship, allow interest to accrue while the borrower is in school.

1 Suppose Mae borrows $5,000 for tuition from a lender who charges interest at an annual rate of 6% compounded monthly.

a. What formula gives the balance of her loan at any time x months later?

b. What is the outstanding loan balance after:

    i. 1 month?

    ii. 2 months?

    iii. 1 year?

c. What is the *effective annual interest rate* of the loan?

2 Suppose that Ava borrows $5,000 for tuition from a lender who charges interest at an annual rate of only 3% compounded monthly.

a. What formula gives the balance of her loan at any time x months later?

b. What is the outstanding loan balance after:

    i. 1 month?

    ii. 2 months?

    iii. 1 year?

c. What is the *effective annual interest rate* of the loan?

3 If a student borrows money at the start of college, repayment might not occur until as much as 5 years later. Consider again the student who borrows $5,000 at an annual percentage rate of 6% compounded monthly. What formulas show the balance on this student's loan after:

a. 1 year?

b. 2 years?

c. t years, where $t < 5$?

4 Write and evaluate exponential expressions with base e that give approximations for the following calculations.

a. $\left(1 + \frac{0.07}{12}\right)^{12}$

b. $\left(1 + \frac{0.2}{50}\right)^{50}$

c. $\left(1 + \frac{0.3}{25}\right)^{25(2)}$

5 Museums and private collectors pay huge prices for famous works of art, and the values of those investments often increase more rapidly than general inflation rates. Suppose that a museum buys a famous painting for $25 million and that the value of that painting is projected to increase at an annual rate of 15% with nearly continuous compounding.

  a. What function gives the projected value of the painting at any time t years after its purchase?

  b. How long should the museum expect it to take before the value of the painting doubles?

6 Elite athletes are often given medicines to help their muscles recover quickly from the strains of competition. However, some of those medicines are illegal since they are seen as artificial performance enhancers. Suppose that a competitor in Iron-Man races is given a rubdown with a cream containing a banned steroid and that the steroid absorbed into his body is metabolized at a rate of 12% per day.

  a. What function gives the percentage of the original concentration of the steroid remaining after d days?

  b. How long will it take for the athlete's body to metabolize the steroid to a level that is only 1% of the original concentration?

7 The population of the United States was about 280 million in the year 2000 and it was estimated at 309 million in 2010. Suppose that the U.S. population is increasing exponentially.

  a. What function with rule in the form $P(t) = P_0(e^{kt})$ matches the population figures for 2000 and 2010? Show how to find that rule by using the given information and the natural logarithm function $\ln x$.

  b. For what value of t is $P(t) = 350$ million? Show how to find that value by algebraic reasoning using the natural logarithm function $\ln x$.

8 Strontium-90 is a dangerous radioactive element that is produced by nuclear weapons tests and absorbed by plants and animals. It has a half-life of about 29 years.

  a. What is the decay constant for strontium-90?

  b. How long will it take 50 mg of strontium-90 to decay to 1% of that amount?

  c. How long will it take 100 mg of strontium-90 to decay to 1% of that amount?

  d. Show that the time for any amount A of strontium-90 to decay to 1% of A is independent of the value of A.

9 The woolly mammoth is an extinct mammal that resembled the modern elephant, but lived in colder climates of Europe, Asia, and North America from 3 million to 10,000 years ago. The distinctive curved tusks of mammoths could reach a length of over 15 feet and samples of those tusks are found often in Northern Asia and Canada.

In 2006, fishermen in Siberia uncovered the remains of a nearly complete woolly mammoth skeleton—backbone, skull, teeth, and tusks.

   a. If carbon dating showed that the level of radioactive carbon ^{14}C in the mammoth remains was only 5% of that in living matter, how long ago did this particular woolly mammoth die? (Recall that the function $R(t) = e^{-0.000121t}$ compares the level of radioactive carbon ^{14}C in an object t years after it was last alive to what would be expected in living matter.)

   b. What level of radioactive carbon ^{14}C would indicate that the mammoth died 8,000 years ago?

10 Express each of the following exponential functions with an equivalent rule in the form $y = A(e^{kx})$.

   a. $y = 10(2^x)$ **b.** $y = 10(1.5^x)$ **c.** $y = 20(0.5^x)$

11 Use the change of base formula to find good decimal approximations for each of these logarithms. In each case, check your answer by using the general definition of $\log_a b$.

   a. $\log_2 90$ **b.** $\log_5 600$ **c.** $\log_{0.5} 10$

12 Express each of these equations in equivalent form that gives y as a function of x. Then identify the function family to which it belongs.

   a. $\log y = 3 \log x + 1.5$

   b. $\ln y = 1.5 \ln x - 2$

   c. $\ln y = 2x + 5$

13 Solve each of these equations for x.

   a. $\log (x + 1) = 4 \log 2 + \log 3 - 3 \log 5$

   b. $\log (x + 2) = \log (3x - 5)$

   c. $\ln (x + 5) = \ln (3x + 1) + 2$

   d. $\ln (5x) + \ln (2x + 1) = 3$

14 Packaged foods often include special cooking instructions for high altitudes because liquids boil at lower temperatures where there is lower atmospheric pressure. The function $P(h) = 76e^{-0.118h}$ gives normal atmospheric pressure in centimeters of mercury at any location h kilometers above sea level. Another way of stating this is that the atmosphere at height h supports a column of mercury $P(h)$ centimeters high.

a. A marker on the steps of the capitol building in Denver, Colorado indicates that the "Mile High City" is 5,280 feet (1.609 km) above sea level. What does this information tell about the normal atmospheric pressure in Denver?

b. La Paz, Bolivia is the highest national capital in the world. The normal atmospheric pressure in La Paz supports a column of mercury that is 49.5 centimeters high. What is the elevation of La Paz above sea level?

c. Because atmospheric pressure is also an indication of the density of air, many mountain climbers carry and use bottled oxygen on their ascents. If a climber needs supplemental oxygen when atmospheric pressure (and thus oxygen content) is below 45% of conditions at sea level, at what elevation should the climber start using bottled oxygen?

d. The atmospheric pressure on the shore of the Dead Sea supports a column of mercury that is 79.83 cm high.

   i. What is the elevation of the Dead Sea relative to sea level of the oceans?

   ii. On the shore of the Dead Sea does water boil at a higher or lower temperature than near sea level of the oceans? Explain.

   iii. Will food cook faster, slower, or in the same time in boiling water on the shore of the Dead Sea than on the shore of the oceans? Explain your reasoning.

15 Have you ever noticed that pizza right out of the oven seems too hot to eat, but that it seems to cool quickly? One of the main factors contributing to how long it takes for pizza to cool is the difference between the temperature of the pizza and that of the room in which it is served. *Newton's Law of Cooling* says just that. In fact, it says that the temperature of the heated object decreases exponentially toward the temperature of the surroundings in a way that can be modeled by the equation

$$T(t) = T_{env} + (T_0 - T_{env})e^{rt}.$$

Here T_0 is the initial temperature of the object, T_{env} is the temperature of the surrounding environment (which generally can be assumed to be unchanged by introduction of a relatively small hot or cold object), and t is the number of minutes of cooling.

a. Suppose you cook a pizza at 400°F. After letting it cool for 5 minutes in a room with air temperature of 75°F, it cools to 165°F. What does this information tell about the value of r in the function $T(t)$?

b. If after the cooling period, it takes you 3 minutes to eat the first slice of pizza, what is the temperature of the last bite of that slice?

c. If your friend shows up 15 minutes after the pizza is taken out of the oven, what is the temperature of the pizza when he gets there?

d. When does the temperature of the pizza fall below 100°F?

CONNECTIONS

16 Write a *NOW-NEXT* rule or a recursive formula that models the pattern of change in each situation.

a. The balance due on a credit card account starts at \$2,500 and is charged an interest penalty on the outstanding balance with an annual interest rate of 18% compounded monthly. Assume that no monthly payments or new charges are made.

b. The balance of a bank savings account starts at \$500, earns interest at an annual interest rate of 3% compounded monthly, and is increased also by monthly deposits of \$25.

c. Population of an elephant herd in a large game preserve starts at 10,000, increases by 5% per year, and is reduced by 250 animals each year due to harvesting by native people living in the preserve.

17 The formula $f(n) = \left(1 + \dfrac{1}{n}\right)^n$ defines a sequence of real numbers, one for every positive integer value of n.

a. List the first 5 terms in the sequence. Round terms to the hundredths place.

b. Explain why the sequence of approximations to e is not an arithmetic sequence.

c. Explain why the sequence of approximations to e is not a geometric sequence.

18 Consider the function $n(x) = \dfrac{1}{\sqrt{2\pi}} e^{\frac{-x^2}{2}}$. This function is a variation of the exponential function $y = e^x$.

a. Produce a graph of $n(x)$. (Be careful to enter the function rule following order of operations conventions carefully!)

b. Identify the line of symmetry of the graph of $n(x)$ and explain why the rule for the function determines that line of symmetry.

c. Identify the probability distribution function whose graph has shape very similar to the graph of $n(x)$. What does the symmetry of the graph reveal about the mean and median of that distribution?

19 In earlier units focused on polynomials, you used those easily calculated expressions to model interesting graph patterns. It turns out that there is a way to approximate values of e^x with suitable polynomials as well. Consider the polynomial $p(x) = 1 + x + \frac{1}{2}x^2 + \frac{1}{6}x^3 + \frac{1}{24}x^4 + \frac{1}{120}x^5$.

a. Evaluate $p(x)$ and e^x for the values in the table below.

x	−1.0	−0.5	0.0	0.5	1.0	1.5	2.0	2.5	3.0
p(x)	0.367								
e^x	0.368								

b. Compare $p(4)$ to e^4 and $p(5)$ to e^5. What does this and your work in Part a suggest about the polynomial approximation of e^x?

c. Describe the pattern in coefficients of the polynomial $p(x)$. Then use the pattern to write a polynomial $q(x)$ of degree eight having the same coefficient pattern.

d. Compare $q(4)$ to e^4 and $q(5)$ to e^5 and see if you can explain why these pairs of calculations agree better than those in Part b.

20 In the 18th century, the Swiss mathematician Leonhard Euler discovered a surprising connection between the functions $y = e^x$, $y = \sin x$, and $y = \cos x$. The connection involves complex numbers. **Euler's formula** states that $e^{i\theta} = \cos \theta + i \sin \theta$, provided that θ is measured in radians.

a. Show that $e^{i\pi} + 1 = 0$. This statement, known as *Euler's identity*, is amazing because it relates five of the most important numbers in mathematics—e, π, i, 1, and 0.

b. Euler was depicted on a German stamp issued in his honor as illustrated at the right. The stamp shows another formula for which Euler is famous, $e - k + f = 2$. What relationship is represented by that formula?

21 Show that $y = a(b^x)$ is equivalent to $\log y = \log a + x \log b$ for $a > 0, b > 0$.

a. Explain why the logarithmic equation is linear in form.

b. If you think of $\log y$ as a function of x and graph the ordered pairs $(x, \log y)$:

  i. what is the slope of the graph of that function?

  ii. what is the vertical intercept of the graph?

22 Show that $y = ax^b$ is equivalent to $\log y = \log a + b \log x$ for $a > 0, x > 0$.

a. Explain why this logarithmic equation is linear in form.

b. If you think of $\log y$ as a function of $\log x$ and graph the ordered pairs $(\log x, \log y)$:

  i. what is the slope of the graph of that function?

  ii. what is the vertical intercept of the graph?

23 Many people find the following pitch enticing.

> *Need some cash to tide you over until the next payday?*
> *We'll loan up to $500 for two weeks.*

For example, in the Commonwealth of Virginia during 2006, over 400,000 people took out over 3.5 million payday loans with total value of over $1.3 billion. (**Source:** Anita Kumar, Pressure Mounts on Va. Payday Lenders, *The Washington Post*, Monday, December 3, 2007, page B5.)

What unsuspecting customers of payday loan stores often fail to realize is that the fee of "only" $15 per $100 loaned amounts to a very large annual percentage interest rate.

a. If you pay $15 upfront to borrow $100 for two weeks, what is the annual percentage rate?

b. In response to complaints by consumer advocates, the U.S. Congress passed a bill limiting to 36% the annual interest rates charged on payday loans to military personnel. What does this limit imply about the cost to military personnel of borrowing $100 for two weeks?

24 How do the results of your work on problems of Investigation 1 illustrate the importance of checking the *effective annual interest rate* when considering savings or credit card plans that advertise only *annual percentage rates* and compounding periods?

25 Look back at your work for Applications Tasks 1 and 2.

a. Suppose in Applications Task 1, Mae borrows $8,000. What is the effective annual interest rate in this case?

b. Suppose in Applications Task 2, Ava borrows $10,000. What is the effective annual interest rate in this case?

c. Based on your work in Applications Tasks 1 and 2 and your answers to Parts a and b, what appears to be true about the initial balance of a loan and the effective annual interest rate? Why does this make sense?

26 Recognizing simple forms of symbolic expressions is a very useful skill in mathematics. The solution to $e^x = 20$ is $x \approx 3$. Use this information to quickly solve the following equations, without the use of technology.

a. $e^{x+1} = 20$ **b.** $e^{3x} = 20$

c. $e^{0.5x} = 20$ **d.** $x \ln 20 = 15$

27 What are the advantages and limitations of each form in which exponential growth and decay functions can be expressed—using $y = a(b^t)$ with b unique to each application problem or using $y = a(e^{rt})$ with the same exponential base in all problems?

28 Common logarithms use 10 as the base and natural logarithms use e as the base. How can you convert the common logarithm of a number into the natural logarithm of the number and vice versa?

29 One of the most commonly used rules in solution of equations says (informally), "With the exception of dividing by zero, you can always do the same thing to both sides of an equation."

a. What are the "same things" you would do to both sides of an equation in the form $a(e^{rt}) = b$ on the way to solving for t? What is the result?

b. What are the "same things" you would do to both sides of an equation in the form $\ln (cx + d) = \ln (ax + b)$ on the way to solving for x? What is the result?

c. What, if any, cautions must be exercised when "doing the same thing to both sides" of such equations?

EXTENSIONS

30 In Course 3, Unit 7, *Recursion and Iteration*, you investigated *finite* series. The irrational number e can be defined as an *infinite* series of rational numbers.

$$e = \sum_{n=0}^{\infty} \frac{1}{n!}$$

The symbol ∞ represents infinity. For convenience, $0!$ is defined be 1.

a. The first three terms of this infinite series are *partial sums*:

$$1, 1 + 1, 1 + 1 + \frac{1}{2}, \text{ or } 1, 2, 2.5.$$

Write the next two terms of the series.

b. How many terms of the series are needed to produce the decimal approximation of e equal to that displayed by your calculator?

31 Work on problems of Investigation 1 suggested that, for large values of n, the expression $\left(1 + \frac{1}{n}\right)^n \approx e$. This fact was then used to derive the approximations

$$\left(1 + \frac{r}{n}\right)^n \approx e^r \qquad \text{and} \qquad \left(1 + \frac{r}{n}\right)^{nt} \approx e^{rt}.$$

a. Calculate $\left(1 + \frac{5}{10}\right)^{10}$ and e^5. Explain the discrepancy of the two results.

b. Calculate $\left(1 + \frac{0.05}{10}\right)^{10}$ and $e^{0.05}$. Explain why these results are so much closer than the pair of results in Part a.

c. Calculate $\left(1 + \frac{3}{25}\right)^{25(2)}$ and $e^{3(2)}$. Explain the discrepancy.

d. Calculate $\left(1 + \frac{0.03}{25}\right)^{25(2)}$ and $e^{0.03(2)}$. Explain why these results are so much closer than the pair of results in Part c.

32 At a typical fondue party, guests gather around small pots of heated chocolate and dip in strawberries for coating. Another popular method for fondue uses "hot rocks" for cooking.

A rock is heated in an oven for several hours. When the rock is removed from the oven, it is hot enough to cook meats and heat cheeses for dining. The pattern of change in the rock's temperature can be modeled by the equation

$$T(t) = T_{env} + (T_0 - T_{env})e^{rt}.$$

Here T_0 is the initial temperature of the rock, T_{env} is the temperature of the cooking environment, and t is the number of minutes of cooling.

a. Suppose that the rock starts at a temperature of 550°F and is used in a room with an air temperature of 75°F. If the rock cools to 250°F in the first 15 minutes, what equation will model the rock's cooling?

b. Suppose that the stone has to be at least 185°F to cook meat. If you want to cook meat over this hot rock, when does the temperature fall below 185°F?

c. After a 45-minute dinner, you want to melt some chocolate for dessert. What is the temperature of the hot rock at that time? Do you think that is hot enough to melt chocolate?

33 Capacitors are electronic devices that store electrical charge and then release it very rapidly when needed, such as to power the flash of a camera. One common way to model the buildup of charge in a capacitor uses functions in the general form $V_C(t) = V_B(1 - e^{rt})$, where V_B represents voltage of the battery providing charge to the capacitor and t represents charging time in seconds.

a. Suppose that a 6.0-V battery is used to charge a capacitor that has an initial voltage of zero. If the capacitor has a voltage of 4.2 V after 0.75 seconds, what is the parameter r in the charging function?

b. What will be the voltage on the capacitor after 1.5 seconds?

c. When is the voltage on the capacitor 30% that of the battery?

d. Will V_C ever be greater than V_B?

(t)David Chasey/Getty Images, (b)Dimitri Vervitsiotis/Getty Images

34 Suppose that an experiment produced the following data showing how charge on a flash camera capacitor (in volts) builds up over time (in seconds):

t	0	1.0	2.0	3.0	4.0	5.0	6.0
V	0.00	3.16	4.32	4.75	4.90	4.96	4.99

Recall what you learned in Unit 1 *Families of Functions* to explain why the function that models the buildup of charge in a capacitor has the form used in Task 33.

a. Plot a graph of the (t, V) data and identify the apparent asymptote for the data points.

b. Describe a transformation of the voltage data that produces a pattern that can be modeled well by an exponential function $y = a(b^t)$.

c. Express the function in Part b with an equivalent rule that uses e as the exponential base.

d. Use the result of Part c and the inverse of the data transformation process in Part b to derive the rule for the function $V(t)$ that matches the pattern of change in the experimental (t, V) data.

35 Using the equation $V_C(t) = V_B(1 - e^{rt})$ (Extensions Task 33), what is the voltage of the capacitor as a proportion of the battery's voltage at the following times:

a. $t = \left|\dfrac{1}{r}\right|$?   **b.** $t = \left|\dfrac{2}{r}\right|$?   **c.** $t = \left|\dfrac{3}{r}\right|$?

d. What proportion of the maximum charge V_B does the capacitor gain from $t = 0$ to $t = \left|\dfrac{1}{r}\right|$? From $t = \left|\dfrac{1}{r}\right|$ to $t = \left|\dfrac{2}{r}\right|$? From $t = \left|\dfrac{2}{r}\right|$ to $t = \left|\dfrac{3}{r}\right|$?

e. If this process continues indefinitely, will the capacitor ever have the same voltage as the battery that is charging it? Why or why not?

f. Electricians, engineers, and physicists consider a capacitor fully charged at $t = \left|\dfrac{5}{r}\right|$. What percentage of the battery's charge does the capacitor have at that time?

36 Not only does a capacitor charge exponentially (Extensions Task 33), it also loses voltage exponentially. Consider the following table of values for loss of charge on a capacitor.

t (in sec)	Voltage (in volts)
0.00	8.8
0.02	6.9
0.04	5.4
0.06	4.2
0.08	3.3
0.10	2.6

a. Find an exponential regression model for this data pattern.

b. Convert the model to an equivalent form $V(t) = V_0 e^{rt}$.

c. The flash on a camera operates by storing charge on a capacitor and releasing it suddenly when the flash goes off. A camera flash only lasts about 0.005 seconds, so it has to discharge most of its voltage quickly. Find values of V_0 and r for a capacitor that releases nearly 3 volts of charge in the time allowed for a camera flash.

37 Evaluate each expression without using technology.

a. $(\log_4 243)(\log_3 64)$

b. $(\log_8 125) \div (\log_4 5)$

REVIEW

38 Sketch a graph of each function and state the domain and range of the function.

a. $f(x) = 3(2^x)$

b. $g(x) = 3\left(\frac{1}{2}\right)^x$

c. $h(x) = 6(0.5^x) + 3$

d. $y = -(3^x)$

e. $y = -(3^x) - 4$

f. $y = -2\left(\frac{1}{2}\right)^x$

39 Without using technology-produced tables or graphs, solve each equation.

a. $10^x = 73$

b. $5(10^x) = 100$

c. $10^{x+5} = 10^{2x+1}$

d. $3(10^{x+2}) = 3.69$

40 Rewrite each expression in a simpler equivalent form.

a. $-4x^8 \cdot 3x^4$

b. $\dfrac{10x^{12}}{22x^3}$

c. $(x^4)^{-3}$

d. $2x^3(6x^5)^2$

e. $5x^{-1}x^{\frac{3}{2}}$

41 Match each expression in Column 1 with an equivalent expression in Column 2. You may use entries in Column 2 more than once or not at all.

Column 1	Column 2
a. $\log 10^x$	**i.** $\log x - 10$
b. $10^{\log x}$	**ii.** x
c. $\log\left(\frac{x}{10}\right)$	**iii.** $10 \log x$
d. $\log 10x$	**iv.** $\log x + 10$
e. $\log x^{10}$	**v.** $\log x + 1$
f. $\log 1$	**vi.** $\log x - 1$
	vii. 0

42 Solve each equation for y.

a. $x^2 + 3y = 7$ **b.** $\dfrac{y^2}{x} = 20$ **c.** $-3x + 10y^2 = 12$

d. $10^y = x$ **e.** $\log y = 3x$

43 Silvia has designed a pattern that she is going to paint on one wall of her bedroom. The pattern contains circles of different sizes. She wants the 5 smallest circles to have a radius of 8 inches.

a. If the paint she plans to use to paint the circles costs 10¢ per square foot, how much will the paint cost for the 5 smallest circles?

b. She wants the next size circle in her pattern to have an area that is 4 times the area of the smallest circle. What should the radius of the next size circle be?

c. Silvia wants to draw a border around some of the circles using glow-in-the-dark paint. If the border is 2 inches wide, how many square inches of paint will she need to outline the 5 smallest circles?

44 Rewrite each expression in the form $a + bi$ with a and b real numbers. Find exact values for a and b whenever possible.

a. $(3(\cos 20° + i \sin 20°))^3$

b. $2(\cos \pi + i \sin \pi) + 4\left(\cos \dfrac{\pi}{2} + i \sin \dfrac{\pi}{2}\right)$

c. $\dfrac{12\left(\cos \dfrac{\pi}{4} + i \sin \dfrac{\pi}{4}\right)}{3\left(\cos \dfrac{3\pi}{4} + i \sin \dfrac{3\pi}{4}\right)}$

45 For each statement, prove that it is an identity, or provide a counterexample to show it is not an identity.

a. $\dfrac{(\cos \theta)(\tan \theta + \sin \theta)}{\sin \theta} = 1 + \cos \theta$

b. $\csc 2\theta = \csc \theta + \csc \theta$

46 Write parametric equations for each of the following paths of point P.

a. P is moving in a direction of 140° at a constant velocity of 10 m/sec.

b. P is moving on a circle with radius 5 and completes 20 counterclockwise revolutions per second.

c. P is moving on an ellipse that intersects the x-axis at $x = \pm5$ and the y-axis at $y = \pm7$. The point begins at (5, 0) and completes one revolution per minute.

Linearization and Data Modeling

Predicting growth or decline of populations is an important task of the government agencies that plan for services to citizens. The table and plot below give the population (in millions) of the United States in census years from 1900 to 2010.

United States Census												
Census Year	1900	1910	1920	1930	1940	1950	1960	1970	1980	1990	2000	2010
Population (in millions)	76	92	106	123	132	151	179	203	227	249	281	309

Source: Census of Population and Housing, U.S. Census Bureau

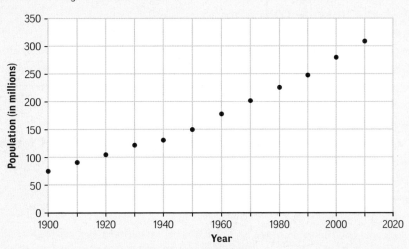

Glowimages/Getty Images

Study the data in the table and plot on the previous page. Think about the type of function that would best model the pattern in United States population growth.

a Does the population growth appear to be linear, quadratic, cubic, exponential, or some other pattern? Explain.

b You could explore the answer to Part a by using calculator or computer data modeling tools. What do you know about the algorithms that are used to fit linear functions to data? To fit nonlinear functions?

To use modeling tools wisely, it helps to understand how they work. The basic idea behind many standard algorithms used to fit nonlinear functions to data is to transform the data to a linear pattern, fit a linear function to the transformed data, and then use the inverse of the transformation to find the appropriate nonlinear function. In this lesson, you will learn ways that logarithmic transformations are particularly useful in the data modeling process.

INVESTIGATION 1

Assessing the Fit of a Linear Model

At first glance, it appears that the pattern of growth in U.S. population for 1900 to 2010 could be represented by a linear function. The linear function $y = 2.1x + 61$ matches the data quite well. However, there is a pattern in the differences between actual and predicted population values suggesting that a linear function is not the best possible fit for the data.

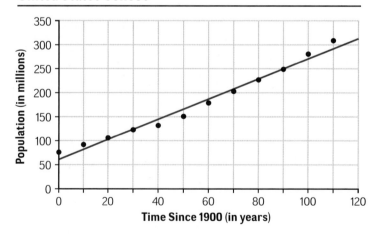

United States Census

As you work on the problems of this investigation, look for answers to this question:

How can analysis of residuals be used to assess the goodness of fit of linear functions to data patterns?

1   Complete the third row of a copy of the (*time, population*) data table to show the population predicted by the function $y = 2.1x + 61$. Then complete the fourth row showing the differences between actual and predicted population values for each year.

Time Since 1900 (in years)	0	10	20	30	40	50	60	70	80	90	100	110
Population (in millions)	76	92	106	123	132	151	179	203	227	249	281	309
Predicted Population												
Actual — Predicted												

Recall from the Course 2 *Regression and Correlation* unit, that the values obtained by calculating (*actual population − predicted population*) are called **residuals**.

a. Make a **residual plot** with *time since 1900* on the *x*-axis and *residual* on the *y*-axis. Describe any pattern that you see in that plot.

b. Explain how the pattern in the residual plot suggests that the (*time since 1900, population*) data are not modeled well by a linear function.

2   The following table shows data relating stride length to leg length for a sample of 20 people from the Cherry Creek Walking Club.

Leg Length (in cm)	Stride Length (in cm)
50	106
52	100
55	108
56	118
58	110
60	110
62	120
64	127
67	138
69	125
70	145
71	140
73	130
74	169
75	152
76	140
81	160
82	142
83	175
88	158

a. Produce a scatterplot of the (*leg length, stride length*) data and then graph the regression line.

b. Explain how the regression line does or does not support the conclusion that the relationship between stride length and leg length is linear.

c. Plot the residuals: *actual stride length − predicted stride length* for the 20 people.

d. Explain how the pattern in the residual plot does or does not support the conclusion that the relationship between stride length and leg length is linear.

3 BASE jumping is an extreme sport in which participants jump from buildings (B), antennas (A), bridge spans (S), and Earth structures (E). They fall freely for a time before opening their parachutes. In October of every year since 1980, daredevil parachutists have traveled to Fayetteville, West Virginia, for Bridge Day and a chance to jump off the New River Gorge Bridge.

In 2007, a crowd of 165,000 spectators came to watch over 400 BASE jumpers flirt with injury and possible death in 850 jumps. The New River Gorge Bridge is 876 feet above the water, and the most daring jumpers fall for about 7 seconds and 650 feet until opening their parachutes.

A news story about the event included data relating time in free fall and approximate speed of typical divers.

Free-fall Speed								
Time (in seconds)	0	1	2	3	4	5	6	7
Speed (in miles per hour)	0	10.9	31.4	51.8	70.9	84.5	94.1	100.9

Source: Eli Saslow, A Heightened Chance of Death, *The Washington Post*, Sunday, November 4, 2007 (A15)

a. Produce a scatterplot of the (*time, speed*) data and then graph the regression line.

b. Explain how the plot shows that speed is probably not a linear function of time in free fall.

c. Produce a residual plot. Explain how the pattern in the residuals illustrates the pattern in the graph of the data and the regression equation in Part a.

4 The next table shows data relating calories to fat in hamburgers sold by a variety of national fast-food chains.

Hamburger Nutrition			
Company	**Burger**	**Fat (in grams)**	**Calories**
Hardee's	Small Hamburger	11	270
	1/3 lb. Original Thickburger®	52	810
Wendy's	Jr. Hamburger	10	250
	Dave's Hot 'N Juicy 1/4 lb. Single	34	600
Burger King	Hamburger	8	240
	WHOPPER® Sandwich	35	630
McDonald's	Hamburger	9	250
	Quarter Pounder® w/Cheese	26	520
	Big Mac®	29	550
Carl's Jr.	Kid's Hamburger	10	280
	Famous Star® w/Cheese	37	670

Source: www.wendys.com (March 2013); www.mcdonalds.com (April 2013); www.bk.com (March 2013); www.hardees.com (May 2007); www.carlsjr.com (November 2012).

a. Plot the (*fat, calories*) data and then on the plot, graph the linear regression equation. Explain how the plot does or does not suggest that there is a linear relationship between the variables.

b. Produce a residual plot: *actual calories − predicted calories*. Then explain how the pattern in those numbers does or does not suggest that there is a linear relationship between the variables.

5 If you fly over the Great Plains region of the United States in summer, you will notice that many parcels of farmland appear to be large discs of green punctuating a dry brown landscape. These green discs of crops like soybeans or corn are created by irrigation systems that pivot like radii around the center of a circle. Suppose that data in the following table give cost for fertilizing a sample of such circular fields as a function of the radius of the field.

Fertilizing Cost for Circular Fields									
Radius (in m)	20	25	40	60	65	75	80	100	110
Fertilizing Cost (in $)	125	170	500	900	1,100	1,700	2,000	3,100	3,600

a. Plot the (*radius, cost*) data and then on the plot, graph the linear regression equation.

b. Plot the residuals that compare actual and predicted cost for the fields of given radii. Then explain how the plot in Part a and the pattern of residuals do or do not support use of a linear model in this situation.

c. Why does it make sense that the relationship between fertilizing cost C and field radius r might be nonlinear? What family of functions seems most likely to provide an accurate model?

SUMMARIZE THE MATHEMATICS

In work on the problems of this investigation, you learned how to judge the validity of linear functions to model data patterns.

a What are residuals and residual plots?

b Think about how you can use a residual plot to evaluate the appropriateness of a linear model.

  i. What does a residual plot look like when it supports use of a linear model?

  ii. What does a residual plot look like when it supports use of a nonlinear model?

Be prepared to explain your ideas to the class.

 CHECK YOUR UNDERSTANDING

When the driver of a car sees danger on the road ahead, the car will travel some distance before the driver is able to apply the brakes and some additional distance before the car actually comes to a stop.

Road tests under various conditions are likely to produce data like those in the following table.

Braking and Stopping Distances as a Function of Speed								
Speed (in miles per hour)	10	20	30	40	50	60	70	80
Distance until Braking (in feet)	7	15	20	29	36	45	51	58
Distance until Stopping (in feet)	5	17	37	65	105	150	205	265

   a. Assess the fit of linear functions to the data relating speed and braking distance.

       i. Plot the (*speed, distance until braking*) data and then on the plot, graph the linear regression equation that best fits the data.

       ii. Make a residual plot.

      iii. Explain how the residual plot does or does not provide evidence that the relationship between braking distance and speed is linear.

   b. Use a procedure similar to that outlined in Part a to decide whether a linear model is appropriate for the relationship between stopping distance and speed.

INVESTIGATION 2

Log Transformations

If you suspect that the relationship between two variables can be represented well by an exponential function, you still have to find the values of a and b so that the rule $f(x) = ab^x$ produces (x, y) values matching the data pattern.

As you work on the problems of this investigation, look for an answer to this question:

> *How can logarithms be used to "straighten" the graphs of exponential data patterns and help find rules for the underlying functions?*

To see how the log transformation process works, it helps to begin with analysis of some well-defined exponential data patterns.

1 Suppose that a medical experiment beginning with a single bacterium shows a pattern in which the number of bacteria doubles every hour.

   a. Complete entries in a table like that on the next page that shows the pattern of change in the number of bacteria and in the logarithms of bacteria counts.

Bacteria Growth I								
Time t (in hours)	0	1	2	3	4	5	10	20
Bacteria Count y	1	2	4					
Log of Bacteria Count (log y)								

Explain how the pattern of entries in row two of the table shows that the bacteria count *is not* a linear function of time, but the pattern of entries in row three shows that the logarithm of the bacteria count *is* a linear function of time.

b. What is the linear function that gives log of bacteria count (log y) as a function of time (t)?

c. Write the equation in Part b in an equivalent form that expresses y as a function of t and see how well this function fits the original bacteria count data.

d. Use algebraic reasoning to show that the equation in Part c is equivalent to $y = 2^t$ as expected.

2 Now suppose that another experiment begins with a bacteria count of 25 and that the count doubles every hour.

a. Complete entries in a table to show the pattern of change over time in bacteria count and log of bacteria count. Then explain how the pattern of entries in row three shows that log of bacteria count is a linear function of time.

Bacteria Growth II						
Time t (in hours)	0	1	3	5	10	20
Bacteria Count y	25	50	200			
Log of Bacteria Count (log y)						

b. What linear function gives log of bacteria count (log y) as a function of time (t)?

c. Write the equation in Part b in an equivalent form that expresses y as a function of t and see how well this function fits the original bacteria count data.

d. Use algebraic reasoning to show that the equation in Part c is equivalent to $y = 25(2^t)$ as expected.

In Problems 1 and 2, you did not really need the logarithmic transformation procedure to find the exponential functions that fit the data patterns. But in real scientific studies, the data will not be so orderly and the appropriate model will not be known. For example, look once more at the U.S. population data reproduced below and plot of the data on the next page.

United States Census												
Time Since 1900 (in years)	0	10	20	30	40	50	60	70	80	90	100	110
Population (in millions)	76	92	106	123	132	151	179	203	227	249	281	309

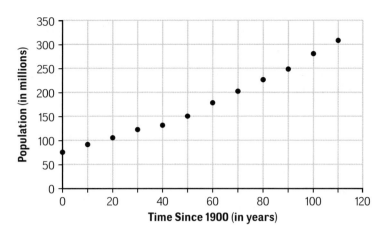

3 Examine the residual plot below to determine if a linear function is an appropriate model for the (*time since 1900, population*) data. What can you conclude? Explain your reasoning.

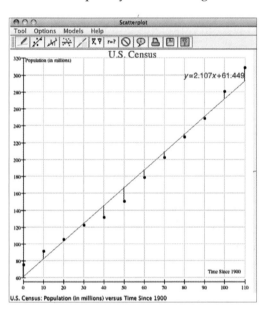
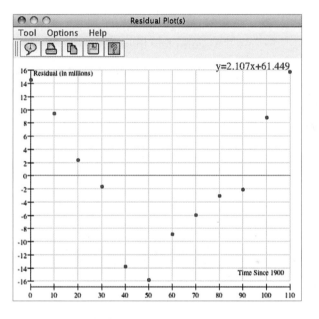

a. Now create a scatterplot of the (*time since 1900, log (population)*) data and find the linear regression equation that shows log (*population*) as a function of time *t*. Round to five decimal places. Then create a residual plot and assess whether a linear relationship is a good model for the (*time since 1900, log (population)*) data. What can you conclude? Explain your reasoning.

b. Write the linear equation that gives log of population as a function of time in an equivalent form that gives a rule for $P(t)$. Use a residual plot to assess the fit of this function to the data.

4 The basic procedure used in Problems 1–3 can also be followed using the natural logarithm function to linearize the data pattern and give a base-*e* exponential function that fits the original data.

a. Refer to the U.S. population data table on the previous page. Find the linear regression equation that gives ln (*population*) as a function of time.

b. Use the result of Part a to find an expression for $P(t)$ in the form $P_0 e^{rt}$.

c. Show that the rule for $P(t)$ in Part b is equivalent to that derived from the same data in Problem 3.

CHECK YOUR UNDERSTANDING

Think about the inverse relationship between exponents and logarithms as you complete these tasks.

a. If $y = ab^x$ and $\log y = 3x + 4$, what are the values of a and b?

b. The table below shows some (x, y) values for an exponential growth pattern. Use the "linearizing" strategy developed in this investigation to estimate the values of a and b for a rule in the form $y = ab^x$.

x	1	3	5	10	20
y	6	13	30	230	13,301

INVESTIGATION

Log-Log Transformations

In some situations where you seek a function relating values of two variables, the experimental data values that you have to work with are of varying magnitudes and the distribution is often skewed. This makes plotting a graph of the full data set very difficult. It also makes it hard to find patterns in the relationship between the variables.

For example, the orbits of planets in our solar system are ellipses with the Sun located a bit off center. The table at the top of the following page gives data on the length of the longer axes of those ellipses and the time required for the planets to complete one full orbit.

©Digital Vision/PunchStock

Planet Orbits

Planet	Half Length of Longer Axis x (in millions of miles)	Orbit Time y (in Earth days)
Mercury	36	88
Venus	67	224
Earth	93	365
Mars	142	686
Jupiter	484	4,344
Saturn	887	10,768
Uranus	1,784	30,660
Neptune	2,795	60,225

A plot of the (x, y) data suggests a nonlinear relationship, but it is not at all obvious what that relationship might be.

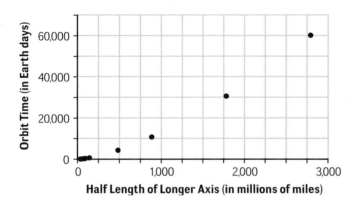

As you work on the problems of this investigation, look for answers to these questions:

How can logarithmic transformations of data reveal patterns relating variables when the values of the variables are of different magnitudes and have a skewed distribution?

How can taking logarithms of both independent and dependent variable values aid in finding the relationship between those variables?

1 About 400 years ago, the astronomer Johannes Kepler (1571–1630) studied data on planetary motion that had been collected by his predecessor Tycho Brahe (1546–1601). Kepler proposed a power function in the form $y = ax^b$ for the relationship between orbit time and orbit axis length. The question is, "What are the parameters a and b?"

a. You have seen earlier that logarithmic transformations can be used to linearize exponential data patterns. This suggests that if the power relationship between orbit time and orbit axis length is quadratic, an inverse transformation using $\sqrt{y}$ might linearize the data. Why does this data transformation make sense?

Johannes Kepler

b. Complete the third row of the following table to show values produced by a square root transformation of the orbit time data.

Half Length of Longer Axis x (in millions of miles)	36	67	93	142	484	887	1,784	2,795
Orbit Time y (in Earth days)	88	224	365	686	4,344	10,768	30,660	60,225
Square Root of Orbit Time $\left(\sqrt{y}\right)$								

c. Study the $(x, \sqrt{y}\,)$ data and provide evidence that this *is not* a linear relationship.

d. What do your results in Parts a and b suggest about the right form for Kepler's power function? Compare your answer with that of your classmates. Resolve any differences.

2 Exponential functions offer another source of models for nonlinear patterns like those in the planetary orbit data. You have seen earlier that if one variable is an exponential function of the other, then performing a logarithmic transformation of y values linearizes the data pattern.

a. Complete the third row of the following table showing values produced by the transformation log y.

Half Length of Longer Axis x (in millions of miles)	36	67	93	142	484	887	1,784	2,795
Orbit Time y (in Earth days)	88	224	365	686	4,344	10,768	30,660	60,225
log of Orbit Time (log y)								

b. Study the $(x, log\ y)$ data and provide evidence that this *is not* a linear relationship.

c. What do your results in Parts a and b suggest about the possibility of the planetary orbit data being represented well by an exponential function?

3 Now try a new third kind of data transformation in search of a linear pattern.

a. Complete a table like the following to show values produced by the log x-log y transformations.

Half Length of Longer Axis x (in millions of miles)	36	67	93	142	484	887	1,784	2,795
Orbit Time y (in Earth days)	88	224	365	686	4,344	10,768	30,660	60,225
log of Half Length of Longer Axis (log x)								
log of Orbit Time (log y)								

b. Study the $(log\ x, log\ y)$ data and provide evidence that this is a linear relationship.

c. Use linear regression to find an equation in the form $\log y = a + b \log x$.

d. Write the regression equation from Part c in an equivalent form that expresses y as a function of x, with parameters a and b. Explain how the results confirm and give details about Kepler's model.

4 What is now known as Kepler's third law states that the square of the time of a planet's orbit y is proportional to the cube of half the length of the longer axis of that orbit x.

a. Write an equation that expresses Kepler's third law, using k to represent the constant of proportionality.

b. Use algebraic reasoning to show how the result of your work in Problem 3 is equivalent to Kepler's third law and explain how it also determines the constant of proportionality $k \approx 0.16$, when units of measurement are millions of miles for distance and Earth days for time.

5 In previous study of inverse variation, you studied the way that light and sound intensity diminish with distance from the source. Consider the following table of values that compares distance from a light source and illuminance (light per unit area).

Light Intensity

Distance (in m)	Illuminance (in lux)
0.5	100
1	25
1.5	11.11
2	6.25
2.5	4
3	2.78
3.5	2.04
4	1.56

a. Graph the data and predict the function family that will model the relationship between the variables.

b. Find a data transformation that seems to linearize the data. Then use the transformed data, linear regression, and further algebraic reasoning to find the resulting equation for illuminance as a function of distance from a light source.

c. Explain why the resulting equation relates the variables in a way that seems reasonable. That is, it gives the pattern of change in illuminance one would expect as distance from the light source increases.

SUMMARIZE THE MATHEMATICS

In this investigation, you learned how to use log-log transformations to "linearize" data patterns and find rules for functions that fit those data patterns.

a When does it make sense to use a log-log transformation of data before seeking a relationship between the variables?

b How can the linear regression equation relating log y to log x be transformed to express y as a function of x?

c Suppose that $y = ax^b$ and log $y = m$ log $x + c$. What are the connections between $a, b, m,$ and c?

Be prepared to explain your ideas to the class.

 CHECK YOUR UNDERSTANDING

The following table gives data about planetary orbits with half the length of the longer axis in millions of kilometers and orbit time in Earth years.

Orbits of Planets								
Planet	Mercury	Venus	Earth	Mars	Jupiter	Saturn	Uranus	Neptune
Half Length of Longer Axis (in millions of km)	57.9	108.2	149.6	227.9	778.4	1,426.7	2,871.0	4,498.3
Orbit Time (in Earth years)	0.241	0.615	1.00	1.88	11.9	29.5	84.0	165

Source: NASA Planetary Fact Sheets; nssdc.gsfc.nasa.gov/planetary/planetfact.html

a. Transform these data using a log-log transformation.

b. Provide evidence that the transformed data are related in a linear pattern, and find the linear regression equation for that relationship.

c. Use the result of Part b and what you know about logarithms, exponents, and equations to express orbit time in Earth years as a function of half length of longer axis in millions of kilometers.

d. Compare the function derived in Parts a–c to that derived in Problem 3 and explain why the results agree with Kepler's third law.

1 The following table shows data relating sodium and fat in hamburgers sold by a variety of national fast-food chains.

Hamburger Nutrition			
Company	Burger	Fat (in grams)	Sodium (in mg)
Hardee's	Small Hamburger	11	560
	1/3 lb. Original Thickburger®	52	1,720
Wendy's	Jr. Hamburger	10	600
	Dave's Hot 'N Juicy™ 1/4 lb. Single	34	1,220
Burger King	Hamburger	8	460
	WHOPPER® Sandwich	35	980
McDonald's	Hamburger	9	480
	Quarter Pounder® w/Cheese	26	1,100
	Big Mac	29	970
Carl's Jr.	Kid's Hamburger	10	580
	Famous Star® w/Cheese	37	1,210

Source: www.wendys.com (May 2013); www.mcdonalds.com (May 2013); www.bk.com (April 2013); www.hardees.com (May 2007); www.carlsjr.com (November 2012).

a. Plot the (*fat, sodium*) hamburger data and then on the plot, graph the linear regression equation for the relationship of sodium to fat. Then explain how the plot does or does not confirm the validity of a linear model relating the variables.

b. Plot the residuals *actual sodium − predicted sodium*. Does the pattern in those numbers confirm the validity of a linear model relating the variables? Explain.

2 Satellite radio services give their subscribers access to a wide selection of radio stations anywhere in the country. The first licenses for satellite broadcasting were granted in 1992. By 2007, there were about 14 million subscribers to the two largest U.S. providers of satellite digital radio, XM and Sirius. Data in the following table show the number of Sirius customers in millions at the end of each quarter year from the beginning of 2004 to the end of 2006.

Satellite Radio Subscribers (in millions)												
Time After Jan. 2004 (in quarter years)	0	1	2	3	4	5	6	7	8	9	10	11	12
Subscribers	0.3	0.4	0.5	0.7	1.0	1.4	1.8	2.1	3.3	4.0	4.7	5.0	6.0

Source: satellitestandard.blogspot.com/2007/01/subscriber-beakdowns.html

  a. Plot the given data and then on the plot, graph the linear regression equation for the relationship of number of Sirius subscribers to number of quarter years since January 2004.

  b. Calculate and plot the residuals comparing actual numbers of subscribers and the numbers predicted by the regression equation.

  c. Explain how the pattern in the residual plot does or does not support use of a linear model for the Sirius subscriber data.

3 Radioactive isotopes of many chemical elements are known to decay exponentially, but at a variety of rates. Suppose that data in the following table give experimental measurements of decay over time for one such radioactive chemical.

Radioactive Isotope							
Time (in days)	0	1	3	5	7	10	20
Amount Left (in grams)	5	4.8	4.3	3.9	3.5	3.0	1.8

  a. Complete a third row of the table with values equal to log (*amount left*), and make a scatterplot of the (*time, log (amount left)*) data.

  b. Find the linear regression equation for the relationship between log (*amount left*) and time. Then calculate and plot the residuals and explain how that information does or does not confirm linearity of the relationship between log (*amount left*) and time.

  c. Use the results of your work in Part b to find the rule in the form $L = ab^t$ for an exponential function that relates amount left L to time t. Explain what the values of a and b tell about the pattern of decay for this substance.

4 The next table shows the cumulative public debt of the United States government at five-year intervals since 1970.

U.S. Public Debt								
Time Since 1970 (in years)	0	5	10	15	20	25	30	35	40
Debt (in billions of $)	389	577	930	1,946	3,266	4,974	5,674	7,933	13,203

Source: U.S. Department of the Treasury, The Public Debt Online.

  a. Why is it reasonable to expect the public debt to grow exponentially?

  b. Produce a scatterplot of the (*year, log (debt)*) data. Does the pattern in that plot confirm the conjecture that U.S. public debt is growing exponentially? Explain.

c. Use the method developed in Investigation 2 to find the rule for a possible exponential model of the (*year*, *debt*) relationship and compare the shape of that function's graph to a scatterplot of the actual data. Explain how the match of the function graph and scatterplot does or does not seem to confirm the conjecture that public debt has been growing exponentially over the past 40 years.

5 When the AIDS disease was first discovered, scientists were very interested in the rate at which it might be spreading in various populations. The following table gives the number of deaths from AIDS in the United States for the years 1981 through 1994.

AIDS Fatalities in the U.S.							
Year	1981	1982	1983	1984	1985	1986	1987
Deaths	128	460	1,501	3,497	6,961	12,056	16,336
Year	1988	1989	1990	1991	1992	1993	1994
Deaths	21,040	27,691	31,402	36,307	40,516	42,992	46,050

Source: United States Centers for Disease Control and Prevention, *HIV/AIDS Surveillance Report*, Year-End Edition 1995.

a. In the late 1980s, people were worrying about whether the number of deaths from AIDS was increasing exponentially. Does that appear to be the case for 1981 through 1989? Decide by making a suitable transformation of the number of cases from 1981 through 1989 and analyzing the resulting scatterplot.

b. Now consider the data for all the years from 1981 through 1994. Does the number of deaths from AIDS appear to be increasing exponentially from 1981 through 1994? Explain your reasoning.

c. In 1992, the definition of AIDS was expanded to include more symptoms. Would this fact affect your answer in Part b? If so, how?

d. The Center for Disease Control and Prevention estimates that in the United States, 49,897 AIDS-related deaths occurred in 1995, 37,359 occurred in 1996, and 21,437 occurred in 1997. What might have caused the pattern of increase in deaths from AIDS to change in the years since 1995?

6 While working on a geometry project, Carisa, Tory, and Amelia were searching for a formula that would give surface area of a sphere as a function of its radius. They tried covering several different spheres with grid paper to make estimates of the surface area. Data from their experiments are given in the next table.

Radius *r* (in cm)	2	5	8	12	15	20
Surface Area *A* (in cm²)	50	300	800	1,800	2,800	5,000

a. How do the values in this table suggest that the relationship between surface area and radius is not linear?

McGraw-Hill Education

b. Apply a log-log transformation to the data and find the linear regression equation that gives log (*surface area*) as a function of log (*radius*).

c. Plot the residuals to show that the relationship between log (*surface area*) and log (*radius*) is linear.

d. Write the equation from Part b in an equivalent form that shows surface area *A* as a function of radius *r*.

e. How close is the formula for the surface area of a sphere derived from data analysis to the theoretical $A = 4\pi r^2$?

7 In *Core-Plus Mathematics* Course 2, you studied the effect of gravity by timing the trip of a ball rolling down an inclined ramp like that pictured here.
The following table gives experimental data relating distance traveled down the ramp to time from one such experiment.

Time (in seconds)	1	2	3	4	5
Distance (in meters)	0.5	2.1	4.4	8.1	12.6

a. How do the values in this table suggest that this relationship between distance and time is not linear?

b. What do you recall from earlier work that suggests a quadratic power relationship between distance and time?

c. Use a log-log transformation to find a power function that fits the given data. Then calculate and plot residuals to assess the validity of that function as a model of the given data pattern.

8 The altimeters of airplanes use atmospheric pressure outside the plane to estimate altitude above sea level. Some sample data relating pressure in pounds per square inch to altitude in miles are given in the next table.

Flight Data								
Altitude (in miles)	0	1	2	3	4	5	6	7
Pressure (in psi)	14.7	11.7	9.4	7.5	6.0	4.8	3.9	3.1

a. Do either log or log-log transformations of the flight data produce a linear pattern? If so, use the transformed data to find an appropriate power or exponential model for the relationship between pressure and altitude.

b. Use exponential and power regression tools to find a model for the given (*altitude, pressure*) data. Compare the regression function to that derived by applying linear regression to suitably transformed altitude and pressure data.

9 Consider the table of values at the right.

a. What equation do you get by performing a linear regression on the data?

b. Use that equation to predict the value of log y and then y, if log x is 3.43.

c. Write the equation from Part a in an equivalent form that expresses y as a function of x.

d. Use the equation from Part c to predict the value of y if x is $10^{3.43} = 2{,}691.53$.

e. Compare the results from Parts b and d and explain why they are equal or different.

log x	log y
0.08	1.06
0.12	1.10
0.25	1.18
0.45	1.31
0.57	1.38
0.95	1.63
1.10	1.73

10 The following table gives brain and body weights from samples of a variety of species of mammals.

Animal Brain and Body Weights		
Animal	**Brain Weight** (in g)	**Body Weight** (in kg)
bat	0.936	0.028
squirrel	3.97	0.183
skunk	10.3	1.7
grey monkey	66.6	4.55
walrus	1,126	667
horse	618	462
chimpanzee	440	56.7
tiger	302	209
African elephant	5,712	6,654
blue whale	6,800	58,059

Scientific theory suggests that the relationship between body and brain weights should be modeled well by a power function with rule in the form $y = a(x^b)$. Check this conjecture in several ways.

a. Use the power regression feature of a data analysis tool to find what is supposedly the best-fitting power function.

b. Use a log-log transformation of the given data and check to see if the result is an approximately linear relationship. Complete the modeling by finding the linear function relating log (*body weight*) to log (*brain weight*) and transforming that equation to an equivalent form that expresses body weight as a function of brain weight directly. Then compare the result with the function derived in Part a.

c. Compare the predictions of the functions from Parts a and b to the actual data and explain whether you think that the scientific theory is supported by the data or not.

d. Brain and body weight data for a typical human would be something like (1,400, 75). How well do these data match the pattern suggested by either of your functions from Parts a and b?

11 Data in Applications Task 5 show the rate at which annual deaths from AIDS increased in the United States between the start of the epidemic in about 1981 and 1994. Your work on that task should have suggested that an exponential function would not match the pattern of change in annual numbers of deaths.

CPMP-Tools provides data on the *total* number of deaths from AIDS in the United States from 1981–1994. Some epidemiologists have suggested that a power function might be a good model for the growth of deaths due to AIDS in the United States. Use what you have learned about how to apply log-log transformations of data to derive a best-fit power function to see if that claim makes sense. If not, explain why.

12 When looking at the following graph and table, it is difficult to tell if the values come from an equation of the form $y = Ax^2$, $y = Ae^{rx}$, $y = Ax^b$, or possibly some other form.

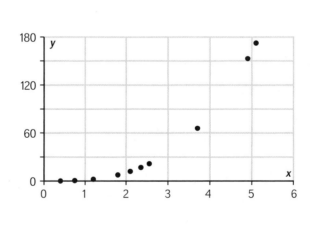

x	y
0.4	0.1
0.75	0.55
1.2	2.2
1.8	7.6
2.1	12.0
2.35	16.9
2.55	21.6
3.7	65.8
4.9	152.9
5.1	172.4

a. Experiment with transformations of the tabled x and y values until you get a linear pattern.

b. Find the linear regression equation that expresses the linearized pattern and then write the resulting equation in equivalent form to express y as a function of x.

CONNECTIONS

When you transform data by taking logarithms, you expect some properties of the data distribution to change, but some other key properties to remain unchanged. Connections Tasks 13–16 ask you to apply the same kind of thinking to other familiar transformations.

13 Suppose that the distribution of quiz scores in one world history class has mean 7.5 and standard deviation 1.4. Explain how the mean and standard deviation change in the following cases.

   a. All scores are multiplied by 10 to make the scale run from 0 to 100 instead of 0 to 10.

   b. Each score is increased by 1 point to correct for an error in wording of one item.

14 Suppose that the heights of students in a physical education class are measured in centimeters and the results show a distribution with mean 160 and standard deviation 8. If the student heights were measured again, this time in inches, what mean and standard deviation would you expect for the distribution of the resulting measurements? (One centimeter is approximately 0.4 inches.)

15 Geometric transformations are often used to enlarge or shrink photographs, designs, and plans for buildings or manufactured objects. Suppose that a polygonal shape is enlarged by the size transformation with rule $(x, y) \rightarrow (3x, 3y)$. Explain how this transformation will change the following properties of the polygon.

   a. Perimeter

   b. Area

   c. Measures of angles

16 Consider the following set of typical weights for several familiar animals.

Animal	gerbil	guinea pig	raccoon	moose	elephant
Weight (in lbs)	0.25	1.5	21	800	14,000

   a. Calculate the mean and standard deviation of this sample of animal weights.

   b. Calculate the square root of each weight. Then compare the mean and standard deviation of the square root of weight data to the same statistics for the real weights.

   c. Calculate the logarithm of each weight and compare the mean and standard deviation of the log of weight data to the same statistics for the real weights.

17 Use a log transformation to decide which of the following sets of values were generated by an exponential function. If an exponential function seems appropriate, find the rule for the best-fit function. If an exponential function does not seem appropriate, explain how the residual plot of the transformed values justifies that conclusion.

a.

x	0	2	4	6	8	10
y	3	7	15	35	75	175

b.

x	5	7	11	14	17	24
y	6	4.8	3.1	2.3	1.7	0.8

c.

x	0	3	6	9	12	15
y	4	8	22	45	75	115

18 Numbers in the following table show how surface area and volume are related for 5 spheres of different sizes. Because both quantities are power functions of the radius of the sphere, it is reasonable to expect that the relationship between them is also a power function of some sort.

Surface Area A (in cm²)	12.57	50.27	113.1	201	314
Volume V (in cm³)	4.19	33.51	113.1	268.2	524

a. Produce a scatterplot of log (*surface area*) and log (*volume*) values. What can you conclude about these transformed data? Explain your reasoning.

b. Use the results from Part a and linear regression to find an equation relating log (*volume*) to log (*surface area*).

c. Use the result from Part b to express volume V as a function of surface area A. Then check the result by applying power regression to the original values for surface area and volume.

d. Substitute the expressions $4\pi r^2$ and $\frac{4}{3}\pi r^3$ for area and volume in the equation of Part c. Then apply algebraic reasoning to show why that equation is valid for all positive values of r.

19 Use log and log-log transformations and residual plots to decide which of these tables of values were probably generated by power functions, exponential functions, or neither of those familiar types.

a.

x	1	2	3	4	5	7	10
y	5.0	6.7	7.5	8.0	8.3	8.8	9.0

b.

x	1	2	3	4	5	7	10
y	12.5	31.3	80	200	500	3,000	50,000

c.

x	1	2	3	4	5	7	10
y	5.0	3.3	2.5	2.0	1.7	1.3	0.9

d.

x	1	2	3	4	5	7	10
y	10	14	17	20	22	26	32

REFLECTIONS

20 Suppose that you were to use population data over the past 50 years to project future population growth. If both power and exponential functions gave fairly good fits to past data, how might choosing the "wrong" function type lead to serious errors of prediction into the next 50 years?

21 One of the most effective strategies for solving a problem is to recall methods used to solve similar problems and to use those solution strategies to attack the new problem. In what sense is linearization of data an application of that strategy?

22 Sometimes in a scatterplot, some of the points are clustered together near the origin. You have seen how taking a log-log transformation can spread out the points in a scatterplot so you can better see them. It has been said that in this case, the wrong scales were used for the original measurements. What would a person who said this mean?

23 Suppose that you suspect that two variables are related by a quadratic function with rule in the form $y = ax^2 + bx + c$, where b and c are not both equal to 0. Why will a log-log transformation fail to produce a linear pattern from which values of a, b, and c can be estimated accurately?

EXTENSIONS

24 Students in a science class at a school wondered about the relationship between the radius and mass of a water-filled balloon. They filled several balloons with water and then measured the mass of water used in each case. They came up with data in the following table.

Radius (in cm)	6	8	10	12
Mass (in grams)	900	2,100	4,200	7,200

a. Why does it make sense to expect a cubic relationship like $M = ar^3$ between mass and radius of a water-filled sphere?

b. Add a row to the data table with entries cube root of mass and explain how the pattern in that row suggests a linear relationship between radius and cube root of mass.

c. What linear equation in the form $y = mx$ fits the pattern relating radius and cube root of mass fairly well?

d. Use your result from Part c to find the relationship between mass and radius.

e. The exact formula for volume of a sphere is $V = \frac{4}{3}\pi r^3$. The mass of one cubic centimeter of water is approximately one gram. How do the results of your work on Parts a–d compare with what the volume formula and the mass of water say should happen?

25 The next table and plot show how world population has grown since the year 1650. It has the kind of shape that makes you think of exponential growth.

a. Calculate log of population values and make a plot of (*year*, *log (population)*) data.

b. What does this result tell you about the accuracy of an exponential model for world population growth over the past 360 years?

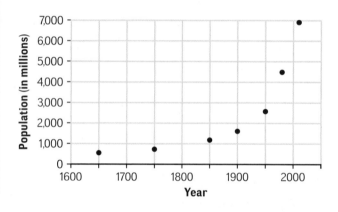

World Population

Year	Population (in millions)
1650	550
1750	725
1850	1,175
1900	1,600
1950	2,564
1980	4,478
2010	6,892

26 In *Core-Plus Mathematics* Course 2, you studied the relationship between slope of inclined ramps and time it takes for a ball or skateboard to roll down the ramp. For a fixed ramp length, the slope is determined by the height to which one end of the ramp is elevated.

Suppose that the following data give run time t (in seconds) for a skateboarder on a ramp that is 16 feet long and tested at various ramp heights h (in feet).

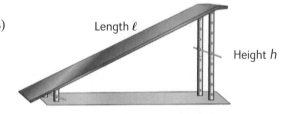

Length ℓ

Height h

Ramp Height h (in feet)	2	4	6	8	10	12
Run Time t (in seconds)	2.82	2.0	1.63	1.41	1.26	1.15

a. Try log and log-log transformations of the given data to see which, if any, leads to linearization.

b. Use the transformed data that is most linear to produce a linear regression equation. Then use algebraic reasoning to find a function that expresses t as a function of h.

c. In earlier study of motion down inclined planes, you saw that run time was related to length and height of the ramp by the formula $t = \dfrac{L}{4\sqrt{h}}$. How do the results of your work in Parts a and b agree with or challenge this formula?

27 Suppose that a rectangular box has dimensions 1 foot $\times$ 2 feet $\times$ 3 feet.

a. What are the volume and surface area of that box?

b. Complete a copy of the following table to show volume and surface area for a variety of boxes that are similar to the 1-2-3 box with various scale factors k.

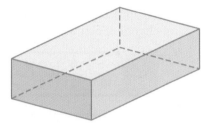

Scale Factor k	2	3	4	5	10	20
Surface Area A (in ft²)						
Volume V (in ft³)						

c. Use methods developed in this lesson to find the relationship between volume and surface area of any rectangular box that is similar to the 1-2-3 box. Since it is generally somewhat simpler to calculate the volume of a box than its surface area, look for an equation that expresses area as a function of volume.

d. How can you explain the particular parameter values in the function rule that you came up with in Part c? That is, can you see any intuitive or formal explanation for why those values are what they are? (*Hint:* Replace A by $22k^2$ and V by $6k^3$ in the equation of Part c and use algebraic reasoning to show that the resulting equation is true for any value of k.)

28 The strength of an electric field is given by the equation $E = \dfrac{k}{r^2}$. Where k is a constant for a charged particle, and r is the distance between the particle and the electric field sensor.

a. Would a graph of the points ($\log r$, $\log E$) be linear? Why or why not?

b. If so, what are the values of a and b for the equation $\log E = b \log r + a$?

29 When experimental data includes zeroes or negative values, log transformations are not directly applicable. But you can often adapt strategies from the *Families of Functions* unit to solve such problems. For example, consider the following table of (x, y) values.

x	1	2	3	5	8	10	12	15
y	−2	−11	−31	−112	−362	−632	−998	−1,743

a. What preliminary transformation of the given values will yield a data set (x, y^*) to which the methods of this lesson can be applied?

b. Study the pattern in the transformed values and use a further log or a log-log transformation to get a linear relationship between the transformed variables. Then use linear regression to show how y^* depends on x.

c. Then "undo" the preliminary transformation in Part a to get a function relating y and x, and check the fit of that function.

REVIEW

30 Determine if each table of values shows a linear or an exponential pattern of change. Then find a function rule that matches each table of values.

a.

x	−5	0	5	10	15
f(x)	10	5	0	−5	−10

b.

x	2	3	4	5	6
f(x)	12	36	108	324	972

c.

x	−4	−1	2	5	8
f(x)	1.5	3	4.5	6	7.5

d.

x	−2	0	2	4	6
f(x)	640	160	40	10	2.5

31 The graph below shows a set of 6 points and two possible linear models for the pattern in those points. Which line is the better fit to the points? Explain your reasoning using sum of squared errors calculations.

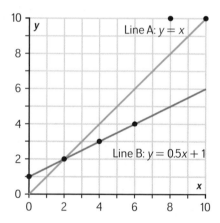

32 Solve each equation and check your answers.

a. $\dfrac{x + 5}{x} = 10$

b. $2x - 7 = \dfrac{8}{x - 3}$

c. $\dfrac{3x + 1}{x - 2} = \dfrac{x + 4}{2x + 6}$

d. $|x^2 + 3| = 12$

e. $\dfrac{1}{x} + \dfrac{3}{x + 2} = 2$

33 Find equations or function rules for relationships that meet these conditions.

   a. Line parallel to $3x + 6y = 12$ and passing through $(3, 2)$

   b. Line perpendicular to $3x + 6y = 12$ and passing through $(3, 2)$

   c. Cubic polynomial with zeroes of 4, 6, and –3 and y-intercept of 24

   d. Sine function with maximum value of 10, minimum value of 4, and period 2π

34 For each of these functions: (1) describe the domain and range; (2) decide whether the function has an inverse; and (3) for those functions that do have inverses, find a rule for the inverse.

   a. $f(x) = 4.5x$ **b.** $g(x) = x^2 - 4.5$

   c. $h(t) = 3t + 5$ **d.** $k(s) = \dfrac{5}{s} + 1$

35 Determine all radian solutions for each of the following equations.

   a. $-4 \sin t = 2$ **b.** $\tan (t + 5) = 3$

   c. $\cos^2 x + 4 \cos x + 3 = 0$

36 Use properties of common logarithms to write each of these expressions in a different but equivalent form.

   a. $\log 3x$ **b.** $\log x^3$

   c. $\log \left(\dfrac{x}{3}\right)$ **d.** $\log 3 + \log x$

37 For each representation of a function below, determine the value of $f^{-1}(4)$.

   a.

x	1	2	3	4
f(x)	1	3	4	6

   b.

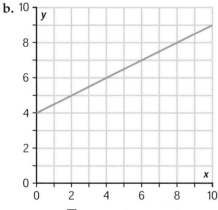

   c. $f(x) = \sqrt{x} + 1$

   d. $f(x) = \log x$

38 Enrollment at Middle State University is expected to increase by 50 students per year. In 2014, the University has 12,780 students. The number of faculty at Middle State University is a function of student enrollment and can be approximated by the rule $F(s) = \frac{s}{30} + 25$, where s is the number of students at Middle State.

a. Write a function rule $S(x)$ that expresses the number of students at Middle State as a function of the number of years x since 2014.

b. Determine each of the following and indicate what it tells you about enrollment at Middle State University.

   i. $S(10)$

   ii. The value of x such that $S(x) = 13{,}180$

c. Determine each of the following and indicate what it tells you about the number of faculty at Middle State University.

   i. $F(13{,}180)$

   ii. The value of s such that $F(s) = 450$

   iii. $F(S(7))$

d. Write a rule for $F(S(x))$ and explain what it tells you about Middle State University.

39 Consider the rectangular prism and square pyramid shown below.

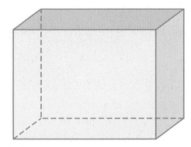 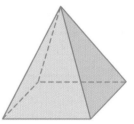

For each shape, complete the following.

a. Without tracing, make a careful sketch of the polyhedron.

b. How many symmetry planes does the shape have? Describe or draw each one.

c. How many axes of symmetry does each shape have? Describe or draw each one and then identify the angles of rotation associated with each axis of symmetry.

d. Imagine a plane that intersects the shape and is parallel to its base. Is the polygon formed by the intersection of the plane and the polyhedron congruent to, similar to, or neither congruent to nor similar to the base of the polyhedron?

Looking Back

The lessons, investigations, and specific problems of this unit challenged you to recall what you have learned about exponential and base-10 logarithmic expressions, equations, and functions in order to extend those ideas and skills to include the mathematically important exponentials with base e and the corresponding natural logarithms. You then applied those new tools to extend your skill in modeling patterns in scientific data.

The tasks in this final lesson of the unit give you a chance to review and consolidate your skill and understanding of exponential and logarithmic functions and data modeling.

1 Suppose you wanted to put money in a bank savings account and saw the following advertisement from an aggressive bank seeking deposits.

> ## We Make Your $ Work
> ### 4% APR Compounded 24/7/365

It appears that this bank will pay interest in a way that is compounded every hour of the day, every day of the week, and every week of the year.

a. What expression shows how to calculate the value of a $500 investment in such an account at any time h hours following your deposit?

b. How many compounding periods will occur in a year, and to what value will the original $500 deposit grow in that time?

c. What is the effective annual interest rate of the savings account?

d. How can the approximate value of the account after one year be calculated by evaluating an expression involving the number e? Calculate the value in this way and compare it to your result from Part b.

e. How can the approximate value of the account after t years be calculated by evaluating an expression involving the number e? Use the expression to estimate the value of the account after 5 years.

2   In the Fall of 2006, the attention of the world was drawn to London, England where a former Russian internal security agent, Alexander Litvinenko, died an agonizing death by poisoning from the radioactive substance Polonium-210. Since 97% of the world production of that rare substance occurs at a closely guarded facility in Russia, many people suggested that Litvinenko's death was an assassination by political enemies.

> **Who Killed Alexander Litvinenko?**
> Bob Simon Reports On A Real-Life Deadly Spy Mystery
>
> By Daniel Schorn
>
> The story would be fit for a spy novel if it weren't so implausible. A Russian ex- KGB agent turns against the Kremlin and flees Moscow. He continues his attacks from exile in London, until he is poisoned with a rare radioactive isotope and dies a slow painful death.
>
> As correspondent Bob Simon reports, this is the real life story of Alexander Litvinenko, the first-ever victim [of] nuclear terrorism.

Source: www.cbsnews.com/stories/2007/01/05/60minutes/main2333207.shtml

Polonium-210 is an extremely toxic substance—an amount as small as one gram could theoretically kill tens of millions of people. (**Source:** Peter Finn, Crime Sets Off Alarm Bells, *The Washington Post*, January 7, 2007, Section A pages 1 and 16.)

a. Polonium-210 has a half-life of 138 days. Based on this fact, what function gives the part of an original 1-gram amount remaining t days after it is manufactured?

b. The Russian factory that makes Polonium-210 (by irradiating Bismuth-209) sells about 85 grams per year, but it actually makes much more, to allow for the rapid decay once the radioactive material is produced. How much of an 85-gram production run will still be radioactive after t days?

c. Suppose that the factory stores each production run for later sale and distribution. How much time will have elapsed since a recent production run if an amount of 50 grams remains?

d. How long would it take 1 gram to decay to a still potent 1 microgram (0.000001 grams)?

3 Evaluate each of the following expressions.

   a. $\log_7 49$

   b. $\log_7 1$

   c. $\log_7 23$

4 Write each equation as an equivalent equation that expresses y as a function of x.

   a. $\log y = 5 + 3x$

   b. $\ln y = 3.2 - 5 \ln x$

5 When police investigate the scene of an automobile accident, they look for skid marks and use the length of those marks to estimate the speed at which the car was traveling. The following table shows data from experiments with a test car.

   a. Plot the data and make a conjecture about the kind of function that will accurately represent the relationship between car speed in miles per hour and skid mark length.

Skid Mark Length ℓ (in feet)	5	17	37	65	105	150	205	265
Speed s (in miles per hour)	10	20	30	40	50	60	70	80

   b. Explore various data transformations to find one that appears to linearize the data pattern.

   c. Apply linear regression to express the relationship in the transformed data set.

   d. Use algebraic reasoning and the result from Part c to express s as a function of ℓ.

   e. Estimate the speed at which an automobile was traveling prior to an accident if the length of its skid marks is 180 feet.

6 Plot I below shows the pattern of change in the number of reported cases of a disease over a seven-year period. Least squares regression lines for the transformed data are shown in Plots II and III.

I

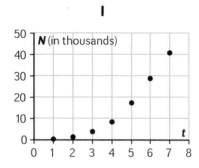

II

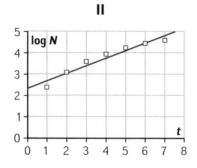

III

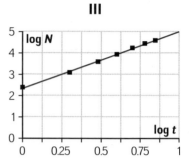

a. The points (1, 2.6953) and (4, 3.7651) are on the regression line for Plot II. Using this information, write a rule expressing N as a function of t.

b. The points (0, 2.3607) and (0.5, 3.6882) are on the regression line for Plot III. Using this information, write a rule expressing N as a function of t.

c. Which of the functions in Parts a and b provides the better fit to the data in Plot I? Why?

SUMMARIZE THE MATHEMATICS

In this unit, you investigated a variety of situations that led to use of power, exponential, and logarithmic functions as models of data patterns.

a What is the approximate value of the number e and how can that value be approximated by use of a formula?

b What is the natural logarithm function? What are the domain and range of the function?

c What kinds of problem situations can be modeled well by use of an exponential function with base e?

d What is a residual plot? How can such a plot be used to evaluate the appropriateness of a linear model?

e Why would you transform given data sets in the process of finding good function models for the patterns in those data sets?

f What data transformations are useful in case the underlying relationship in a situation is likely an exponential function? A general power function?

Be prepared to share your responses and reasoning with the class.

 CHECK YOUR UNDERSTANDING

Write, in outline form, a summary of the important mathematical concepts and methods developed in this unit. Organize your summary so that it can be used as a quick reference in future units.

6

Surfaces and Cross Sections

In your previous *Core-Plus Mathematics* studies, you learned methods for representing three-dimensional figures such as cylinders, prisms, pyramids, and other polyhedra in two dimensions. Often three-dimensional objects such as the terrain of a planet like Mars—currently being explored by the *Curiosity* rover—are less uniform in shape. How to represent more complex three-dimensional objects in two dimensions on a computer monitor or paper is a fundamental problem in geometry and its applications.

In this unit, the focus will be primarily on creating and reasoning with representations of surfaces of three-dimensional objects such as those of a distant planet and of cross sections of three-dimensional objects such as those that are so useful in medical imaging.

In the following two lessons, you will extend your skills in visualizing surfaces and cross sections and representing those objects, both geometrically and algebraically, through the use of three-dimensional coordinates.

LESSONS

1 Three-Dimensional Representations

Represent three-dimensional surfaces with contour diagrams and identify and sketch cross sections of three-dimensional objects, including conic sections. Identify and sketch graphs of conic sections represented algebraically and write equations matching graphs of conic sections.

2 Equations for Surfaces

Extend ideas of coordinate representation and methods in two dimensions to three dimensions. Visualize, describe, and sketch surfaces represented by equations and identify and sketch surfaces of revolution and cylindrical surfaces.

Three-Dimensional Representations

Hikers regularly use *contour diagrams* of trails to get an overall picture of the terrain, showing where the mountains are and where the flat regions are. The map at the right shows a section of the Presidential Range of the White Mountains National Forest in New Hampshire. Trails are shown by heavier lines that include the name of the trail. The thin lines are *contour lines*. If a hiker walks along a contour line, she stays at the same elevation. On this map, adjacent contour lines differ in elevation by 100 ft.

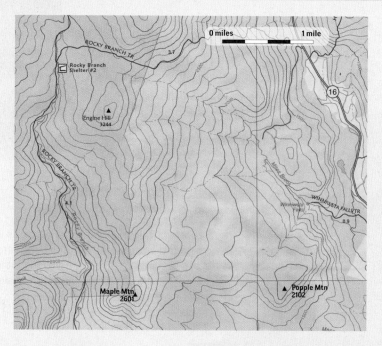

(t)Pixland/AGE fotostock, (b)©Appalachian Mountain Club Books

Maps showing contour lines are one way to represent three-dimensional objects in two dimensions. Other methods you have previously studied include orthographic (face-view) drawings and oblique drawings. In this unit, you will learn how to represent and analyze three-dimensional objects using contour diagrams, using cross sections, and using coordinates and equations.

INVESTIGATION 1

Using Data to Determine Surfaces

NASA (the National Aeronautics and Space Administration) has sent men to the Moon and spacecrafts to Mars and Venus. In preparing to land on the Moon, a critical concern was identifying a safe location for landing the spacecraft. In the case of Earth's moon, the surface was visible, so relatively flat sites could be located.

However, in seeking a flat landing site on Venus, the problem was more difficult because the surface was obscured by clouds. Orbiting NASA spacecraft used radar to determine the elevations of many points on the surface of Venus. These elevations were then used to make contour diagrams of portions of the surface. The diagrams were used to identify possible landing sites. During the 1978 Pioneer mission to Venus and the 1990 Magellan mission, nearly the entire surface of Venus was mapped using radar altimetry.

As you work on the problems of this investigation, look for answers to these questions:

How can you use altitude data to make and interpret a contour diagram of a surface?

How can you use coordinates to identify locations in space?

Exploration 1 You can simulate the radar altimeter mapping of Venus and other planets by using a partially-filled shoe box containing a hand-molded surface inside it and having grid paper on the lid. The molded surface represents a region of the surface of Venus. A bamboo skewer can be used as a radar probe. You will also need grid paper, a centimeter ruler, and a marking pen.

Calibrate your radar probe by making a mark 1 cm from the blunt end. Continue marking and numbering centimeter intervals on the entire skewer. If the box you are using does not have holes in its lid at grid marks, you will need to make holes using the sharp end of the skewer before collecting your data.

1 You are now ready to collect data. Keep the shoe box closed.

    Step 1. Keep the radar probe perpendicular to the box top as you insert it into a hole near a corner of the box top. Continue the gentle insertion until the probe touches the surface inside the box. Note the closest centimeter mark on the probe above the box top.

    Step 2. On your grid paper, record the number of the centimeter mark determined in Step 1.

    Repeat these steps for each hole in the box top. Your grid paper should represent the grid paper on the box top as if you had been allowed to write directly on the lid.

2 Now, using your data, make a contour diagram of the terrain. Follow the procedure illustrated with the sample data below. Do not open the shoe box and look at the terrain inside.

    Step 1. For your data, what number represents the highest peak? For the data below, a cluster of 1s shows the highest peak. It is best to start drawing the contour around either the highest peak or lowest valley. The example below begins with a contour at 1.5 cm, as shown by the loop around the 1s. Now draw contour lines every 2 cm, that is for 3.5, 5.5, 7.5, and 9.5 cm. The open contour drawn around the 10 represents the 9.5 cm contour. It is open because there may be more 10s beyond the edge.

10	8	7	5	4	6	5
8	6	5	5	5	3	3
2	3	4	5	3	2	2
2	2	3	3	2	1	3
2	2	3	3	1	1	3
6	5	4	3	2	2	4

Step 2. Work outward from the chosen peak or valley. In this example, working outward from the 1s loop, you need to draw contour lines for 3.5 cm. One of these lines is drawn in the diagram below. Why does the contour line at the lower-left corner extend between the 2 and 6?

```
10 / 8   7   5   4   6   5

    8   6   5   5   5   3   3

    2   3   4   5   3   2   2

    2   2   3   3   2 / 1 \ 3

    2   2   3   3 ( 1   1 ) 3

    6   5   4 \ 3   2   2   4
```

Step 3. Continue drawing the remaining contour lines; these should look like those on the diagram below.

```
10 / 8 / 7 / 5   4 \ 6 / 5

  8 / 6 / 5   5   5 / 3   3

    2   3 \ 4   5 / 3   2   2

    2   2   3   3   2 / 1 \ 3

    2   2   3   3 ( 1   1 ) 3

    6 \ 5   4 \ 3   2   2 ( 4
```

3 Now examine your contour diagram.

a. How is a region of small numbers related to a region of larger numbers? Explain why this makes sense.

b. Should the contour lines drawn for two numbers intersect? Explain.

c. Based on your data, describe the terrain in the box. Then open the box and compare your description of the terrain to the actual surface.

Exploration 2 The following altitude data were collected by the Pioneer mission to Venus. The location is a region near the Venusian equator. For the purpose of this exploration, altitude data are collected at points 90 feet apart and then rounded to the nearest multiple of 10 ft.

Stocktrek/Corbis

4 Examine these data and then on a copy of the data draw a contour diagram with contours at 10-ft altitude intervals starting at 725 ft.

720	780	820	830	830	830	830	840	840	840	850	850	850
750	800	820	830	830	830	830	830	840	840	840	850	850
790	820	830	830	830	830	820	830	830	840	840	840	840
820	830	830	820	820	810	810	820	830	840	840	840	840
840	840	830	810	800	800	800	810	830	840	840	840	840
840	840	820	800	790	780	780	800	820	840	850	850	860
830	820	800	780	770	770	770	790	820	850	860	870	890
800	790	780	770	770	770	770	790	820	860	890	910	930
780	770	760	760	770	770	770	790	820	860	910	940	970
760	760	760	760	770	770	780	790	820	860	920	960	980
750	760	760	760	770	780	790	790	810	860	920	970	990

Source: nssdc.gsfc.nasa.gov

a. Using your contour diagram, locate and justify possible landing sites for a space probe.

b. If the captain wishes to land at a high elevation, where would you suggest landing?

c. What can you infer about the terrain by the spacing between the lines?

d. Using the *CPMP-Tools* geometry custom app "Contour," select the NASA Pioneer data set. Compare your contour diagram to the one provided by this software. What does the coloring of the computer-produced diagram show?

5 Discuss the similarities and differences between the two contour diagrams you made in Explorations 1 and 2. For what contexts might a representation such as that in Exploration 1 be most helpful? In Exploration 2?

6 A portion of the altitude data from the bottom-left corner of the surveyed region on Venus is shown on the coordinate system below. Adjacent grid points are 90 ft apart.

How could you describe each of the following lettered locations on the planet using three numbers (coordinates) that give its position in feet, where the third coordinate is altitude?

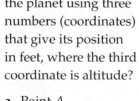

a. Point A

b. Point B

c. Point C

d. Point D

e. Point E

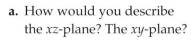 **A Three-Dimensional Coordinate System** The three-number descriptions of a point in three-dimensional space found in Problem 6 are *rectangular coordinates* of the point.

Rectangular coordinates of a point are determined by measuring the directed perpendicular distance from the point to each of three mutually perpendicular planes called **coordinate planes**. The *origin O* is the intersection of the three coordinate planes. The intersections of the pairs of planes are the *coordinate axes*. The axes are usually called the *x*-, *y*-, and *z*-axes. The labels *x*, *y*, and *z* indicate the positive direction on each axis.

It is customary to represent a three-dimensional rectangular coordinate system with the *y*-axis horizontal and the *z*-axis vertical as shown. The positive *x*-axis appears to come forward out of the page and creates the appearance of depth, or the third dimension. Coordinates of points are given in the order (x, y, z).

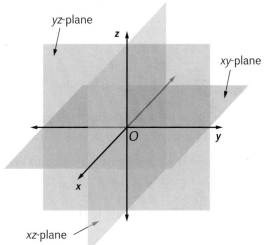

a. How would you describe the *xz*-plane? The *xy*-plane?

b. Into how many regions do these three coordinate planes separate space?

c. Describe the location of all points with:

   i. positive *x*-coordinates.

   ii. negative *z*-coordinates.

   iii. positive *y*- and negative *x*-coordinates.

8 Care is needed in plotting and interpreting points in three-dimensional space. When plotting points, it is helpful to show the "path" to the point as indicated in the diagram at the right.

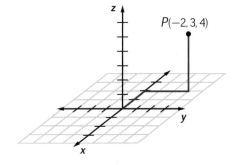

a. Plot and draw the paths to points $P(-2, 3, 4)$ and $Q(2, 3, 4)$ on a copy of the three-dimensional coordinate system shown.

b. What should appear to be true about the plot of these two points?

c. On a copy of a three-dimensional coordinate system, plot and label each of the following points.

   $A(0, 2, 3)$ $B(3, 2, 4)$ $C(3, -2, -4)$

   $D(-5, 3, 5)$ $E(-4, -7, -2)$

d. Describe in general how you would plot a point $A(a, b, c)$.

In this investigation, you examined how to draw contour diagrams from altitude data and considered how to use three coordinates to identify locations in space.

a Describe how contour lines can be drawn to describe a smooth three-dimensional surface.

b Describe how coordinates can be used to specify the location of a point in three-dimensional space.

c The altitude data for Venus (Problem 4, page 398) were given every 90 feet. If the x- and y-axes are placed so they intersect at the bottom-left corner of the data set, give the three coordinates of a possible landing site for a spacecraft mission to Venus.

Be prepared to share your group's ideas and reasoning with the class.

 CHECK YOUR UNDERSTANDING

The height (in feet) of a surface, represented by the z values in the table below, was measured for the 49 lattice points in a square region: $-3 \leq x \leq 3$, $-3 \leq y \leq 3$.

a. Use these data to plot the positive z values on an xy-coordinate plane. Then construct a contour diagram for the height z of the surface above the xy-plane. Draw contour lines at 1-ft intervals, starting at 9 ft. Use your contour diagram and patterns in the data to describe the shape of the mapped surface.

		y						
		−3	**−2**	**−1**	**0**	**1**	**2**	**3**
	−3	8.5	10	11	11.5	11	10	8.5
	−2	10	12	13.5	14	13.5	12	10
	−1	11	13.5	15	15.5	15	13.5	11
x	**0**	11.5	14	15.5	16	15.5	14	11.5
	1	11	13.5	15	15.5	15	13.5	11
	2	10	12	13.5	14	13.5	12	10
	3	8.5	10	11	11.5	11	10	8.5

b. On a three-dimensional coordinate system, plot the point that has z-coordinate of 16 in the contour diagram. Also plot the point that has x-coordinate of -2 and y-coordinate of -3.

Visualizing and Reasoning with Cross Sections

As you saw in Investigation 1, contour diagrams are developed by sampling heights of a terrain at specific intervals. The contour lines represent horizontal cross sections of the terrain. The representation only approximates the true characteristics of the terrain. However, for many people who use contour diagrams, such as hikers, forest rangers, deep sea divers, and ship navigators, the approximate nature of the information is adequate. They want to know generally what the land or sea bottom near them looks like.

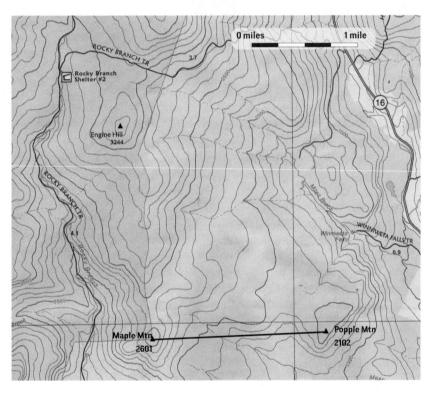

A contour diagram can be used to create a *topographic profile* of a region that also shows the variation in the altitude of the land surface.

As you work on the problems of this investigation, look for answers to these questions:

How can you interpret and draw a profile of a region?

How can horizontal and vertical cross sections of a three-dimensional figure be used to determine and sketch the figure?

Topographic Profiles Understanding what contour maps represent can be enhanced through the construction of a topographic profile. A **topographic profile** is a vertical cross-sectional view along a line segment drawn across a portion of a map. In other words, if you could slice vertically through a section of earth, pull away one half, and look at it from the side, the surface would be a topographic profile. Not only does constructing a topographic profile help hikers visualize terrain, it is very useful for geologists when analyzing numerous problems about land formations.

1 Suppose you wanted to hike from the summit of Maple Mountain to the summit of Popple Mountain following a path shown by the line segment drawn near the bottom of the map on the previous page. Consecutive contour lines on this map represent 100-ft elevation differences. Consider the vertical cross section of the region indicated by this segment, called a **relief line**.

    a. Along the relief line, what can you say about the land surface if the contour lines are close together? If they are far apart?

    b. The section of the map with the relief line between Maple Mountain and Popple Mountain is shown below. Using the following procedure, you can make a topographic profile to help visualize the terrain of the hike.

        Step 1. Place a sticky note along the relief line as shown. Make tic marks at the endpoints of the segment. Write the elevation of the starting and ending points next to their tic marks.

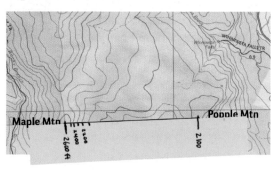

        Step 2. Next, make a tic mark at places the paper crosses a contour line on the map. Write the elevations for the contour lines below their tic marks on the sticky note.

        Step 3. Now, place your marked sticky note along a horizontal axis on grid paper. Draw vertical lines for the starting and ending elevations of the topographic profile as shown at the right. Label the vertical axis with units appropriate for your elevation data.

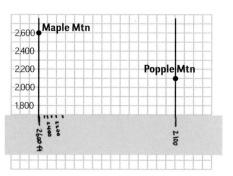

        Step 4. Plot points representing the elevation data and connect them to visualize the terrain of your hike. Discuss the shape of the terrain with your classmates.

    c. How could you add information to the topographic profile to represent the distance along the trail?

2 A group of divers earning a certification is doing a compass run from a rocky beach to a buoy in Lake Superior. The goal of the exercise is to maintain their heading in spite of terrain and current. Contour lines on this map differ by a depth of 10 feet. Sketch a profile of the lake bottom along their path. What purpose do you think the buoy is serving?

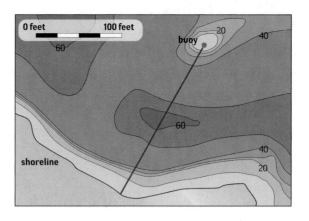

3 **Cross Sections** Just as a contour diagram can help you visualize portions of the surface of Earth or other planets, **cross sections**—the intersections of planes with physical or mathematical objects—can reveal important features of those objects.

a. The contour diagram at the right shows the series of horizontal cross sections at 2-in. intervals for an object. What polyhedron might this contour diagram represent?

b. Now sketch a circular cone and a triangular pyramid with each shape resting on its base. Assuming that the height of each shape is 8 in. and that contour lines are based on horizontal cross sections at 2-in. intervals, draw the corresponding contour diagrams.

c. How would the contour diagrams for the cone and pyramids change if each shape were resting on the vertex instead of the base?

d. Look back at the map on page 401. Locate Engine Hill and the three closest contour lines to the summit. How is this contour map similar to and different from the contour diagrams formed by horizontal cross sections in this problem?

4 Consider a cylinder with its base in the xy-plane as shown below.

a. Describe the cross sections of the cylinder if the intersecting planes are parallel to the xy-plane.

b. Could any other three-dimensional object have cross sections parallel to its bases exactly like those you described in Part a? If so, describe the object.

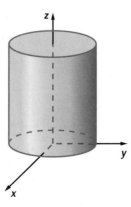

c. Describe cross sections of the cylinder if the intersecting planes are parallel to the yz-plane.

d. How could you display or record both the horizontal and vertical cross sections so that they could be used to identify and sketch the cylinder? Compare your method to those of your classmates. Resolve any differences.

5 The **circular paraboloid** at the right is the three-dimensional analog of a parabola. It was formed by rotating a parabola with vertex at the origin about the z-axis.

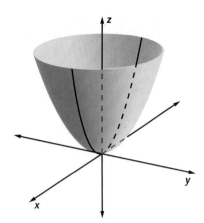

a. Sketch five cross sections of this object determined by equally spaced planes parallel to the xy-plane.

b. Repeat Part a for planes parallel to the yz-plane.

c. How could you label and display the cross sections so that they could be used to identify and sketch the object?

The displays you prepared in Problems 3–5 were made using horizontal or vertical cross sections determined by a series of parallel planes at uniform intervals. It is standard practice to record the series of horizontal cross sections with the one corresponding to the highest point drawn first and subsequent cross sections placed to the right. For the series of vertical cross sections, the first-drawn cross section is the one made by the vertical plane nearest the viewer; subsequent cross sections will be those obtained by intersections with vertical planes at uniform intervals further from the viewer.

6 Consider the following ordered horizontal and vertical cross sections of three-dimensional objects. Use the cross sections to help you describe each object.

a. Horizontal cross sections:

Vertical cross sections:

b. Horizontal cross sections:

Vertical cross sections:

c. Vertical cross sections for another object with the same horizontal cross sections as in Part a:

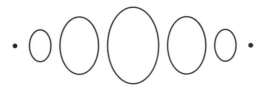

d. Horizontal cross sections:

Vertical cross sections:

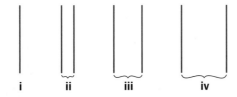

i   ii   iii   iv

SUMMARIZE THE MATHEMATICS

In this investigation, you explored how topographic profiles are created from maps displaying contour lines. You also explored how analysis of both vertical and horizontal cross sections of an object can help you better understand the object.

a   Explain how to make a topographic profile.

b   How is a profile similar to, and different from, a vertical cross section of a surface?

c   Explain how cross sections of a three-dimensional object can be used to figure out what the object looks like.

Be prepared to share your ideas with the entire class.

 CHECK YOUR UNDERSTANDING

Successive horizontal cross sections of a three-dimensional object are given at the right. Cross sections are made at 2-cm intervals beginning at the top of the object.

a. Sketch an object that has the given horizontal cross sections.

b. Make sketches of several differently shaped vertical cross sections of the object.

Conic Sections

Cross sections of three-dimensional objects are two-dimensional objects. As you observed in previous investigations, different plane figures may result when a plane and an object in space intersect. An important class of two-dimensional figures are those that can be obtained from the intersection of a plane and a right circular *double cone* as shown below. Such intersections are called **conic sections**, or simply **conics**.

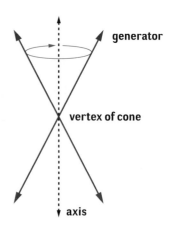

A double cone is produced by rotating a line called a *generator* about a fixed line called the *axis* of the cone, that intersects the generator at the vertex of the cone.

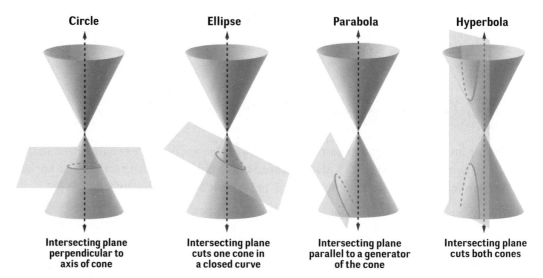

Circle	Ellipse	Parabola	Hyperbola
Intersecting plane perpendicular to axis of cone	Intersecting plane cuts one cone in a closed curve	Intersecting plane parallel to a generator of the cone	Intersecting plane cuts both cones

The various intersections of a plane with the surface of a double cone are familiar curves. In Unit 2, *Vectors and Motion*, you used parametric equations of circles to model the motion of rotating CDs, parametric equations of ellipses to model the motion of orbiting satellites, and parametric equations of parabolas to model the path of a kicked soccer ball. In Unit 5 of Course 3, *Polynomial and Rational Functions*, you saw that the graph of the inverse power model $y = \frac{1}{x}$ had the shape of a hyperbola with the x-axis as horizontal asymptote and the y-axis as vertical asymptote. In this investigation, you will explore other representations, characterizations, and applications of the conic sections.

As you work on the following problems, look for answers to these questions:

> *What is the general x-y form of an equation of each conic section?*
>
> *How can you rewrite the general form of a particular conic in a form from which it is easier to sketch its graph?*

Each of the conics shown on the previous page can also be described as a set (or **locus**) of points satisfying a specified distance condition. In this investigation, you will revisit circles and parabolas from the locus-of-points perspective and derive general formulas for ellipses and hyperbolas. In Extensions Task 33, you will examine the connection between the definition of conics as cross sections of double cones and the definition of conics as *loci* that satisfy a distance condition.

1 **Representing a Circle** Recall that a **circle** is the set of points in a plane that are at a fixed distance r, called the radius, from a given point O, called the center of the circle. In Course 2 Unit 3, *Coordinate Methods*, you used this definition to develop the **standard form of the equation of a circle** with center (h, k) and radius r, $(x - h)^2 + (y - k)^2 = r^2$.

   a. Write, in standard form, the equation of a circle with center $(-3, 2)$ and radius 5. Then rewrite your equation in expanded form $ax^2 + by^2 + cx + dy + e = 0$.

   b. The graph of a quadratic equation like $x^2 + y^2 + 6x + 10y + 11 = 0$, in which the coefficients of x^2 and y^2 are equal (1, in this case), is a circle. You can use your understanding of the factored form of a perfect square trinomial, $a^2x^2 + 2abx + b^2 = (ax + b)^2$, to rewrite such equations in standard form, revealing the center and radius of the circle.

      i. Explain why the following equations are equivalent. In the case of the second equation, also explain why 9 and 25 were chosen to be added to both sides.

$$x^2 + y^2 + 6x + 10y + 11 = 0$$
$$(x^2 + 6x + 9) + (y^2 + 10y + 25) = -11 + 9 + 25$$
$$(x + 3)^2 + (y + 5)^2 = 23$$

      ii. What are the center and radius of the circle represented by the third equation?

   c. Use the method above, called *completing the square*, to determine the center and radius of the circle with equation $x^2 + y^2 + 12x - 2y + 21 = 0$. Then sketch a graph of the circle.

2 **Representing a Parabola** In Unit 3, *Algebraic Functions and Equations*, you considered a **parabola** as the set of points in a plane equidistant from a fixed line, called the *directrix*, and a point not on the line, called the *focus*.

   a. In the diagram below, a parabola is shown where the directrix line has been chosen as $y = -8$ on a coordinate grid. What point on the diagram represents the focus?

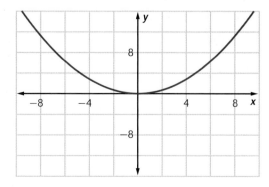

b. What expression represents the distance from any point $P(x, y)$ to the focus?

c. What expression represents the distance from any point $P(x, y)$ to the directrix?

d. What equation represents the condition that the distance from any point $P(x, y)$ on the parabola to the focus is always equal to the distance from that point to the directrix?

e. Rewrite the equation you derived in Part d in an equivalent expanded form that does not involve a radical.

3 Now generalize your reasoning in Problem 2.

a. Find the equation of a parabola with focus $F(0, c)$ and directrix $y = -c$.

  i. Where is the vertex located?

  ii. For what values of c is the vertex a maximum point? A minimum point?

b. Find the equation of a parabola with focus $F(c, 0)$ and directrix $x = -c$.

  i. Explain why your equation is not an equation of a function of the form $y = f(x)$.

  ii. Where is the vertex of this parabola located?

  iii. How is this parabola similar to, and different from, the parabola in Part a?

c. Find the equation of the parabola with directrix $x = 2$ and focus $(-2, 0)$. Sketch your parabola.

4 In your previous work, you saw that graphs of quadratic functions $y = ax^2 + bx + c$, $a \neq 0$, were also parabolas. You also learned that the **vertex form of the equation of a parabola** is $y = a(x - h)^2 + k$, and that equivalent forms of a quadratic function rule reveal different information about its graph.

a. What information can you deduce about the graph of a parabola by examining its expanded symbolic form, $y = ax^2 + bx + c$?

b. What information can you deduce about the graph of a parabola by examining the vertex form, $y = a(x - h)^2 + k$?

c. Because different forms of the function rule for a parabola allow you to easily determine different information about the parabola, it is important to be able to transform a function rule from one form to the other. Rewrite each rule below in vertex form. Then describe the graph as completely as possible.

  i. $y = x^2 + 6x + 8$

  ii. $y = -2x^2 + 12x - 17$

  iii. $y = 5x^2 - 10x + 9$

Parabolic surfaces (surfaces found by rotating a parabola about its line of symmetry or axis) have useful reflective properties. In the case of parabolic headlight reflectors as on the left below, the bulb is placed at the focus for the high beam and in front of the focus for the low beam. When light is projected from the focus, the light will be reflected in rays parallel to the axis and, in the case of high-beam headlights, parallel to the road surface.

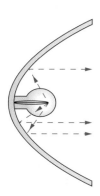

The opposite principle is used in satellite receiving dishes such as those above on the right. In this case, parallel television or radio waves are reflected from the parabolic surface and concentrated at the focal point where the electronic receiver is located.

5 **Representing an Ellipse** An **ellipse** is the set of points in a plane for which the sum of the distances from two fixed points, called the *foci* (plural of focus), is a constant.

a. To draw an ellipse, take a piece of string about 6 inches long and tie the ends together. Place two thumbtacks into a stiff piece of cardboard about 2 inches apart, and loop the string around them. Use a sharp pencil with its point on the cardboard to form a triangle from string as shown below. Keeping the string taut, draw a curve on the cardboard.

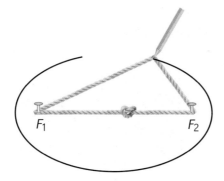

b. How must the distance between the foci compare to the length of the tied string? Explain your reasoning.

c. If you use the same length string but change the distance between the foci, how will the new ellipse compare to your first ellipse?

6 Now consider how to write an equation that represents an ellipse. Suppose the foci of an ellipse are at $F_1(-4, 0)$ and $F_2(4, 0)$, and that for any point $P(x, y)$ on the curve, $PF_1 + PF_2 = 10$.

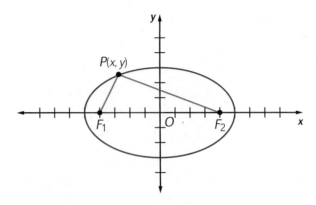

a. Complete this derivation of the equation of the ellipse with foci given above. Give reasons for each statement and provide intermediate statements with reasons where needed.

$$PF_1 = \sqrt{(x + 4)^2 + y^2} \text{ and } PF_2 = \sqrt{(x - 4)^2 + y^2}$$

So, $PF_1 + PF_2 = \sqrt{(x + 4)^2 + y^2} + \sqrt{(x - 4)^2 + y^2} = 10.$

$$\vdots$$

$$(x + 4)^2 + y^2 = 100 - 20\sqrt{(x - 4)^2 + y^2} + (x - 4)^2 + y^2$$

$$\vdots$$

$$4x - 25 = -5\sqrt{(x - 4)^2 + y^2}$$

$$16x^2 - 200x + 625 = 25(x^2 - 8x + 16 + y^2)$$

$$\vdots$$

$$9x^2 + 25y^2 = 225$$

$$\text{or } \frac{x^2}{25} + \frac{y^2}{9} = 1$$

b. Consider the geometry of this ellipse.

i. What are the lines of symmetry for the ellipse?

ii. The intersection of the lines of symmetry is the *center O* of the ellipse. What are the coordinates of the center of the ellipse?

iii. The segment of the horizontal (or vertical) symmetry line with endpoints on the ellipse is called the *horizontal (or vertical) axis*. What is the length of the horizontal axis? The vertical axis?

c. How would your last two equations in Part a and your answers to Part b change if the foci were $F_1(0, -4)$ and $F_2(0, 4)$?

7 Now consider the more general case. Suppose the foci of an ellipse are $F_1(-c, 0)$ and $F_2(c, 0)$, and that for any point $P(x, y)$ on the ellipse, the sum of the distances to the foci is $2a$, where $a > c > 0$.

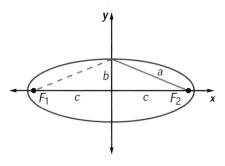

a. Explain why the x-intercepts are $(-a, 0)$ and $(a, 0)$.

b. Explain why the y-intercepts are $(0, -b)$ and $(0, b)$, where $b = \sqrt{a^2 - c^2}$.

c. Use the distance formula and reasoning as in Problem 6 to show that the equation of the ellipse can be written in the form $\dfrac{x^2}{a^2} + \dfrac{y^2}{b^2} = 1$.

d. Next consider the geometry of this ellipse.

    i. What is the center of the ellipse?

    ii. What are the lines of symmetry?

    iii. What is the length of the horizontal axis? The vertical axis?

e. How would you modify the equation in Part c if the line through the foci is parallel to the x-axis and the center is at a point with coordinates (h, k)?

8 The equation you wrote in Problem 7 Part e is called the **standard form of the equation of an ellipse** with center (h, k), $2a$ the length of the horizontal axis, and $2b$ the length of the vertical axis.

a. Write in standard form the equation of an ellipse with center $(3, -5)$, horizontal axis of length 12, and vertical axis of length 20. Then write your equation in expanded form.

b. For each equation, write the equation in standard form and decide if it represents a circle or an ellipse. Then sketch the graph labeling the center and radius (for circles) or the length of the horizontal and vertical axes (for ellipses).

    i. $9x^2 + 4y^2 - 36 = 0$ **ii.** $3x^2 + 3y^2 - 108 = 0$

    iii. $36x^2 + 9y^2 - 216x = 0$ **iv.** $4x^2 + 25y^2 + 24x + 50y - 39 = 0$

The ellipse, like the parabola, has interesting and useful reflective properties. Any sound or light wave initiated at one focus will be reflected to the other focus. You can experience this phenomenon in "whispering galleries." Statuary Hall in the U.S. Capitol building, shown at the right, is an elliptical chamber. A person whispering at one focus can be easily heard by another person standing at the other focus, even though the person cannot be heard at many places between the foci!

This reflective principle of an ellipse is also used in lithotripsy, a procedure for treating kidney stones. The patient is submerged in an elliptical tank of warm water with the kidney stone positioned at one focus. High-energy shock waves generated at the other focus are reflected and concentrated on the stone, pulverizing it.

In the case of an ellipse, the sum of the distances from any point on the curve to the foci is a constant. A different two-dimensional curve, the **hyperbola**, is obtained when the absolute value of the *difference* of the distances from the foci is constant.

9 **Representing a Hyperbola** Suppose the foci of a hyperbola are $F_1(-c, 0)$ and $F_2(c, 0)$, with $c > 0$. Consider points $P(x, y)$ such that the absolute value of the difference of the distances between $P(x, y)$ and the foci is a constant $2a$, where $a > 0$.

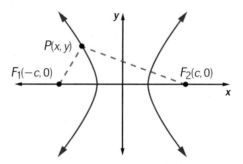

a. If point P is not on the segment connecting the foci, explain why $F_1P + 2c > F_2P$ and $c > a$.

b. Show that the equation of the hyperbola can be written in the form $\frac{x^2}{a^2} - \frac{y^2}{c^2 - a^2} = 1$ by supplying the reasoning for each step in the derivation of this equation. Provide intermediate steps with reasons as needed.

By definition, $|F_1P - F_2P| = 2a$. So, $F_1P - F_2P = 2a$ or $F_1P - F_2P = -2a$. Consider first the case of $F_1P - F_2P = 2a$:

$$\sqrt{(x + c)^2 + y^2} - \sqrt{(x - c)^2 + y^2} = 2a$$

$$\vdots$$

$$(x + c)^2 + y^2 = 4a^2 + 4a\sqrt{(x - c)^2 + y^2} + (x - c)^2 + y^2$$

$$\vdots$$

$$cx - a^2 = a\sqrt{(x - c)^2 + y^2}$$

$$c^2x^2 - 2ca^2x + a^4 = a^2x^2 - 2ca^2x + a^2c^2 + a^2y^2$$

$$\vdots$$

$$x^2(c^2 - a^2) - a^2y^2 = a^2(c^2 - a^2)$$

$$\frac{x^2}{a^2} - \frac{y^2}{c^2 - a^2} = 1$$

c. Why does the equation for the hyperbola turn out to be the same for the case of $F_1P - F_2P = -2a$?

d. Explain why $c^2 - a^2 > 0$. Since $c^2 - a^2$ is positive, it is convenient to represent $c^2 - a^2$ by the positive number b^2. Then the equation in Part b can be written as $\frac{x^2}{a^2} - \frac{y^2}{b^2} = 1$.

e. Each continuous part of the hyperbola is called a *branch of the hyperbola*. The points on the hyperbola that lie on the segment connecting the foci are called the *vertices* of the hyperbola.

  i. Explain why the vertices of the hyperbola in Part c are $(-a, 0)$ and $(a, 0)$.

  ii. Explain why that hyperbola has no y-intercepts.

  iii. What are the lines of symmetry?

10 Now consider more carefully some specific hyperbolas.

  a. For the hyperbola with equation $9x^2 - 4y^2 = 36$, what are the intercepts? The lines of symmetry?

  i. Explain why the lines $y = \frac{3}{2}x$ and $y = -\frac{3}{2}x$ are asymptotes of this hyperbola.

  ii. Sketch the hyperbola by first sketching its asymptotes.

  b. Use asymptotes to help you sketch the graph of $25y^2 - 4x^2 = 100$.

11 Explain how you know that an equation of the form

$$\frac{(x-h)^2}{a^2} - \frac{(y-k)^2}{b^2} = 1 \text{ or}$$

$$\frac{(y-k)^2}{a^2} - \frac{(x-h)^2}{b^2} = 1$$

is an equation of a hyperbola.

  a. What are the lines of symmetry of the graph of each equation?

  b. The equations above are called the **standard forms of the equation of a hyperbola** with center (h, k) and constant difference $2a$.

  i. Write the equation $9x^2 - 16y^2 + 36x + 32y - 124 = 0$ in standard form.

  ii. Describe the graph as completely as possible and make a sketch using the asymptotes as guides.

Hyperbolas are not as commonly seen as circles, ellipses, and parabolas in physical objects or in paths of objects, but one place you can see them is on the wall of a room behind the lamp shade of a lighted lamp. Hyperbolas do play important roles in astronomy and particle physics. Also, the LORAN (LOng RAnge Navigation) navigational system is based on locating a vessel at the intersection of two hyperbolas.

SUMMARIZE THE MATHEMATICS

In this investigation, you examined conic sections in terms of their algebraic and geometric representations and properties.

a Describe each of the conics in terms of the intersection of a plane and a double cone.

b Conics whose symmetry lines are parallel to the *x*-axis and/or *y*-axis can be expressed in several forms. How can you tell by examining the coefficients of an equation of the form $Ax^2 + By^2 + Cx + Dy + E = 0$, where *A* and *B* are not both zero, whether the equation is that of a circle, an ellipse, a parabola, or a hyperbola?

c How can you rewrite an equation like that in Part b so that it is easier to sketch the graph of the curve?

d Write the standard form equation(s) of each of the following conics and explain as completely as possible what that form allows you to conclude about the graph. Assume each conic has the *x*- and/or *y*-axes as symmetry lines.

   i. Circle   **ii.** Ellipse

   iii. Parabola   **iv.** Hyperbola

e How would you modify the equations in Part d if the lines of symmetry were $x = h$ and $y = k$, where *h* and *k* are not both 0?

Be prepared to share your descriptions and reasoning with the class.

 CHECK YOUR UNDERSTANDING

Identify the conic section represented by each equation and write the equation in standard form. Then use properties of the conic to sketch a graph.

a. $x^2 - 6x + y^2 - 40 = 0$

b. $4x^2 - y^2 - 16 = 0$

c. $3x^2 + 12x + 5y^2 + 30y + 42 = 0$

d. $y^2 - 20x = 0$

1 Study the portion of a contour map shown below. Describe as precisely as you can the terrain along the 3.5-mile White Ledge Trail including the highest and lowest altitudes. Follow the trail clockwise, beginning at White Ledge Campground.

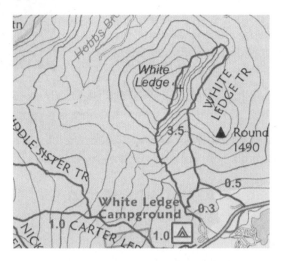

2 The diagram below shows the contours of the temperature along one wall of a heated room throughout one winter day, with time indicated as on a 24-hour clock. The room has a heater located at the left-most corner of the wall, and there is one window in the wall. The heater is controlled by a thermostat several feet from the window. (**Source:** Adapted from *Multivariable Calculus, Preliminary Edition* by William McCallum, Deborah Hughes-Hallett, Andrew Gleason, et al., New York: Reprinted by permission of John Wiley & Sons, Inc., 1996.)

a. Where is the window? When was it open?

b. Why do you think that the temperature at the window at 5 P.M. (17 hours) is less than at 11 A.M. (11 hours)?

c. When was the heat on?

d. To what temperature do you think the thermostat is set? How do you know?

e. How far is the thermostat from the left-most corner of the wall?

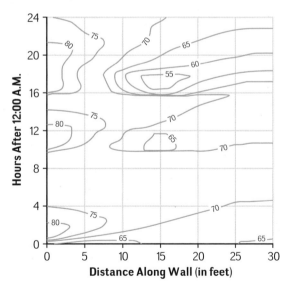

3 The altitude data below (in feet) were collected by the Pioneer probe of equatorial Venus, with adjacent probe points 90 feet apart.

a. Use the data to construct a contour diagram of the region. To simplify the contour construction, use one contour line to enclose all values within a 50-ft interval, beginning with 400–449 feet. Two contour lines have been drawn. One encloses a region of altitudes from 400 to 449 feet and the other a region of altitudes from 450 to 499 feet.

b. Are there possible landing sites in this region? If so, identify and justify your site choice(s). If not, explain why there is no appropriate site.

c. Sketch three vertical cross sections determined by the altitude data in the left, middle, and right columns of the chart.

473	424	449	523	585	623	631	616	587	572	575	565	533	512	504
526	507	530	570	625	659	663	658	619	587	587	576	525	507	511
591	606	611	623	658	666	666	661	636	619	624	617	551	527	531
612	635	660	696	697	675	672	663	652	655	663	649	591	565	565
623	702	788	809	760	707	703	673	653	657	659	651	613	584	580
642	735	852	868	835	744	716	696	671	652	638	634	628	611	591
633	699	869	943	936	795	736	723	676	651	628	628	642	642	621
628	698	903	1,028	998	845	761	747	682	651	633	634	646	661	695
606	651	807	965	973	856	752	719	682	649	634	633	658	708	753
582	595	668	824	877	838	740	696	668	630	629	641	701	763	769
551	567	610	747	834	803	719	686	672	638	631	639	672	729	753
516	519	529	630	696	671	639	650	648	645	645	627	616	665	724
495	503	504	538	547	543	566	615	624	637	654	653	674	704	736

Source: nssdc.gsfc.nasa.gov

4 A team of engineers at Compudesign is designing an air cooling system for a metal plate in a new model computer. Without the system, the temperatures (in °C) on the plate can be represented by the following contour diagram.

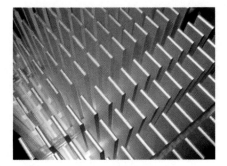

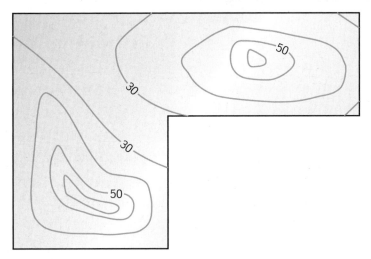

The planned cooling system will consist of two stiff, thin metal fins attached in such a way that they stand perpendicular to the plate. Heat from the plate will travel up into the fins and a fan will blow air across them, cooling the system. The engineers have decided that one fin will be placed along the line shown below.

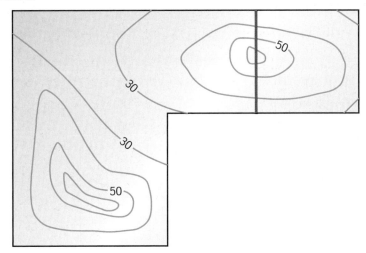

a. Why do you think the engineers chose this location for the fin?

b. Draw a graph of the temperature along the base of the fin as a function of distance from the top of the plate.

c. Choose a location for the second fin and draw a graph of the temperature along the base of the fin similar to that in Part b. What do points on your *x*-axis represent?

Jim Laser/CPMP

5 An Air Canada plane is flying in a straight line from Thunder Bay, Ontario, to New York, NY. The following contour diagram shows the barometric pressure in millibar (mbar) on the day of the flight. The pressure is assumed to be unchanging throughout the day.

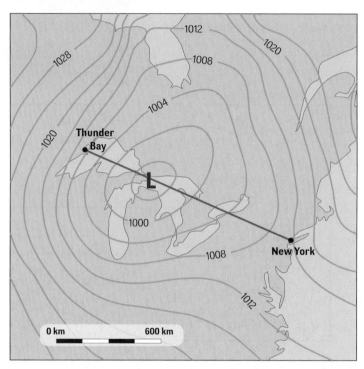

a. Draw a graph of the barometric pressure experienced by the plane as a function of distance from Thunder Bay.

b. Mark a point on your graph representing the location of the "L" (the lowest pressure).

6 One way to build a sense of points in three dimensions is to consider points plotted on isometric dot paper. Study the three-dimensional coordinate system at the right that shows the location of points A, B, and C in space, each defined by three coordinates (x, y, z). Think of the x- and y-axes as defining the floor of your room and the z-axis pointing straight up. (You might want to compare the diagram at the right to a corner of your room.)

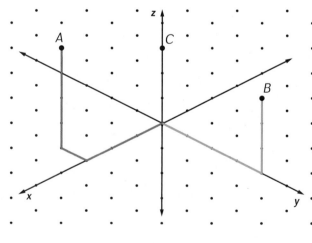

a. Following the marked trails for points A and B, what are the coordinates for these points?

b. Alanna has proposed that the coordinates for point C are $(-2, -2, 1)$. What do you think about her suggestion?

7 Just as rectangles are often replaced by nonrectangular parallelograms to give the impression of depth in two-dimensional drawings, circles are often replaced by ovals or ellipses in three-dimensional drawings. Consider the sphere shown below.

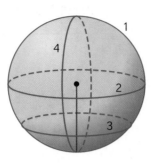

a. Which numbered circular cross sections are drawn as circles? Why?

b. Which numbered circular cross sections on the sphere are congruent? Under what condition will two horizontal or vertical cross sections be congruent?

c. Suppose the diameter of the sphere is 10 cm and that horizontal cross sections are taken at 1-cm intervals. Find the radius of each cross section and sketch the corresponding contour diagram.

d. How would a contour diagram of the bottom hemisphere of the sphere differ from the contour diagram of the sphere? From the contour diagram of the top hemisphere?

e. How does the contour diagram for a sphere differ from that of a cone standing on its base?

8 Assume that the height of each object below is 8 cm. Sketch a diagram for each object that shows horizontal cross sections at 2-centimeter intervals.

a.

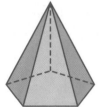

b.

c.

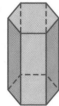

d.

9 Conic sections can be characterized by **Moiré patterns**—patterns formed by two intersecting sets of concentric circles. Graph paper with intersecting sets of concentric circles is sometimes called *conic graph paper*. In the conic graph paper below, the radius of the smallest circles is 1 unit and the centers, F_1, F_2, of the nonconcentric circles are 10 units apart.

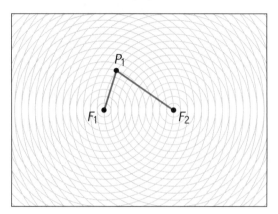

a. Explain why $F_1P_1 + F_2P_1 = 16$ in the diagram above.

b. On a piece of conic graph paper:

- Label the centers of the circles F_1 and F_2.

- Label the point P_1 as in the diagram above.

- Find 20 points P such that $F_1P + F_2P = 16$.

- If you were to connect all points P such that $F_1P + F_2P = 16$, what type of figure would you draw? How do you know? What are the points F_1 and F_2?

c. Use conic graph paper to draw an ellipse with horizontal axis of length 20. Can more than one such ellipse be drawn? Explain.

10 Conic graph paper, as described above, is also helpful when drawing hyperbolas. The radius of the smallest circles is 1 unit and the centers of the nonconcentric circles are 10 units apart.

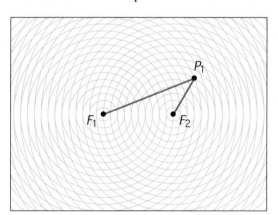

a. On a sheet of conic graph paper, label points F_1, F_2, and P_1 as in the previous diagram.

   i. Verify that $F_1P_1 = 14$ and $F_2P_1 = 6$.

   ii. Locate all points P such that $F_1P = 14$ and $F_2P = 6$. How many different points satisfy these conditions?

b. Find and mark all points P such that:

   i. $F_1P = 16$ and $F_2P = 8$.

   ii. $F_1P = 8$ and $F_2P = 16$.

c. Plot 20 other points P such that $|F_1P - F_2P| = 8$ and then connect your points.

d. Explain why the figure you drew in Part c is a hyperbola.

e. Use conic graph paper to draw a hyperbola with vertices that are 4 units apart and that has foci that are 10 units apart.

11 Identify the conic section represented by each equation and write the equation in standard form. Then draw a sketch of the figure and label lines of symmetry and center as applicable.

a. $81x^2 - 36y^2 = 2{,}916$

b. $5y^2 - 10y - x + 9 = 0$

c. $3x^2 + 3y^2 - 51 = 0$

d. $y^2 - x^2 - 8y - 20 = 0$

e. $4x^2 + y^2 - 8x + 6y + 9 = 0$

12 Write an equation describing each conic section shown below.

a.

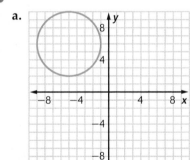

b.

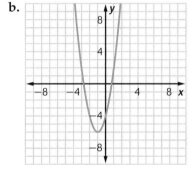

c.

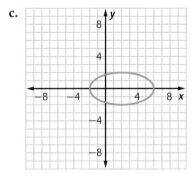

d.

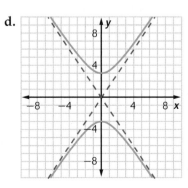

CONNECTIONS

13 Draw a three-dimensional coordinate system.

    a. Using the origin as a vertex, draw a cube with 4 units on a side with its edges along the positive x-, y-, and z-axes. Label and find the coordinates of each vertex.

    b. Draw another 4-unit cube with edges along the negative x- and y-axes and along the positive z-axis. Label and find the coordinates of each vertex.

14 High cholesterol resulting in clogged arteries was once thought to be the major underlying cause of heart attacks. Yet, half of all heart attack victims have cholesterol levels that are normal or even low. Research at Boston's Brigham and Women's Hospital suggests that inflammation, as measured by elevated levels of C-reactive protein, is another important independent trigger.

    Study the three-dimensional bar graph at the right that shows the relative risk of cardiovascular problems by levels of cholesterol and C-reactive protein.

Explain how each of the following findings is indicated in this plot.

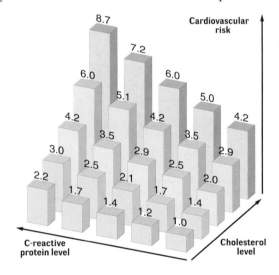

    a. A person with the highest combination of cholesterol and C-reactive protein has almost 9 times the risk as someone with the lowest combination.

    b. A person with C-reactive protein in the top quintile (fifth) has twice the risk as someone in the lowest quintile with the same cholesterol level.

    c. High cholesterol levels seem to contribute more to the risk of cardiovascular problems than do high C-reactive protein levels.

15 Use the "Slicing or Unfolding Polyhedra" custom app to explore different possible cross sections formed by a plane intersecting a rectangular prism. Write a summary, with sketches, of your findings.

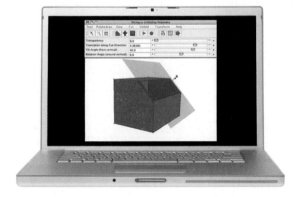

MaRoDee Photography/Alamy

16 Look back at the map on page 401. Suppose you are standing at Winniweta Falls and looking at the summit of Maple Mountain. What is the angle of elevation of your line of sight?

17 Position vectors are not limited to two dimensions. For example, three friends, Cristina, Marisol, and Rina, are in the same biology class. Cristina's scores on the first three biology tests can be thought of as a three-dimensional vector $\vec{t_1} = (84, 67, 90)$. Marisol's score vector is $\vec{t_2} = (68, 72, 86)$ and Rina's is $\vec{t_3} = (76, 74, 92)$.

a. Write $\frac{1}{3}(\vec{t_1} + \vec{t_2} + \vec{t_3})$ in coordinate form. What is the meaning of this vector relative to the friends' biology test scores?

b. Suppose the first test is over less important material and the third test is the final exam, so the tests are weighted according to the vector $\vec{w} = (25\%, 30\%, 45\%)$. Compute the following dot products and explain the meaning of each relative to the friends' biology test scores.

  i. $\vec{w} \cdot \vec{t_1}$

  ii. $\vec{w} \cdot \vec{t_2}$

  iii. $\vec{w} \cdot \vec{t_3}$

  iv. $\vec{w} \cdot \left[\frac{1}{3}(\vec{t_1} + \vec{t_2} + \vec{t_3}) \right]$

18 In Investigation 3, you derived the equation of an ellipse from the locus-of-points definition. In Unit 2, *Vectors and Motion*, you used parametric equations to model elliptical motion. The standard form of the ellipse can also be derived from the parametric equations representation.

a. Sketch the ellipse represented by the following parametric equations.

$$x = 2 \cos t \qquad y = 5 \sin t \qquad 0 \le t \le 2\pi$$

  i. What is the length of the horizontal axis? Of the vertical axis?

  ii. Sketch and describe the ellipse represented by the following equations.

$$x = 2 \cos t + 5 \qquad y = 5 \sin t + 8 \qquad 0 \le t \le 2\pi$$

  iii. Write parametric equations for the ellipse with center (h, k) having horizontal and vertical lines of symmetry, with $2a$ the length of the horizontal axis and $2b$ the length of the vertical axis.

b. Use the Pythagorean identity $\sin^2 t + \cos^2 t = 1$ to show that the ellipse described by the parametric equations you wrote in Part aiii also satisfies the following equation:

$$\frac{(x-h)^2}{a^2} + \frac{(y-k)^2}{b^2} = 1$$

19 In Connections Task 18, you used the parametric equations representation of an ellipse with center (h, k) to derive the standard form of the equation of an ellipse. Now reverse this process and use a different Pythagorean identity to find a parametric equations representation for a hyperbola with equation of the form $\frac{x^2}{a^2} - \frac{y^2}{b^2} = 1$.

20 From your work in Investigation 3, you know that the set of points whose coordinates (x, y) satisfy the equation $\frac{x^2}{2} - \frac{y^2}{2} = 1$ is a hyperbola.

a. Use a graphing calculator or graphing software to produce a graph showing the shape of the hyperbola for $-10 \le x \le 10$.

b. What lines appear to be asymptotes for the graphs in Part a?

c. What geometric transformation appears to map the graph of $y = \frac{1}{x}$ onto that of $\frac{x^2}{2} - \frac{y^2}{2} = 1$? (See Extensions Task 35 to prove that your idea works.)

21 Consider the ellipse with equation $\frac{x^2}{9} + \frac{y^2}{4} = 1$.

a. What equation would define an ellipse that is congruent to the ellipse $\frac{x^2}{9} + \frac{y^2}{4} = 1$ but is centered at $(5, 2)$ and has axes of symmetry $x = 5$ and $y = 2$?

b. What are the coordinates of points where the axes of symmetry intersect the ellipse in Part a?

22 Examine the following diagram of an ellipse.

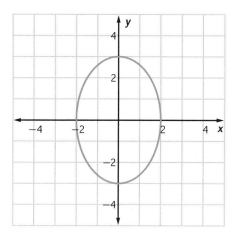

a. Write the equation representing this ellipse.

b. Use the diagram to estimate the area of that ellipse.

c. A circle with radius 1 has coordinate equation $x^2 + y^2 = 1$. Show algebraically, that the transformation $(x, y) \rightarrow (2x, 3y)$ maps the unit circle onto the ellipse with equation $\frac{x^2}{4} + \frac{y^2}{9} = 1$.

d. Use your answer to Part c and the area formula for a circle to devise a way of calculating the *area of the ellipse*. Compare the result of your area calculation with the estimate you made in Part a and explain why the two are (probably) somewhat different.

e. Recall that the general coordinate equation for an ellipse centered at the origin with x-intercepts $(\pm a, 0)$ and y-intercepts $(0, \pm b)$ is $\frac{x^2}{a^2} + \frac{y^2}{b^2} = 1$. What general formula for the area of such an ellipse is suggested by your work in Parts c and d?

23 In Investigation 1, you were introduced to a three-dimensional rectangular coordinate system. It is often helpful to picture a three-dimensional coordinate system in terms of a room you are in. Think of the origin of the coordinate system as a corner at floor level where two walls meet.

  a. Describe the x-, y-, and z-axes in this context.

  b. What would points with negative coordinates correspond to in this context?

24 Find an example of how contour diagrams are used to communicate the temperature patterns found across the United States and Canada. How is color sometimes used to indicate the higher temperature regions and the lower temperature regions?

25 Maps displaying contour lines are informative but do not necessarily convey a complete picture of a region. Here are two contour lines at 40-foot elevation intervals. The horizontal distance between the contours is approximately 300 yards.

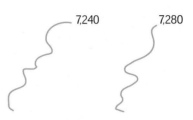

  a. Could there be a region between these contours that is higher than 7,280 ft? Explain.

  b. Sketch several possible side views of the 300 yards between these contours.

  c. How does the existence of a trail across the region between contours eliminate some possible configurations of the terrain?

26 A beam of light from a pen light or flashlight forms a cone. When the beam of light is projected onto a flat surface, the image is a conic section. In a darkened room, using a piece of cardboard as a plane and a pen light or flashlight, explore how to place the cardboard relative to the light source to produce each of the four conic sections. Write a brief summary of your findings.

27 Try to visualize each of the following possibilities. You can use the "Slicing Cones, Cylinders, and Spheres" custom app to aid your thinking as necessary.

  a. If possible, describe or sketch a plane whose intersection with a double cone is:

   i. a point.

   ii. a line.

   iii. a pair of intersecting lines.

  b. If possible, describe or sketch a plane whose intersection with a cylinder is:

   i. a point.

   ii. a line segment.

   iii. a rectangle.

c. If possible, describe or sketch a plane whose intersection with a sphere of radius r is:

    i. a point.   **ii.** a circle with radius $\frac{1}{2}r$.   **iii.** an oval.

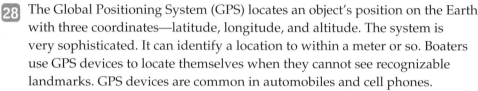

EXTENSIONS

28 The Global Positioning System (GPS) locates an object's position on the Earth with three coordinates—latitude, longitude, and altitude. The system is very sophisticated. It can identify a location to within a meter or so. Boaters use GPS devices to locate themselves when they cannot see recognizable landmarks. GPS devices are common in automobiles and cell phones.

    Research and write a short report on the capabilities and underlying mathematical features of GPS. Include some information on an application of GPS, other than in automobiles, boats, and cell phones.

29 Select one of the following projects to learn more about the usefulness of contour diagrams.

a. Consult your state's Department of Natural Resources or another agency that has responsibility for parks, lakes, and rivers to locate topographical maps of regions in your state. Choose one such map for a land region and one for a lake and describe the region and lake on the basis of the data included in the map. Describe also the contour intervals used in these maps.

b. If a local weather forecasting service is nearby, contact it and ask for copies of maps describing the atmospheric pressure patterns over the United States on a particular day. Learn how these pressure maps are used to assist in determining wind velocity and direction. Make a sketch of the pressure data. Write a report describing how pressure maps are used in the weather forecasting business.

30 Research information on the process of CAT scans and MRIs (magnetic resonance imaging) and the interpretation of their images. Write a brief report of your findings, including illustrations. Explain how the ideas are related to your work with cross sections in Investigation 2.

31 You have seen that sketching a hyperbola can be aided by first sketching the asymptotes and then using symmetry of the curve as in the diagram below.

a. If a point (not a vertex) with coordinates (p, q) is on the hyperbola with equation $\frac{x^2}{a^2} - \frac{y^2}{b^2} = 1$, use symmetry to identify three other points that are on the graph.

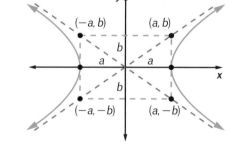

b. Provide an argument that the asymptotes of the hyperbola with equation $\frac{x^2}{a^2} - \frac{y^2}{b^2} = 1$ are $y = \pm\frac{b}{a}x$.

c. Provide an argument that the asymptotes of the hyperbola with equation $\frac{y^2}{a^2} - \frac{x^2}{b^2} = 1$ are $y = \pm\frac{a}{b}x$.

32 Recently, scientists at NASA observed a satellite with a mass of about 800 kg moving through the solar system at an approximate initial distance of 2.5×10^8 km from the Sun (at its farthest, Mars is roughly this distance from the Sun). They determined that the orbit of the satellite is a conic with equation

$$\frac{(x-c)^2}{a^2} + \frac{y^2}{a^2-c^2} = 1,$$

where the x and y are measured in kilometers. In these coordinates, the Sun is at the origin and the satellite initially lies on the x-axis. The constants a and c can be found from the initial speed v (in kilometers per second) of the object and a constant ε (called the *eccentricity* of the orbit) as follows:

$$\varepsilon = \sqrt{1 - (3.818 \times 10^{-3})v^2 + (3.548 \times 10^{-6})v^4}$$

$$a = \frac{(4.709 \times 10^5)v^2}{1 - \varepsilon^2} \qquad c = \frac{(4.709 \times 10^5)\varepsilon v^2}{1 - \varepsilon^2}$$

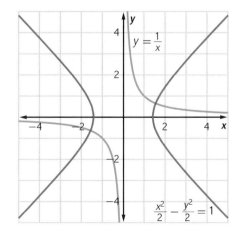

a. The orbital speed of the Earth is approximately 30 km/sec. What kind of orbit will the satellite have if its initial speed is $v = 30$ km/sec?

b. Sketch a graph of the orbit from Part a and describe the long-term behavior of the satellite.

c. What kind of orbit will the satellite have if its initial speed is $v = 35$ km/sec?

d. Sketch a graph of the orbit from Part c and describe the long-term behavior of the satellite.

33 Access the article, "Can You Really Derive Conic Formulae from a Cone?" by Gary S. Stoudt at www.maa.org/publications/periodicals/convergence/can-you-really-derive-conic-formulae-from-a-cone. Write a report on the derivation of the coordinate equation of one of the conics from the Greek definition in terms of a double cone.

34 Use symbolic reasoning to rewrite the parametric equations

$$x = 4 \sec t \qquad y = 5 \tan t \qquad 0 \leq t \leq 2\pi$$

as a single equation in variables x and y. Describe as completely as possible the curve described by the equation. Then use technology to confirm your reasoning.

35 In Connections Task 20, you discovered that it looks like the graph of $y = \frac{1}{x}$ could be rotated about the origin through an angle of 45° in the clockwise direction to coincide with the graph of the equation $\frac{x^2}{2} - \frac{y^2}{2} = 1$.

a. Check that the matrix $M = \begin{bmatrix} \frac{1}{\sqrt{2}} & \frac{1}{\sqrt{2}} \\ -\frac{1}{\sqrt{2}} & \frac{1}{\sqrt{2}} \end{bmatrix}$

can be used to calculate images for points under the required rotation by finding the product of that matrix and the sample points $\begin{bmatrix} 0 \\ 1 \end{bmatrix}$, $\begin{bmatrix} 1 \\ 1 \end{bmatrix}$, and $\begin{bmatrix} 1 \\ 0 \end{bmatrix}$.

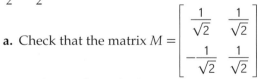

b. The graph of the function $y = \frac{1}{x}$ consists of all points with coordinates in the form $\left(x, \frac{1}{x}\right)$. Find the matrix product $\begin{bmatrix} \frac{1}{\sqrt{2}} & \frac{1}{\sqrt{2}} \\ -\frac{1}{\sqrt{2}} & \frac{1}{\sqrt{2}} \end{bmatrix} \begin{bmatrix} x \\ \frac{1}{x} \end{bmatrix}$.

c. The result of the calculation in Part b is a matrix in the form $\begin{bmatrix} A \\ B \end{bmatrix}$.

Use algebraic reasoning to show that $A^2 - B^2 = 2$. Explain how this result proves the conjecture about the connection between the two hyperbolic graphs.

36 Suppose C is a circle with center at the origin and radius r. Find an equation for the set of all points $P(x, y)$ that are equally distant from the circle C and the x-axis.

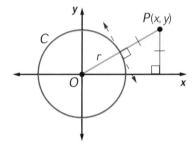

REVIEW

37 Rewrite each expression in expanded form.

a. $(4x + k)^2$

b. $(6a + 2b)^2$

c. $\left(\frac{x}{3} + \frac{y}{6}\right)^2$

d. $(3x - 2)(4x^2 + 2x - 7)$

38 Consider the line $3x + y = 6$.

a. Draw the graph of this line on a coordinate system.

b. Write an equation for a line that does not intersect the given line. Add a graph of your line to the graph in Part a.

c. Write an equation for a line that intersects the line in Part a in exactly one point. Add a graph of this line to your graph from Part a.

d. Write an equation for a line that intersects the line in Part a at infinitely many points.

39 Solve each equation. Check your solutions.

a. $\sqrt{x - 16} = 8$

b. $\sqrt{2x + 6} = x - 1$

c. $\sqrt{x + 6} = \sqrt{2x + 1}$

d. $\sqrt{4x} - \sqrt{x} = 2$

40 For each rational function:

- state the domain.
- determine all vertical, horizontal, or oblique asymptotes.
- sketch a graph of the function.

a. $f(x) = \dfrac{x + 2}{x^2 - 4}$

b. $g(x) = \dfrac{x^2 - 2x}{x + 1}$

c. $h(x) = \dfrac{x^2 + 3x}{x^2 + 7x + 6}$

41 For each pair of expressions, determine the value of k that will make the expression in Column II equivalent to the expression in Column I.

Column I	**Column II**
a. $x^2 + 8x + 16$	$(x + k)^2$
b. $x^2 + 6x - 10$	$(x + 3)^2 + k$
c. $2x^2 + 20x + 5$	$2(x + k)^2 - 45$

42 Determine the length of the line segment connecting each pair of points.

a. $A(3, 7)$ and $B(-12, 7)$ **b.** $X(9, -6)$ and $Y(-2, -6)$

c. $C(0, 0)$ and $D(4, 10)$ **d.** $H(-3, 12)$ and $K(2, 9)$

43 Consider the quadrilateral $ABCD = \begin{bmatrix} 4 & -1 & -4 & 1 \\ 1 & 4 & -1 & -4 \end{bmatrix}$.

a. Prove that quadrilateral $ABCD$ is a square.

b. What are the coordinates of the intersection point of the diagonals?

c. Determine the equation of the circle that circumscribes the square.

d. Determine the equation of the circle that is inscribed in the square.

44 Recall that the equation of a circle with radius r centered at (h, k) has equation $(x - h)^2 + (y - k)^2 = r^2$.

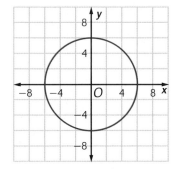
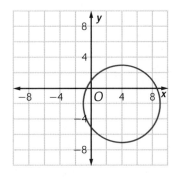

a. Suppose point A is on the left circle above so that $\overline{OA}$ and the positive x-axis form an angle with radian measure $\frac{2}{5}$. In what quadrant is point A located? Find the approximate coordinates of point A.

b. What is the equation of the circle on the left above? On the right?

c. What are the y-coordinates of all points on the circle above on the left with x-coordinate of 4? Above on the right with x-coordinate of 4?

Equations for Surfaces

In the previous lesson, you represented surfaces of planets and familiar objects using contour diagrams and cross sections. You also explored the use of three-dimensional coordinates. Computer models for designing cooling towers like the one above are based on coordinate representations. The images are often produced from information about cross sections derived from the equations of the surfaces of the shapes.

You can use your understanding of coordinate representations of two-dimensional figures to guide your thinking about coordinate representations of related three-dimensional objects.

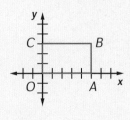 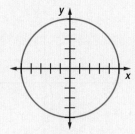 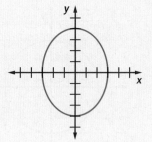

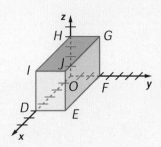 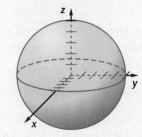 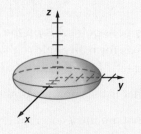

©ImageState/age fotostock

THINK ABOUT THIS SITUATION

Use the diagrams on the previous page to help you think about possible similarities and differences between two-dimensional and three-dimensional objects and their coordinate representations. The scales on the axes are 1 unit.

a What are the coordinates of point *B*, a vertex of the rectangle? What are the coordinates of point *J*, a vertex of the rectangular prism? How was your reasoning similar in each case?

b What are the coordinates of the point symmetric to point *B* with respect to the *y*-axis? With respect to the *x*-axis? What do you think are the coordinates of the point symmetric to point *J* with respect to the *xz*-plane? With respect to the *xy*-plane? Explain your reasoning.

c What is true about all points on the line determined by points *B* and *C*? What is the equation of the line? What is true about all points on the plane determined by points *I*, *D*, and *E*? What do you think is the equation of the plane?

d How would you determine the length of $\overline{AC}$? How might you determine the length of $\overline{DG}$?

e What is the equation of the circle? Of the ellipse? What equation do you think would describe the *sphere*? The *ellipsoid*?

In this lesson, you will extend important ideas and reasoning strategies involving coordinates in two dimensions to coordinates in three dimensions.

INVESTIGATION 1

Relations Among Points in Three-Dimensional Space

In Lesson 1, you explored how to represent points in space with *x*-, *y*-, and *z*-coordinates. The *x*-axis and the *y*-axis determine the *xy-plane* as shown below. Similarly, the *x*-axis and the *z*-axis determine the *xz-plane*, and the *y*-axis and *z*-axis determine the *yz-plane*.

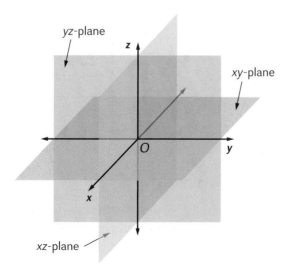

As you work on the problems of this investigation, look for answers to these questions:

How are graphs and equations of planes parallel to a coordinate plane in three dimensions similar to graphs and equations of lines parallel to a coordinate axis in two dimensions?

How are the formulas for calculating distance between points and midpoints of segments in three dimensions similar to the corresponding formulas in two dimensions?

How is the equation of a sphere with center at the point (j, h, k) similar to the equation of a circle with center at the point (j, h)?

1 **Equations for Special Planes** You can determine the equations of the *coordinate planes* and planes parallel to them by reasoning from analogous cases in a two-dimensional coordinate system.

a. Think about the form of equations of horizontal and vertical lines in a two-dimensional (x, y) coordinate system.

   i. What is the equation of the x-axis? Of the y-axis? Why do these equations make sense?

   ii. What are the equations of the lines 5 units away from and parallel to the x-axis?

   iii. What are the equations of the lines 10 units away from and parallel to the y-axis?

b. Now think about the form of equations of horizontal planes in a three-dimensional (x, y, z) coordinate system.

   i. Which coordinate plane is represented by the equation $z = 0$? Why does this make sense?

   ii. What is the equation of the plane 4 units above and parallel to the xy-plane?

   iii. What is the equation of the plane 6 units below and parallel to the xy-plane?

   iv. The sketch below shows one of the two planes described in part iii. Reproduce this diagram and then sketch the other plane. Use different colors for each plane. Label both planes.

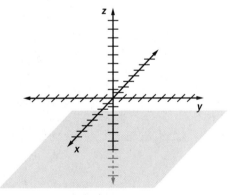

c. What is the equation for the xz-plane? What are the equations of the planes 5 units away from and parallel to the xz-plane? Sketch and label all three planes on the same coordinate system.

d. What is the equation for the yz-plane? What are the equations of the planes 12 units away from and parallel to the yz-plane? Sketch and label those two planes on the same coordinate system.

2 In Course 2 Unit 3, *Coordinate Methods*, you derived a formula for computing distances between pairs of points in a coordinate plane. If $P(x_1, y_1)$ and $Q(x_2, y_2)$ are points in a coordinate plane, then

$$PQ = \sqrt{(x_1 - x_2)^2 + (y_1 - y_2)^2}.$$

You used this distance formula to help derive equations for the conic sections in Lesson 1.

In the Think About This Situation at the beginning of this lesson, you considered how you might compute the distance between points in three-dimensional space. Compare your ideas with the following approach suggested by one class in Winston-Salem, North Carolina. Their approach is based on the rectangular prism at the beginning of this lesson. The scales on the axes are 1 unit.

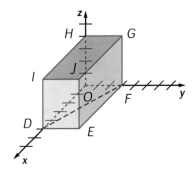

To find the length of $\overline{DG}$, the students suggested drawing $\overline{DF}$ and then using the Pythagorean Theorem twice.

a. Why is $\triangle DOF$ a right triangle? What is the length of $\overline{DF}$?

b. Why is $\triangle DFG$ a right triangle? What is the length of $\overline{DG}$?

c. What is the length of $\overline{OJ}$ in the diagram above?

3 **Distance Formula: From Two to Three Dimensions** Now use the diagram below to derive a formula for finding the distance PQ if $P(x_1, y_1, z_1)$ and $Q(x_2, y_2, z_2)$ are points in three-dimensional space and the figure shown is a rectangular prism. You will find it helpful to label the coordinates of points S and R.

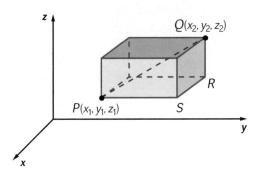

a. In the diagram, point Q is above point P. Does your distance formula hold if point Q is below point P? Explain your reasoning.

b. Use the distance formula you developed to find the distance between the points $A(1, 2, 3)$ and $B(6, -2, -7)$.

c. Does the formula you derived hold for any two points that lie in a plane parallel to a coordinate plane? Explain your reasoning.

4 **Equations for Spheres** Recall that a circle is the set of points in a plane at a given distance from a fixed point, its center. In three-dimensional space, the set of points at a given distance from a fixed point is a **sphere**. Like a circle, a sphere is determined by its center (the fixed point) and its radius (the given distance).

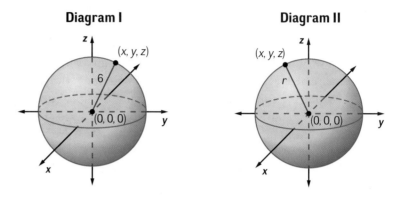

Diagram I   **Diagram II**

a. Find the equation for a sphere with center at the origin and radius 6 as shown in Diagram I.

b. What are the *x*-, *y*-, and *z*-intercepts of the sphere in Part a? Do the coordinates of these points satisfy your equation?

c. Using Diagram II, derive an equation for a sphere with center at the origin and radius *r*.

d. How would you modify the equation in Part c if the center of the sphere had coordinates (j, h, k) and radius *r*? Compare your ideas with those of your classmates. Resolve any differences.

e. The equation derived in Part c represents the surface of the sphere. How would you represent algebraically the solid ball enclosed by the sphere?

5 The pictures below illustrate just a few of the many applications of a solid sphere.

Ball bearings

Bocce balls

Replacement ball-and-socket joints

a. What physical property of spheres makes them ideal for all kinds of uses that involve smooth or rotating movement?

b. The surface areas of a cubic storage tank and of a spherical tank are each 60 m². Which tank has the greatest volume?

c. What research question or conjecture does your answer to Part b suggest? Discuss with your classmates how you might investigate your conjecture or question.

6 **Midpoint Formula: From Two to Three Dimensions** The distance formula in three-dimensional space is a generalization of the distance formula in two dimensions. Now, investigate how to generalize the formula for the midpoint of a segment in a plane to that of a segment in three dimensions.

a. If $U(x_1, y_1)$ and $V(x_2, y_2)$ are points in a plane, what are the coordinates of the midpoint of $\overline{UV}$? Suppose P with coordinates (x_1, y_1, z_1) and Q with coordinates (x_2, y_2, z_2) are points in space. Reasoning by analogy, conjecture an expression for the midpoint of $\overline{PQ}$.

b. Now consider the points $A(5, 6, 12)$ and $B(1, -4, 4)$.

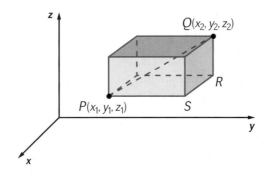

   i. Use your knowledge of computing the midpoint of a segment in a plane to calculate the midpoint of $\overline{BC}$. Then use your result to calculate the midpoint of $\overline{AB}$.

   ii. Use the distance formula to verify that the point you found is the midpoint of $\overline{AB}$.

c. Use the diagram at the right to derive a formula for the midpoint M of $\overline{PQ}$. Compare your formula for the midpoint of $\overline{PQ}$ with that of your classmates. Resolve any differences.

7 Now pull together the ideas you developed in this investigation. Consider the points $S(-3, 1, 5)$ and $T(2, 4, -3)$ in three-dimensional space.

a. Find the equation of the plane containing point S and parallel to the xy-plane.

b. Find the equation of the plane containing point T and parallel to the yz-plane.

c. Find ST.

d. Find the coordinates of the midpoint of $\overline{ST}$.

e. Find the equation of the sphere with center at the origin and containing point T.

f. Find the equation of the sphere with center at point S and containing point T.

SUMMARIZE THE MATHEMATICS

In this investigation, you extended important ideas involving coordinates in two dimensions to coordinates in three dimensions.

a In three-dimensional space, how are the equations and graphs of planes parallel to a coordinate plane similar to, and different from, the equations and graphs of lines parallel to a two-dimensional coordinate axis?

b Describe how the formula for the distance between two points in a three-dimensional coordinate system is similar to, and different from, the distance formula for two points in a two dimensional coordinate system.

c Describe how to find the coordinates of the midpoint of a segment in three dimensions.

d How is the equation of a sphere with center at the origin and radius r similar to, and different from, the equation of a circle with center at the origin and radius r?

e How is the equation of a sphere with center at (a, b, c) and radius r similar to, and different from, the equation of a circle with center (a, b) and radius r?

Be prepared to share your ideas and reasoning with the entire class.

 CHECK YOUR UNDERSTANDING

In the diagram below, vertex T of the rectangular prism has coordinates $(12, 3, 6)$ and O is the origin.

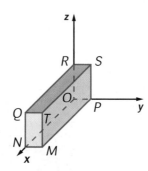

a. Determine the coordinates of the remaining vertices.

b. Write equations for the planes that contain the faces of the prism.

c. Use right triangles to explain why $(PQ)^2 = (PM)^2 + (MN)^2 + (NQ)^2$.

d. Calculate PQ using the equation in Part c and then using the distance formula.

e. Find the midpoints of $\overline{OT}$ and $\overline{RM}$. What can you conclude?

f. Does the sphere with equation $x^2 + y^2 + z^2 = 200$ contain the prism? Explain.

The Graph of $Ax + By + Cz = D$

In two dimensions, two coordinates locate a point, and equations in two variables can be used to specify lines or curves. For example, the graph of $2x - 3y = 12$ is a line and the graph of $\frac{x^2}{25} - \frac{y^2}{10} = 1$ is a hyperbola.

In three-dimensional space, three coordinates locate a point, and equations in three variables can be used to specify surfaces. The surface may be flat, as in the case of a plane, or curved, as in the case of a sphere. The challenge, often, is to figure out what the surface looks like given its equation and to make a sketch of it, or to interpret the reasonableness of a technology-produced graph.

As you work on the problems of this investigation, look for answers to this question:

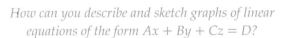

> *How can you describe and sketch graphs of linear*
> *equations of the form $Ax + By + Cz = D$?*

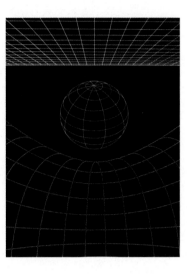

1. The equation $Ax + By + Cz = D$ in three dimensions is analogous to the equation $Ax + By = C$ in two dimensions. Based on this analogy, what do you think the graph of $Ax + By + Cz = D$ could be? Explain your reasoning.

2. Now consider a specific instance of the equation $Ax + By + Cz = D$, namely $2x + 4y + 3z = 12$.

   a. Reasoning by analogy from your previous work in a two-dimensional coordinate system, what are the points at which the surface represented by $2x + 4y + 3z = 12$ intersects the x-, y-, and z-axes (called the *intercepts*)? Plot these points on a three-dimensional coordinate system.

   b. Cross sections of the surface formed by its intersection with the coordinate planes or with planes parallel to the coordinate planes provide useful information about the nature of the surface. The first cross sections to check are those made with the coordinate planes. These cross sections are called **traces**.

      i. Explain why you can find the equation of the trace of the surface in the xy-plane by setting $z = 0$ and examining the resulting equation.

      ii. Find the equation of the xy-trace for $2x + 4y + 3z = 12$. What is the shape of the trace?

      iii. Find the equation and describe the graph of each of the other two traces.

      iv. Plot the portion of each of the traces between the x- and y-axes, the y- and z-axes, and the x- and z-axes.

   c. Based on your information about the traces of the graph of $2x + 4y + 3z = 12$, what do you think the surface is? Why? Revise your sketch to better display this surface.

d. Additional information about a surface can be obtained by examining the equations of cross sections found by setting x, y, or z to a constant value.

    i. Explain how setting $z = 8$ generates a cross section of $2x + 4y + 3z = 12$ that is in a plane parallel to one of the coordinate planes. Describe the location of the cross section.

    ii. Find the equation of the cross section determined by setting $z = 8$. What is the shape of this cross section? How is this cross section related to the xy-trace?

    iii. Describe the cross section formed by setting $x = 5$. By setting $y = 2$.

    iv. Does the information you found by examining the equations of these cross sections confirm or change your thinking about the surface described by $2x + 4y + 3z = 12$? Explain.

Reasoning like you did in Problem 2 should suggest that a linear equation in x, y, and z is a **plane**.

3 Discuss how the graph of the plane shown below is similar to and different from your sketch of $2x + 4y + 3z = 12$. Try to find the equation for this plane using the connection between an equation and points on its graph.

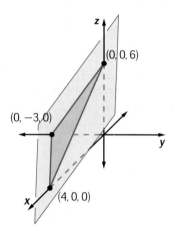

4 Illustrate how you could use the points where the graph of $2x + 3y + 2z = 6$ intersects the x-, y-, and z-axes to quickly sketch the surface on a three-dimensional coordinate system. Explain why your method works.

5 Sketch the graph of $3x + 5z = 15$ on a three-dimensional coordinate system. Describe the surface.

6 Consider the general equation $Ax + By + Cz = D$ for each set of conditions below. Describe the graph of each equation. Identify the intercepts, the traces, and the cross section at $z = 4$.

a. A, B, C, and D are all nonzero.

b. A, B, and D are nonzero, but $C = 0$.

c. $A = B = 0$, while C and D are nonzero.

7 Now consider how you can work backward from information about traces to the equation of a surface. Suppose a plane has traces with equations $x + y = 8$, $x + 4z = 8$, and $y + 4z = 8$. Sketch the plane and find an equation for the plane.

8 As you have seen, equations of planes in three dimensions are similar to equations of lines in two dimensions. An important characteristic of lines is that they are straight; they have constant slope. Planes have an analogous characteristic: they are flat. But what makes them flat? Consider the portion of the graph of $2x + 3y + 6z = 6$ shown below. Scales are 1 on each axis.

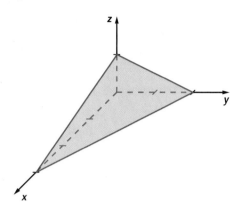

a. How can you quickly check the reasonableness of this graph?

b. Now imagine starting at the point $(0, 0, 1)$ and walking around on this plane. What can you say about the "slope" of your walk if you stay directly above the x-axis? The y-axis?

c. Suppose you walk on the plane along a path with constant y-coordinate. What can you say about the "slope" of your path? Explain your reasoning.

d. Suppose you walk on the plane along a path with constant x-coordinate. What can you say about the "slope" of your path in this case? Why does this make sense?

e. What can you say about the "slope" of walks on the plane along paths with the z-coordinate constant?

9 In your previous work in two dimensions, you saw that two lines could intersect in 0, 1, or infinitely many points. Consider the analogous situation in three dimensions in the case of planes with equations $2x + 3y + 4z = 12$ and $3x + 2y - z = 6$.

a. Sketch the planes and describe their intersection.

b. What are the possible ways the graphs of two nonequivalent equations of the form $Ax + By + Cz = D$ can intersect?

c. Under what conditions would the graphs of $A_1x + B_1y + C_1z = D_1$ and $A_2x + B_2y + C_2z = D_2$ be parallel?

d. Is it possible for the graphs of three equations of the form $Ax + By + Cz = D$ to intersect in 0 points? One point? Infinitely many points? Explain.

The equation $Ax + By + Cz = D$, where not all A, B, and C are zero, is a linear equation in x, y, and z.

a What is the graph of an equation of this form?

b How can you quickly sketch the graph?

c What is the nature of each cross section of the graph of $Ax + By + Cz = D$?

d How are the equations of planes parallel to coordinate planes special cases of $Ax + By + Cz = D$?

Be prepared to explain your ideas to the class.

 CHECK YOUR UNDERSTANDING

Sketch the graph of each equation on a three-dimensional coordinate system.

a. $2x - y + 3z = 6$

b. $x + y - 2z = 30$

c. $x - 4y = 8$

INVESTIGATION 3

Surfaces Defined by Nonlinear Equations

In Investigation 2, you found that your understanding of linear equations made sketching planes defined by linear equations of the form $Ax + By + Cz = D$ much easier. Similarly, to sketch surfaces defined by nonlinear equations, you can draw on your understanding of curves, particularly the conics, in a coordinate plane.

As you work on the following problems, look for answers to these questions:

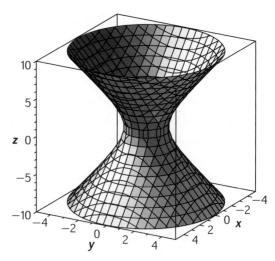

What strategies are helpful in sketching surfaces and judging the reasonableness of technology-produced graphs defined by nonlinear equations in three variables?

How can you describe symmetries of three-dimensional figures?

1 Re-examine the ellipse and corresponding surface, called an **ellipsoid**, from the beginning of this lesson.

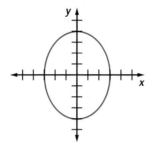 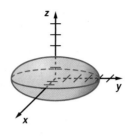

a. Is the ellipse symmetric with respect to the x-axis? With respect to the y-axis? Explain your reasoning.

b. If a point $A(a, b)$ is on the ellipse, name the coordinates of at least two other points on the ellipse.

c. The ellipsoid above is *symmetric with respect to the xz-plane*. If a point $A(a, b, c)$ is on the ellipsoid, then the point $A'(a, -b, c)$ is also on the ellipsoid.

  i. Is the ellipsoid *symmetric with respect to the xy-plane*? If so, what are the coordinates of the point that is symmetric to point A with respect to the xy-plane?

  ii. Is the ellipsoid *symmetric with respect to the yz-plane*? If so, what are the coordinates of the point that is symmetric to point A with respect to the yz-plane?

2 Now consider the algebraic representation for the ellipse above. Scales on the axes are 1.

a. Explain why an equation for the ellipse is $\frac{x^2}{9} + \frac{y^2}{16} = 1$.

b. Note that when you substitute $-x$ for x in the equation in Part a, you get $\frac{(-x)^2}{9} + \frac{y^2}{16} = 1$ or $\frac{x^2}{9} + \frac{y^2}{16} = 1$, which is the same as the original equation.

  i. Why can you use this fact to conclude that the graph of the equation is symmetric with respect to the y-axis?

  ii. How could you determine that the graph of $\frac{x^2}{9} + \frac{y^2}{16} = 1$ is symmetric with respect to the x-axis by only reasoning with the symbolic form?

3 In the Think About This Situation at the beginning of this lesson, you may have conjectured that an equation for the ellipsoid is $\frac{x^2}{9} + \frac{y^2}{16} + \frac{z^2}{4} = 1$. Explain why this graph-equation match makes sense in terms of each of the following.

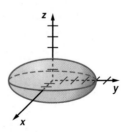

a. x-, y-, and z-intercepts

b. Cross sections determined by the coordinate planes and planes parallel to the coordinate planes

   i. What are the equation and shape of the cross section of the graph of $\dfrac{x^2}{9} + \dfrac{y^2}{16} + \dfrac{z^2}{4} = 1$ determined by the yz-plane? The xz-plane? The xy-plane?

   ii. What are the equation and shape of the cross section determined by setting $z = 2$? By setting $y = 2$? By setting $x = 1$?

c. Symmetry

   i. What symmetry of the graph of $\dfrac{x^2}{9} + \dfrac{y^2}{16} + \dfrac{z^2}{4} = 1$ is implied by the fact that $\dfrac{x^2}{9} + \dfrac{y^2}{16} + \dfrac{(-z)^2}{4} = \dfrac{x^2}{9} + \dfrac{y^2}{16} + \dfrac{z^2}{4} = 1$?

   ii. How could you test for 180° rotational symmetry of the graph of $\dfrac{x^2}{9} + \dfrac{y^2}{16} + \dfrac{z^2}{4} = 1$ about the x-axis by reasoning with the equation itself?

   iii. How would you test for symmetry with respect to the yz-plane by reasoning with the equation itself?

4 Now consider in more detail 180° rotational symmetry in two and in three dimensions.

a. Does the ellipse on page 441 appear to have 180° rotational (*half-turn*) symmetry about the origin? Explain.

b. How could you determine that the graph of $\dfrac{x^2}{9} + \dfrac{y^2}{16} = 1$ has half-turn symmetry about the origin by only reasoning with the symbolic form?

c. The ellipsoid in Problem 3 has *half-turn symmetry about the z-axis*. If a point $A(a, b, c)$ is on the ellipsoid, then the point $A'(-a, -b, c)$ is also on the ellipsoid.

   i. Does this ellipsoid have half-turn symmetry about the x-axis? If so, what are the coordinates of the point that is symmetric to point A under this half-turn?

   ii. Does this ellipsoid have half-turn symmetry about the y-axis? Explain your reasoning.

   iii. How could you test for half-turn symmetry of the graph of $\dfrac{x^2}{9} + \dfrac{y^2}{16} + \dfrac{z^2}{4} = 1$ about the x-axis by reasoning with the equation itself?

d. The paraboloid at the right is defined by the equation $x^2 - y + 4z^2 = 0$.

   i. Does the paraboloid have half-turn symmetry about the y-axis? How could you test for half-turn symmetry about the y-axis by reasoning with the equation itself?

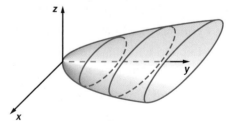

   ii. Does the paraboloid have half-turn symmetry about the x-axis? How could you test for half-turn symmetry about the x-axis by reasoning with the equation itself?

5 Shown below is a technology-produced graph of $x^2 + y^2 - z = 0$. Horizontal cross sections (and perhaps a contour diagram) can be used to explain why the graph is correct. That is, it is a bowl shape rather than a cone.

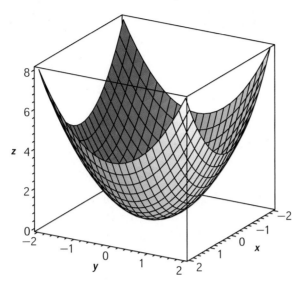

a. The horizontal cross sections are determined by planes with equations of the form $z = c$. Find the equations for the horizontal cross sections at intervals of $c = 0, 1, 2, 3,$ and 4 units.

b. Describe the sequence of graphs of the equations you found in Part a.

c. What does this tell you about the shape of the graph and its correctness?

d. Explain why vertical cross sections of the graph of $x^2 + y^2 - z = 0$ are determined by equations $y = c$ or $x = c$.

e. Find the equation of several vertical cross sections at 1-unit intervals. Explain how those equations are revealed in the shape of the graph above.

6 Use horizontal cross sections at 1-unit intervals (and perhaps a contour diagram) to help you explain why the graph of $x^2 + y^2 - z^2 = 0, z \geq 0$, is a cone. Sketch a graph of the cone.

7 Use analysis of intercepts, cross sections, and symmetry to help you match each surface with one of the following equations.

I

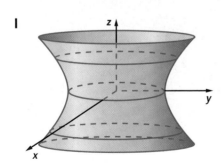

II

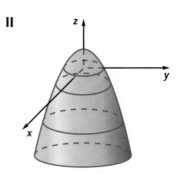

III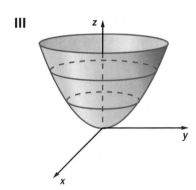

IV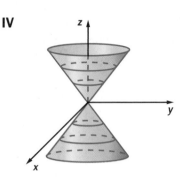

a. $x^2 + y^2 - z^2 = 0$

b. $x^2 + 4y^2 - z^2 = 4$

c. $x^2 + 2y^2 + 3z = 6$

d. $x^2 + y^2 - z = 0$

8 Use analysis of intercepts, cross sections, and symmetry to help you visualize and sketch surfaces given by each of the following equations. Describe each surface. Compare your surfaces and descriptions with those of your classmates and resolve any differences.

a. $x^2 - y + z^2 = 1$

b. $z^2 - x^2 - y^2 = 16$

Graphing surfaces in three dimensions can be time consuming and difficult. Three-dimensional graphs are more easily produced using computer software with three-dimensional graphing capability. Another advantage of these technologies is that they may permit you to view the surface from different angles. Software with implicit graphing capabilities can graph equations such as $4x + 5y = 7$ in two dimensions and equations such as $x^2 - y + z^2 = 1$ (Problem 8) in three dimensions.

Computer-produced graphs of surfaces such as that in Problem 5 and below are produced by drawing cross sections parallel to two coordinate planes, creating a *mesh*. The fineness of the mesh is determined by how close adjacent cross sections are to each other. In a "smooth" rendering of a surface by computer graphic software as shown below, color and shading are used to fill in the quadrilaterals of the mesh and then a *smoothing algorithm* is applied so the surface appears realistic.

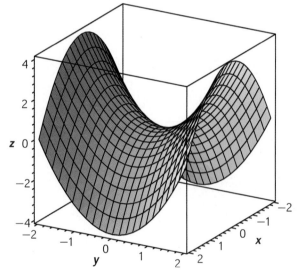

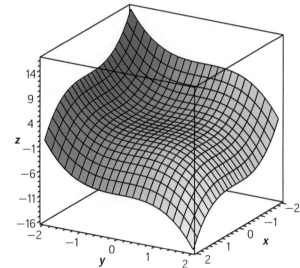

9 Three-dimensional graphing software typically graphs equations of the form $z = f(x, y)$. This notation means that z is a function of two variables, x and y.

a. The technology-produced display below shows the graph of $z = f(x, y) = \sqrt{1 - x^2 + y}$. Explain differences between this graph and the one you sketched in Problem 8 Part a.

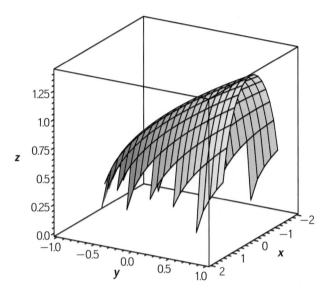

b. Being able to judge the reasonableness of technology-produced graphs is as important when working in a three-dimensional coordinate system as when working in a two-dimensional coordinate system. Explain why it is reasonable that the graph shown below is that of $x^2 - y^2 - z = 0$.

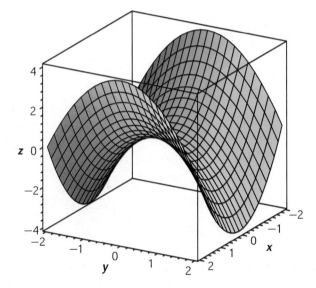

c. Use three-dimensional graphing software to graph each of the following equations. Describe the general shape of each surface. Experiment with viewing the surface from different angles.

   i. $z = \dfrac{x^2y - y^2x}{400}$ **ii.** $z = 2x - 3y + 3$

   iii. $z = \sqrt{25 - x^2 - y^2}$ **iv.** $z = \dfrac{x^2y^2 - y^2x}{400}$

CHECK YOUR UNDERSTANDING

Sketch and describe each of the following surfaces. Then check your graphs using three-dimensional graphing software.

a. $(x - 2)^2 + y^2 + (z - 1)^2 = 25$

b. $x^2 + y^2 + z = 4$

INVESTIGATION 4

Surfaces of Revolution and Cylindrical Surfaces

Graphs of equations in three variables are surfaces. Some of these surfaces can be generated by rotating (or revolving) a curve about a line, sweeping out a **surface of revolution**. The line about which the curve is rotated is called the **axis of rotation**. A table leg or lamp base turned on a lathe has a surface of revolution. A potter using a potting wheel makes surfaces of revolution.

Some common surfaces can be thought of as surfaces of revolution. As you work on the problems of this investigation, look for answers to this question:

What strategies can be used to identify and sketch surfaces of revolution and cylindrical surfaces?

1 Sketch a graph of $x^2 + y^2 = 25$ in the xy-plane of a three-dimensional coordinate system.

    a. Imagine rotating the circle about the y-axis. What kind of surface is formed?

    b. Would you get the same surface if the circle was rotated about the x-axis? Explain your reasoning.

    c. Write the equation for this surface of revolution.

2 The paraboloid from Lesson 1 Problem 5 (page 404) is reproduced in Diagram I. Write an equation of a curve in the yz-plane that would produce a similar surface when rotated about an axis. What is the axis of rotation?

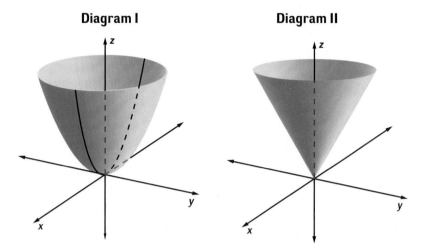

Diagram I	**Diagram II**

3 Describe, and illustrate with a sketch, how a cone with vertex at the origin (Diagram II) can be generated as a surface of revolution from a two-dimensional figure.

4 Next consider the segment determined by the points $P(0, 5, 1)$ and $Q(0, 5, 6)$.

    a. Imagine rotating the segment about the z-axis. What kind of surface is formed? Draw a sketch of the surface.

    b. If $A(x, y, z)$ is a point on the surface of revolution, what conditions must be satisfied by x, y, and z?

    c. Develop an equation for this surface of revolution.

In the next four problems, you will examine surfaces formed by other familiar curves rotated about the x-, y-, or z-axis.

5 Consider the curve $y - x^2 = 0$ in the xy-plane. Imagine the curve revolving about the y-axis to generate a surface.

    a. Describe the traces of this surface.

    b. Describe the cross sections parallel to the xz-plane.

    c. Sketch the surface.

6 Next consider the curve $z - y^4 = 0$ in the yz-plane.

    a. Sketch the surface generated by rotating this curve about the z-axis.

    b. Describe the cross sections formed by planes parallel to the xy-plane and give reasons for your answers.

c. To find an equation for this surface, choose a point $P(x, y, z)$ on the surface. Then P is on a circular cross section with radius r.

    i. Explain why $x^2 + y^2 = r^2$, where r is a function of z.

    ii. Explain why for a point $A(0, y, z)$, $|y| = r$.

    iii. Explain why $z = r^4$.

    iv. Explain why $z = (x^2 + y^2)^2$.

7 Look back at your work for Problem 2.

    a. Use algebraic and geometric reasoning to develop a possible equation for that paraboloid.

    b. Based on your work in Part a, what can you conclude about the equation for the surface in Problem 5?

8 Now consider a line through the origin in the yz-plane that makes an angle of θ, $0° < \theta < 90°$, with the positive z-axis. Generate a surface by rotating the line about the z-axis.

    a. Make a sketch of the line and the surface generated. Describe the surface.

    b. Describe the cross section of the intersection of the surface and a plane perpendicular to the z-axis, other than the xy-plane.

    c. To find the equation of this surface, you can use reasoning similar to that used in Problems 6 and 7. Let $P(x, y, z)$ be any point on the surface.

        i. Explain why $x^2 + y^2 = r^2$, where r is a function of z.

        ii. Explain why $r = |z| \tan \theta$.

        iii. Explain why $x^2 + y^2 = k^2 z^2$, where $k^2 = \tan^2 \theta$.

    d. What are the equations of the traces of this surface?

    e. Use the equation of the surface to determine the shapes of cross sections parallel to the yz- and xz-planes.

In Problem 4, you formed a cylinder by rotating a segment perpendicular to the xy-plane about the z axis. More generally, a **cylindrical surface** can be generated by moving a line along the path of a plane curve keeping the line at a fixed angle to the curve so that as the line moves, it is always parallel to its original position. The horizontal cross sections will all be congruent, but do not need to be circles as in pipes and cans. They can have a variety of shapes; think of cookie cutters or roof gutters. In addition, cylindrical surfaces need not have a closed cross section such as a circle or triangle, but can be open like a parabola or a hyperbola.

Sketches of three cylindrical surfaces are shown below, one closed and the other two open.

9 Sketch a graph of $y - x^2 = 0$ in the xy-plane and then consider the following actions.

- Select a line parallel to the z-axis (perpendicular to xy-plane) intersecting the graph.

- Imagine sweeping out a surface by moving the line along the curve $y - x^2 = 0$, keeping the line parallel to the z-axis at all times.

The surface is a **parabolic cylindrical surface**. Its equation is $y - x^2 = 0$ since z takes on all values for every (x, y) pair satisfying $y - x^2 = 0$.

a. Describe the cross sections parallel to the xy-plane.

b. Describe the cross sections parallel to the other coordinate planes.

c. Sketch the parabolic cylindrical surface on a three-dimensional coordinate system.

10 Sketch graphs of the following surfaces on a three-dimensional coordinate system. Check your graphs using three-dimensional graphing software.

a. The **logarithmic cylindrical surface** with equation $z = \log y$

b. The **hyperbolic cylindrical surface** with equation $\dfrac{z^2}{4} - \dfrac{y^2}{9} = 1$

SUMMARIZE THE MATHEMATICS

In this investigation, you examined how to generate surfaces in three-dimensional space.

a Describe two ways that plane curves can be used to generate surfaces.

b How can you use cross sections parallel to a coordinate plane to identify a surface of revolution?

c Describe a general procedure to develop an equation for a surface of revolution.

d What kind of surface of revolution is defined by an equation with only two variables? What is the effect of the omitted variable? How is this seen in intercepts and cross sections at intercepts?

e How does the generation of a cylindrical surface differ from that of a surface of revolution?

Be prepared to share your descriptions and thinking with the class.

 CHECK YOUR UNDERSTANDING

Consider surfaces that can be generated using an ellipse.

a. Sketch the graph of $\dfrac{x^2}{4} + \dfrac{y^2}{9} = 1$ in the xy-plane of a three-dimensional coordinate system.

b. Sketch the surface of revolution generated by rotating $\dfrac{x^2}{4} + \dfrac{y^2}{9} = 1$ about the y-axis. Then find the equation of the surface of revolution.

c. Sketch the cylindrical surface with equation $\dfrac{x^2}{4} + \dfrac{y^2}{9} = 1$.

d. Compare the equations of the surfaces in Parts b and c and summarize the information revealed by the two different forms.

APPLICATIONS

1 In this lesson, you saw that geometric ideas such as distance, shape, and symmetry in a two-dimensional coordinate model had analogous representations in a three-dimensional coordinate model. Complete a table like the one below, which summarizes some of the key features of a three-dimensional coordinate model.

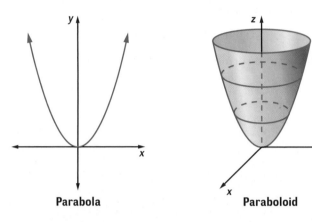

Parabola   Paraboloid

Geometric Idea	Two-Dimensional Coordinate Model	Three-Dimensional Coordinate Model
Point	Ordered pair (a, b) of real numbers	
Plane	All possible ordered pairs (x, y) of real numbers	
Distance Between Two Points	For points $A(x_1, y_1)$ and $B(x_2, y_2)$, $AB = \sqrt{(x_1 - x_2)^2 + (y_1 - y_2)^2}$	
Midpoint of a Line Segment	For points $A(x_1, y_1)$ and $B(x_2, y_2)$, midpoint of $\overline{AB}$ is $M\left(\dfrac{x_1 + x_2}{2}, \dfrac{y_1 + y_2}{2}\right)$	
Reflection Symmetry	Across the x-axis $(x, y) \rightarrow (x, -y)$ Across the y-axis $(x, y) \rightarrow (-x, y)$	Across the xz-plane $(x, y, z) \rightarrow$? Across the yz-plane $(x, y, z) \rightarrow$? Across the xy-plane $(x, y, z) \rightarrow$?
Half-Turn Symmetry	About $(0, 0)$ $(x, y) \rightarrow (-x, -y)$	About the z-axis About the y-axis About the x-axis
Conics	Circle $x^2 + y^2 = r^2$ Ellipse $\dfrac{x^2}{a^2} + \dfrac{y^2}{b^2} = 1$ Parabola $y = ax^2$	Sphere $x^2 + y^2 + z^2 = r^2$ Ellipsoid Paraboloid

2 Triangle PQR has vertices $P(1, 2, 3)$, $Q(5, 4, 1)$, and $R(-1, 6, 5)$.

    a. Draw $\triangle PQR$ in a three-dimensional coordinate system.

    b. What kind of triangle is $\triangle PQR$?

    c. Find, plot, and label the coordinates of the midpoints of each side.

    d. Is $\triangle PQR$ similar to a triangle with the midpoints as vertices? Justify your response.

3 The table below gives ratings for six laptops by the editors of a magazine. A rating of 5 is the highest rating in each category.

	Hard Drive Size	Cost	Battery Life
Brand A	4	5	4
Brand B	3	5	5
Brand C	3	3	5
Brand D	2	5	5
Brand E	4	2	3
Brand F	3	4	3

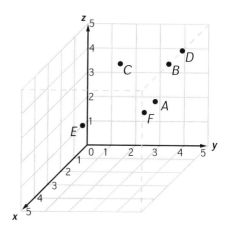

    a. The three-dimensional plot at the right provides a graphical model for the ratings shown in the table. What does each axis represent?

    b. Rank the laptops by adding the three ratings for each.

    c. Explain how the distance formula might be used to rank the laptops. Then rank them using your described method. Are the rankings the same or different than the ones in Part b? Which method would you recommend and under what conditions?

    d. Suppose you use your laptop primarily while sitting at a desk. In this case, its battery life is not as important to you and the cost is very important. How might you modify your method from Part b to take this into consideration?

    e. The magazine also included ratings for the laptop weight. How might you modify the distance formula to take this fourth rating into account?

4 Sketch graphs of each of the following equations on a three-dimensional coordinate system.

   a. $x + 2y + z = 8$ **b.** $x + y + 3z = 3$

   c. $4x - 2y + z = 4$ **d.** $2x + 3y - z = 2$

   e. $2x - 3z = 6$ **f.** $|x| = 5$

5 Use analysis of intercepts, cross sections, and symmetry to help you match each equation with one of the following surfaces.

     I   **II**   **III**

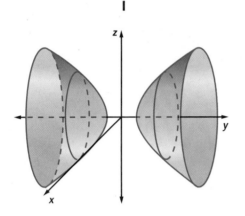 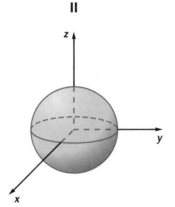 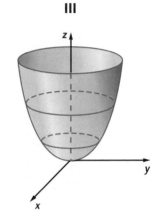

   a. $x^2 + y^2 + z^2 = 9$

   b. $\dfrac{y^2}{4} - \dfrac{x^2}{9} - \dfrac{z^2}{9} = 1$

   c. $4x^2 + 9y^2 - 36z = 0$

6 Sketch and name the surfaces given by the following equations. Verify your sketches using three-dimensional graphing technology.

   a. $x^2 + y^2 = 9$ **b.** $x^2 + 2y^2 + 3z^2 = 12$

   c. $2x - 3y + z - 6 = 0$ **d.** $\dfrac{x^2}{4} + \dfrac{y^2}{9} - \dfrac{z^2}{16} = 1$

7 A surface is generated by rotating about the z-axis a line that makes a 45° angle with the z-axis and contains the origin.

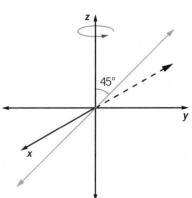

   a. Describe the traces.

   b. Describe the cross sections parallel to each of the three coordinate planes.

   c. Make a sketch of the surface.

   d. Derive an equation for this surface.

8 Spheres are the three-dimensional analogue of circles in a plane. In this lesson, you discovered that the equation of a sphere with center at the point (j, h, k) and radius r is similar to that for a circle with center at the point (h, k) and radius r.

   a. Describe as completely as possible the surface with equation

$$(x - 3)^2 + (y - 2)^2 + (z + 5)^2 = 36.$$

   i. Find the equations of the curves (if any) where the surface intersects each coordinate plane.

   ii. Find the points (if any) where the surface intersects each coordinate axis.

   b. Translate the sphere with equation $x^2 + y^2 + z^2 = 16$ so that its center is at the point $P(4, -2, 3)$.

   i. Sketch both spheres.

   ii. Write the equation for the translated sphere in expanded form.

   c. Describe as completely as possible the surface with the equation

$$x^2 + y^2 + z^2 - 2x - 8z = 8.$$

9 Sketch the graph of $y^2 - 4z^2 = 16$ in the yz-plane of a three-dimensional coordinate system.

   a. Sketch the surface of revolution generated by rotating $y^2 - 4z^2 = 16$ about the z-axis. Then find the equation of the surface of revolution.

   b. Sketch the cylindrical surface with equation $y^2 - 4z^2 = 16$.

   c. Compare the equations of the surfaces in Parts b and c and summarize the information revealed by the two different forms.

10 The diameter of the midsection of the cooling tower shown below is 70 meters. The tower is about 160 meters tall.

Develop an equation whose graph is a good mathematical model of the cooling tower.

CONNECTIONS

11 Look back at the table you completed for Applications Task 1. Now consider the similarities and differences between graphs of linear inequalities in two dimensions and linear inequalities in three dimensions.

    a. In a two-dimensional coordinate model, how would you graph $4x + 6y \leq 12$? How would you describe the graph?

    b. In a three-dimensional coordinate model, how would you graph $4x + 6y + 3z \leq 12$? How would you describe the graph?

12 For each of the following equations, describe the surface and identify the x-, y-, and z-intercepts, symmetries, and traces.

    a. $x^2 + y^2 - 2z = 0$ **b.** $x^2 + y^2 + z^2 = 9$

    c. $\dfrac{x^2}{6} + \dfrac{y^2}{4} + \dfrac{z^2}{3} = 1$ **d.** $x^2 + y^2 - z^2 = 0$

    e. $x^2 + y^2 - z^2 = -1$ **f.** $6x - 5y + 10z = 30$

13 In Unit 1, *Families of Functions*, you saw that changing the rule of a function of a single variable by adding or multiplying by a constant transformed its graph in predictable ways.

A graph of the paraboloid $z = x^2 + y^2$ is shown at the right. Describe the graph of each of the following functions, and describe its relationship to the original paraboloid. Check your predictions using three-dimensional graphing technology.

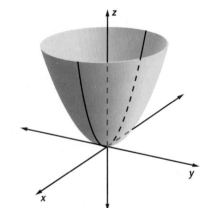

    a. $z = x^2 + y^2 + 5$

    b. $z = 3 - x^2 - y^2$

    c. $z = x^2 + (y - 2)^2$

    d. $z = 5x^2 + 5y^2$

14 The disposable drinking cup shown below can be thought of as a surface of revolution with a bottom. The cup has a 5-cm diameter bottom, 7-cm diameter top, and has height 9 cm.

    a. Beginning with an appropriate segment in three-dimensional space, describe how the surface can be generated.

    b. The volume of the cup can be approximated by cutting it horizontally into sections each with a height of 0.5 cm, approximating each section with a cylindrical disc, and then summing the volumes of the discs. Approximate the volume in this manner.

c. Another way to generate the cup surface is to make the rotating segment part of a line through the origin. In this case, the volume of the cup is the difference of the volumes of two cones. Find the volume in this manner.

d. Compare the values of the volume found in Parts b and c. Describe how you could improve your approximation in Part b.

15 It can be shown that the volume of any three-dimensional figure bounded by a quadratic surface and two parallel planes (like those pictured in Applications Task 10 and Connections Task 14) can be calculated using the **prismoidal formula**

$$V = \frac{B + 4M + T}{6} \cdot h,$$

where B is the area of the cross section at the base, M is the area of the cross section at the middle, T is the area of the cross section at the top, and h is the height of the figure.

a. Use the prismoidal formula to calculate the volume of the cup in Task 13. Compare your answer to the approximation you calculated in Part b of that task.

b. Use the prismoidal formula to derive familiar formulas for the volumes of the following three-dimensional figures.

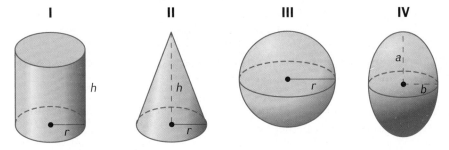

REFLECTIONS

16 Suppose you are standing at the point $S(4, 5, -3)$. North is in the negative x-direction, east is in the positive y-direction, and up is in the positive z-direction.

a. At what coordinates will you be standing after you move north 6 units, up 5 units, west 8 units, down 3 units, east 10 units, and south 4 units? Call this point A.

b. After completing the trip in Part a, will you be looking up or down at the point $T(-2, 5, 1)$?

c. After completing the trip in Part a, will you be closer to point S or point T?

17 In this lesson, you found reasoning by analogy helpful in extending ideas of two-dimensional space to three-dimensional space. In a science course, you may have learned that Einstein's theory of special relativity involves a four-dimensional model (x, y, z, t). In that model, x, y, and z represent the usual coordinates of three-dimensional space and t represents the time coordinate.

Think about how you could generalize the following ideas to four-dimensional space that is key to Einstein's theory of relativity.

a. Distance between points $P(x_1, y_1, z_1, w_1)$ and $Q(x_2, y_2, z_2, w_2)$

b. Midpoint of $\overline{PQ}$ for the points in Part a

c. Four-dimensional analogue of a sphere with center at $(0, 0, 0, 0)$ and radius r

d. In two-dimensional space, a square can be formed by translating a segment of length S a distance of S units in a direction perpendicular to the segment, and then connecting corresponding vertices. In three-dimensional space, a cube can be found by translating a square of side length S by a distance of S units in a direction perpendicular to the plane of the square, and then connecting the corresponding vertices. How could you form a **hypercube**, the four-dimensional analogue of a cube? How many vertices would the hypercube have? How many edges?

18 Sketch graphs of $z = 0$, $z = 5$, and $z = -2$ on the same three-dimensional coordinate system. Describe how the graphs are related.

19 In sketching a surface, why is it important to examine other cross sections, in addition to the traces?

20 Must the graph of a linear equation in three variables have a trace in each of the three coordinate planes? Explain.

21 How does knowing the shape of the surface defined by $x^2 + y^2 + z = 4$ (described in the Check Your Understanding on page 446) help you quickly recognize the shape of the surface represented by $x + y^2 + z^2 = 4$?

22 How might a plane be generated as a surface of revolution?

23 *Reasoning by analogy* is another important mathematical habit of mind.

a. What is meant by reasoning by analogy?

b. Give two examples where reasoning by analogy was helpful in this lesson.

24 The coordinates of the midpoint of $\overline{AB}$ where $A(x_1, y_1, z_1)$ and $B(x_2, y_2, z_2)$ are $\left(\dfrac{x_1 + x_2}{2}, \dfrac{y_1 + y_2}{2}, \dfrac{z_1 + z_2}{2}\right)$. Suppose M is on $\overline{AB}$ and divides $\overline{AB}$ so that $AM = \dfrac{m}{n}AB$, where $0 < m < n$. Find the coordinates of this point of division. Check that when $\dfrac{m}{n} = \dfrac{1}{2}$, the coordinates are those of the midpoint of $\overline{AB}$.

25 A system of two linear equations in two variables represents two lines in a plane. The two lines may intersect in no points, one point, or infinitely many points. You have used several strategies to solve such systems, including the inverse-matrix method, linear-combinations method, substitution method, and graphing. In this task, you will consider the interpretation and solution of systems of linear equations in three variables.

  a. What are the possible intersections for the graphs of two linear equations in three variables?

  b. What are the possible intersections for the graphs of three linear equations in three variables?

  c. How would you solve the following system of equations by extending the inverse-matrix method to three dimensions? Solve the system and check your solution.

Entrance to Royal Ontario Museum

$$
\begin{array}{ll}
(1) & x + y + z = 1 \\
(2) & 2x - y + z = 0 \\
(3) & x + 2y - z = 4
\end{array}
$$

  d. Solve the system in Part c by extending the linear-combination method for two dimensions. Begin by combining equations (1) and (3) and equations (2) and (3) to eliminate the z-variable. Then solve the resulting system of two variables. Complete your solution and compare it to that obtained in Part c.

  e. Solve the system in Part c by first using the linear-combination method to eliminate the y-variable.

26 Solve each of the following systems by using the linear-combination method and then check your results using the inverse-matrix method or the "Solving Linear Systems" custom app in *CPMP-Tools*.

  a. $\begin{cases} x + 2y + 2z = 5 \\ x - 3y + 2z = -5 \\ 2x - y + z = -3 \end{cases}$

  b. $\begin{cases} x + y + z = 6 \\ 2x + 2y + 2z = 12 \\ 2x + 3y + 2z = 8 \end{cases}$

27 Find the equation of the plane through $P(2, 0, 0)$, $Q(0, 5, 0)$, and $R(0, 0, -8)$.

28 Many of the surfaces you have sketched in this unit have descriptive names. Most have three symbolic forms, each related to a particular axis, and have combinations of quadratic terms, linear terms, and constants. Match each surface with the appropriate equation below. (One surface matches two equations.) Then explain why your match makes sense.

Sphere

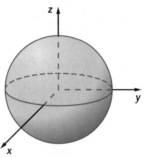

Elliptic Paraboloid

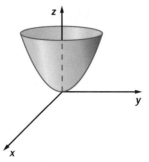

Double Cone

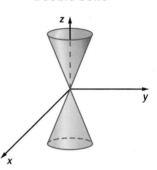

Ellipsoid

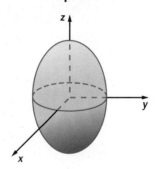

Hyperboloid of one sheet

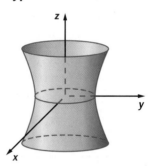

Hyperboloid of two sheets

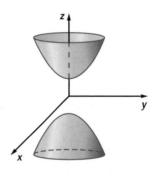

Hyperbolic Paraboloid

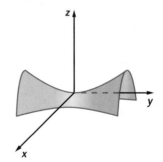

a. $\dfrac{x^2}{a^2} + \dfrac{y^2}{b^2} + \dfrac{z^2}{c^2} = 1$

b. $x^2 + y^2 + z^2 = r^2$

c. $\dfrac{x^2}{a^2} + \dfrac{y^2}{a^2} - \dfrac{z^2}{b^2} = 0$

d. $\dfrac{x^2}{a^2} + \dfrac{y^2}{b^2} - \dfrac{z^2}{c^2} = 1$

e. $\dfrac{z^2}{c^2} - \dfrac{x^2}{a^2} - \dfrac{y^2}{b^2} = 1$

f. $\dfrac{x^2}{a^2} + \dfrac{y^2}{b^2} - z = 0$

g. $\dfrac{y^2}{b^2} - \dfrac{x^2}{a^2} - z = 0$

h. $\dfrac{x^2}{a^2} + \dfrac{y^2}{a^2} + \dfrac{z^2}{a^2} = 1$

29 In Course 3 of *Core-Plus Mathematics*, you used two-dimensional coordinate methods to represent and solve linear programming problems. Many linear programming problems, such as the one below, give rise to an objective function and constraint inequalities that require use of more than two variables.

There are different types of surfboards for surfing, each of which is individually manufactured. A small surfboard manufacturer in Hawaii produces three different types of surfboards: shortboard, funshape, and longboard. Each board goes through a process of (1) forming and shaping the outer shell; (2) laminating and fin adding; and (3) sanding and final finishing. The number of hours for each surfboard type is summarized in the following table.

Hours Required to Manufacture Surfboards			
	Forming and Shaping	Laminating and Fin Adding	Sanding and Final Finishing
Shortboard	2	1	3
Funshape	3	2	3
Longboard	4	1	4

During the summer season, the surfboard manufacturer spends at most 600 hours on forming and shaping, 275 hours on laminating and fin adding, and 725 hours on sanding and final finishing.

a. Explain how the linear inequality $2x + 3y + 4z \leq 600$ represents the forming and shaping constraint for this surfboard manufacturer.

b. Describe and sketch the region formed by this linear inequality. Then explore the use of the "Linear Programming" custom app in *CPMP-Tools* to graph this constraint inequality. Select "3D" from the View menu before entering the constraint inequality. Compare and resolve any differences between your sketch and the computer-produced graph.

c. Write the inequality for the laminating and fin adding time constraint. On a new set of coordinate axes, sketch the region formed by this constraint inequality.

d. Make a conjecture about the shape of the region that satisfies *both* constraint inequalities. Check your conjecture using the "Linear Programming" custom app.

e. Write the inequality for the sanding and final finishing time constraint. Use the custom app to produce the region that satisfies the three inequalities in Parts b, c, and d. What does this region of space represent in terms of surfboard manufacturing?

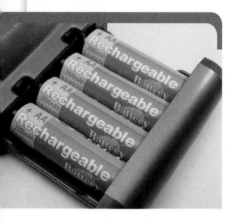

f. The small surfboard manufacturer makes a profit on each shortboard of $150, on each funshape surfboard of $160, and on each longboard of $200. Its goal is to maximize its profit.

    i. What algebraic rule shows how to calculate total profit *P* for the season? Enter this objective function into the "Linear Programming" custom app.

    ii. Describe the graph of the objective function. Explain why this makes sense.

    iii. Click and drag the graph of the objective function across the feasible region. What is the greatest profit the surfboard manufacturer can expect? How many of each type of surfboard should be manufactured?

30 A researcher involved in a study to determine the life of a new model battery, measured in terms of charge-discharge cycles, found that battery life *Z* was a function of the charge rate *X* and the temperature *Y* where it was used. The function was

$$Z = 262.58 - 55.83X + 75.50Y + 27.39X^2 - 10.61Y^2 + 11.50XY.$$

Use three-dimensional graphing software to examine enough views of the surface to be able to describe the effect of charge rate and temperature on battery life.

31 Study this technology-produced graph of $z = 2^{-(x^2 + y^2)}$.

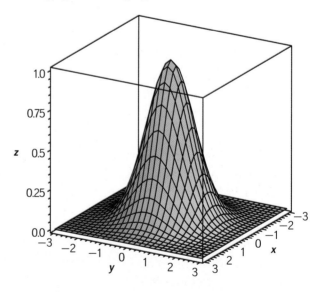

Judge the reasonableness of the graph by analyzing the given function rule.

a. Why must the graph lie entirely above the *xy*-plane?

b. What is the *z*-intercept?

c. Why does the value of *z* get smaller as you move away from the origin of the *xy*-plane in any direction?

d. What must be true about the horizontal cross sections of the graph?

e. Why does the graph flatten out as it gets closer to the *xy*-plane?

f. What symmetry does the graph have?

32 How can you quickly sketch the graph of $8x - 3y = 36$ in a coordinate plane? Make a quick sketch.

33 Given the equation $6x + 7y = 10$ in an xy-coordinate plane, determine a second equation so that the system of equations has:

a. no solution.

b. one solution.

c. infinitely many solutions.

34 Consider the ellipse with equation $\frac{x^2}{16} + \frac{y^2}{8} = 1$.

a. Verify that the point $(2, \sqrt{6})$ is on the ellipse.

b. Use symmetry to find the coordinates of three other points on the ellipse.

c. Determine the coordinates of the x- and y-intercepts of the ellipse.

35 Without using technology, describe the graph of each pair of parametric equations on $0 \leq \theta \leq 2\pi$. Then check your work using technology.

a. $x = 4 \cos \theta$
   $y = 4 \sin \theta$

b. $x = 2 \cos \theta - 2$
   $y = 2 \sin \theta + 4$

c. $x = 6 \cos \theta$
   $y = 2 \sin \theta$

36 Prove or disprove that the statement $\sin \theta\, (\sin \theta \tan \theta + \cos \theta) = \tan \theta$ is an identity.

37 Determine the solutions to each of the following equations.

a. $(x - 2)(2x^2 + x - 3) = 0$

b. $3|3x + 9| = 15$

c. $\log (x - 4) = 2$

d. $3e^{\frac{x}{2}} = 30$

e. $8 \cos \theta = 4$

f. $\frac{-4}{x + 2} + 3 = x$

38 In which of the following accounts is it better to invest?

Account A: pays 8% annual interest compounded quarterly
Account B: pays 7.6% annual interest compounded continuously

39 If $\sin \theta = -\frac{5}{13}$ and θ is in Quadrant III, find exact values for each of the following.

a. $\cos \theta$

b. $\tan \theta$

c. $\csc \theta$

d. $\sec \theta$

e. $\sin^2 \theta + \cos^2 \theta$

f. $\csc^2 \theta - \cot^2 \theta$

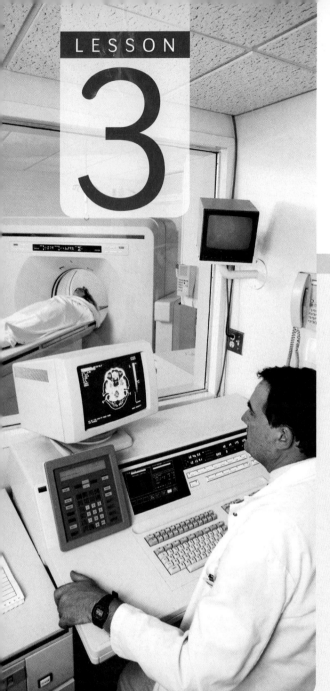

LESSON
3

Looking Back

In this unit, you developed skill in interpreting and representing surfaces in three dimensions. In the process, you investigated the usefulness of contour diagrams and cross sections as means to represent and analyze complex three-dimensional shapes. Analysis of cross sections of a double cone led to an alternate way of thinking about special curves you had previously studied—the circle, parabola, ellipse, and hyperbola.

The introduction of a three-dimensional coordinate system enabled you to explore and formalize analogs of important geometric ideas in two dimensions including distance, symmetry, and graphs of equations. Although three-dimensional graphing software enables you to produce graphs of three-dimensional surfaces, being able to visualize the cross sections and intercepts is essential in interpreting the images and judging their reasonableness. In this final lesson, you will review and consolidate your understanding of these key concepts and skills.

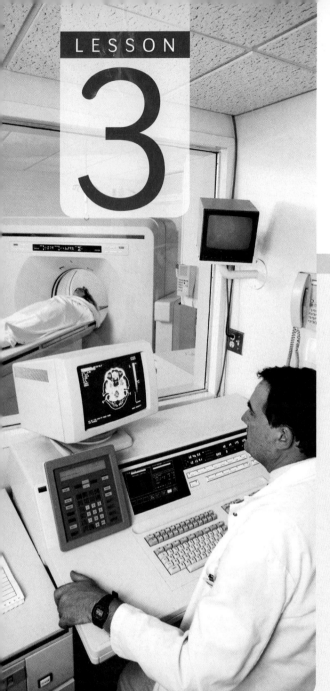
©Royalty-Free/Corbis

1 The York Pond Trail and the Kilkenny Ridge Trail in the Mahoosuc Range of the White Mountains National Forest meet near Willard Notch. Use the map below to help complete the following tasks. On this map, contour lines represent 100-ft elevation differences.

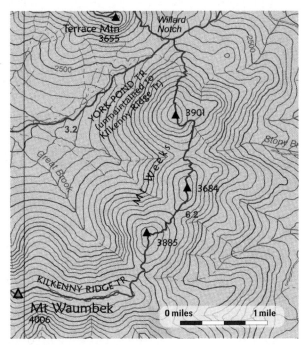

a. What do the three sets of closed curves in the center portion of the map represent?

b. Describe the terrain along the Kilkenny Ridge Trail from Mt. Waumbek to Willard Notch. Include the length of the section of trail and high and low altitudes.

c. Describe the terrain along the York Pond Trail.

d. Visualize a relief line from the Mt. Weeks peak with altitude 3,885 feet to the peak with altitude 3,684 feet. Sketch the vertical cross section along this line.

e. Explain why the contour lines in this map and those in other contour diagrams in this unit never cross each other.

2 Identify the conic section represented by each equation or graph below.

- In the case of an equation, rewrite the equation in standard form and use properties revealed by the equation to sketch the graph.

- In the case of a graph, write the corresponding equation.

a. $9x^2 - 16y^2 = 144$

b. $2x^2 + 2y^2 - 8x + 12y - 36 = 0$

c.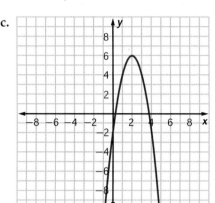

d.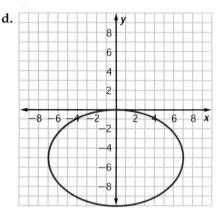

3 Complete the following tasks in the context of a three-dimensional coordinate system.

a. Suppose you are standing above the xy-plane at the point $P(5, 8, 3)$ and looking at the point $Q(1, -3, 5)$. Are you looking up or down? How far is point Q from point P?

b. Describe the set of points whose distance from the y-axis is 4. What equation describes this set of point(s)?

c. Which of the points $A(3, 1, -2)$, $B(2, -3, 0)$, or $C(-10, 0, 0)$ is closest to the yz-plane? Which point(s) are on the xy-plane?

d. Find the equation of a sphere with radius 5 centered at the origin. What is the equation of a congruent sphere with center at $(1, -3, 2)$?

e. For the sphere in Part d with center at $(1, -3, 2)$, is the point $(-2, 2, -2)$ on the sphere, in its interior, or in its exterior? Explain.

4 Consider the plane with equation $6x + 4y + 3z = 24$.

a. Sketch the plane.

b. How would you describe the horizontal cross sections of this plane?

c. Write the equation of a plane parallel to the given plane.

d. Write the equation of a plane that intersects the given plane at its yz-trace.

5 A surface has the equation $x^2 + 4y^2 + 16z^2 = 64$.

a. Find the equations of the traces.

b. Find the intercepts.

c. Describe all symmetries.

d. What curves do planes perpendicular to the z-axis form with the surface?

e. Sketch the surface. Compare your sketch to the graph produced using three-dimensional graphing technology.

6 Use analysis of intercepts, cross sections, and symmetry to help you match each surface with one of the equations that follow.

I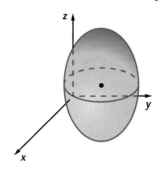

II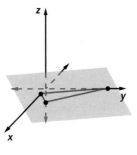

III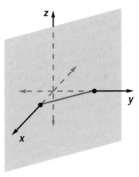

IV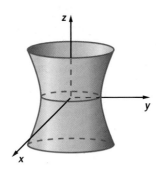

V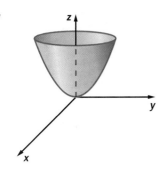

VI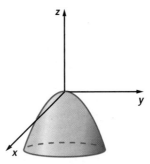

a. $4x^2 + 9y^2 = 36z$

b. $36x^2 + 16y^2 - 9z^2 = 144$

c. $-(x^2 + y^2) = 16z$

d. $12x + 2y - 9z = 36$

e. $9(x - 3)^2 + 4(y - 6)^2 + 2(z - 4)^2 = 36$

f. $6x + 4y = 24$

7 Which of the surfaces in Task 6 can be interpreted as surfaces of revolution about a coordinate axis?

a. What is the axis of rotation?

b. What is the equation of the generating two-dimensional figure?

CHECK YOUR UNDERSTANDING

Write, in outline form, a summary of the important mathematical concepts and methods developed in this unit. Organize your summary so that it can be used as a quick reference in future units.

7

Concepts of Calculus

One of the most important roles of mathematical calculation and reasoning is describing and predicting patterns of change in the world around us. Calculus is the branch of mathematics that provides ideas and methods for analyzing rates of change in the many different families of functions that are used to model cause-and-effect and change-over-time relationships between quantitative variables.

In this unit, you will explore important problems requiring analysis of rates of change and develop understanding of the basic calculus ideas that are useful in solving those problems.

The key ideas will be developed in two lessons.

LESSONS

1 Introduction to the Derivative
Develop understanding of ways to describe instantaneous rates of change for important functions and the phenomena that they model.

2 Introduction to the Definite Integral
Develop understanding of mathematical ideas for measuring the cumulative effect of continuous change in quantitative variables.

Introduction to the Derivative

In Course 4 and in previous courses of the *Core-Plus Mathematics* program, you studied several important families of functions—linear, exponential, power, quadratic, trigonometric, polynomial, rational, and logarithmic. You found that members of each family had closely related patterns in tables, graphs, and symbolic rules.

In several of the units, you experimented with and analyzed various aspects of the motion of bungee jumpers. The following graph shows a record of (*time*, *height*) data from such a jump. You can learn a lot about that jump by studying the shape of the graph and the coordinates of particular data points.

Flight of a Bungee Jumper

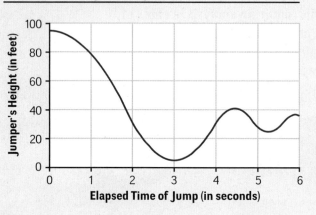

Think about how you could use information in the graph to answer questions like these.

a When was the jumper in free fall, and when was the bungee cord pulling at him?

b How long is the un-stretched bungee cord?

c How will the jumper's *velocity* change over time, and how is that pattern of change shown by the shape of the graph?

d When was the jumper traveling at maximum *speed* downward? At maximum *speed* upward?

e How could you estimate the jumper's *average velocity* during the first two seconds of the jump?

f How could you estimate the jumper's *velocity* at any instant in time?

The branch of mathematics called *differential calculus* provides ideas and techniques for answering questions like those above. Work on the problems of this lesson will give you an introduction to that very important subject.

Rates of Change from Graphs and Tables

The basic information provided by the bungee jump graph is the jumper's height at any time in the up-and-down trip. You have probably realized that by looking at the shape of the graph carefully, it is also possible to make inferences about the velocity of the jumper's motion.

As you work on the problems of this investigation, look for answers to the following general question:

> *How can data in a table or graph be used to describe*
> *the rate of change in the dependent variable?*

Walk That Graph Graphs of (*time*, *position*) data are often produced by radar and sonar devices that send out radio and sound waves and detect echoes as the waves bounce off target objects in the sky, on the ground, and underwater.

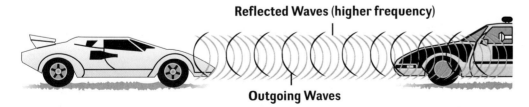

Reflected Waves (higher frequency)

Outgoing Waves

1 The key to both radar and sonar detection is measuring the time it takes a radio or sound wave to travel from the transmitter to a distant object and back to the receiver.

a. How do you think radar and sonar devices convert measures of elapsed time into estimates of distances to target objects?

b. How could measures of time and distance be used to estimate velocity of moving objects?

2 There are now handheld sonar devices, called motion detectors, that connect to your graphing calculator or computer. If you aim such a detector at someone walking toward or away from the device, it will produce a sequence of (*time, distance*) data pairs and convert them into a graph.

For each of the following (*time, distance*) graphs:

- explain in detail how you would walk toward and/ or away from a motion detector to produce the given graph pattern.

- test your ideas with actual walks in front of a motion detector or at least in front of classmates who have thought about the same challenge.

a.
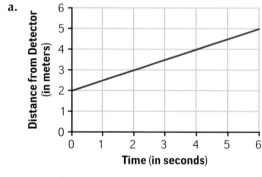

b.
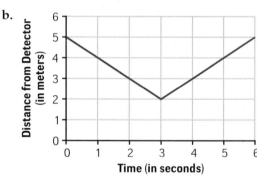

c.
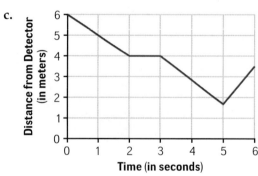

d.
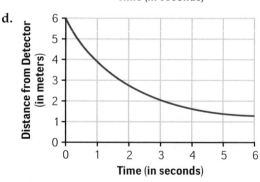

e.
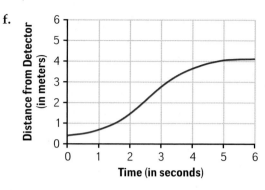

f.

3 Use your experience from work on Problem 2 to formulate guidelines for interpreting the shape of a (*time*, *position*) graph.

    a. What graph shape indicates motion at a constant speed?

    b. What graph shape indicates motion at an increasing speed? At a decreasing speed?

    c. What graph shape indicates change in direction of motion?

    d. What graph feature shows when motion is occurring at the greatest speed in a time interval? At the slowest speed in a time interval?

Instantaneous Velocity The bungee jumper graph, reproduced below, shows that it took 3 seconds for the jumper to fall 90 feet, giving an average speed of 30 feet per second. But you have learned enough about falling objects to know that the jumper is not falling at the same speed throughout that 3-second segment of the ride.

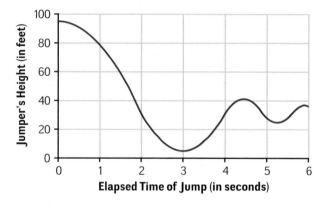

4 Data in the following table match the first segment of the bungee jumper graph.

Time (in seconds)	0.0	0.5	1.0	1.5	2.0	2.5	3.0
Height (in feet)	95.0	90.0	80.0	60.0	30.0	10.0	5.0

    a. What do the data suggest about the time when the bungee cord began pulling at the jumper and about the length of the unstretched bungee cord?

    b. Estimate the jumper's velocity at the time 1.0 second into the jump. What unit of measure did you use in reporting the velocity estimate?

c. Use the data given in the next table to find a more accurate estimate of the jumper's velocity at the time 1.0 second into the jump.

Time (in seconds)	0.4	0.6	0.8	1.0	1.2	1.4	1.6
Height (in feet)	91.4	88.4	84.6	80.0	74.6	68.4	61.4

d. What do your estimates in Parts b and c suggest about the jumper's velocity at the time exactly 1.0 second into the jump? This is called his **instantaneous velocity**.

5 When students in a Maine class were asked to estimate jumper velocity at 2.5 seconds, they used different strategies and came up with different results.

a. Which of these methods makes most sense to you? Be prepared to explain why you think your choice is better than the others.

i. $\dfrac{5 - 10}{3.0 - 2.5} = -10$ feet per second

ii. $\dfrac{10 - 30}{2.5 - 2.0} = -40$ feet per second

iii. $\dfrac{5 - 30}{3.0 - 2.0} = -25$ feet per second

b. Why do the negative velocity numbers make sense in this situation?

c. Use the data given in the next table and each strategy illustrated in Part a to find better estimates of the jumper's velocity at the time exactly 2.5 seconds into the jump.

Time (in seconds)	2.2	2.3	2.4	2.5	2.6	2.7	2.8
Height (in feet)	20.2	16.2	12.8	10.0	7.8	6.2	5.2

d. What instantaneous velocity at the time 2.5 seconds into the jump is suggested by the estimates in Parts a and c?

Rates of Change in the Bungee Business For customers of a bungee jump attraction, the thrill is in the up and down ride. For operators of the bungee business, the goal is making a profit. Suppose that the graph at the right shows projected daily profit as a function of price charged per jump.

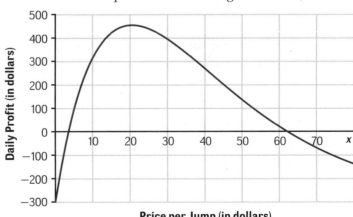

6 How would you describe the overall pattern of change in projected daily profit that is shown in the graph?

7 Use data in the next tables to estimate the instantaneous rate of change in projected daily profit if the price per jump is $10 and if it is $30. In each case, explain what the estimated instantaneous rate of change tells about profit prospects for the bungee business.

a.

Price per Jump (in $)	7	8	9	10	11	12	13
Daily Profit (in $)	195	240	280	315	345	370	390

b.

Price per Jump (in $)	27	28	29	30	31	32	33
Daily Profit (in $)	423	414	405	394	383	372	360

8 What unit did you use in reporting the instantaneous rate of the change estimates?

9 How is the difference between your two estimates for instantaneous rate of change in Problem 7 shown in the graph of (*price per jump, daily profit*) data?

SUMMARIZE THE MATHEMATICS

In this investigation, you explored ways to use graphs and data to analyze average and instantaneous rates of change in variables.

a Suppose that a given graph provides (*time, distance*) data for a moving object.

   i. How does the shape of the graph show when the velocity of the object is increasing, decreasing, or constant over an interval of time?

   ii. How can you calculate average velocity of the object over a specific time interval?

  iii. How can you estimate instantaneous velocity of the object at a specific point in time?

b Suppose that a given graph provides (*price, profit*) data for a business with one main product.

   i. What does the shape of the graph show about the rate of change in profit as price increases?

   ii. How can you estimate the instantaneous rate of change in profit at a particular price, and what does that value tell about the way profit changes in response to change in price?

Be prepared to explain your thinking and strategies to the class.

✓ CHECK YOUR UNDERSTANDING

The bungee jumper graph from the beginning of this lesson is reproduced here, along with some sample data about the trip from the bottom of the first drop to the top of the first rebound.

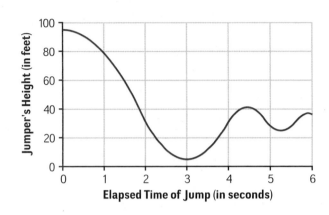

Time (in seconds)	Height (in feet)
3.00	5
3.25	7
3.50	12
3.75	19
4.00	31
4.25	39
4.50	40
4.75	37

a. Estimate the jumper's average velocity over the first rebound in the trip.

b. Compare that average velocity to your best estimates of the instantaneous velocity at these times in the ride:

  i. 3.25 seconds

  ii. 4.00 seconds

  iii. 4.50 seconds

INVESTIGATION 2

Rates of Change from Function Rules

You know from prior experience that having an algebraic rule for a function makes it much easier to create tables and graphs of function values and to answer questions about the function. As you work on the problems of this investigation, look for answers to the following questions:

How can function rules be used to make good estimates for instantaneous rates of change in the quantities that those functions represent?

How does the instantaneous rate of change in a function f(x) give meaning to the idea of slope for a curved graph?

Speed of Falling Bodies In previous courses, you studied the motion of athletes diving from high platforms. One of the most spectacular of such events occurs in Mexico where cliff divers drop off rocky outcroppings into ocean bays.

If one of these divers drops from a spot that is 30 meters above the water, his fall to the water will be modeled well by the function $h(t) = 30 - 4.9t^2$, where height is in meters and time is in seconds.

1 Use the function $h(t)$ to answer these questions as accurately as possible.

    a. How long will it take the diver to reach the surface of the water?

    b. What will be the diver's average speed from takeoff to hitting the water?

    c. How will the diver's speed change during flight? How is that change shown in the shape of a (*time, height*) graph?

2 One of the most interesting questions about the diver's flight is how fast he will be traveling when he hits the water. Devise and use a strategy for making a good estimate of that speed. Then compare your method and results to those of others in your class. Resolve any differences.

3 Now consider the problem of estimating the diver's speed at several other points in his dive. Try several different ways of producing good estimates.

    a. What is his speed exactly 1 second into the dive?

    b. What is his speed exactly 2 seconds into the dive?

    c. What is his speed just as he takes off from the cliff?

4 How are the instantaneous speed results from Problems 2 and 3 illustrated by the shape of the graph for $h(t)$?

5 The questions about the cliff diver refer to speed, not velocity. Why does that make sense in this context?

Rates of Change in Populations Suppose that a laboratory experiment uses a fruit fly population that doubles every five days. If the initial population contains 100 flies, the number at any time t days into the experiment would be modeled by the function with rule $P(t) = 100(2)^{\frac{1}{5}t}$, or $P(t) = 100(2)^{0.2t}$.

6 Use the rule for $P(t)$ to answer these questions about growth of the experimental population.

   a. What is the average growth rate of the population (flies per day) from day 0 to day 25?

   b. Give estimates for the instantaneous growth rate in the experimental population on day:

   i. 0   **ii.** 5   **iii.** 10

   iv. 15   **v.** 20   **vi.** 25

   c. Produce a graph of the function $P(t) = 100(2)^{0.2t}$. Then explain how the shape of that graph illustrates the differences between average and instantaneous rates of change calculated and estimated in Parts a and b.

Rates of Change in Periodic Functions Many interesting variable quantities such as alternating electrical current, tides in ocean harbors, hours of daylight, or sound waves from musical instruments change in repeating patterns as time passes. For example, there are places on Earth where ocean tides change the water depth in harbors by as much as 50 feet every six hours.

Low Tide   **High Tide**

7 Suppose that water depth in a tidal harbor is given by the function $D(t) = 9 \sin 0.5t + 15$. The depth D is measured in feet and the time t is measured in hours.

   a. Based on what you know about variation in the graph of the sine function, what overall pattern of change would you expect in water depth for $0 \le t \le 24$?

   b. Use the function rule to calculate the water depth when $t = 0, 3, 6, 9,$ and 12.

   c. Use the information from Part b to calculate the average rate of change in water depth (feet per hour) in the following time intervals.

   i. $0 \le t \le 3$ **ii.** $3 \le t \le 6$

   iii. $6 \le t \le 9$ **iv.** $9 \le t \le 12$

   d. Calculate the average rates of change in water depth for $0 \le t \le 6$ and for $6 \le t \le 12$. Then produce a graph of $D(t)$ and on a copy of the graph mark the points for $t = 0, 6,$ and 12. Explain how this graph shows that the average rates of change in water depths you calculated give misleading information about the overall pattern of tidal change in the intervals from $t = 0$ to $t = 6$ and from $t = 6$ to $t = 12$.

8 Now focus on the rate at which harbor water depth is changing at some specific times.

   a. Estimate the instantaneous rate of change in water depth at these times.

       i. $t = 0$ **ii.** $t = 1.5$

       iii. $t = 3$ **iv.** $t = 4.5$

       v. $t = 6$ **vi.** $t = 7.5$

   b. Sketch a graph of $D(t)$ for $0 \le t \le 12$ and explain how the pattern of instantaneous rates of change in Part a is illustrated by the shape of the graph.

9 Study the water depth graph from Problem 8 and a table of values for $D(t)$ to identify some times when you think the tide will be changing as described below.

   a. moving in most rapidly

   b. moving out most rapidly

   c. changing most slowly

Slope of a Curve The calculations used to estimate instantaneous rates of change in height of the bungee jumper and cliff diver, the population in a fruit fly experiment, and the depth of water in a tidal harbor should have reminded you of the formula for slope of a straight line

$$\frac{\Delta y}{\Delta x} = \frac{y_2 - y_1}{x_2 - x_1}.$$

You also discovered that rate of change estimates can be used to explain the shape of the graphs for the various functions involved. To see why it makes sense to connect slopes of straight lines and slopes of curves, use a graphing tool to look closely at the graphs of several nonlinear functions in small neighborhoods around specific points.

10 Start by graphing $f(x) = \frac{1}{x}$. Select a window that allows you to trace to $(1, 1)$. (On some calculators, the **Zdecimal** feature can be used to achieve this.) Then zoom in on the graph in a series of windows centered at the point $(1, 1)$.

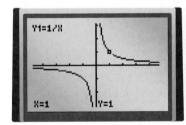 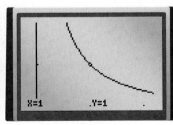 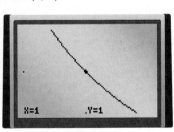

                                     Zoom 1   **Zoom 2**

   a. What do you notice as your displays magnify the graph near the point $(1, 1)$?

b. Estimate the instantaneous rate of change in $f(x)$ at $x = 1$. Then explain why the result of your graph magnification in Part a makes it sensible to think of this rate of change as the *slope of the curve* at (1, 1).

c. Analyze the graph of $f(x)$ in the same way near other specific points like $(-2, -0.5)$ or $(1.6, 0.625)$ to see if the connection between slopes of straight lines, instantaneous rates of change, and slopes of curves apply there as well.

11 Explore the connection between slopes of straight lines, instantaneous rates of change, and slopes of curves by examining the graphs of other nonlinear functions. For each function and point described below:

- graph the function with a window centered at the indicated point.

- magnify the graph by zooming in on the indicated point.

- estimate the instantaneous rate of change in the function at the indicated point and explain why it does or does not make sense to describe that value as the slope of the curve at that point.

a. $g(x) = 20(0.5^x)$ at (3, 2.5)

b. $h(x) = \cos x$ at $\left(\frac{\pi}{2}, 0\right)$

c. $j(x) = \sin x$ at $\left(\frac{\pi}{2}, 1\right)$

SUMMARIZE THE MATHEMATICS

In this investigation, you explored the rates of change in speed of falling cliff divers, populations of experimental fruit flies, and tides in ocean harbors. Suppose you were to model some other situation with a function $y = f(x)$.

a How would you use the rule for $f(x)$ to discover the overall pattern of change in $f(x)$ values over an interval of its domain $x_0 \le x \le x_1$?

b How would you use the rule for $f(x)$ to calculate the average rate of change of $f(x)$ over the interval $x_0 \le x \le x_1$?

c How would you use the rule for $f(x)$ to estimate the instantaneous rate of change of $f(x)$ at any specific point $(a, f(a))$?

d In what sense does the instantaneous rate of change in $f(x)$ at any specific point $(a, f(a))$ describe the slope of the graph of $f(x)$, even when $f(x)$ is a nonlinear function?

Be prepared to share your methods and reasoning with the class.

If you want to chill a drink quickly on a hot day, you can place the bottle or can into the freezer compartment of a refrigerator. The temperature of such a drink as a function of cooling time might be modeled well by

$$T(m) = 60e^{-0.06m} + 10,$$

where temperature is measured in degrees Fahrenheit and time is measured in minutes. A graph of this function is shown at the right.

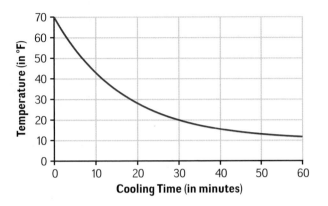

a. Find the average rate of change in drink temperature over the 60 minutes shown in the graph.

b. Calculate good estimates of the rate at which drink temperature is changing when:

  i. $m = 10$

  ii. $m = 30$

  iii. $m = 50$

c. How are the differences between the overall average rate of change calculated in Part a and the three instantaneous rates of change estimated in Part b shown by the shape of the graph?

d. Does it make sense to describe the instantaneous rates of change in Part b as the slopes of the graph at the points $(10, 43)$, $(30, 20)$, and $(50, 13)$? Why or why not?

INVESTIGATION 3

The Derivative

In Investigations 1 and 2, you calculated average rates of change for a variety of functions and used those functions to estimate instantaneous rates of change at specific points. The rate at which a function $f(x)$ is changing at some particular point $x = a$ in its domain is called the **derivative of f(x) at a**. The derivative is indicated with the notation $f'(a)$ that is read "*f* prime of *a*" or "the derivative of *f* at *a*."

As you work on the problems of this investigation, look for answers to these questions:

What general procedure shows how to find $f'(x)$ for any function $f(x)$ and any value of x?

How is the derivative of a function related to the shape of its graph?

What are some methods for finding derivatives in familiar function families?

1 Based on your work in the preceding investigations, what strategy would you recommend for finding the derivative of $f(x)$ when given a particular function and a specific value $x = a$? Consider two cases.

    a. $f(x)$ is given only by a graph or a table of values.

    b. $f(x)$ is defined by a rule involving algebraic, exponential, and/or trigonometric expressions.

2 Study the function $f(x)$ defined by this graph.

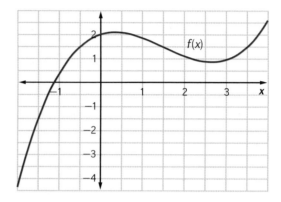

    a. For what values of x is $f'(x)$ positive? Negative? Equal to 0?

    b. Compare the following derivative values to see which is the larger in each pair. Use only visual estimation (no calculation) to come up with your answers.

        i. $f'(-2)$ and $f'(0)$

        ii. $f'(0)$ and $f'(1)$

        iii. $f'(-2)$ and $f'(3)$

    c. How would you summarize, for someone unfamiliar with the concept of derivative, what $f'(a)$ tells about the shape of the graph of $f(x)$?

Finding Derivatives Efficiently You have probably noticed that problems calling for good estimates of instantaneous rates of change in functions require a fair amount of calculation. Furthermore, finding the derivative at one point in the domain of a function does not seem to tell you anything about derivatives at other points. It will probably not surprise you to learn that mathematicians have discovered methods that give $f'(x)$ for any value of x.

3 The rules for some derivative functions can be determined by simply thinking about what the derivative of a function means.

    a. Suppose that $f(x) = k$ for all values of x.

        i. What will the graph of $f(x)$ look like?

        ii. What is the rate of change in $f(x)$ as x increases?

        iii. What rule tells the value of $f'(x)$ for any value of x?

b. Suppose that $g(x) = mx + b$ for all values of x.

    i. What will the graph of $g(x)$ look like?

    ii. What is the rate of change in $g(x)$ as x increases?

    iii. What rule tells the value of $g'(x)$ for any value of x?

    iv. What derivative formula does your answer to part iii suggest in the special case $g(x) = 3x + 5$? In the case $h(x) = -5x + 7$? In the case $j(t) = 4.3t$?

To find rules for derivatives of nonlinear functions, it helps to have a definition of $f'(x)$ that makes the process of calculating instantaneous rates of change explicit. There are two common ways that mathematicians define the **derivative of a function**:

$$(1) \quad f'(a) = \lim_{h \to 0} \frac{f(a + h) - f(a)}{h} \qquad\qquad (2) \quad f'(a) = \lim_{x \to a} \frac{f(x) - f(a)}{x - a}$$

The two definitions can be expressed in words as follows.

    (1) The derivative of f at a is the *limit as h approaches 0* of the quotient formed by dividing the difference $f(a + h) - f(a)$ by h.

    (2) The derivative of f at a is the *limit as x approaches a* of the quotient formed by dividing the difference $f(x) - f(a)$ by the difference $x - a$.

These two definitions can be interpreted geometrically by the following graphs that show the connection between slopes of straight lines and curves.

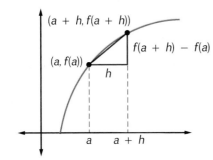 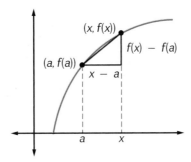

4   Discuss the derivative definitions with your teacher and classmates to answer these questions.

    a. What do you think the phrases "limit as h approaches 0" and "limit as x approaches a" mean?

    b. How well do the two definitions match the strategies that you developed in order to find good estimates for instantaneous rates of change in Investigations 1 and 2?

    c. How would you respond in each case to students who have suggested these alternative definitions for the *derivative* of a function?

      i. $f'(a) = \lim_{h \to 0} \dfrac{f(a) - f(a - h)}{h}$

      ii. $f'(a) = \lim_{h \to 0} \dfrac{f(a + h) - f(a - h)}{2h}$

      iii. $f'(a)$ is the slope of the graph of $f(x)$ when you are close to the point $(a, f(a))$.

5 Now consider what is perhaps the simplest nonlinear function, $q(x) = x^2$. You could try to discover a rule for $q'(x)$ by estimating the numeric derivative of $q(x)$ at several values of x and searching for a pattern in those results. It turns out that in this case, you can deduce the derivative of $q(x) = x^2$ by some fairly simple algebraic reasoning.

a. Study the following reasoning and provide justification for each step.

$$q'(x) = \lim_{h \to 0} \frac{q(x + h) - q(x)}{h} \tag{1}$$

$$= \lim_{h \to 0} \frac{(x + h)^2 - x^2}{h} \tag{2}$$

$$= \lim_{h \to 0} \frac{(x^2 + 2xh + h^2) - x^2}{h} \tag{3}$$

$$= \lim_{h \to 0} 2x + h \text{ (if } h \neq 0) \tag{4}$$

$$= 2x \tag{5}$$

$$\text{So, } q'(x) = 2x. \tag{6}$$

b. Test the formula $q'(x) = 2x$ by evaluating the derivative formula at several specific points and comparing the information provided to the graph of $q(x)$ and numerical estimates for $q'(x)$ at those same points.

c. Prove that the derivative of $p(x) = -x^2$ is $p'(x) = -2x$.

6 Suppose that you know the rates of change in two functions $f(x)$ and $g(x)$ at every point of their common domain.

a. What relationship would make sense for the derivative of the function $f(x) + g(x)$ and the separate derivatives of $f(x)$ and $g(x)$? Compare your answer with your classmates and resolve any differences.

b. What do your answers to Part a of this problem and Problems 3 and 5 above suggest about the derivative of a quadratic function $s(x) = x^2 + bx + c$?

c. Once again, test your answer to Part b by comparing the derivative values produced by that formula to the graph of $s(x)$ and numerical estimates for $s'(x)$ in some specific cases—for example:

i. $s(x) = x^2 + 5x - 3$ at $x = 2$ and $x = -3$

ii. $s(x) = -x^2 + 3x + 2$ at $x = 4$ and $x = 0$

7 Suppose that you know the rate of change in a function $f(x)$ at every point in its domain.

a. What relationship would make sense for the derivative of the function $af(x)$, $a \neq 0$, and the derivative of $f(x)$? Compare your answer with others and resolve any differences.

b. What do your answers to Part a of this problem and to Problems 5 and 6 above suggest about the derivative of the general quadratic function $p(x) = ax^2 + bx + c$, $a \neq 0$?

c. Once again, test your answer to Part b by comparing the derivative values produced by the formula to the graph of the given function and to numerical estimates of $p'(x)$ in some specific cases—for example:

    i. $p(x) = 3x^2 + 5x - 3$ at $x = -2$ and $x = 1$

    ii. $p(x) = -0.5x^2 + 4x + 2$ at $x = 0$ and $x = 6$

Tools for Finding Derivatives Developers of calculators and computer software have written algorithms that perform the repeated calculations required to find derivatives at specific points by numeric approximation. Developers of modern computer algebra systems (CAS) have written programs that use calculus methods for producing derivative functions, so all you have to do is enter the original function and ask the computer to find the related derivative function.

For example, many graphing calculators have *numeric derivative* routines that make the kind of estimates for $f'(x)$ you have provided by looking at the

$$\textbf{difference quotient } \frac{f(x+h) - f(x)}{h}.$$

Most computer algebra systems have symbolic derivative routines that use rules for functions to produce rules for derivative functions. They often use notation for derivatives based on the historic form $\frac{dy}{dx}$.

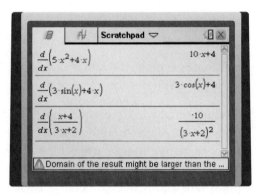

Calculating numerical
approximations for derivatives

Calculating derivative formulas for functions
involving polynomial, circular, and rational
function family members.

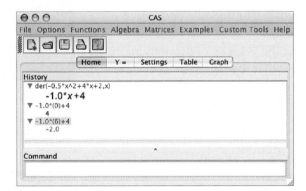

Calculating the derivative formula for the polynomial function
in Problem 7 Part cii and evaluating at $x = 0$ and $x = 6$

8 Test the results produced by a numeric derivative routine by comparing them to estimates using the difference quotient formula $\dfrac{f(x + 0.1) - f(x)}{0.1}$ in the following situations.

   a. Recall that the cliff diver's height (in meters) is a function of time (in seconds) with rule $h(t) = 30 - 4.9t^2$ (page 475). How fast is he falling at the point that is 2 seconds into his dive?

   b. Recall that the population of experimental fruit flies is a function of time in days with rule $P(t) = 100(2^{0.2t})$ (page 475). How fast is that population growing at a time 15 days after the start of the experiment?

   c. Recall that the depth of water (in feet) in a tidal harbor is a function of time in hours with rule $D(t) = 9 \sin 0.5t + 15$ (page 476). How fast is the water depth changing when $t = 5$?

9 Use a CAS to find derivative rules for the functions in Problem 8. Then check the rules by evaluating them at the points of interest and comparing the derivative values obtained in this way to the numeric estimates of derivatives you produced in Problem 8.

   a. What rule gives the cliff diver's speed at any time in his dive, and what is the value of this rule when $t = 2$?

   b. What rule gives the rate of change in fruit fly population on any day, and what is the value of this rule when $t = 15$?

   c. What rule gives the rate of change in tidal water depth at any time in the day, and what is the value of this rule when $t = 5$?

10 In your initial work with quadratic functions in *Core-Plus Mathematics*, you analyzed the flight of many different projectiles—from baseballs, basketballs, and soccer balls to pumpkins and water balloons. In each case, the main questions were about how high above the ground the objects were at various key times.

   Introduction of the notion of a derivative from calculus makes it easy to answer questions about the velocity of those flying objects. Suppose that the height in feet of a pumpkin shot from a "punkin' chunker" is given at any time t seconds by $h(t) = -16t^2 + 160t + 30$.

   a. At what time will the pumpkin reach its maximum height, and when will it return to the ground?

   b. What function gives the vertical component of velocity for the flying pumpkin at any time t?

   c. What is the vertical component of velocity of the pumpkin:

      i. when it leaves its chunker?

      ii. when it reaches its maximum height?

      iii. when it returns to the ground?

CHECK YOUR UNDERSTANDING

The distance traveled by an inline skater going down a hill is given by a function $d(t) = at^2 + bt$, where distance is in meters and time in seconds. The constant a depends on the slope of the hill and the constant b depends on the skill of the skater. Suppose that for a particular hill and skater, the value of a is 1.5 and the value of b is 2.

a. What is $d'(t)$ and what does it tell you about this situation?

b. How fast will the skater be going after 2 seconds?

c. How fast will the skater be going after covering 32 meters?

INVESTIGATION **4**

From Function Graph to Derivative Graph

You can learn a lot about the rates at which variables change by studying the shape of function graphs. As you work on the problems of this investigation, look for answers to the following question:

> *How can the graph of the derivative of a function be sketched from*
> *information obtained by analysis of the graph of the function itself?*

iStockphoto.com/JOsefino

1 In big city subway systems, the trains might travel 2,000 to 3,000 meters between stations and cover that distance in only 2 to 3 minutes. The next graph shows distance as a function of time for one such leg between stations.

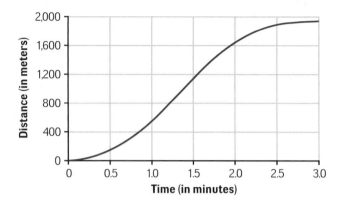

Use information from the graph to answer these questions.

a. For what time interval(s) is the speed of the train:

   i. increasing?

   ii. constant?

   iii. decreasing?

   In each case, explain how the shape of the graph provides the required information.

b. When did the train appear to be moving at top speed and approximately what was that speed?

c. If the function $d(t)$ gives train distance (in meters) as a function of time (in minutes) since leaving a station, what does the derivative $d'(t)$ tell about the train?

d. Which of the following graphs best matches the pattern of change in $d'(t)$ over the interval $[0, 3]$? Explain why your choice is better than the other two.

Graph I

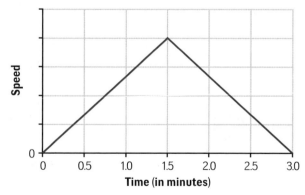

Graph II

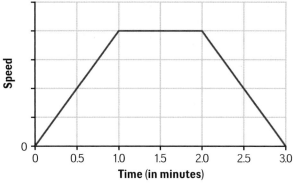

Time (in minutes)

Graph III

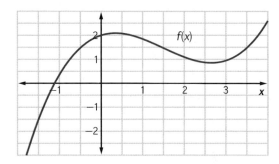

Time (in minutes)

2 There are some points on the graph of a function $f(x)$ that are especially helpful in creating a graph of $f'(x)$. Study the shape of the graph below and think about other graphs that you have seen in earlier investigations to come up with answers for the following questions about graphs of functions and their derivatives.

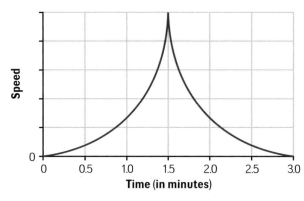

a. If a function has a local maximum or local minimum at $(a, f(a))$, what can you infer about the value of $f'(a)$? What does that tell about the graph of $f'(x)$?

b. What information from the graph of $f(x)$ tells where the graph of $f'(x)$ is positive? Where it is negative?

c. What information from the graph of $f(x)$ tells where the graph of $f'(x)$ is increasing? Where it is decreasing?

d. What information from the graph of $f(x)$ tells where the graph of $f'(x)$ will have a local maximum or local minimum value?

3 Study each of the following functions and their graphs. Without using graphing technology, complete each of the following tasks. Then check your ideas by calculating and graphing numerical estimates of the derivative function.

- Write a description of how the derivative changes as the input variable x increases. Explain where the derivative is positive, where it is zero, and where it is negative. Also explain where the derivative is increasing and decreasing.

- Estimate the derivative at several points by considering the slope of the graph.

- Use your analysis to sketch a graph of the derivative function on a copy of the corresponding function graph.

- Identify the kind of function whose graph is likely to match the derivative.

a. A graph of $a(x) = 0.5x$ with window $-5 \leq x \leq 5$ and $-3 \leq y \leq 3$

b. A graph of $f(x) = x^2 - 2$ with window $-4 \leq x \leq 4$ and $-4 \leq y \leq 10$

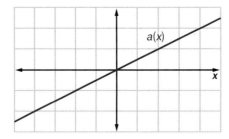
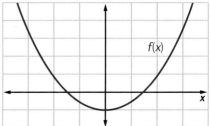

c. A graph of $g(x) = 1.5^x$ with window $-3 \leq x \leq 7$ and $-2 \leq y \leq 10$

d. A graph of $h(x) = \cos x$ with window $-\pi \leq x \leq \pi$ and $-2 \leq y \leq 2$

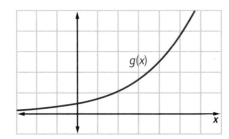
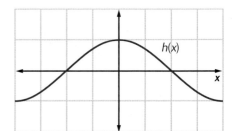

e. A graph of $j(x) = 0.5x^3 - 8x$ with window $-5 \leq x \leq 5$ and $-15 \leq y \leq 15$

f. A graph of $\ell(x) = \ln x$ with window $-1 \leq x \leq 9$ and $-2 \leq y \leq 4$

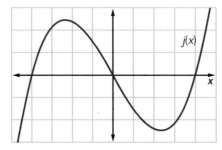
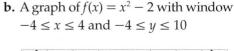

✔ CHECK YOUR UNDERSTANDING

The graph below shows the growth of a fish population after stocking of a new lake. Time t is measured in years and the fish population is measured as a percent of the maximum capacity of the lake.

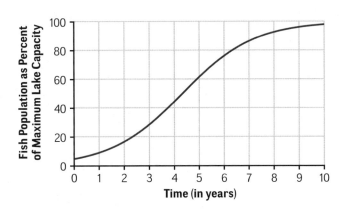

a. On a copy of the graph of $P(t)$, sketch a graph of $P'(t)$.

b. The function $P(t) = \dfrac{100}{1 + 20(0.5)^t}$ models the growth of the fish population. Compare the graph of $P'(t)$ that you sketched with that produced by a CAS. What adjustments, if any, do you need to make to your sketch?

c. Explain what the pattern in the graph of $P'(t)$ tells about the pattern of change in the fish population over the 10-year period following stocking of the lake.

APPLICATIONS

1 When you turn on your oven, it gradually heats up from room temperature to whatever temperature you set. Suppose that you set the oven control to 450°F for baking a pie.

a. Describe the patterns of heating illustrated by the following graphs.

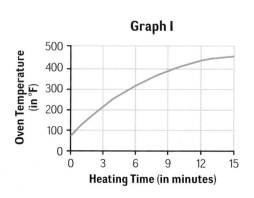

Graph I

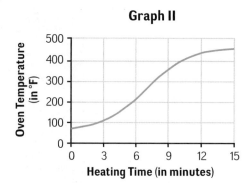

Graph II

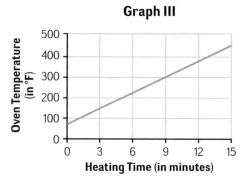

Graph III

b. Shown below are three tables of (*time, temperature*) data that you might get by watching an oven warm up from room temperature to 450°F. Match these tables to the graphs in Part a. Explain how the different patterns of temperature change are illustrated in the matching graphs.

Data Pattern A							
Heating Time (in minutes)	0	1	2	3	4	5	6	7
Temperature (in °F)	70	95	120	145	170	195	220	245
Heating Time (in minutes)	8	9	10	11	12	13	14	15
Temperature (in °F)	270	295	320	345	370	395	420	445

Data Pattern B								
Heating Time (in minutes)	0	1	2	3	4	5	6	7
Temperature (in °F)	70	80	90	115	140	170	210	260
Heating Time (in minutes)	8	9	10	11	12	13	14	15
Temperature (in °F)	310	350	380	405	430	440	445	450

Data Pattern C								
Heating Time (in minutes)	0	1	2	3	4	5	6	7
Temperature (in °F)	70	125	170	210	250	280	310	335
Heating Time (in minutes)	8	9	10	11	12	13	14	15
Temperature (in °F)	360	380	400	415	430	440	445	450

c. Estimate the rate of heating predicted by each graph and table at a time exactly 5 minutes after the oven is turned on. Be sure to explain the units in which those rates of heating would be measured and reported.

2 Among the most familiar problems concerning distance, time, and rate of motion are those that involve balls that have been thrown, hit, or kicked. For example, if a soccer ball is kicked into the air, its height will be a function of time in flight. This function might be given by graphs or tables like those below.

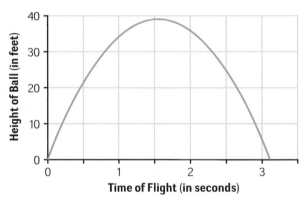

Time of Flight (in seconds)	0	0.25	0.50	0.75	1.00	1.25	1.50	1.75	2.00	2.25	2.50	2.75	3.00
Height of Ball (in feet)	0	11.5	21.0	28.5	34.0	37.5	39.0	38.5	36.0	31.5	25.0	16.5	6.0

a. Use the table and graph to estimate both the speed and vertical velocity of the moving ball at times 0.5 seconds and 2.5 seconds into its flight.

b. Explain the difference between speed and vertical velocity for the ball and how that difference shows up in the graph and the table of the (*time, height*) relation.

3 Suppose $d(t)$ gives the distance in meters of a walker from a motion detector as a function of time t in seconds. Write algebraic expressions that show how you would calculate or estimate:

   a. the average velocity of the walker between $t = 1$ and $t = 5$.

   b. the average velocity of the walker over the interval $0 \leq t \leq 6$.

   c. the average velocity of the walker over the interval $a \leq t \leq b$.

   d. the instantaneous velocity of the walker at $t = 3$ seconds.

   e. the instantaneous velocity of the walker at $t = a$ seconds.

4 Suppose the daily operating profit (in dollars) for a movie theater is a function of the number of tickets sold with rule $P(x) = 8.5x - 3,500$ for $0 \leq x \leq 1,000$.

   a. Find the rate at which theater profit is changing when $x = 100$, $x = 500$, and $x = 800$. Give the unit in which those rates should be expressed.

   b. How will the answers to Part a be reflected in the shape of a graph for $P(x)$?

5 Suppose the height of a soccer kick (in feet) is a function of time in flight (in seconds) with rule $h(t) = -16t^2 + 50t$ for $0 \leq t \leq 4$.

   a. Estimate the vertical velocity of the ball when $t = 0.5$, $t = 1.5$, and $t = 2.0$ seconds. Explain how the differences in those values are shown by the shape of the graph of $h(t)$.

   b. About how fast will the ball be traveling when it hits the ground?

6 In Washington, D.C., the number of minutes between sunrise and sunset varies throughout the year according to the function rule

$$S(d) = 180 \sin (0.0172d - 1.376) + 720.$$

In this rule, d represents the day of the year (with January 1 as $d = 1$).

   a. Produce a graph of $S(d)$ for $0 \leq d \leq 365$. Explain what the shape of that graph tells you about the number of minutes between sunrise and sunset throughout the year.

   b. March 21, June 21, September 21, and December 21 are the approximate first days of spring, summer, fall, and winter. In a non-leap year, what day numbers correspond to these dates?

   c. Estimate the instantaneous rate of change in time between sunrise and sunset on the four special dates in Part b. Explain how those rates are illustrated in the graph of $S(d)$.

   d. At what time(s) of the year are days growing "longer" or "shorter" most rapidly? When are they growing most slowly? What are those rates of change?

7 Suppose that under laboratory conditions, an initial population of 150 bacteria doubles every hour.

   a. What function gives the number of bacteria after t hours?

   b. Calculate the predicted number of bacteria after 3, 6, 9, and 12 hours.

   c. Calculate the average growth rate of the bacteria population between $t = 0$ and $t = 12$.

   d. Estimate the instantaneous growth rate of the bacteria population at $t = 3$, $t = 6$, $t = 9$, and $t = 12$.

   e. Compare the results from Parts c and d and explain how the differences can be seen in the shape of a graph of the growth function.

8 Consider again the function $P(x) = 8.5x - 3{,}500$ giving daily profit for a movie theater as a function of the number of tickets sold.

   a. What is the rule for $P'(x)$?

   b. What does the rule for $P'(x)$ tell about the rate at which profit changes as the number of tickets sold increases?

   c. What is the shape of the graph of $P'(x)$? Why does that make sense?

9 Consider again the function $h(t) = -16t^2 + 50t$ giving height of a soccer kick as a function of time.

   a. What is the rule for $h'(t)$?

   b. How does the shape of the graph of $h'(t)$ relate to the graph of $h(t)$ and to the motion of the ball?

10 Suppose that the height of a suspension cable above a bridge surface (in meters) is given by $h(x) = 0.001x^2 - 0.5x + 75$, where x is the distance in meters from the left tower to the right tower.

   a. What is the rule for $h'(x)$?

   b. Sketch graphs of $h(x)$ and $h'(x)$ for $0 \le x \le 500$.

   c. Find values of x that meet each of the following conditions and explain what those derivative values tell about the height of the suspension cable.

   i. $h'(x) > 0$ **ii.** $h'(x) < 0$ **iii.** $h'(x) = 0$

11 Consider the function $g(x)$ with graph given at the right.

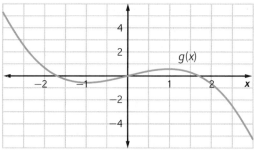

a. Study the graph of $g(x)$ to identify local maximum/ minimum points and intervals where $g(x)$ is increasing or decreasing.

b. Use the information from Part a to sketch the shape of the graph of $g'(x)$ for $-3 \leq x \leq 3$.

12 Consider again the function $S(d) = 180 \sin (0.0172d - 1.376) + 720$ that gives the number of minutes between sunrise and sunset on any day of the year in Washington, D.C.

a. Study a graph of $S(d)$ to identify points or intervals that meet these conditions:

    i. times of maximum sunlight

    ii. times of minimum sunlight

    iii. times when sunlight time is increasing

    iv. times when sunlight time is decreasing

b. Use the information from Part a to sketch the shape of the graph of $S'(d)$ for $0 \leq d \leq 365$.

CONNECTIONS

13 Because small portions of a curve often look like straight line segments, it is usually possible to sketch the graph of a curve by plotting a sequence of points and connecting those points in order. Test this idea with the function $f(x) = x^2$.

a. Calculate $f(-5), f(0)$, and $f(5)$ and plot the corresponding points on a coordinate system. Then connect those points in order with line segments.

b. Calculate $f(-3)$ and $f(3)$ and plot the corresponding points on the same diagram. Then connect all five plotted points in order (from $x = -5$ to $x = 5$) with line segments of a second color.

c. Calculate $f(-1)$ and $f(1)$ and plot the corresponding points on the same diagram. Then connect all seven plotted points in order with line segments of a third color.

d. Calculate $f(x)$ for $x = -4, -2, 2$, and 4, and plot the corresponding points on the same diagram. Connect all eleven points in order with yet a fourth color.

e. Explain how the progression of line plots illustrates the connection between finding rates of change for nonlinear functions, slopes of curved graphs, and slopes of linear functions.

f. Check your work by entering the eleven (x, x^2) pairs in your calculator data lists or computer software data table, making a connected plot of those data points, and graphing the function $f(x) = x^2$ over that plot.

The word "limit" that is used in defining the derivative of a function appears in many other mathematical situations. Tasks 14–16 illustrate a few of those applications of the limit idea.

14 The number sequence $1, \frac{3}{2}, \frac{7}{4}, \frac{15}{8}, \ldots, 2 - \left(\frac{1}{2}\right)^n$ converges to the limit 2 because for any specified margin of error, it is always possible to find a value of n such that all terms in the sequence beyond that term will be closer to 2 than that margin of error.

   a. Find n so that all terms of the sequence beyond the nth are within 0.001 of the limit 2.

   b. Find n so that all terms of the sequence beyond the nth are within 0.0001 of the limit 2.

15 In Unit 5, *Exponential Functions, Logarithms, and Data Modeling*, you discovered that e can be calculated by evaluating the expression $\left(1 + \frac{1}{n}\right)^n$ for ever larger values of n. In the language of this unit, we would say that $\lim_{n \to \infty} \left(1 + \frac{1}{n}\right)^n = e$.

   a. If you let $m = \frac{1}{n}$, then what does m approach as n approaches infinity?

   b. What would you conjecture to be $\lim_{m \to 0} (1 + m)^{\frac{1}{m}}$? Make a table of values of the expression $(1 + m)^{\frac{1}{m}}$ for $m = 0.01$, $m = 0.001$, $m = 0.0001$, and $m = 0.00001$. Explain what those values say about your conjecture of the limit of that expression as m approaches 0.

16 Consider the behavior of the rational function $f(x) = \frac{x^2 - 1}{x - 1}$ near $x = 1$.

   a. Can you calculate the value of $f(1)$?

   b. What is the $\lim_{x \to 1} f(x)$ coming from the right?

   c. What is the $\lim_{x \to 1} f(x)$ coming from the left?

   d. Sketch a graph and create a table of values for $f(x)$. Though $f(1)$ is undefined, what do the table and graph suggest about the limit of $f(x)$ as x approaches 1?

   e. If $x \neq 1$, then $f(x)$ can be simplified. What is the limit as x approaches 1 for the simplified function?

17 The value of the derivative of a function at a point tells you how the function is changing at that point.

   a. What can you tell about the way that a function f is changing near a if you know:

      i. $f'(a) > 0$?

     ii. $f'(a) < 0$?

    iii. $f'(a) = 0$?

   b. Explain how your conjectures about properties of $f'(x)$ will be shown in a graph of the original function $f(x)$.

18 In Course 3 Unit 5, *Polynomial and Rational Functions*, you examined the graphs of rational functions that approached an asymptote.

a. Sketch a graph of the function $f(x) = \dfrac{x^2}{x+1}$ for $-10 \leq x \leq 10$.

b. To examine the end behavior of the function, we want to know how the function behaves for large values of x. Divide x^2 by $x + 1$ and use the result to explain the pattern in the graph of $f(x)$ for large values of x.

c. As x approaches infinity, is $f(x)$ ever less than $x - 1$? For what value of x is the difference between $f(x)$ and $x - 1$ less than 0.001?

19 In Course 3 Unit 7, *Recursion and Iteration*, you studied geometric sequences and series with terms in the following patterns, where $r \neq 1$.

Sequence: $1, r, r^2, r^3, r^4, \ldots, r^{n-1}, \ldots$

Series: $1, 1 + r, 1 + r + r^2, 1 + r + r^2 + r^3, 1 + r + r^2 + r^3 + r^4, \ldots,$
$1 + r + r^2 + r^3 + r^4 + \cdots + r^{n-1}, \ldots$

a. For what values of r does a geometric sequence approach a finite limit as n increases? How is that limit related to the value of r?

b. For what values of r does a geometric series approach a finite limit as n increases?

c. Recall the formula $S_n = \dfrac{r^n - 1}{r - 1}$ gives the sum of the first n terms in the above geometric sequence or the nth term in the associated geometric series. How do this formula and your answer to Part a combine to give a formula for the limit of a geometric series as n increases, in the cases when a finite limit exists?

20 In Investigation 3, you used a CAS to find a derivative formula given a function rule. In Investigation 4, you used the graph of a function to sketch a graph of its derivative function.

Use what you know about the shape of the graphs of exponential functions, power functions, and the sine function to explore the derivative formulas for these three function families. First sketch or visualize the derivative functions to formulate a conjecture. Then check your conjecture using a CAS.

REFLECTIONS

21 Think about the significance of the derivative of a function in modeling applied problem situations.

a. Suppose that $h(t)$ gives the height (in feet) of a golf ball at various times (in seconds) in its flight.

  i. What are the measurement units for $h'(t)$?

  ii. What does $h'(t) = 0$ tell about the velocity of the ball at the time of maximum height and why is that result reasonable?

b. When an earthquake strikes somewhere in the United States, news about the event spreads quickly. Suppose that $N(t)$ gives the number of people (in millions) who have heard the news t hours after the quake.

  i. What are the measurement units of $N'(t)$?

  ii. What can you infer about spread of the news if $N(4) = 60$ and $N'(4) = 6.5$?

c. When deep-sea divers return to the surface, they need to rise slowly to adapt to the change in pressure or else risk serious injury or even death. Suppose that a diver working at a depth of 175 feet begins ascent to the surface so that her depth is given by $D(t)$ at time t minutes later.

  i. What are the measurement units of $D'(t)$?

  ii. What can you infer about the diver's ascent if $D(5) = 150$ and $D'(5) = -30$?

22 Select one of the modeling functions in Investigation 2. Pose your own question involving instantaneous rate of change of that function in the chosen context. Illustrate how you could use the difference quotient $\dfrac{f(x + h) - f(x)}{h}$ and a spreadsheet to help answer your question. How could you improve the accuracy of your estimate?

23 The process of finding the derivative of a function is called **differentiation**. The credit for inventing differentiation and, more generally, the main ideas of calculus is shared by Isaac Newton and Gottfried Leibniz. At roughly the same time, but in different countries and using different symbolic notation, each independently obtained similar results.

Isaac Newton

Gottfried Leibniz

Leibniz realized the importance of good notation in solving problems. For a function $y = f(x)$, Leibniz devised the notation $\dfrac{dy}{dx}$ to denote the "rate of change of y as x changes" or "the derivative of y with respect to x." As you saw in Investigation 3, this notation is still used today, along with the notation $f'(x)$.

a. Sketch the graph of a function $y = f(x)$. On your sketch, show a geometric interpretation of the notation $\dfrac{dy}{dx}$.

b. What advantages, if any, do you see for using Leibniz's notation for the derivative?

24 After completing Lesson 1, Lynette said to her teacher, "When I estimate the instantaneous rate of change for a function, I am really just finding an average rate of change." Do you agree or disagree with Lynette? Explain your response.

25 In the Check Your Understanding task following Investigation 4, you studied the pattern of growth in a fish population, from the time a few fish were stocked in a new lake to the time the fish population approached maximum capacity for the lake's food supply. The population function rule was $P(t) = \dfrac{100}{1 + 20(0.5^t)}$. Following is a graph of the derivative of this function, $P'(t)$.

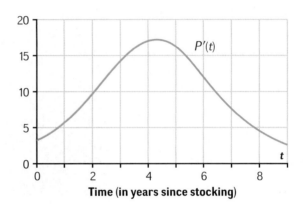

Time (in years since stocking)

a. How should the vertical axis be labeled in words?

b. According to this graph, when is the derivative a maximum?

c. What does the answer to Part b tell about growth rate of the fish population?

d. What would the answer to Part b tell about the graph of $P(t)$ itself?

26 Use the **Zoom** capability of your graphing calculator or CAS to examine "close up" the graphs of $f(x) = x^3$ and $g(x) = |x|$ near $x = 0$. What seems to be true about the derivative of each function at 0?

EXTENSIONS

27 Everyone who drives a vehicle has to stop suddenly sometimes. The time and distance it takes to stop depend on several variables. For a typical car on a dry asphalt road, the stopping distance (in feet) is a function of speed (in miles per hour) with a quadratic rule like $d(s) = 0.04s^2 + 0.75s$.

This function is actually the sum of two separate functions, each related to parts of the stopping problem.

- $b(s) = 0.75s$ gives the distance traveled before you manage to press the brake pedal.

- $c(s) = 0.04s^2$ gives the distance traveled from the time you press the brake pedal until you stop the car.

a. Use a calculator or spreadsheet to make a table showing how the three functions change as the car's speed changes from 0 to 100 mph in steps of 10 mph.

Speed	$b(s)$	$c(s)$	$d(s)$
0	0	0	0
10	⋮	⋮	⋮
⋮			

b. What do each of the columns of the table in Part a tell about stopping distance for a car at various speeds?

c. How does the rate of change in total stopping distance relate to the rate of change in braking distance and the rate of change in distance traveled after braking?

d. Use a calculator or spreadsheet to make another table showing approximations to the derivatives of the three functions involved in stopping a car.

28 In Unit 1, *Families of Functions*, you investigated transformations of graphs of basic functions and how those transformations are represented in symbolic rules.

a. How does the rate of change of $f(x) = x^2$ at $x = 1$ compare to the rate of change of $g(x) = x^2 + 5$ at $x = 1$?

b. Why does your finding in Part a make sense graphically?

c. Would your answer to Part a be the same for other values of x? Why or why not?

d. In general, what do you think is true about the rates of change at corresponding points of two functions whose graphs are related by a vertical translation? Prove your assertion using algebraic reasoning and the definition of the derivative.

29 Repeat Task 28 for the case of vertical stretches by first comparing the rate of change of $f(x) = x^2$ at $x = 4$ and the rate of change of $h(x) = 3x^2$ at $x = 4$.

30 Imagine a ball rolling down each of the following ramps.

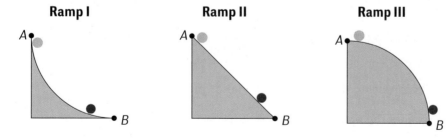

a. Which ramp provides the longest path for the ball? The shortest? Explain your reasoning.

b. On which ramp is the ball traveling fastest when it is one-quarter of the way down the ramp? The slowest? Explain your reasoning.

c. For which ramp do you think the speed of the ball at the end of the ramp will be the greatest? The least? Explain your thinking.

d. For which ramp do you think the ball will reach the bottom quickest? Slowest?

e. For each of the ramps, sketch what you think would be a graph of the speed of the ball as a function of the distance traveled.

31 You can also look at rules for exponential functions from a limits perspective. From work on Connections Task 15, you know that $\lim_{n \to \infty} \frac{1}{n}$ is 0, so you would expect that $\lim_{n \to \infty} A^{\frac{1}{n}} = A^0$.

a. If $A > 0$, What does $A^{\frac{1}{n}}$ represent for $n = 2$? For $n = 3$?

b. If $A = 64$, calculate $A^{\frac{1}{n}}$ for $n = 1, 2, 3,$ and 6. For what value of n will $64^{\frac{1}{n}} < 1.01$?

c. If $A > 1$, describe the behavior of $A^{\frac{1}{n}}$ as n gets larger. Can $A^{\frac{1}{n}}$ ever be less than 1?

d. Describe the behavior of $A^{\frac{1}{n}}$ as n gets larger if $0 < A < 1$. Can $A^{\frac{1}{n}}$ ever be greater than 1?

32 Consider the functions $f(x)$ and $g(x)$ as graphed at the right.

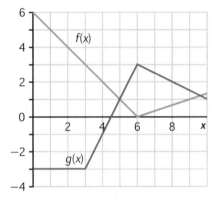

a. If $h(x) = f(x) + g(x)$, then what is $h'(7)$?

b. If $j(x) = f(x) - g(x)$, then what is $j'(5)$?

c. If $k(x) = 2f(x)$, then what is $k'(2)$?

REVIEW

33 Find the slope, y-intercept, x-intercept, and an equation for each line described below.

a. Contains $(-1, 3)$ and $(1, -1)$

b. Contains $(-1, -3.8)$ and $(-5, -6.6)$

c. Is perpendicular to the line $y = \frac{2}{3}x + 1$ and has x-intercept at $(-2, 0)$

d. Is parallel to the line $3x - y = 7$ and contains the origin

34 Consider the polynomial function $f(x) = (2x - 1)(x^2 - 10)$.

a. What is the y-intercept of $f(x)$?

b. Determine all solutions to $f(x) = 0$.

c. For what values of x is $f(x) > 0$?

d. For what values of x is $f(x)$ increasing?

35 Sketch graphs of functions $f(x)$ that model the following patterns of change. For each function graph, describe the type of function rule that would match the pattern you sketched.

  a. As x increases, $f(x)$ increases at a constant rate.

  b. As x increases, $f(x)$ decreases at a constant rate.

  c. As x increases, $f(x)$ increases at a rate that is gradually increasing.

  d. As x increases, $f(x)$ increases at a rate that is gradually decreasing.

  e. As x increases, $f(x)$ decreases at a rate that is gradually decreasing.

36 The average high temperature in degrees Fahrenheit in Dallas, Texas, can be approximated by $H(d) = 21 \sin (0.016d - 1.64) + 75$, where d is the day number of the year with January 1 being day 1.

  a. What are the maximum and minimum average high temperatures in Dallas, Texas?

  b. Evaluate $H(50)$ and explain what it tells you.

  c. On what days of the year is the average high temperature in Dallas greater than 75°F?

37 Rewrite each rational expression in a simplified form. Then determine for what values of the variable, if any, the original and simplified expressions are *not* equivalent.

  a. $\dfrac{6x + 9}{3}$

  b. $\dfrac{(x + 5)^2 - x^2}{5}$

  c. $\dfrac{2t - 12}{t - 6}$

  d. $\dfrac{8ab + 4b}{12b}$

  e. $\dfrac{2x - 5}{4x^2 - 25}$

  f. $\dfrac{8d^2 + 22d + 15}{8d + 12}$

38 Recall that a *piecewise function* is a function whose rule is different on different parts of the domain. An example of a simple piecewise function is the **Heaviside step function**:

$$H(x) = \begin{cases} 0, & \text{if } x < 0 \\ 1, & \text{if } x \geq 0 \end{cases}$$

  a. Graph $H(x)$ on the interval $[-2, 2]$.

  b. The Heaviside step function is often used to model situations that involve abrupt changes. For example, a particular flashlight, when turned on, emits a cone of light with brightness 150 candela. If the flashlight is off for all times less than zero and turned on at $t = 0$, make a graph that indicates the brightness for any time t.

  c. If $L(t)$ is the brightness of the light at any time t, how is $L(t)$ related to $H(t)$?

  d. Draw a graph of the function $150H(t - 3)$. What kind of situation could this function be used to model?

  e. Draw a graph of the function $150H(t - 1) - 150H(t - 3)$. What kind of situation could this function be used to model?

39 Solve each equation.

   a. $1{,}000e^{0.05t} = 1{,}500$

   b. $5e^x - 40 = e^x$

   c. $\ln(x + 1) = 3.7$

   d. $\log(25x) - \log(x - 1) = 2$

40 Prove whether or not each statement is an identity.

   a. $\sin^3 \theta + \sin \theta \cos^2 \theta = \tan \theta \cos \theta$

   b. $\csc 2\theta = 2 \csc \theta$

41 Consider all real and complex number solutions to these equations.

   a. Find the product of the two solutions to $3x^2 - 5x + 25 = 0$.

   b. Find the sum of all solutions of $(x + 3)(x^2 + x + 4) = 0$.

42 Carmela is training for a triathlon. Currently when she is training, she averages the following distance/time for each event.

      Biking: 15 mph
      Swimming: 0.5 miles in 20 minutes
      Running: 8 minutes per mile

Assume that she can maintain these speeds throughout her workout.

   a. What is Carmela's swimming speed in mph?

   b. What is her running speed in mph?

   c. Suppose that she runs for 45 minutes, bikes for 40 minutes, and swims for 30 minutes.

      i. How far will she have traveled?

      ii. What is her average speed for the entire workout?

43 In the following, $i^2 = -1$. Rewrite each product as a polynomial with complex coefficients.

   a. $(7x + 2i)^2$

   b. $(9x - 4i)^2$

   c. $(x - (2 + 5i))^2$

Introduction to the Definite Integral

Some motion detectors are designed to produce tables and graphs of (*time, speed*) data rather than (*time, distance*) data. For example, police radar scanners focus on the speed of approaching cars, and baseball "radar guns" report the speed of pitched balls.

When automobile companies promote their new cars, one feature mentioned often is acceleration or "pickup." A standard statistic for measuring pickup is the time it takes to go from 0 to 60 miles per hour (88 feet per second).

The following graph shows the performance of one car during such an acceleration test.

Acceleration Test

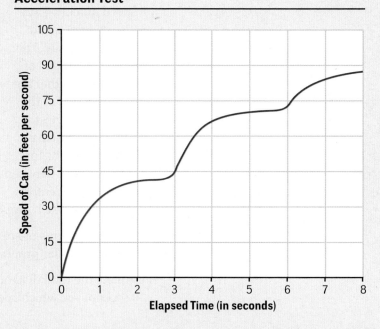

Lesson 2 | Introduction to the Definite Integral **503**

Steve Allen/Getty Images

Work on the problems of this lesson will develop ideas of *integral calculus* needed to answer questions like those above.

INVESTIGATION 1

What is the Total?

If a function tells the position of a moving object at any time, the derivative of that function tells the instantaneous rate of change in position, or velocity, at any time. But the graph of speed for an accelerating car poses the inverse problem—given a function that tells velocity at any time, how can we find the position?

As you work on the problems of this investigation, look for answers to this question:

How can the information provided by a rate of change graph be used to calculate total change over some time interval?

1 Unlike cars, motorcycles, or airplanes, human walkers and runners can reach top speed quickly. They can also change speed and direction easily. The result will be graphs of (*time, speed*) data that are simpler than that of the 0–60 acceleration test.

   Study the following (*time, speed*) graphs for students walking between classes at Plano Central HighSchool. For each graph:

- describe (in terms of speed and acceleration) how you would walk to produce a similar graph.

- estimate the total distance the walker would have covered in the 10 seconds for which speeds are shown on the graph.

a. Patricia's Walk

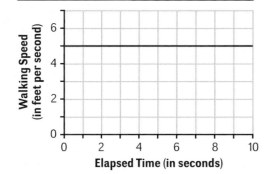

b. Raymond's Walk

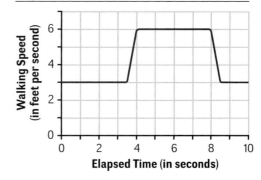

c. Fraser's Walk

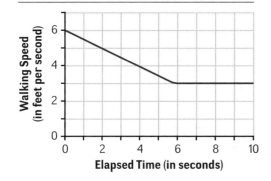

d. Ayana's Walk

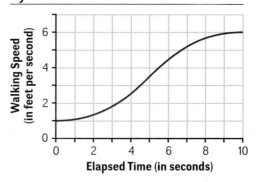

2 People, cars, planes, and baseballs are not the only things in the world for which rate of change is an important measured statistic. Electricity, water, natural gas, oil, and computer data flow through a variety of "pipelines." The flow rates are monitored by gauges that report in units like gallons, cubic meters, or megabytes per second.

Study the following graphs that show recorded flow rates for pumps used to irrigate fields of a large farm. In each case, estimate the total amount of water pumped onto the fields in the indicated time intervals. Compare your results with those of your classmates and resolve any differences.

a.

Graph I

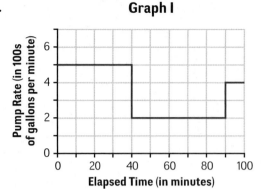

y-axis: Pump Rate (in 100s of gallons per minute)

x-axis: Elapsed Time (in minutes)

b.

Graph II

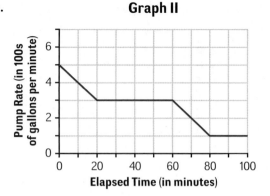

y-axis: Pump Rate (in 100s of gallons per minute)

x-axis: Elapsed Time (in minutes)

c.

Graph III

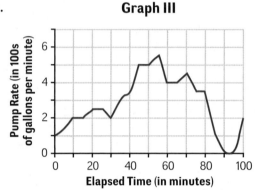

y-axis: Pump Rate (in 100s of gallons per minute)

x-axis: Elapsed Time (in minutes)

SUMMARIZE THE MATHEMATICS

In this investigation, you were given information about the speed of a moving object or the flow rate of a substance like irrigation water and used that information to calculate a total distance traveled or total water used.

a How can you use a speed graph to estimate distance traveled by a moving object in some interval of time? Why does your proposed strategy provide the required result?

b How can you use a flow rate graph to estimate total flow in some interval of time? Why does your proposed strategy provide the required result?

Be prepared to explain your ideas to the class.

When an airplane touches the runway after a flight, it has to slow down fairly quickly from its landing speed that might be about 150 mph or 220 feet per second to a taxiing speed of 20 mph or about 30 feet per second.

The speed graph for such a plane might look like this.

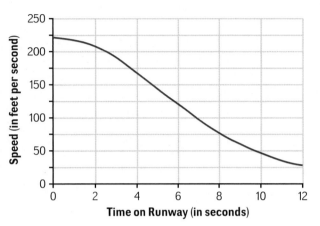

Use the graph to estimate the distance traveled by the plane during:

a. the first 6 seconds after landing.

b. the time from 6 to 12 seconds after landing.

INVESTIGATION **2**

Velocity and Net Change

In work on the problems of Investigation 1, you developed strategies for using speed graphs to estimate distance traveled and rate graphs to estimate total flow through a pipeline. In each of the problems, the motion or flow was in one direction. But cars and people move both forward and backward, and pumps can be used to move fluids in both directions through a pipeline.

As you work on the problems of this investigation, look for answers to these questions:

> *How can graphs indicate both rate and direction of change in a quantity?*
>
> *How can such directional rate graphs be used to calculate the net change in position of a moving object or the net flow in a pipeline?*

Directed Motion To accurately describe the rate of change in position of a moving object, both speed and direction of motion need to be specified. These two attributes—speed and direction—define the term **velocity**. In situations involving motion toward and away from a sensor, it is customary to indicate speed in one direction with positive numbers and speed in the other direction with negative numbers.

1 Suppose that the following simplified graph shows the velocity of a person walking toward and away from a motion detector during a 10-second period.

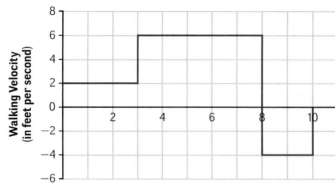

a. How would you walk toward and away from the detector in order to replicate the given graph? Compare your description with that of others and resolve any differences.

b. How far from the motion detector would you be after 10 seconds if you began your walk:

   i. 5 feet from the detector?   **ii.** 3 feet from the detector?

   iii. 12 feet from the detector?

c. What is the net change in distance from the motion detector that results from each 10-second walk described in Part b?

d. What is the total distance you covered in each 10-second trip described in Part b?

2 Here is another graph showing the velocity of a person walking toward and away from a motion detector during a 10-second period.

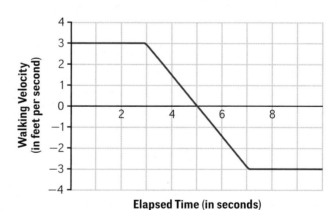

a. How would you walk toward and away from the detector in order to replicate the given graph? Compare your description with that of others and resolve any differences.

b. How far from the motion detector would you be after 10 seconds if you began your walk:

   i. 6 feet from the detector?   **ii.** 4 feet from the detector?

   iii. 9 feet from the detector?

c. What is the net change in distance from the motion detector that results from each 10-second walk described in Part b?

d. What is the total distance you covered in each 10-second trip described in Part b?

Underground Water Supplies In some parts of the United States, water for human consumption, manufacturing processes, and agriculture is taken directly from major rivers that collect rainfall and snowmelt from large watershed areas. In other parts of the country, water is pumped from wells that reach deep into *aquifers* that are underground reservoirs.

The Edwards Aquifer is one of the largest underground reservoirs in the world, collecting water from an area of 6,000 square miles on the Edwards Plateau in Texas. Rainwater runs off into streams that disappear underground into the aquifer. While the Edwards Plateau can experience some of the highest rainfall intensities in the world, it also goes through periods of severe drought. Some regions of the plateau did not see any rain between 1988 and 1996, while other regions experienced record rainfalls of 22 inches in three hours.

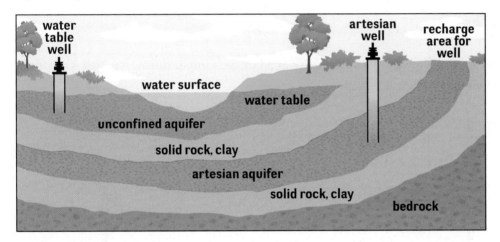

3 The following graph shows the rate of change in aquifer volume during a recent year.

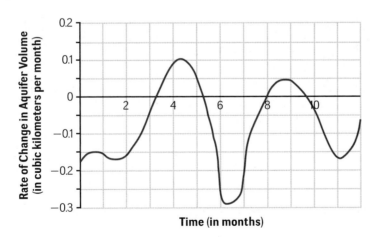

a. How would you describe the pattern of inflow and outflow of the aquifer over the year?

b. Estimate the net change in water volume of the aquifer from the beginning to the end of the year in question. Be prepared to show and explain the calculations used in producing your estimate.

CHECK YOUR UNDERSTANDING

The graph below shows the velocity of a bungee jumper during the first 8 seconds of a jump.

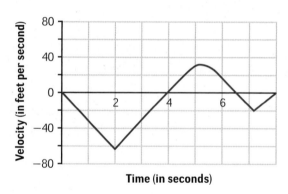

a. What pattern of change in direction and speed of the jumper is shown by the graph? When is he falling downward? Bouncing upward? Gaining speed? Losing speed?

b. Suppose that the jumper starts from a height of 130 feet above a safety net. Estimate his height above the net:

    i. after 4 seconds.

    ii. after 6.5 seconds.

    iii. after 8 seconds.

c. What is the total distance traveled by the jumper in the 8-second time period?

Karl Weatherly/Getty Images

The Definite Integral

If you look back at the problems of Investigations 1 and 2, you will see an important relationship between net change in a varying quantity and the areas of regions bounded by its rate of change graph and the x-axis. For example, the following diagram shows how to calculate change in position of a person walking away from and toward a motion detector.

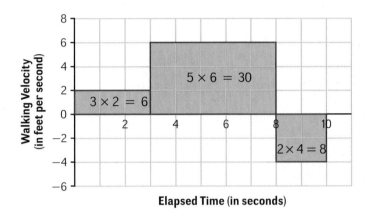

To find net change in distance from the detector, it makes sense to calculate $(3 \times 2) + (5 \times 6) - (2 \times 4)$. On the diagram, that is equivalent to finding a sum and difference of areas. The calculus idea that exploits this connection to solve problems is the *definite integral*.

As you work on the problems of this investigation, look for answers to these questions:

What does it mean to find the definite integral of a function $f(x)$ over an interval $[a, b]$?

What strategies are useful for finding approximate and exact values of definite integrals?

1 You have seen rate functions with quite irregular graphs like that below showing rate of change in a city water reservoir. In such cases, you have seen that it makes sense to use a step function approximation to the irregular graph.

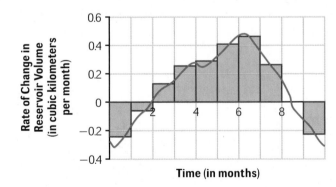

a. Use the approximating step function shown in the graph to estimate net change in reservoir volume for the 10-month period. Record the calculations used to produce that estimate.

b. Explain how the areas of the "bars" on the graph contribute to the estimate of net change in reservoir water volume.

The key point illustrated in the two preceding examples is that the net change in a quantity can be calculated using the graph of its rate of change function to find the difference:

(area of regions above the x-axis) − *(area of regions below the x-axis)*

For a function $f(x)$ defined over an interval $[a, b]$, this difference of areas is called the **definite integral of $f(x)$ from a to b**.

The diagram below illustrates the general strategy used to estimate definite integrals:

(1) The interval from a to b is divided into smaller intervals of length Δx.

(2) In each subinterval, you choose a point x so that $f(x)$ is representative of the function on that interval.

(3) Then calculate the product $f(x)\Delta x$ corresponding to the area (positive or negative) of a rectangle above or below the x-axis.

(4) Approximate the integral from a to b by summing the various $f(x)\Delta x$ products.

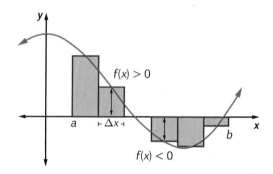

The definite integral of $f(x)$ from a to b is commonly indicated by the notation

$$\int_a^b f(x)\, dx.$$

This notation for the definite integral is a reminder of the process by which integrals can be approximated—with a *sum of products* in the form $f(x)\Delta x$ that represent areas of rectangles. The symbol $\int$ is like an elongated "S" for "sum."

2 When the graph of a function $f(x)$ is made up of familiar geometric shapes, you can use your knowledge about the areas of shapes involved to find exact values of definite integrals for $f(x)$. For example, suppose that the following (*time, velocity*) graph models the way someone walked toward and away from a motion detector.

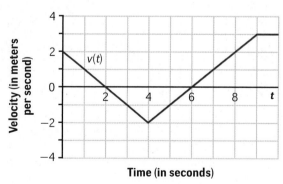

a. Evaluate these definite integrals and explain what the results tell about the walker's distance from the motion detector.

i. $\int_0^5 v(t)\, dt$ **ii.** $\int_5^{10} v(t)\, dt$ **iii.** $\int_0^{10} v(t)\, dt$

b. What is the total area between the graph of $v(t)$ and the t-axis and what information about the walker's trip is given by that number?

3 When the rate of change function is defined by an algebraic rule with a curved graph, there are many useful procedures for making good estimates of the definite integral. For example, the diagram below shows the graph of $f(x) = -x^2 + 2x + 4$ on the interval $[0, 3.5]$. Six rectangles have been drawn to make a rough estimate of $\int_0^3 f(x)\, dx$.

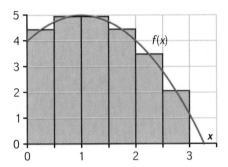

a. The total area of the six rectangles can be estimated as
$4.4(0.5) + 4.9(0.5) + 4.9(0.5) + 4.4(0.5) + 3.4(0.5) + 1.9(0.5) = 11.95$.
What do the separate terms in the sum represent and how are they determined by the function rule for $f(x)$?

b. Do you think that the approximation of 11.95 is larger than, smaller than, or exactly equal to the value of the integral $\int_0^3 f(x)\, dx$? Explain your reasoning.

c. How could you refine the general strategy to get a more accurate approximation for the definite integral? Try your idea and compare the result to the first estimate of 11.95.

4 The next diagram shows a graph of the function $h(x) = \dfrac{4}{x+1}$ on the interval $[0, 5]$.

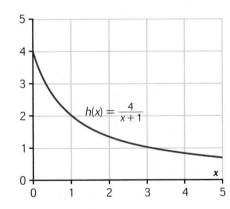

Outline a strategy for estimating $\int_0^5 h(x)\, dx$ and then use that strategy.

a. Record the calculations required by your strategy and the approximation that results.

b. Explain why you think your estimate is greater than, less than, or exactly equal to the value of the integral of $h(x)$ from 0 to 5.

c. Explain how you would proceed if it were necessary to find a better approximation.

5 The next diagram shows a graph of the function $g(x) = x^3 - 4x$.

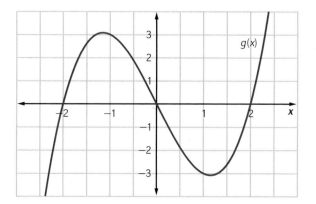

Use an approximation strategy with $\Delta x = 0.5$ to estimate values for these definite integrals.

a. $\int_{-2}^{0} g(x)\, dx$

b. $\int_{0}^{2} g(x)\, dx$

c. $\int_{-2}^{2} g(x)\, dx$

6 Most graphing calculators and computer algebra systems offer tools for evaluating definite integrals. They often use notation as shown below. Syntax varies depending on the tool.

The **nInt** command produces a numerical estimate (which may, in some cases, be an exact value) for the given definite integral over the specified interval. The CAS command ∫ provides an exact value.

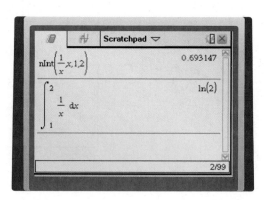

Use technology tools to check the approximations you found in Problems 3–5.

In this investigation, you discovered the connection between problems that require accumulation of change at variable rates and the calculus concept of definite integrals.

a What does $\int_m^n f(x)\,dx$ tell about the function $f(x)$?

b How can you calculate an approximation of $\int_m^n f(x)\,dx$? How could you improve the accuracy of your estimate?

c How can you calculate the exact value of $\int_m^n f(x)\,dx$?

Be prepared to explain your ideas to the class.

 CHECK YOUR UNDERSTANDING

The following graph shows velocity of a golf ball when it is tossed upwards and then bounces several times on a sidewalk. Positive velocity indicates upward motion and negative velocity indicates downward motion.

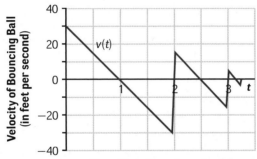

a. What does each segment of the graph tell about the motion of the ball?

b. If the ball is set in motion by a chip shot at ground level, what will be its position 1 second later? 2 seconds later? 2.5 seconds later? 3 seconds later?

c. If the velocity function of time is indicated by $v(t)$, estimate the following definite integrals and explain what each tells about the motion of the ball.

   i. $\int_0^2 v(t)\,dt$ **ii.** $\int_0^3 v(t)\,dt$ **iii.** $\int_1^3 v(t)\,dt$

APPLICATIONS

1 The following graph shows the reported speed of a cross-country skier during a 2.5-hour trip.

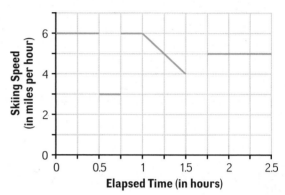

a. Describe the up and down contours of a ski trail that might lead to such a speed graph.

b. How far did the skier travel in the first hour? In the second hour?

c. How far did the skier travel in the time shown on the graph?

2 At the beginning of this lesson, you studied the acceleration test graph reproduced below.

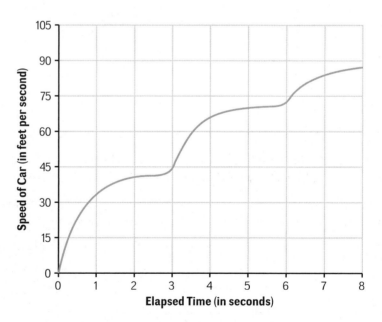

a. Estimate the total distance traveled by the car during the 8 seconds of the 0–60 mph test.

b. Explain a strategy that you could use to make a more accurate estimate from the same graph.

c. Test your strategy and compare your new estimate of total distance traveled with that in Part a.

3 A well-known church in Rockville, Maryland, has four large windowed arches that can be modeled by the function $h(x) = 30 - 0.1x^2$ with measurements in feet.

Use the window outline and grid shown to estimate the total area of glass used in each arch. Show the calculations used in making your estimate and explain how you could improve the accuracy of the estimate.

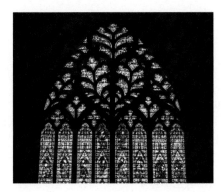
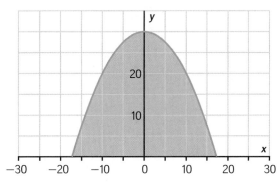

4 For commercial airlines, one important variable factor in their profit picture is the price of jet fuel for their planes. When fuel prices rise, the airlines tend to lose money every day; when fuel prices fall, they tend to make money every day.

Suppose that the following graph shows variation in profit per day for an airline during one twelve-month period.

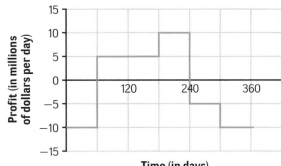

a. What pattern of change in the price of jet fuel is suggested by the profit graph?

b. What was the net profit of the airline:

   i. over the first 120 days of the year?

   ii. over the first 240 days of the year?

   iii. over the first 360 days of the year?

   Be prepared to explain how you calculated your answers to each question.

5 The next graph shows the rate at which water depth changes (in feet per hour) in the tidal water of an ocean harbor during one full period of the tide.

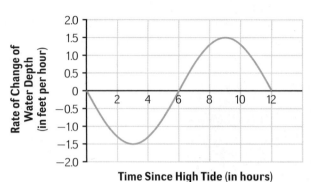

a. Explain what the pattern in the graph tells about change in water depth over each of these time intervals:

    i. $0 \leq t \leq 3$ **ii.** $3 \leq t \leq 6$ **iii.** $6 \leq t \leq 9$ **iv.** $9 \leq t \leq 12$

b. Use the graph to find what you believe is an accurate approximation for the net change in water depth in the harbor over the first six hours of the period shown.

c. Use the symmetry of the graph and your result from Part b to estimate the net change in water depth over the time interval $6 \leq t \leq 12$.

d. What is the approximate net change in water depth over the time interval $0 \leq t \leq 12$?

6 The function $h(x) = 30 - 0.1x^2$ describes the parabolic church arches in Applications Task 3.

a. Write the definite integral that will provide the area of the window in that arch.

b. Use a calculator or computer tool to evaluate the integral in Part a and compare the result to your approximation in Task 3.

7 The function $f(t) = -1.5 \sin 0.5t$ matches the tidal flow rate graph in Applications Task 5.

a. Write the definite integral that will provide the net change in water depth during the first six hours of the time shown on the graph in Task 5.

b. Use a calculator or computer tool to evaluate the integral in Part a and compare the result to your approximation in Task 5.

c. Use a calculator or computer tool to evaluate an integral giving the net change in water depth over the interval $0 \leq t \leq 12$ and compare the result to your approximation in Task 5.

8 Sketch a graph of $g(x) = x^2 - 4x + 3$ on the interval $[0, 5]$. Then evaluate $\int_0^5 g(x)\,dx$ using the following two strategies. Compare the two results and explain why they differ.

Strategy I: Using a step function approximation to $g(x)$ with $\Delta x = 0.5$

Strategy II: Using a calculator or computer integration tool

9 Sketch a graph of the function $f(x) = 1 + x^2$ for the interval $[0, 3]$.

a. Calculate these estimates of $\int_0^3 f(x)\,dx$ and explain the difference between them.

  i. $f(0)(1) + f(1)(1) + f(2)(1)$

  ii. $f(1)(1) + f(2)(1) + f(3)(1)$

b. Calculate these estimates of $\int_0^3 f(x)\,dx$ and explain the difference between them.

  i. $f(0)(0.5) + f(0.5)(0.5) + f(1)(0.5) + f(1.5)(0.5) + f(2)(0.5) + f(2.5)(0.5)$

  ii. $f(0.5)(0.5) + f(1)(0.5) + f(1.5)(0.5) + f(2)(0.5) + f(2.5)(0.5) + f(3)(0.5)$

c. Explain the difference between the estimates of $\int_0^3 f(x)\,dx$ in Parts a and b.

d. Use a calculator or computer tool to evaluate $\int_0^3 f(x)\,dx$. Give a possible explanation for the difference between that result and the estimates in Parts a and b.

e. Will $\int_0^4 f(x)\,dx$ be larger, smaller, or the same as $\int_0^3 f(x)\,dx$? How can you tell without actually evaluating either integral?

10 Shown below is a graph of the function $f(x) = 6 - x^2$.

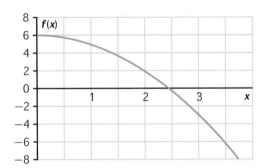

a. Will $\int_0^1 f(x)\,dx$ be larger or smaller than $\int_0^2 f(x)\,dx$? How can you decide without evaluating either integral?

b. Will $\int_0^2 f(x)\,dx$ be larger or smaller than $\int_0^4 f(x)\,dx$? How can you decide without evaluating either integral?

11 Shown below is the graph of a circle with radius 1 centered at the origin of a coordinate system.

a. What is the area of each small square in the coordinate grid?

b. What is the total area of the squares that lie completely within the circle?

c. What is the total area of the squares needed to cover the circle?

d. What is the exact area of the circle?

12 Now consider the problem of finding the area of a circle using ideas of calculus.

a. What is the equation for a circle of radius 1 and center $(0, 0)$?

b. Use the equation from Part a to find the rule for a function $s(x)$ whose graph is the upper half of the unit circle.

c. Use the function from Part b to write a definite integral that gives the area of the upper half of the unit circle.

d. Use a calculator or computer integration tool to evaluate the integral from Part c and compare the result to what your knowledge of the area formula for circles predicts. What might explain any differences?

13 The following graph shows the standard normal probability distribution $z(x)$ on the interval $[-3, 3]$. To find the probability of an outcome between $x = m$ and $x = n$, where $m < n$, you need to find $\int_m^n z(x)\, dx$.

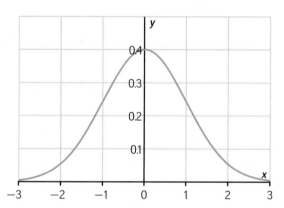

Use the graph to estimate probabilities that a standard normal random variable will fall in each of these intervals:

a. between 0 and 1

b. between 1 and 2

c. between 2 and 3

d. between −1 and 1

e. between −2 and 2

f. between −3 and 3

14 The following diagrams show two steps in finding an approximation for $\int_1^5 f(x)\,dx$.

Diagram I

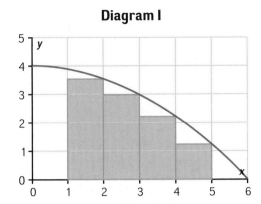

Diagram II

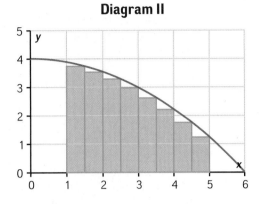

a. Which diagram will lead to the more accurate approximation? Why?

b. What next step would lead to an even more accurate approximation?

c. What would it mean to say that $\int_m^n f(x)\,dx$ is the limit of approximations that are sums of terms in the form $f(x)\Delta x$ as $\Delta x \to 0$?

15 The next diagrams show a circle with radius 1 and an ellipse with x-intercept $(a, 0)$ and y-intercept $(0, b)$. Rectangles have been drawn to show how one might approximate the areas of the circle and the ellipse.

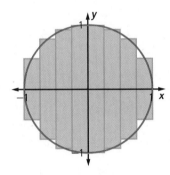
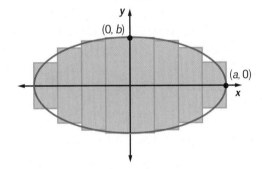

a. What transformation $(x, y) \to (\_\_, \_\_)$ would map the circle onto the ellipse?

b. How would the areas of the rectangles covering the ellipse compare to the corresponding rectangles covering the circle?

c. What formula for area of an ellipse with semi-major axis length a and semi-minor axis length b is suggested by the answer to Part b?

REFLECTIONS

16 One description of the derivative is that it solves problems about nonlinear functions using linear approximations. In what sense is that same statement true of the definite integral?

17 The process of finding the derivative of a function is often called *differentiation*; the process of finding the definite integral is called *integration*. Why do these two words seem to fit the processes that are occurring in each case?

18 Look back at Applications Task 9 (page 519). How is the integral estimation strategy used similar to, and different from, that used in Investigation 3? What, if any, are its benefits?

19 An important part of calculus is devoted to developing rules for calculating exact values of definite integrals for many different kinds of functions. Based on your work of this lesson, what do you see as the benefits and costs of developing rules for calculating integrals and the benefits and costs of using numeric approximation algorithms and computer algebra system integration tools?

EXTENSIONS

20 In 1960, the Aral Sea in the former Soviet Union was one of the largest lakes in the world, with water volume of about 1,100 km³. Fed by two rivers, the Amu and the Syr, with no outflow other than evaporation, its level was stable.

In 1954, Soviet government officials decided to construct a system of canals that would divert water from the Amu and the Syr rivers to irrigate agriculture in the Central Asian desert. As the irrigation increased, water flow to the Aral Sea was reduced and the lake began to shrink dramatically.

1973

1987

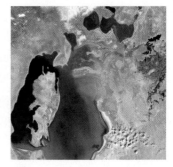
1999

Source: U.S. Geological Survey

    a. Describe, in terms of inflow and evaporation, the pattern of change for Aral Sea volume shown by the graph at the top of the following page.

U.S. Department of the Interior U.S. Geological Survey

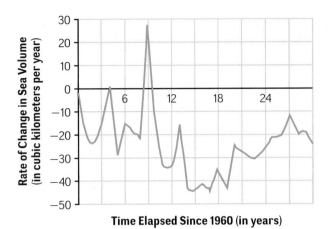

Time Elapsed Since 1960 (in years)

b. What was the approximate volume of water in the Aral Sea in 1966? In 1972? In 1978? In 1984? In 1990?

c. If the pattern of change in Aral Sea water volume continued as shown in the graph, what volume would you predict for 1996, 2006, and 2016? Compare your predictions with current data available in an atlas or Internet sources.

21 In this lesson, you estimated area between curves and the x-axis by covering the region with rectangles. Another method for approximating such areas uses *trapezoids* as shown in the next diagram where $f(x) = -x^2 + 5x + 1$.

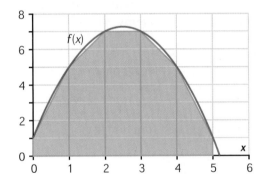

a. Estimate $\int_0^5 f(x)\, dx$ by finding the areas of the indicated trapezoids.

b. Compare the trapezoidal estimate for area to one obtained by using an approximation method that uses five rectangles with $\Delta x = 1$.

c. Compare the trapezoidal estimate for area to one obtained by use of a calculator or computer integration tool.

The close connection between area and the definite integral is useful in deducing important properties of the integration process. Apply that connection in Tasks 22–27.

22 Sketch a graph of the function $f(x) = x$ and use it to complete the following tasks.

a. Evaluate these integrals by finding areas of corresponding regions between the graph of $f(x)$ and the x-axis in each case.

i. $\int_0^1 f(x)\,dx$ ii. $\int_0^3 f(x)\,dx$

iii. $\int_0^6 f(x)\,dx$ iv. $\int_0^k f(x)\,dx,\, k > 0$

b. Next consider functions with rules in the form $g(x) = mx$ for various values of m. What formula shows how to evaluate $\int_0^k g(x)\,dx$ for any given nonzero values of m and k? Explain how you know that your answer is correct.

23 Now consider the case of more general linear functions with rules in the form $h(x) = mx + b,\, m \neq 0$.

a. Evaluate these integrals by finding areas of corresponding regions between the graph of $h(x)$ and the x-axis in each case.

i. $\int_0^1 (x + 2)\,dx$ ii. $\int_0^3 (x + 2)\,dx$

iii. $\int_0^6 (x + 2)\,dx$ iv. $\int_0^k (x + 2)\,dx$

b. What formula shows how to evaluate $\int_0^k (x + b)\,dx$ for any given values of b and k? Explain how you know that your answer is correct.

c. What formula shows how to evaluate $\int_0^k (mx + b)\,dx$ for any given values of m, b, and k? Explain how you know that your answer is correct.

24 For any number $k > 0$, $\int_0^k x^2\,dx = \dfrac{k^3}{3}$. Use the given formula, a graph of $y = x^2$, and the connection between integration and area to evaluate these integrals.

a. $\int_0^k (x^2 + 1)\,dx$ **b.** $\int_0^k (x^2 + 5)\,dx$ **c.** $\int_0^k (x^2 + c)\,dx$

25 Shown below is the graph of a function $h(x)$.

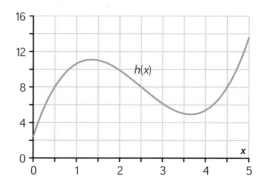

Using the fact that $\int_0^5 h(x)\,dx \approx 43$:

a. what would you expect for $\int_0^5 [h(x) + 7]\,dx$ and why?

b. what would you expect for $\int_0^5 [7h(x)]\,dx$ and why?

26 For the function graphed in Task 25, $\int_0^3 h(x)\, dx \approx 26$.

    a. What value would you expect for $\int_3^5 h(x)\, dx$?

    b. In general, what relationship would you expect between $\int_a^b h(x)\, dx$, $\int_b^c h(x)\, dx$, and $\int_a^c h(x)\, dx$ when $a < b < c$? Why does this make sense?

27 Use a computer algebra system to explore patterns in these families of definite integrals.

    a. $\int_a^b x^n\, dx$ for $n = 1, 2, 3, 4, \ldots$

    b. $\int_a^b (x^2 + x + c)\, dx$, $\int_a^b (x^3 + x^2 + x + c)\, dx$, $\int_a^b (x^4 + x^3 + x^2 + x + c)\, dx$, $\ldots$

    c. What do the results of your work on Part b suggest about the relationship between $\int_a^b f(x)\, dx$, $\int_a^b g(x)\, dx$, and $\int_a^b [f(x) + g(x)]\, dx$ for any combination of functions and intervals of integration? Explain why you believe your conjecture is correct.

REVIEW

28 For their summer vacation, the Chavez family rented a houseboat on Lake Mead. One day they started at Boulder Harbor and traveled 5 nautical miles (nm) on a path with direction 50°. They had lunch at that spot.

    a. If Boulder Harbor is located at the origin of a coordinate system, what are the coordinates of the spot where they had lunch?

    b. After having lunch, they turned and traveled 3 nm on a route with direction 255° to a nice swimming spot. How far and in what direction should they travel to go directly back to Boulder Harbor when they finish swimming?

    c. Determine the magnitude and direction for a direct trip from Boulder Harbor to the swimming spot.

29 Consider the functions $f(x) = \frac{1}{x^2} + 1$, $g(x) = 3 \cos x$, and $h(x) = -2e^x$.

   a. Sketch a graph of each function and state the domain and range of each function.

   b. How many solutions are there to each of the following equations? Justify your responses.

      i. $f(x) = h(x)$

      ii. $h(x) = g(x)$

30 Let $f(t) = \frac{1}{2}t + \frac{3}{4}$.

   a. Solve $f(t) = 11$.

   b. Evaluate $f(f(4))$.

   c. Write a symbolic rule for $f(f(t))$.

31 Find all solutions to each equation.

   a. $4 \sin \theta = -2$

   b. $\tan^2 \theta + 2 \tan \theta + 1 = 0$

   c. $\sqrt{x + 4} = x - 2$

   d. $1.3e^{x+2} = 123.5$

32 Write equations or function rules for each of the following.

   a. Circle with center $(3, 5)$ and containing the origin

   b. Parabola with vertex $(-2, 5)$ and y-intercept at $(0, 3)$

   c. The number of bacteria in an experiment if the initial quantity was 108 and after 3 hours, there were 270 bacteria. Assume that the growth is exponential.

   d. The perpendicular bisector of the segment with endpoints $A(6, 2)$ and $B(4, 8)$

33 Rewrite each expression as a single algebraic fraction in simplest form.

   a. $t^2 + t^{-2}$

   b. $\frac{a}{b} + \frac{b}{a}$

   c. $\frac{d^2}{4c} \div \frac{c^4}{d^{-1}}$

   d. $\frac{y+1}{y} - \frac{1}{y^2}$

34 Consider the rectangular prism shown below.

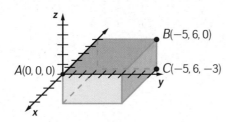

a. Sketch the prism on your paper and identify the coordinates of all remaining vertices.

b. Find AB.

c. Give the equations of two vertical planes containing points B and C.

d. What is the equation of the vertical plane containing points A and B?

35 The graph of a function $f(x)$ is shown below.

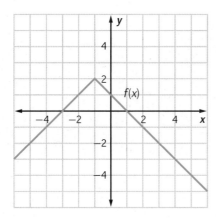

Sketch and label a graph of each of the following functions. Then identify the x- and y-intercepts of each graph.

a. $g(x) = -f(x)$

b. $h(x) = 2f(x)$

c. $j(x) = f(x - 3)$

d. $v(x) = f(x) + 3$

Looking Back

In this unit, you explored two basic concepts and related techniques that calculus offers for studying rates of change in variables. You learned how to use *derivatives* to measure the instantaneous rate of change of a function and *definite integrals* to measure accumulation of change. The tasks of this lesson give you an opportunity to review and apply those two big ideas.

1. When a skydiver jumps out of an airplane that is several thousand feet above the ground, the distance and speed of her flight are modeled better by functions in the exponential family than the quadratic and linear functions that describe motion of simpler shapes over shorter distances. For example, distance (in feet) and speed (in feet per second) might be given by:

$$d(t) = 120t - 540(1 - 0.8^t)$$
$$s(t) = 120(1 - 0.8^t)$$

   a. Use the distance function and the difference quotient $\dfrac{d(t + 0.1) - d(t)}{0.1}$ to estimate $d'(0)$, $d'(5)$, and $d'(10)$ and compare those estimates to $s(0)$, $s(5)$, and $s(10)$.

   b. Sketch graphs of $d(t)$ and $s(t)$ and explain the relationship between the patterns of change in the two graphs.

c. The following diagram shows one way to estimate the area between the
   t-axis and the graph of the speed function $s(t)$. Use that strategy to estimate
   the following definite integrals and explain what each result tells about
   motion of the skydiver.

   i. $\int_0^{10} s(t)\, dt$

   ii. $\int_1^5 s(t)\, dt$

   iii. $\int_5^9 s(t)\, dt$

   iv. $\int_9^{10} s(t)\, dt$

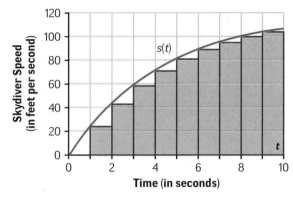

d. Use the distance function $d(t)$ to check your answers in Part c. Explain
   any differences between results of the two strategies and how you could
   modify the estimation strategy of Part c to produce more accurate results.

e. Show the instructions needed to calculate the definite integrals in Part c
   with use of a CAS integration tool, and compare the results to the estimates
   in Part c and the calculations using the distance function in Part d.

2 For objects in motion near the
surface of the Moon, the
relationship between time
and height is quite accurately
modeled by a quadratic
function, because there is
negligible atmosphere to
provide air resistance.

   For example, when astronaut
Alan Shepard hit a golf ball
during a Moon walk in 1971,
the height of the ball (in feet)
could have been modeled by
the rule $h(t) = -2.7t^2 + 30t$.

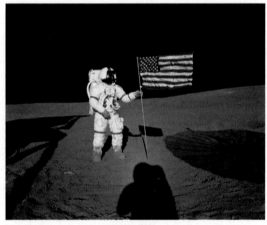

Alan Shepard on the Moon in 1971

a. Produce a graph of $h(t)$ and give coordinates of the points representing
   the ball's maximum height and return to the surface.

b. Find the rule for $h'(t)$ and produce a graph of that function from the time
   the ball was hit until the time it returned to the Moon's surface. Explain
   how the information provided by this derivative graph is related to that
   in the height graph of Part a.

3 Compare the speed graph for the skydiver (Task 1) to the velocity graph for the golf ball hit on the Moon (Task 2). Explain how the differences in the graphs tell the different stories about motion of the skydiver and the golf ball.

4 When some interesting event occurs in your community, news about it can spread in several different ways. If the event is covered by radio, television, newspaper, or Internet reporters, the story will soon appear in those media. If the event is not reported in the news media, word might still be spread by informal contacts among people.

   For example, suppose that on some winter weekend, there is a breakdown in your school's heating system, making it necessary to close the school on Monday. Word of this school closing could spread in several ways.

   a. Which of the following graphs seems most likely to model spread of the school closing news if it is reported on local media? Which if the word is spread only from person to person in the school community?

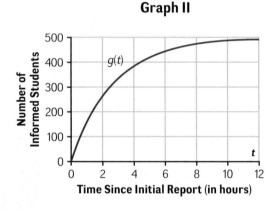

Graph I

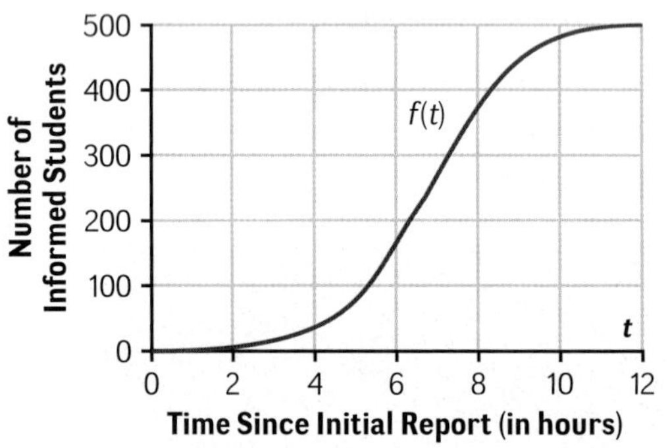

Graph II

   b. The rules that match the graphs in Part a are:

   Graph I: $f(t) = \dfrac{500}{1 + 499(0.4^t)}$

   Graph II: $g(t) = 500 - 499(0.7^t)$

   Estimate values of $f'(t)$ and $g'(t)$ for $t = 0, 4, 6, 8,$ and 12. Then use those values and the graphs of $f(t)$ and $g(t)$ to sketch graphs of the two derivative functions.

   c. Explain how the two derivative graphs show different patterns in the rate at which news of the school closing spreads in the community.

5 Suppose that a skateboarder rides back and forth on the half-pipe ramp shown below.

a. If the side view of the ramp is a semicircle with diameter 30 feet, what is the length of a trip from one side to the other?

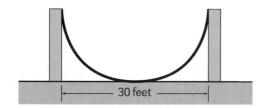

b. Sketch a graph that you believe shows the probable relationship between time and distance traveled for two complete round trips from left to right and back.

c. Sketch a graph that you believe shows the probable relationship between time and skateboarder velocity over the two round trips. Explain the relationship of this graph and the graph in Part b. Assume motion away from the start point is positive velocity and motion back toward the start point is negative velocity.

d. Identify points on the velocity graph corresponding to maximum distance from the starting point.

e. Identify sections of the graph that represent motion away from the start and sections that represent motion back toward the start.

6 Find rules and sketch graphs of functions $f(x)$ with the following properties.

a. The derivative of $f(x)$ is $2x + 4$.

b. $f(x)$ is linear and its definite integral from 1 to 5 is 0.

c. $f(x)$ is linear and its definite integral from 1 to 4 is 3.

d. $f(x)$ is a trigonometric function and its definite integral from 0 to 2π is 0.

In this unit, you learned how to use the derivative $f'(x)$ and the definite integral $\int_a^b f(x)\,dx$ to analyze important properties of functions and solve problems about the situations those functions model.

a How can you estimate values of $f'(x)$ for any specific value of x?

b How can you calculate exact values of $f'(x)$ for any specific value of x?

c What does the value of $f'(a)$ tell about:

    i. the rate of change in $f(x)$ near $x = a$?

    ii. the shape of the graph near $x = a$?

d What does $f'(x)$ tell about a situation in which $f(x)$ models:

    i. position or distance traveled by a moving object?

    ii. size of a population that changes over time?

    iii. temperature of an object as a function of time?

e What does $\int_a^b f(x)\,dx$ tell about the graph of $f(x)$?

f What does $\int_a^b f(x)\,dx$ tell about a situation in which $f(x)$ models:

    i. velocity of a moving object as a function of time?

    ii. rate of flow in a system as a function of time?

g How can you use the rule and graph for $f(x)$ to estimate $\int_a^b f(x)\,dx$?

Be prepared to explain your ideas to the class.

 CHECK YOUR UNDERSTANDING

Write, in outline form, a summary of the important mathematical concepts and methods developed in this unit. Organize your summary so that it can be used as a quick reference in future units.

Counting Methods and Induction

Counting methods are part of an area of mathematics called *combinatorics*. They are used to answer the question, "How many?" This question arises in everyday life and in more abstract settings. For example, how many boys with GPA greater than 3.0 are in your graduating class? How many Internet addresses (IP numbers) are possible? How many cell phone numbers can have the prefix 919? Or even, how many terms are in the expansion of $(a + b)^6$ and what is the coefficient of a^2b^4 in that expansion?

In this unit, you will develop skill in systematic counting. You will learn concepts and methods that will help you solve counting problems in many contexts. You will also develop skill in combinatorial reasoning and proof by mathematical induction.

These ideas will be developed in the following four lessons.

Systematic Counting

It seems like everywhere you look—in school, at the mall, at movie theaters and concerts, on college campuses, and at work—denim jeans have become a wardrobe staple. According to a recent representative survey of 6,000 consumers aged 13–70, 60% female and 40% male, conducted by Cotton Incorporated's Lifestyle Monitor™, 96% of consumers own denim jeans—seven pairs on average! Consumers in the survey reported wearing denim jeans an average of four days a week.

(**Source:** www.cottoninc.com)

Just Jeans, a store in local malls, specializes in denim jeans and carries several of the more popular brands for both males and females. Maintaining proper inventories requires understanding the market and mathematics.

Consider, for example, that one popular brand of jeans for males is available in six styles: Slim, Bootcut, Straight Leg, Relaxed, Original, and Loose; nine finishes, including stone washed, raw indigo, Texas crude, and vintage; 14 waist sizes, 28"–46" (37", 39", 41", 43", and 45" excepted); and nine inseam lengths, 28"–36".

The same brand also manufactures women's denim jeans. Women's jeans are available in seven styles: Legging Jeans, Skinny, Real Straight, Curvy, Bootcut, Natural, and Long and Lean; two finishes, basic indigo and black-dyed; 11 sizes, 0 (XS), 2 (XS), 4 (S), 6 (S), 8 (M), 10 (M), 12 (L), 14 (L), 16 (XL), 18 (XL), and 20 (XXL); and 12 inseam lengths, 29"–37", including some "half" lengths such as $32\frac{1}{2}$".

You can see that there are many different types of jeans possible! Knowing just how many different types is important for the store manager, who must order and stock all the jeans. In this unit, you will learn how to count in situations like this and many others.

THINK ABOUT THIS SITUATION

Think about situations in which counting is important, and how you might carry out that counting.

a Assuming your school does not require school uniforms, how many students in your mathematics class are wearing jeans today? How many are boys? How many are girls?

b Considering the styles, finishes, waist sizes, and inseam lengths for the male jeans listed on the previous page, how many different male jean options do you think are available from the manufacturer? Explain how you determined your answer.

c Considering the styles, finishes, sizes, and inseam lengths for the brand's female jeans, how many different female jean options do you think are available from the manufacturer? Explain how you determined your answer.

d Think of other situations in which it is important to answer the question, "How many?" For example, what does a typical car license plate look like in the state where you live? Why is it important to know how many different license plates are possible? Give two examples of your own.

As you have seen, answering the question, "How many?" can, at times, be more difficult than you might think. The branch of mathematics that deals with systematic methods of counting is called **combinatorics**. In this unit, you will learn some of the basic concepts, strategies, applications, and reasoning methods important in combinatorics.

INVESTIGATION 1

Methods of Counting

In the Think About This Situation, you brainstormed about denim jeans selection and store inventories involving careful counting, its importance, and some possible counting methods. In this investigation, you will begin a careful analysis of counting concepts, methods, and their applications.

As you work on the problems in this investigation, look for answers to this question:

What are some useful methods for systematic counting?

1 Some Websites, as you may know, require users to have passwords so that access can be controlled and security can be maintained. The number and type of characters allowed in a password can vary.

   a. Suppose a password consists of only two characters: one letter from A to D, followed by one digit from 0 to 2.

      i. How many different passwords are possible? Explain the method you used to get your answer. Describe at least one other method that could be used.

      ii. Do you think this password format is practical? Why or why not?

   b. For one Website, passwords consist of five characters: three digits from 0 to 9, followed by two letters from A to Z.

      i. How many different passwords are possible? Compare your answer with that of your classmates. Resolve any differences.

      ii. Suppose an unauthorized user tries to gain access to the system simply by trying different passwords. If the person can try one password every second, and the system does not cut her off, what is the longest it could take to get into the system? How many minutes? Hours? Days? (For this question, ignore the need to also know the correct username.)

2 Personal Identification Numbers (PINs) used in Automated Teller Machines (ATMs) are similar to computer passwords. You insert your card and then enter the correct PIN in order to get access to your bank account and withdraw or deposit money. A PIN often consists of four digits, 0 to 9. Some ATMs will capture your card if you enter the wrong PIN too many times. Use a counting argument to help explain why this is done.

There are many different strategies you might use to solve counting problems. Make note of different strategies as you solve Problems 3 and 4.

3 Outdoor Adventure Clothing and Gear stocks windproof ski jackets in a different style for men and women. For each of these styles, there are 4 colors (purple, teal, yellow, blue), in each of 3 sizes (small, medium, large).

   a. How many different jacket options do you think are possible?

   b. Students in one class at Pioneer High School proposed the following methods for answering this question. Carefully study each method. Which methods work? Compare your answer with those of your classmates. Resolve any differences.

Ryan McVay/Getty Images

Julia's Method
(tree diagram)

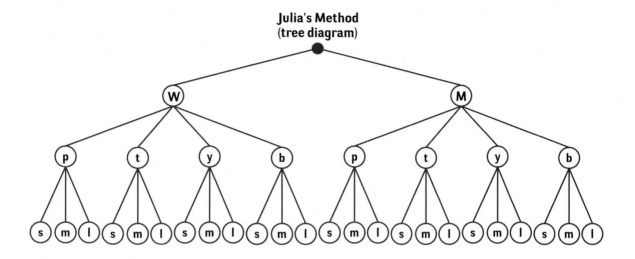

Alicia's Method

M-b-s	W-b-s
M-b-m	W-b-m
M-b-l	W-b-l
M-y-s	W-y-s
M-y-m	W-y-m
M-y-l	W-y-l
M-t-s	W-t-s
M-t-m	W-t-m
M-t-l	W-t-l
M-p-s	W-p-s
M-p-m	W-p-m
M-p-l	W-p-l

Kaya's Method

W-b-l M-p-s W-y-s W-t-s
W-t-m M-y-s M-y-m
M-t-l W-p-s M-p-l W-t-l
W-b-s W-p-l M-y-l
M-b-s W-p-m M-t-m
W-b-m W-y-m M-t-s
M-b-l

Alonzo's Method

The jackets are for men or women.
There are 12 options for Men and
12 options for Women.
So, there are 12 + 12 = 24 different
jacket options.

Monica's Method

2	×	4	×	3	= 24 different
2 choices: men or women		4 choices for color		3 choices for size	jacket options

Samuel's Method
There are 2 genders
plus 4 colors
plus 3 sizes.
So, there are
2 + 4 + 3 = 9
different jacket
options possible.

Sandra's Method

There are 4 colors for
both men and women;
that makes 4 × 4 = 16.
Then there are 3 sizes;
that makes 16 × 3 = 48.
So, there are 48
different jacket
options possible.

c. For those methods that work, describe connections among them. For example, how does Monica's multiplication method relate to Julia's tree diagram? Find connections among the other correct methods.

4 Use methods similar to those in Problem 3 that were correct or invent your own method to solve the following counting problems.

a. Suppose the men's and women's jackets from Problem 3 are available in four sizes—small, medium, large, and extra large and the same four colors as before. How could you modify the correct methods in Problem 3 to determine the number of different jacket options that are now possible? What is the total count now?

b. Suppose that the men's jackets are available in five sizes including an extra, extra large (XXL), while the women's jackets are available in four sizes (and the same four colors as before). If each different jacket will be displayed on a separate hanging rack in the store, how many racks are needed? Explain your method.

c. Suppose a password consists of only two characters—one letter from A to D, followed by the digit 1 or 2. Count the number of passwords in this situation using each correct method from Problem 3.

d. Suppose a password consists of six characters—four digits from 0 to 9, followed by two letters from A to Z. Which of the methods in Problem 3 would be most effective for determining how many different passwords are possible? Why? Use that method to determine the answer.

> **Authentication Required**
>
> Enter username and password for "Your Guest Username" at https://www.wmich.edu
>
> User Name:
>
> |
>
> Password:
>
> ☐ Use Password Manager to remember this password.
>
> (Cancel) (OK)

SUMMARIZE THE MATHEMATICS

In this investigation, you used systematic counting to answer the question, "How many?" in several different contexts.

a List four correct counting methods you have seen or used in this investigation.

b For each method:

- describe how it works.
- give an example of how to use it.
- discuss some advantages and disadvantages of the method.

c Look back at the counting situations you have considered. Find an example where you might use more than one counting method to solve a problem.

Be prepared to explain your methods, examples, and thinking to the class.

The telephone number for a local landline has seven digits, like 472-5555. The first three of these digits are called the *prefix*. As the population or phone use grows in a community, new prefixes may be needed. For example, on October 1, 2011, a fourth prefix, 974, was added for the city of Iqaluit, Nunavut (Canada). Read the notice below from the local telephone company announcing the addition of a new prefix.

Northwestel Introduces New Telephone Number Prefix in Iqaluit

Northwestel customers in Iqaluit will soon be dialing 974 when making local calls. The telephone company will introduce a new phone number prefix in Iqaluit on October 1st.

There are already three prefixes, also known as office codes or NXX numbers, being used in Iqaluit. They are 979, 975, and 222.

Northwestel provides a new prefix in a community when the available numbers begin to be depleted, to ensure that the community does not run out of phone numbers.

Source: nwtel.ca

a. Assuming no restrictions on the digits, how many different phone numbers can be created with the 974 prefix?

b. How many different phone numbers are available in Iqaluit using all four prefixes?

INVESTIGATION 2

Counting Principles

Some of the counting methods in Investigation 1 are based on fundamental principles of counting. A common element of these principles is that they enable you to count without counting!

As you work on the problems in this investigation, look for answers to the following questions:

What are some fundamental principles of counting?

How and when are these principles useful in solving counting problems?

1 **Multiplication Principle of Counting** Monica's reasoning in Problem 3 of Investigation 1 illustrates the *Multiplication Principle of Counting*.

Monica's Method

$$2 \quad \times \quad 4 \quad \times \quad 3 \quad = 24 \text{ different}$$

2 choices: men or women	4 choices for color	3 choices for size	jacket options

In general, if you want to count all the outcomes from a sequence of tasks, count how many outcomes there are from each task and multiply those numbers together.

There are three requirements that must be met in order to use the Multiplication Principle of Counting. It helps to think of a *counting tree* diagram as shown below. If you can picture the counting situation as a counting tree, then the Multiplication Principle probably applies.

Suppose you want to count all the possible outcomes that result from a sequence of tasks.

Requirements to Be Met to Use the Multiplication Principle of Counting

Use the Multiplication Principle if you want to count all the *combined outcomes* from a *sequence* of tasks, where the numbers of outcomes from each task are *independent* of each other.

I *Sequence*—You want to count all the possible outcomes that result from a sequence of tasks. (In a counting tree, the sequence is seen as the different levels of branches in the tree.)

II *Distinct Combined Outcomes*— You want to count all the combined outcomes at the end of the sequence of tasks (at the ends of the final branches of the counting tree). All these combined outcomes must be different and distinct from each other.

III *Independent Number of Outcomes*—The number of outcomes from each task is the same no matter which outcomes happened in previous tasks. (In a counting tree, there are the same number of branches from each node at a given level of the tree.) That is, suppose the first task in a sequence has n_1 outcomes; then for each of these, the second task has n_2 outcomes; and for each of these, the third task has n_3 outcomes; and so on. (In terms of a counting tree diagram, the first level of the tree has n_1 branches, the second level has n_2 branches from each node, the third level has n_3 branches from each node, and so on.)

If these three requirements are met, then the number of possible combined outcomes from the whole sequence of tasks is $n_1 \times n_2 \times n_3 \times \cdots$.

a. Identify two examples from your solutions of previous counting problems where you used the Multiplication Principle of Counting.

b. For each of your examples in Part a, describe how the counting problem satisfies the three requirements above.

2 Examine the structure of each of the following problems in terms of requirements for use of the Multiplication Principle of Counting. Then solve each problem.

a. Consider a password that consists of just two characters. One character is a letter from A to C and the other character is a digit from 0 to 4. The two characters can be in either order: letter-digit or digit-letter. How many such passwords are there?

b. A Website password consists of any 7 letters followed by any 2 digits.

   i. How many passwords are possible?

   ii. Suppose that no letters or digits can be repeated. In this case, how many passwords are possible?

c. Suppose a password consists of 4 letters and 1 digit in any order. Letters can be repeated in a given password. How many passwords are possible?

3 More and more people are using more and more phones. This creates a demand for new telephone numbers. In most of North America, a phone number looks like 641-555-0136. There is a three-digit area code, then a three-digit local prefix, then the final four digits. One way to create more phone numbers is to create more area codes. There are rules for how to create an area code. The rules changed in 1995.

a. Examine the area code rules below from before 1995. How many area codes were possible at that time?

Area Code Rules Before 1995

- The first digit cannot be 0 or 1, since these digits indicate special numbers for the phone company, like calling the operator.

- The second digit must be 0 or 1.

- The third digit can be any digit 0–9.

b. The old area codes were running out fast. So, now the restriction on the second digit has been lifted. The rules today are the same except that the second digit can be any digit 0 to 9. How many different area codes are possible today?

4 Addition Principle of Counting Some counting situations require more than the Multiplication Principle of Counting. Addition also can play an important role. In fact, another counting principle is the **Addition Principle of Counting**—The total number of outcomes among two tasks is the sum of the number of outcomes from each task (minus the number of outcomes that are common to both tasks, if there are any).

a. Identify two examples from your solutions of previous counting problems where you used the Addition Principle of Counting. Explain how the principle was used.

b. Look back at the student counting methods from Problem 3 in Investigation 1. Which student(s) used the Addition Principle of Counting? Explain.

5 Counting problems are sometimes based on information presented in a table. For example, the number of men and women in the U.S. Senate in late 2013 who are Democrats or Republicans is shown in the table below.

	Men	Women	Total
Democrat	37	16	53
Republican	41	4	45
Total	78	20	98

a. Suppose a brochure is to be sent to every U.S. Senator who is a Republican or a woman. How many brochures are to be sent so that every Republican and every woman gets exactly one brochure?

b. Study George's answer to Part a. Do you agree with George's solution? Why or why not?

> I need to count everybody who is a Republican or a woman. There are 45 Republicans and 20 women. So, by the Addition Principle of Counting, the answer is 45 + 20 = 65.

SUMMARIZE THE MATHEMATICS

In this investigation, you studied two fundamental principles of counting—the *Multiplication Principle of Counting* and the *Addition Principle of Counting*.

a State each principle in your own words.

b These two principles are meant to express fundamental common sense ideas about counting. For each principle, explain why it makes sense.

Be prepared to explain your statements and reasoning to the entire class.

Think about how many different automobile license plates can be made under certain constraints. For example, in 2012 in the state of New York, an Empire license plate had seven characters. The first three characters were letters and the last four were numbers.

a. How many different Empire license plates were possible in New York in 2012?

b. Suppose that no letters or numbers can be repeated. How many Empire license plates are possible?

c. Suppose that the three letters are used to represent the county, with MON used for all Empire license plates issued to residents of Monroe County. Do you think this is a good plan? Explain.

d. License plate configurations do not always include all possible letters and numbers. Suppose New York started their lettering with ACA-1000 and counted up, leaving out letters "I," "O," and "Q" to avoid confusion with 1 and 0. How many plates are possible?

e. In some states, for a special fee, you can request a personalized plate. Suppose that in Michigan, you can order a personalized plate with five characters drawn from all possible letters or the digits 0 to 9 with repetitions allowed. How many personalized plates could be issued in Michigan?

f. How many personalized license plates could be issued in your state?

APPLICATIONS

1 Every device on the Internet is assigned a unique number, known as an *IP address*, like 188.165.140.31. If you check the network settings on your computer, you will probably find a number like this. The most commonly used version of IP addresses as of 2012 was IP version 4 (IPv4). Below is a diagram showing an IPv4 address.

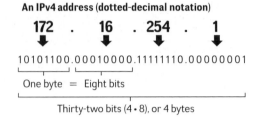

An IPv4 address (dotted-decimal notation)

172 . 16 . 254 . 1

10101100.00010000.11111110.00000001

One byte = Eight bits

Thirty-two bits (4 · 8), or 4 bytes

a. An IPv4 number consists of four decimal numbers separated by dots (which is called "dotted-decimal notation"), like 172.16.254.1 in the figure above. The actual number as stored in a computer is a 32-bit binary number. A "bit" (binary digit) is a 0 or a 1. Thus, an IPv4 number is a string of 32 bits. How many IPv4 numbers are possible, assuming no restrictions on the bits?

b. As with telephone numbers, there are some restrictions on IPv4 numbers. The restrictions get rather technical, but the most common restrictions put limitations on the use of the decimal numbers 0 and 255. The decimal number 0 is represented in binary as 00000000; the decimal number 255 is represented as 11111111. Suppose that the only restrictions for an IPv4 number are that the decimal numbers 0 and 255 are not allowed for the 1st or 4th bytes. In this situation, how many IPv4 numbers are possible?

c. Due to the rapidly growing numbers of Internet users and Internet devices, there are not enough IPv4 numbers. Another type of IP address is IPv6. An IPv6 number has 128 bits. How many times bigger is the number of possible IPv6 numbers than the number of possible IPv4 numbers (assuming no restrictions on bits)?

2 Web site passwords sometimes have very specific requirements. For example, a password for some non-classified U.S. military Web sites in 2010 had requirements similar to the following. The password must contain exactly 10 characters in the following order: 2 upper-case letters, 2 lower-case letters, 2 digits, 2 special characters, and 2 final characters that can be of any of the four previous types. A "special character" must be one of the "shift" characters from the row that includes the numbers (and characters such as the hyphen) on a standard computer keyboard. How many different passwords are possible with these requirements?

3 Suppose you toss a fair coin three times, and make a note of *heads* or *tails* on each toss.

a. Construct a tree diagram showing all possible results.

b. List the ways to get at least two heads.

4 AT&T was the only telephone company in the United States until 1984. This company devised the original telephone numbering policies in the 1940s. The original policies are listed below. Under these restrictions, how many phone numbers were possible?

Original Telephone Numbering Policies

- A phone number consists of ten digits: three digits for the area code, three digits for the local prefix, and four digits for the local number.

- 0 cannot be used as the first digit of an area code or a local prefix, since dialing 0 is reserved for reaching the operator.

- Phone numbers beginning with 1 are reserved for internal use within the telephone system, so 1 cannot be used as the first digit of an area code or a local prefix.

- Early phone numbers included letters, for example PYramid4-1225 instead of 794-1225. In order to be able to dial letters, most numbers on a telephone have letters associated with them. However, to avoid confusing the numbers 0 and 1 with the letters O and I, the numbers 0 and 1 on a telephone do not have associated letters. So, 0 and 1 cannot be the second digit of a local prefix.

- The telephone system used the second digit to distinguish an area code from a local prefix. Since the second digit of a local prefix can be anything *except* 0 and 1, the second digit of an area code *must* be 0 or 1.

- The third digit of an area code can be any number. The third digit of a local prefix can be any number except 0 or 1.

- The four digits of the local number can be anything.

---

5 The "call sign" of a radio station is the set of letters by which it is identified, like KOA or WGBH. The call signs of most regular broadcast radio stations in the United States have 3 or 4 letters.

  a. Generally, the call signs of radio stations west of the Mississippi river have letters starting with K. How many different 3- and 4-letter call signs like this are possible? In your answer, are you assuming that letters can be repeated or not? Whatever you decide, give a brief argument supporting your decision.

  b. If all radio call signs begin with K or W, how many different call signs are possible?

6 Dominos are rectangular tiles used to play a game. Each tile is divided into two squares with a number of dots in each square, as in the figure below.

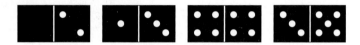

  a. The standard set of dominos has from 0 to 6 dots in each square. How many different standard dominos are possible?

  b. A deluxe set of dominos has from 0 to 9 dots in each square. How many different deluxe dominos are possible?

  c. How many dominos are possible in a set that has 0 to n dots in each square?

7 The original plan for assigning telephone numbers that you investigated in Applications Task 4 was implemented in 1947. At that time, the supply of numbers was expected to last for 300 years. However, by the 1970s the numbers were already starting to run out. So, the numbering plan had to be modified. In this task, you will count the number of different phone numbers that were available in 2012.

  a. For three-digit area codes, the first digit cannot be a 0 or a 1. Assuming no additional restrictions, how many three-digit area codes are possible under this plan?

  b. Certain area codes are classified as "Easily Recognizable Codes" (ERCs). ERCs designate special services, like 888 for toll-free calls. The requirement for an ERC is that the second and third digit of the area code must be the same. The first digit again cannot be a 0 or a 1. How many ERCs are there?

  c. Consider the seven digits after the area code. As with the area code, the first digit of the three-digit local prefix cannot be a 0 or a 1. The remaining six digits for the local number have no restrictions. How many of these seven-digit phone numbers are possible?

  d. Assuming only the 0 and 1 restrictions in Parts a and c, how many ten-digit phone numbers are possible?

8 The password format for a particular Web site has five letters and three digits. How many different passwords are possible in each of the following situations?

  a. The letters follow the digits, and the letters and digits can be repeated.

  b. The letters follow the digits, and the letters and digits cannot be repeated.

  c. The three digits can be anywhere in the password, and the letters and digits can be repeated.

9 The table below shows information about age and income for employees in a manufacturing business.

	$30,000 or Less	More Than $30,000	Total
25 or Younger	19	8	27
Older Than 25	34	31	65
Total	53	39	92

  a. How many employees make $30,000 or less?

  b. How many employees are older than 25 or make more than $30,000?

  c. How many employees are older than 25 and make more than $30,000?

  d. Which of Parts a, b, or c uses the Addition Principle of Counting?

CONNECTIONS

10 Recall that a matrix is a rectangular array of numbers. The 3×2 matrix below has 3 rows and 2 columns.

$$\begin{bmatrix} 0 & 0 \\ 1 & 1 \\ 0 & 1 \end{bmatrix}$$

  a. How many 3×2 matrices that have only 0 or 1 as entries are possible?

  b. How many $n \times m$ matrices that have entries of only 0 or 1 are possible?

11 In Investigation 2, Problem 5, George incorrectly counted the number of brochures needed so that every U.S. Senator who is a Republican or a woman was sent exactly one brochure. A Venn diagram representing this counting task is shown at the right.

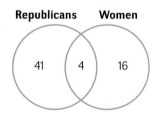

  a. Analyze the diagram by comparing it to the table on page 542. How does the Venn diagram help avoid George's mistake?

  b. Draw a Venn diagram to help count the number of brochures needed to send exactly one to each U.S. Senator who is a Democrat or a man. How many brochures are needed?

12 At many private and some public schools, students are required to wear uniforms. Suppose the boys must wear navy or khaki pants; a short-sleeve or long-sleeve white shirt; and a sweater, vest, or blazer.

a. Explain how the Multiplication Principle of Counting can be used to reason that in any group of at least 13 boys, two will be wearing the same type of uniform.

b. How big would the group need to be to guarantee that three boys would be wearing the same type of uniform?

REFLECTIONS

13 Examine the license plates from Michigan and Minnesota shown below.

Michigan and Minnesota license plates in 2007 and 2006, respectively

a. What appear to be the license identification schemes for the two states?

b. Why do you think the two states have different schemes?

c. How is the license plate identification in your state similar to, and different from, those for Michigan and Minnesota?

14 Recall or listen to the song "The Twelve Days of Christmas." How many gifts have accumulated after the 12th day?

15 Find a counting problem in your daily life or in a newspaper whose solution involves the Multiplication Principle of Counting or the Addition Principle of Counting. Describe the problem and explain how it can be solved.

16 Refer to the background information for the Think About This Situation (page 534). Create and answer a counting question regarding Just Jeans' inventory of jeans from the particular brand mentioned. For example, make assumptions about what styles and sizes of male and female jeans should be stocked and their quantity. The demands for some sizes or styles may be such that they can only be special ordered.

17 Two useful mathematical practices are:

- Make sense of problems and persevere in solving them.

- Reason abstractly and quantitatively.

For each of these practices, give one example from this lesson where you used that practice. Explain how you used the practice.

EXTENSIONS

18 In Applications Task 1, you were introduced to binary numbers. The decimal number 2 is written as 10 in binary notation. The decimal number 3 is written as 11 in binary notation. Explain the conversion of the decimal number 255 to the binary number 11111111.

19 In the book *The Man Who Counted: A Collection of Mathematical Adventures* by Malba D. Tahan, translated by Leslie Clark and Alastair Reid, a story is told of Beremiz Samir, a man with amazing mathematical skills. Read at least the first chapter of this book. Then describe and explain at least one counting feat performed by Beremiz.

20 In Applications Task 6, you counted the number of standard dominos. To play the game of dominos, you take turns trying to place dominos end-to-end by matching the number of dots. For example, for the three dominos pictured below, the 3–5 domino can be placed next to the 1–3 domino, but the 0–2 domino cannot be placed next to either of the other dominos.

a. Is it possible to make a chain of all the dominos in a standard 6-dot set? Justify your answer.

b. In general, is it possible to make a chain of all the dominos in an n-dot set? A vertex-edge graph (see *Core-Plus Mathematics* Course 1, pages 243–246 and page 256) can be used to solve this problem, as outlined below. Let the vertices be the number of dots and let an edge between two vertices, or from a vertex to itself, represent a domino with dots corresponding to the vertices. Then consider paths through the graph model. Construct a graph model like this for domino sets using a few different values of n. Then figure out, and explain, how to use the graph model to determine whether it is possible to make a chain using all the dominos in the set.

21 Think about all possible non-negative integers with their standard ordering; that is, 0, 1, 2, 3, 4, 5, … . What proportion of all non-negative integers contains the digit 3?

22 Fifty-five seniors at Hackett High School were surveyed about their food preferences, with regard to fish, chicken, and beef. Here are the results.

23 like fish
17 like chicken
17 like beef
6 like beef and chicken
8 like beef and fish
10 like chicken and fish
2 like all three

a. Represent this situation with a Venn diagram. Include the appropriate numbers in all regions of the diagram.

b. How many students like chicken and fish, but not beef?

c. How many students like beef and fish, but not chicken?

d. How many students like only beef, and not chicken or fish?

e. How many students do not like any of the three?

REVIEW

23 When using a calculator to multiply large or small numbers, the result may be displayed in scientific notation. Write a standard decimal number and a number written in scientific notation that is equivalent to each expression below.

a. 250^5

b. $(5 \times 10^4)(12 \times 10^6)$

c. 0.02^5

d. $\dfrac{1}{(100)(25^2)}$

24 Without using technology, match each of the function rules below to one of the following graphs. Then describe the reasoning you used to match the function rules and the graphs.

a. $f(x) = \dfrac{x^2 - 9}{x^2 - x - 6}$

b. $g(x) = \dfrac{2}{x^2 - x - 6}$

c. $h(x) = \dfrac{x^2 - x - 6}{2}$

d. $k(x) = \dfrac{x^2 + 2x - 8}{x^2 - x - 6}$

I

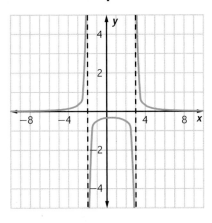

II

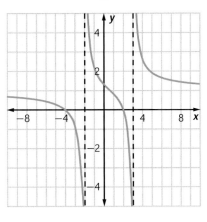

III

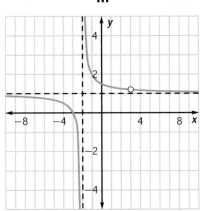

IV

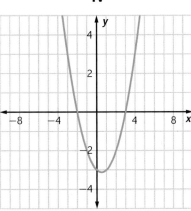

25 Darius has a savings account that earns 2.35% interest compounded continuously. His current account balance is $5,278. Assume that Darius does not withdraw or deposit any money and that he leaves all earned interest in the account.

a. How much interest will Darius earn over the next year?

b. What will his account balance be five years from now?

c. How long will it take for his account balance to reach $10,000?

26 Without the use of technology, rewrite each expression in simplest form.

a. $\dfrac{4(4-10)}{-8}$

b. $\dfrac{(7)(6)(5)(4)}{(3)(2)}$

c. $\dfrac{(10)(9)(8)(7)}{(6)(5)(4)(3)(2)}$

d. $\dfrac{x^{70}}{x^{10}}$

e. $\dfrac{ab^2+b^2}{a+1}$

f. $\dfrac{(n-4)(n-3)(n-2)(n-1)}{(n-2)(n-1)}$

27 Without using the regression features of your calculator or data analysis software, write a function rule that matches each description.

   a. A linear function whose graph contains the points $(1, 7)$ and $(3, -5)$

   b. A quadratic function with vertex $(2, -1)$ and y-intercept $(0, 7)$

   c. A cubic polynomial function with zeroes of $-3, 3,$ and 5

28 Consider a right cone resting on its base. The base of the cone has a radius of 6 in. and the height of the cone is 15 in.

   a. Draw and label a careful sketch of the cone.

   b. Imagine a plane parallel to the base intersecting the cone 5 in. above the base. Give the shape and dimensions of the cross section that is formed by the intersection of the cone and this plane.

   c. Imagine a plane perpendicular to the base and containing the vertex of the cone. Give the shape and dimensions of the cross section that is formed by the intersection of the cone and this plane.

29 Consider the graph of a cubic function $f(x)$ shown below.

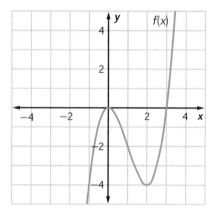

   a. Explain why $f(x)$ has no local maximum or local minimum other than what is shown in the graph.

   b. Order these function values from smallest to largest. Explain your reasoning.

   $$f(-2) \qquad f(1) \qquad f'(-0.5) \qquad f'(2) \qquad f'(4)$$

   c. For what values of x is $f'(x)$ negative?

   d. For what values of x is $f'(x)$ increasing?

   e. The function rule that matches the graph is $f(x) = x^3 - 3x^2$. Use the function rule to estimate $f'(3)$.

30 Find all complex (or real) number solutions to each equation. Give exact values whenever possible.

  a. $x^3 - 2x^2 - 7x - 4 = 0$

  b. $\sqrt{x + 5} = 2x$

  c. $4 \cos x = 2\sqrt{3}$

  d. $\log (2x - 18) = 2$

  e. $4(10^x) - 127 = 250$

31 Suppose that you roll a regular tetrahedron that has the letters F, N, T, and M on its faces and then spin a spinner that is divided into two equal parts with the letter O on one side and the letter Y on the other side.

  a. Make a sample space of all possible two-letter outcomes where the first letter is from the tetrahedron and the second letter is from the spinner. What is the probability of each outcome?

  b. What is the probability that your outcome spells a word?

  c. What is the probability that both of the letters come before S in the alphabet?

  d. What is the probability that your outcome spells a word if you spin an O?

  e. Are the two events of rolling the die and spinning the spinner independent events? Explain your reasoning.

Order and Repetition

Two important issues that often arise in counting problems are *order* and *repetition*. You have already seen these issues in your counting work in Lesson 1, such as whether letters or digits could be repeated in a password or not. Now consider how these issues are treated in the *Education Week* article below.

Take Note: Counting Their Chickens

A national restaurant chain that boasts about its tasty chicken is eating crow after a high school mathematics class cried foul over a television ad.

The ad shows Joe Montana, the National Football League quarterback, standing at the counter at a Boston Chicken restaurant puzzling over side-dish choices when an announcer says that more than 3,000 combinations can be created by choosing three of the restaurant's 16 side dishes.

But Bob Swaim, a math teacher at Souderton Area High School near Philadelphia, and his class did the math and told the Colorado-based chicken chain that there were only 816 combinations.

"We goofed," said Gary Gerdemann, a spokesman for Boston Chicken, explaining that the restaurant had confused "combinations" with "permutations."

"Apparently we didn't listen to our high school math teachers," Mr. Gerdemann said.

The company has, however, listened to Mr. Swaim and corrected its ads. For their eagle eyes, the students were awarded free meals and $500 to expand the math menu at Souderton.

Source: *Education Week.* Vol. 14, No. 20, p. 3

THINK ABOUT THIS SITUATION

Think about the mathematics in the situation involving side-dish choices described at the bottom of the previous page.

a Consider order and repetition in choosing three of the restaurant's 16 side dishes.

   i. Are repetitions allowed in this counting situation? Explain your thinking.

   ii. Do two different orderings count as two different possibilities? Explain.

b The announcer claimed that, "more than 3,000 combinations can be created by choosing three of the restaurant's 16 side dishes." How do you think this number was determined?

c What error in reasoning likely explains the restaurant's inflated claim of more than 3,000 combinations?

d How do you think Mr. Swaim's mathematics class came up with the number of 816 combinations of three side-dish meals?

In this lesson, you will apply the Multiplication Principle of Counting while taking into account order and repetition to obtain formulas for counting *permutations* and *combinations*. More generally, you will learn concepts and methods for counting the number of possible selections from a collection of objects. You will see how these problems are similar to, and different from, counting problems involving a sequence of tasks, which were the focus of Lesson 1.

INVESTIGATION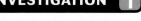

Permutations and Combinations

In the last lesson, you solved counting problems using fundamental counting principles. In this investigation, you will apply the Multiplication Principle of Counting, while taking into account the issues of order and repetition, to count *permutations* and *combinations*. These arise when you make selections from a collection of objects.

As you analyze the counting situations in this investigation, look for answers to the following questions:

Does the situation involve counting the number of selections from a collection of objects?

Are order and repetition important in the situation?

What strategies and formulas are useful in solving counting problems in which order and repetition are important?

Considering Order and Repetition in Counting To help you understand order and repetition in counting, consider the number of different types of groups chosen from a club. Suppose one club is selecting *officers*—President, Vice-President, and Treasurer. Another club is choosing a *committee*.

- Officers—The French Club will select three officers: President, Vice-President, and Treasurer. There are 15 members in the club from which to select these officers.

- Committee Members—The Ski Club will choose three people for a committee. There are 15 members in the club from which to choose these committee members.

Officers Election Ballot	**Committee Election Ballot**
President _____	_____
Vice-President _____	_____
Treasurer _____	_____

1 Think about order and repetition in each of the above situations.

  a. French Club Officers

   i. Is repetition allowed when selecting President, Vice-President, and Treasurer? Explain.

   ii. Do different orderings count as different possibilities when selecting President, Vice-President, and Treasurer? Explain.

  b. Ski Club Committee Members

   i. Is repetition allowed when choosing three members of a committee? Explain.

   ii. Do different orderings count as different possibilities when choosing three members of a committee? Explain.

  c. Which is larger—the number of possible slates of three officers or the number of possible groups of three committee members? Why does that make sense?

2 Now think about the number of different possibilities in each situation.

  a. French Club Officers

  How many different selections for President, Vice-President, and Treasurer are possible? Show the calculation you used to determine how many. Compare your reasoning and answer with that of your classmates. Resolve any differences.

  b. Ski Club Committee Members

  How many different three-person committees are possible? Show the calculation you used to determine how many. Compare your reasoning and answer with that of your classmates and resolve any differences.

3 Below are responses from two students related to the club election problems. Carefully study each student's response.

a. Analyze Amy's reasoning about the number of French Club officer selections.

> There are 15 choices for President. Once the President is chosen, then there are 14 members left who could be chosen for Vice-President. Once the President and Vice-President have been chosen, then there are 13 members left who could be Treasurer. So, there are $15 \times 14 \times 13 = 2{,}730$ different possibilities for the three officers. This counts different orders as different possibilities. For example, ABC is different from BAC.

This reasoning is essentially correct. Compare Amy's reasoning and answer with yours in Problem 2 Part a. If necessary, revise your work.

b. Analyze Latricia's reasoning about the number of Ski Club committees.

> There are 15 choices for the first committee member. This leaves 14 club members who could be chosen as the second committee member, and 13 choices for the third committee member. So far, this is $15 \times 14 \times 13$ possibilities. However, in this situation, order doesn't matter. For example, a committee of ABC is the same as a committee of BAC. In fact, for a committee consisting of A, B, and C, there are 6 different orderings that make the same committee. So, you must divide the first calculation by 6. This gives a final answer of $\frac{15 \times 14 \times 13}{6} = 455$ possible committees.

This reasoning is essentially correct. Compare Latricia's reasoning and answer to yours in Problem 2 Part b. If necessary, revise your work.

4 Whenever an idea is particularly common and important in mathematics, a definition is made to capture that idea. There are a couple of definitions related to order and repetition. Specific mathematical terms are used for what you have been counting in the club situations.

a. A **permutation** is an arrangement in which order matters and repetitions are not allowed.

   i. Explain why counting the number of possible three-officer selections is a permutation problem.

   ii. Look again at the French Club officer selection problem in Problem 3 Part a. You now know this is a permutation problem. It also uses the Multiplication Principle of Counting. Where did Amy use the Multiplication Principle of Counting in her reasoning?

b. A **combination** is an arrangement in which order does *not* matter, that is, different orderings are *not* counted as different possibilities. Also, repetitions are not allowed.

  i. Explain why counting the number of three-person committees is a combination problem.

  ii. The word "combination" in mathematics has a very specific meaning. It means something different than, for example, the combination of a dial locker or bicycle lock. Explain why the combination of a lock is *not* a mathematical combination.

  iii. Look again at the Ski Club committee problem in Problem 3 Part b. You now know that this is a combination problem. The Multiplication Principle of Counting is *not* used in this problem. Latricia correctly says that:

  > I did not use the Multiplication Principle in my reasoning. But it's like I started to use it, and then adjusted.

  Where did Latricia start to use the Multiplication Principle, and at what point did it not work so she adjusted her strategy?

c. In general, for a collection of people or identified objects, are there more permutations or combinations?

  i. Explain why your answer makes sense in terms of order and repetition.

  ii. Explain why your answer makes sense in terms of the computations used for counting permutations and combinations in the previous problems.

Formulas for Counting Permutations and Combinations Formulas can be very helpful in computations and in understanding concepts and connections. There are several different yet equivalent formulas for both combinations and permutations.

Some of the formulas use *factorials*. **Factorial** notation is a compact way of writing certain products of consecutive non-negative integers. For example, $5 \times 4 \times 3 \times 2 \times 1 = 5!$, which is read as "5 factorial." In general, when n is a positive integer, $n! = n \times (n - 1) \times \cdots \times 2 \times 1$. For convenience, $0!$ is defined to be 1.

5 Consider possible formulas for counting the number of permutations of k objects selected from n objects.

a. When Javon was working on Problem 2 Part a, his calculations suggested a general approach.

> To count the number of permutations of *k* objects selected from *n* objects, start with *n* and carry out a factorial-type computation using exactly *k* factors.

Describe how Javon's formula-in-words works for counting the number of French Club officer selections in Problem 2 Part a.

b. Elise proposed the following algebraic formula for counting permutations:

$$P(n, k) = n(n - 1)(n - 2) \cdots (n - k + 1)$$

i. What do you think Elise meant by the notation $P(n, k)$?

ii. Explain how her formula fits the formula-in-words from Part a.

c. Amy proposed a somewhat different formula for counting permutations, where $P(n, k)$ is the number of permutations of k objects selected from n objects.

$$P(n, k) = \frac{n!}{(n - k!)}$$

She described her formula as follows.

> I'm thinking of filling k boxes. In each box, I will put one of n objects. No repetitions are allowed and order matters. To count the number of ways to do this, I start with n possibilities for the first box, then $n - 1$ possibilities for the second box, then $n - 2$ for the next box, and I keep going like that until I've used up all k boxes. So, it's like a factorial but you cut it off after k factors. I can cut it off after k factors by dividing by $(n - k)!$.

i. Use Amy's formula to find the number of French Club three-officer selections from 15 people. Make sure your answer agrees with that in Problem 4 Part a.

ii. Discuss Amy's reasoning with your classmates. Does it make sense to you?

iii. Explain how her formula fits with the formulas in Parts a and b.

d. Use any of the methods and formulas for counting permutations to solve these problems:

i. How many different officer slates for a club are possible if there are four officers (President, Vice-President, Treasurer, and Secretary) selected from 20 club members?

ii. How would your reasoning and answer in part i change if there were 25 club members?

6 Javon's, Amy's, and Latricia's teacher next challenged the class to try to find a general formula for $C(n, k)$, the number of different combinations when choosing k objects from n objects, like choosing three Ski Club committee members from 15 people. Sometimes $C(n, k)$ is read as "n choose k." Working together, with much effort, they proposed the formula below to their class.

$$C(n, k) = \frac{n(n - 1)(n - 2) \cdots (n - k + 1)}{k!}$$

a. Latricia described their thinking as follows:

> Combinations are different than permutations because order doesn't matter. My formula counts permutations and then adjusts. So, I start by counting the permutations—that's the numerator. That gives me too many because it counts different orderings as different possibilities. So, I'll adjust for the fact that order doesn't matter by dividing by all the ways that k things can be ordered—that's the denominator.

i. Discuss Latricia's reasoning with your classmates. Does it make sense to you?

ii. In her description, she claims that k objects can be ordered in $k!$ ways. Explain why this is so.

iii. Use Latricia's formula to find the number of three-person Ski Club committees chosen from 15 people. Make sure your answer agrees with that in Problem 2 Part b.

b. As you know from your previous studies, formulas can be written in different, but equivalent, forms. Below is another common formula for combinations.

$$C(n, k) = \frac{n!}{k!(n-k)!}$$

Explain why this formula is equivalent to the formula:

$$C(n, k) = \frac{n(n-1)(n-2)\cdots(n-k+1)}{k!}$$

c. Alex gave this brief description of how to count combinations.

> *Start with n and carry out a factorial-type computation using exactly k factors. But then you have counted too many, so divide by k!.*

Explain how Alex's description fits with the formulas in Parts a and b.

d. Use any of the methods and formulas for counting combinations to solve these problems:

i. Find the number of different possible four-member committees that can be chosen from a club with 20 members.

ii. How would your answer to part i change if there were 30 club members?

7 Use the methods and formulas you have learned to help solve the following counting problems.

a. A sample of 10 cell phones will be selected from a shipment of 200 phones to test for flaws. How many different samples of 10 can be chosen?

b. Five students in a music class will be chosen to perform individually in a recital. There are 18 students in the class. How many different groups of five students can be chosen to perform?

c. Think again about the recital in Part b. How many different programs for the recital can be created using five students from the class of 18? Each student will perform individually, as in Part b. How is this problem different from Part b?

SUMMARIZE THE MATHEMATICS

Many counting situations can be analyzed in terms of order and repetition. In this investigation, you focused on permutations and combinations.

a The cells in the following table show four possibilities for counting situations in terms of order and repetition. On a copy of this table, fill in two of the cells by writing "permutation" or "combination" in the appropriate cells.

Counting with Order and Repetition		
	No Repetitions	**Repetitions Okay**
Different Orderings Count as Different Possibilities		
Different Orderings Do Not Count as Different Possibilities		

b In the "permutation" and "combination" cells, enter this additional information:

- a relevant example
- a relevant formula, using factorials
- a brief description in words of how to carry out the calculations indicated by the formula

c Explain how the Multiplication Principle of Counting is used in reasoning strategies to derive formulas for counting permutations and combinations.

Be prepared to explain your ideas and examples to the class.

✓ CHECK YOUR UNDERSTANDING

The Union High School marching band has developed a repertoire of 10 music pieces for this semester. Due to time constraints, four pieces can be performed at an upcoming event.

a. How many different collections of four music pieces could the director choose?

b. Four music pieces will be put together to create a program for the event. How many different concert programs are possible?

ThinkStock/SuperStock

Collections, Sequences, This or That

In the previous investigation, you focused on permutations and combinations. In each case, repetition is *not* allowed. In this investigation, you will study counting situations in which repetition *is* allowed.

As you work on the problems of this investigation, look for answers to the following questions:

What are similarities and differences among the four types of problems involving order and repetition?

What are similarities and differences among methods that count the number of selections from a collection of objects (Lesson 2) and methods that count the outcomes from a sequence of tasks (Lesson 1)?

1 Begin by analyzing the following two counting situations that you experience in your daily life. Think of how they are connected to the counting classifications in the table at the bottom of the page.

   a. In which cell of the table below does this "Pick 5" situation belong? Enter this into your copy of the table.

   b. Place this Web site password situation in the appropriate cell of your table.

Counting with Order and Repetition		
	No Repetitions	**Repetitions Okay**
Different Orderings Count as Different Possibilities		
Different Orderings Do Not Count as Different Possibilities		

2 Look back at the article on page 554, "Take Note: Counting Their Chickens." The counting problem in this article is: *How many different three-side-dish orders can be made from the restaurant's 16 side dishes?*

a. The spokesman explained that the restaurant had confused combinations with permutations. However, neither combinations nor permutations are correct in this situation. Explain why.

b. Place this restaurant-side-dish problem in the appropriate cell of your table.

c. The solution given by the Souderton math class is 816. Verify that this is correct. You may need to work hard to do this, but the math students in the article got it and so can you.

Sides
(served after 10:30 A.M. until 8:00 P.M.)

Tomato Bisque	Macaroni and Cheese
Caesar Salad	Black Beans and Rice
House Salad	Sweet Potato Fries
Potato Salad	Cornbread
Green Beans	Mashed Potatoes
Vegetable Medley	Cranberry Relish
Cole Slaw	Cinnamon Apples
Creamed Spinach	Fruit Salad

Apply your understanding of counting methods to help solve the following problems. Permutations and combinations may be involved directly, indirectly, or not at all. You may use any counting principles or methods from Lessons 1 and 2. Be prepared to justify your solutions.

3 How many different 13-card hands can be made from a deck of 52 cards?

4 Information is stored in computers as strings of 0s and 1s (because 0 and 1 can be interpreted as "off" and "on" settings for switches inside the computer). 0 and 1 are called binary digits or *bits*. A sequence of bits is called a *binary string*. For example, 10110 is a binary string with five bits. A binary string with eight bits is called a *byte*. For example, 11010110 is a byte but 010 is not.

   a. How many different bytes are possible?

   b. How many bytes can be created that contain exactly two zeroes?

5 How many different ways are there for win, place, and show (first, second, and third, respectively) positions in a horse race with seven horses?

6 Seven people are running for three unranked positions on the school board. In how many different ways can these three positions be filled?

7 A **set** is any well-defined collection of objects. A *subset* of a set is, roughly, a smaller set inside the set. Precisely, set A is a **subset** of set B if every element of A is an element of B. For example, if $B = \{1, 2, 3\}$, then *some* of the subsets of B are

$$\varnothing, \{1\}, \{1, 3\}, \{2, 3\}, \{1, 2, 3\}.$$

The symbol $\varnothing$ denotes the **empty set**—the set with no elements. The empty set is a subset of every set. The whole set (in this case, $B = \{1, 2, 3\}$) is also considered a subset of itself.

   a. There are more subsets of B than those listed above. List all the subsets of $B = \{1, 2, 3\}$. How many are there?

   b. How many subsets of B have two elements? You can of course just look at your list in Part a to answer this question. In addition, explain how you could use combinations to answer this question.

   c. Suppose a set has five elements. How many subsets of three elements does this set have?

   d. For a set with five elements, how many subsets are there?

SUMMARIZE THE MATHEMATICS

In this investigation, you refined and extended your understanding of counting methods. You considered four types of counting problems related to issues of order and repetition. You solved problems that require several different counting strategies.

a When you are presented with a complex counting problem, what questions should you ask yourself as a start to solving the problem? What might be the next step in your solution strategy?

b In solving counting problems, it is often helpful to detect the underlying structure of the problem as outlined below. Give an example of each of these three types of problem structures.

    i. Count the number of selections from a collection of objects

    ii. Count the number of outcomes from a sequence of tasks

    iii. Count the number of outcomes from one situation or another situation

c What are similarities and differences among the four types of counting problems involving order and repetition?

d Permutations and combinations can be thought of (and defined) in terms of sets and subsets.

    i. Explain why a subset of a set is a combination.

    ii. Explain why an ordered sequence of distinct elements of a set is a permutation.

Be prepared to explain your thinking and examples to the class.

✔ CHECK YOUR UNDERSTANDING

Complete each of the following counting tasks.

a. Four of the 11 members of a championship gymnastics team will be chosen at random to stand in a row on stage during an awards ceremony. Assuming it is more prestigious to stand closer to the podium, how many different arrangements of four gymnasts on stage are possible?

b. A particular bicycle lock has four number dials. Each dial consists of the digits from 0 to 9.

    i. How many different lock combinations are possible? Explain, including any assumptions you are making about how the lock works.

    ii. Are these lock combinations actually "combinations" in the mathematical sense of the word? Are they permutations? Explain.

APPLICATIONS

1 Carlos's hockey team has ten players, not counting the goalie. Five non-goalie players need to be selected for the starting line-up.

   a. How many different starting line-ups (not including the goalie) are possible if positions are not assigned?

   b. How many starting line-ups are possible if positions are assigned?

2 A group of four students is working on a collection of eight simple tasks. They decide they will share the work and each do two of the tasks. How many ways are there for them to divide the work?

3 Major League Baseball teams maintain a 25-man roster and also a 40-man roster. Players on the 25-man roster may play in official games throughout the season. The additional 15 players on the 40-man roster may play in games starting September 1.

   a. How many 9-player batting orders are possible from a 25-man roster? A 40-man roster?

   b. In 2011, world population was estimated at about 7 billion. If all the distinct batting orders for a 25-man roster were written down and distributed equally to the people of the world, how many batting orders would each person receive?

4 A researcher distributes a questionnaire to 80 participants. To gain additional insight into participant responses, 14 of the 80 participants are randomly selected to be interviewed.

   a. How many different ways could she pick 14 participants to interview?

   b. A stack of 500 sheets of paper is about 5 cm high. Assume she writes the names of a single selection of the 14 interview participants on a sheet of paper.

   i. How high would the stack of papers containing all possible selections stand?

   ii. How does your answer compare to the average distance from Earth to the Sun, roughly 150 billion meters?

5 Suhayla is making a music playlist for Jahanna. She has narrowed the possible songs down to 30 and wants to make a mix with exactly 20 songs.

   a. If she carefully chooses the order of the songs, how many possible mixes could she make for him?

b. The total surface area of Earth is nearly 150 million square kilometers. Though sand grains vary considerably in size, a reasonable approximation for the average volume of a grain of sand is 10^{-12} m³. If Suhayla had one grain of sand for each possible playlist, spread out over the land surface area of Earth, how deep would the sand be?

6 Examine the following portion of a television commercial.

Customer:	So what's this deal?
Pizza Chef:	Two pizzas.
Customer:	[Looking towards a four-year-old boy.] Two pizzas. Write that down.
Pizza Chef:	And on the two pizzas choose any toppings—up to five [from the list of 11 toppings].
Customer:	Do you …
Pizza Chef:	… have to pick the same toppings on each pizza? No!
4-Year-Old Boy:	Then the possibilities are endless.
Customer:	What do you mean? Five plus five are ten.
4-Year-Old Boy:	Actually, there are 1,048,576 possibilities.
Customer:	Ten was just a ballpark figure.
Pizza Chef:	You got that right.

a. Do you think the customer's "ballpark figure" is too low? Explain your reasoning.

b. Suppose you order just one pizza and you must choose exactly 5 different toppings from 11 choices. How many different pizzas are possible?

c. Suppose you order just one pizza and you must choose exactly 3 different toppings from 11 choices. How many different pizzas are possible?

d. Suppose you order just one pizza and you can choose from 0 to 5 different toppings. How many different pizzas are possible?

e. In the TV commercial, does the 4-year-old boy have the correct answer? If so, explain how to compute his answer. If not, explain why it is incorrect and determine the correct answer.

f. Belinda reasoned as follows.

> There are 1,024 possibilities for one pizza. Since 2 pizzas are ordered, that makes $(1,024)^2$ possibilities for a two-pizza order. But order does not matter for the two pizzas, so divide by 2. Thus, the correct answer is 524,288.

Explain the error in Belinda's reasoning.

CONNECTIONS

7 Use counting methods to help answer each of the following geometric questions.

   a. Given a set of n points, how many distinct line segments can be formed with two of the n points as endpoints? Does it make any difference if all of the points are not in the same plane? Explain your reasoning.

   b. How many points of intersection are formed by n coplanar lines if no two are parallel and no three intersect in a common point?

   c. Using combinations, explain why the questions in Parts a and b are essentially the same.

8 The notation used for the number of combinations is not completely standardized. One of the most common notations is the one used in this lesson, namely, $C(n, k)$. As you use technology and read other books or Web sites you might see notations like $_nC_k$ or $\binom{n}{k}$. Practice interpreting the different notations by computing each of the following. Compute each by hand and then check your answer using technology.

   a. $C(12, 5)$

   b. $_8C_3$

   c. $\binom{12}{7}$

9 The following formula shows one way that $C(n, k)$ and $P(n, k)$ are related.
$$C(n, k) = \frac{P(n, k)}{k!}$$

   a. Explain this relationship in words (by reasoning about combinations, permutations, order, and repetition).

   b. Prove this relationship using factorials and other formulas.

10 As you learned in this lesson, a general formula for the number of permutations is
$$P(n, k) = \frac{n!}{(n - k)!}.$$
A general formula for the number of combinations is
$$C(n, k) = \frac{n!}{(n - k)!k!}.$$
What restrictions must be placed on n and k for these formulas to make sense?

11 Find a general formula for those problems in which you are counting the number of possible selections of k objects from a collection of n objects when order matters and repetitions are allowed. (An example of a problem of this type is counting the number of different Web site passwords possible if a password consists of six letters and repetitions are allowed.) Enter this formula into the appropriate cell of your Counting with Order and Repetition table.

12 As you have learned, two important issues to consider in counting situations are order and repetition. It can be tricky sometimes to decide how order and repetition are involved.

   a. How do you decide whether order should be considered in counting? Give an example.

   b. How do you decide whether repetitions are involved in a counting situation? Give an example.

13 Combinatorics is sometimes described as "methods for counting without counting." In what sense is this an apt description of the mathematics you have been doing in this lesson?

14 The restaurant-side-dish problem (Investigation 2, page 563) is not a direct combination or permutation problem. As you discovered, you cannot simply refer to the Counting with Order and Repetition table, choose a formula, and compute. Often problems involve several counting ideas. Using the restaurant-side-dish problem as an example:

   a. explain how the Multiplication Principle of Counting is used, or could be used, in solving this problem.

   b. explain how the Addition Principle of Counting is used, or could be used, in solving this problem.

   c. explain how combinations are used, or could be used, in solving this problem.

15 You can often think about a counting problem in several ways. Consider this problem:

> How many different passwords are possible that consist of five lowercase letters, with no letters repeated?

   a. Explain how you can solve this problem using the Multiplication Principle of Counting.

   b. Explain how you can solve this problem using permutations.

16 Suppose you have n objects with which you can sequentially fill k slots. Are there more or fewer possible k-slot sequences if repetitions are allowed when filling the slots? Justify your answer.

17 An important mathematical practice is to: *Construct viable arguments and critique the reasoning of others.* Look back at the work you did in Investigation 1. Find at least two instances where you used this mathematical practice. Explain how you used the practice in each instance.

18 RNA (ribonucleic acid) is a messenger molecule associated with DNA (deoxyribonucleic acid). RNA molecules consist of a chain of bases. Each base is one of four chemicals: U (uracil), C (cytosine), A (adenine), and G (guanine). It is difficult to observe exactly what an entire RNA chain looks like, but it is sometimes possible to observe fragments of a chain by breaking up the chain with certain enzymes. Armed with knowledge about the fragments, you can sometimes determine the makeup of the entire chain. One type of enzyme that breaks up an RNA chain is a "G-enzyme." The G-enzyme will break an RNA chain after each G link. For example, consider the following chain:

AUUGCGAUC

A G-enzyme will break up this chain into the following fragments:

AUUG   CG   AUC

a. What fragments result when a G-enzyme is applied to the following chain?

CGUUGGAUCGAU

b. Unfortunately, the fragments of a broken-up chain may be mixed up and in the wrong order. However, you can use reasoning to figure out the right order. For example, suppose a chain is broken up by a G enzyme and the fragments are out of order. Explain why a fragment that does not end in G must be the last fragment in the chain.

In Parts c–e, you will use information about fragments to reconstruct the complete chain. Suppose you have the fragments of an unknown RNA chain of 10 bases.

c. Suppose the complete RNA chain of 10 bases is broken by a G-enzyme into the following fragments (although not necessarily in this order):

AUG   AAC   CG   AG

How many different ways can these fragments be combined into a complete RNA chain of 10 bases?

d. Another enzyme, the U-C enzyme, breaks up an RNA chain after every U and every C. This enzyme breaks the unknown RNA chain in Part c into the following fragments.

GC   GAAC   AGAU

How many different ways can these fragments be combined into a single RNA chain of 10 bases?

e. So far, just by counting, there are many possible ways to recombine the fragments into a complete 10-base chain. However, if you reason about the fragments resulting from the two enzymes, by examining the fragments in Parts c and d, there is only one possible complete chain. What is the complete chain?

19 In Problem 2 of Investigation 2 (page 563), you solved the restaurant-side-dish problem from the newspaper article. That is, you determined the number of different 3-side-dish selections that can be made from 16 side dishes. The answer is 816. There are several possible solution methods. Read and fill in the missing pieces of the following correct solution.

For simplicity, label the 16 side dishes A through P. Think about an order form that could be used to record someone's order, like the following.

```
A   B  C  D  E  F  G  H  I  J  K  L  M  N  O  P
|XX|  |  |  |  |  |  |  |  |  |  |X|  |  |  |
```

a. To help in the counting process, think about this order form as a sequence of marks, without the letter labels. Each mark is either a dividing line to separate the different side dishes or an "X" to indicate a selection.

   i. What 3-side-dish order is represented by the completed order form above?

   ii. Why are there three Xs?

   iii. Why are only 15 vertical dividing marks needed?

b. You can count the number of orders by thinking about sequences of marks, as follows.

   So, there are 18 total marks. Think about 18 slots. In each of those slots, you will put a mark, either a dividing line or an X.

   You want 3 Xs in this sequence of marks. The sequence of marks thus created will determine an order.

   So to count the number of different orders, you just count the number of ways you can choose 3 of the 18 slots for the Xs. Thus, you are choosing 3 slots from 18 slots; order does not matter in the selection of slots; and none of the 18 slots can be repeated.

   So, the solution is $C(18, 3)$, which is 816.

Go back through this reasoning process and restate it in your own words, so that you understand it and could explain it to someone else.

c. Suppose the restaurant had 10 side dishes and you could choose 4. How many 4-side-dish orders are possible?

d. Suppose you want to select k objects from a collection of n objects, where order does not matter and repetitions are allowed. Explain why the number of possible selections in this case is $C(n + k - 1, k)$. Enter this formula into the appropriate cell of your Counting with Order and Repetition table.

20 Look back at the "Pick 5" part of Problem 1 (page 562) of Investigation 2. How many different "Pick 5" orders are possible?

21 Determine the number of all possible three-number combinations for a dial combination lock with 20 numbers on the dial. The combination is entered by rotating the dial to the right, then left, then right again, but the left number cannot be the same as either of the right numbers.

22 Counting problems may look simple but they often require very careful thinking. Consider this counting problem.

> How many different flags of 8 horizontal stripes contain at least 6 blue stripes if each stripe is colored red, green, or blue?

Here is one student's proposed solution.

> *Each stripe on the flag must be labeled with a color—red, green, or blue. Because at least 6 blue stripes are required, one can begin by choosing 6 stripes to color blue. There are C(8, 6) ways to do this. Once the minimum of 6 blue stripes is fulfilled, the remaining 2 stripes (whichever ones they are) can each be colored in any one of 3 ways (red, green, or blue). Therefore, there are (3)(3) = 9 ways to finish coloring the stripes, giving a total of (28)(9) = 252 different patterns of color on the flag's stripes.*

This solution may seem plausible, but it is incorrect. Explain why it is wrong. Then solve the problem correctly. (**Source:** Annin, Scott A. and Kevin S. Lai. "Common Errors in Counting Problems." *Mathematics Teacher*, 103, no. 6 (February 2010): 402–409.)

23 The graph of and rule for $f(x)$ shown below represent the snowfall rate in inches per hour during a heavy snowstorm.

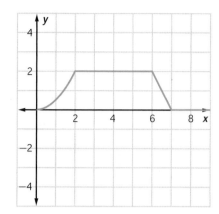

$$f(x) = \begin{cases} 0.5x^2 & \text{for } 0 \leq x < 2 \\ 2 & \text{for } 2 \leq x < 6 \\ 14 - 2x & \text{for } 6 \leq x < 7 \\ 0 & \text{for } 7 \leq x \end{cases}$$

a. Determine the value of $\int_2^7 f(x)\,dx$ and explain what it tells you about the snowfall.

b. Use four rectangles to approximate the number of inches of snow that fell during the first two hours of this storm.

24 Rewrite each expression in standard polynomial form.

a. $(2x - 3)^2$

b. $(a + 4b)^2$

c. $(3x - 5y)^2$

d. $(t - 4)(t^2 + 5t - 3)$

e. $(x + 3)^3$

f. $(5w - 2t)^3$

25 Rewrite each rational expression as a single algebraic fraction in simplest form.

a. $\dfrac{x + 3}{7} - \dfrac{x + 2}{4}$

b. $\dfrac{t^2 + 6t + 8}{t - 2} \cdot \dfrac{t}{t + 4}$

c. $\dfrac{y}{y - 3} + \dfrac{y + 1}{y^2 - 4y + 3}$

d. $\dfrac{x + 4}{x^2 - 2x + 1} \div \dfrac{x^2 + 7x + 12}{x^2 - 1}$

e. $\dfrac{n^2 - 3n + 2}{2} + (n - 1)$

26 For each sequence below, assume that the first term is t_0.

- Is the sequence arithmetic, geometric, or neither?

- Determine a recursive formula for the sequence.

- Determine a function formula for the sequence.

a. 4, 7, 10, 13, …

b. 2, 6, 18, 54, …

c. 3, 4, 7, 12, 19, …

27 Find the area of each polygon described below.

a. An isosceles triangle with legs of length 10 cm and base of length 6 cm

b. An equilateral triangle with sides of length 12 cm

c. A square with diagonals of length $5\sqrt{2}$ in.

d. A regular hexagon with sides of length 1 in.

28 Draw a sketch of each plane described below and then write an equation of the plane in the form $Ax + By + Cz = D$.

a. Is parallel to the xy-plane and contains the point $(1, 2, 3)$

b. Has traces with equations $x + y = 12$, $x + 2z = 12$, and $y + 2z = 12$

c. Intersects the x-axis at $(2, 0, 0)$, the y-axis at $(0, -4, 0)$, and the z-axis at $(0, 0, 5)$

Counting Throughout Mathematics

In the last lesson, you saw that counting problems arise in many different contexts, for example, menu choices, telephone numbers, automobile license plates, Web site passwords, club committees, card games, IP addresses, and even DNA strings. The counting methods you learned are also applied throughout mathematics.

For example, one of the most famous mathematicians of the last century, Paul Erdös, often used clever methods of counting in his research. Erdös published well over 1,000 papers, almost always working with collaborators. In fact, there is a counting number based on working with Erdös, called an Erdös number. If you collaborated with Erdös to write a paper, your Erdös number is 1. If you collaborated with a collaborator, then your number is 2, and so on. For instance, some of the authors of this book have Erdös number 4 and thus others have Erdös number at most 5.

In this lesson, you will apply counting methods to some of the broad areas in which Erdös worked, including probability and discrete mathematics, and also algebra and geometry.

INVESTIGATION 1

Counting and Probability

In your previous studies, you used counting to help calculate probabilities. In this investigation, you will review and extend your understanding of probability and counting.

As you work on the problems in this investigation, look for answers to these questions:

> *How can counting methods be used in determining probabilities?*
>
> *How are the Multiplication Rules for probability similar to, and different from, the Multiplication Principle of Counting?*

Counting is especially useful in determining probability when there are a finite number of outcomes, all of which are equally likely. In this case, the **probability of an event** A can be defined as

$$P(A) = \frac{\text{number of outcomes corresponding to event } A}{\text{total number of possible outcomes}}.$$

Thus, when all the outcomes are equally likely, you can determine the probability of an event by counting the number of outcomes corresponding to the event and dividing by the total number of possible outcomes.

1 Suppose you roll two fair dice, one red and one blue. An *outcome* is the number of spots showing on the top face of each die. For example, (3, 5) denotes the outcome of getting 3 on the red die and 5 on the blue die. Note that (3, 5) is a different outcome from (5, 3).

   a. Are all the outcomes equally likely? What is the total number of possible outcomes?

   b. Consider the *event* of getting a sum of 5 on the two dice. How many outcomes correspond to this event?

   c. What is the probability of getting a sum of 5 when rolling two dice? The probability of getting a sum of 8?

2 In a certain state lottery, a player fills out a ticket by choosing five "regular" numbers from 1 to 45 and one PowerBall number from 1 to 45. The goal is to match the numbers with those drawn at random at the end of the week. The regular numbers are not repeated and they do not have to be in the same order as those drawn. The PowerBall number can be the same as one of the regular numbers.

   a. How many different ways are there to fill out a ticket?

   b. A player wins the jackpot by matching all five regular numbers plus the PowerBall number. This is called "Match 5 + 1." Since all the numbers must match, there is only one way to fill out a ticket that is a "Match 5 + 1" winner. What is the probability of the event "Match 5 + 1"?

   c. A player wins $100,000 by matching the five regular numbers but not the PowerBall number. This is called "Match 5." What is the probability of getting a "Match 5" winner?

   d. A player wins $5,000 for "Match 4 + 1" (match exactly four of the regular numbers plus the PowerBall number). What is the probability of getting a "Match 4 + 1" winner?

   e. Look back at your solutions to Parts a–d. Describe at least two different counting methods from Lessons 1 and 2 that you used in your solutions.

Counting Methods used to Determine Probabilities You used counting to help find the probabilities in Problems 1 and 2. For example, you may have used combinations and the Multiplication Principle of Counting. Think about which counting methods you use as you solve the rest of the problems in this investigation.

3 Suppose you toss a fair coin four times and record the sequence of heads and tails.

a. How many possible outcomes are there? Describe two ways of determining this number. Are the outcomes equally likely?

b. Find the following probabilities.

   i. *P(four heads)*

  ii. *P(exactly one head)*

 iii. *P(at least three heads)*

c. You may recall that the **Multiplication Rule for independent events** states that if A and B are independent events, then $P(A$ and $B) = P(A) \times P(B)$. Show how to calculate the three probabilities in Part b using the Multiplication Rule.

4 Suppose the names of six boys and four girls written on individual slips of paper are placed in a hat. You draw two names, in succession and with replacement. That is, you draw a slip of paper, record the name, return the slip of paper to the hat. Shake the hat to re-mix the slips of paper. Then draw again and record the name.

a. Find the probability that the first name drawn is a girl's name and the second name is a boy's name.

b. Show how you used, or could use, the Multiplication Rule for independent events to find this probability.

c. You now have a multiplication rule for probability and a multiplication principle for counting. They are closely related. Find the probability in Part a using the Multiplication Principle of Counting and the definition of probability given at the beginning of this investigation. (Or just explain how you did it if you have already solved the problem this way.)

Jim Laser/CPMP

5 Suppose again the names of six boys and four girls are written on individual slips of paper and placed in a hat. This time you draw two names *without* replacement. That is, you draw one name, you do *not* return the slip of paper to the hat, then you draw a second name.

   a. Find the probability that the first name drawn is a girl's name and the second name is a boy's name.

   b. Explain why the answer to Part a is *not* $\frac{4}{10} \times \frac{6}{10}$.

   c. Show how you can find the probability in Part a using the Multiplication Principle of Counting and the definition of probability given at the beginning of this investigation.

   d. To find the probability in Part a, you can also use the **General Multiplication Rule** for any two events:

   > If A and B are events, then $P(A \text{ and } B) = P(A) \times P(B \mid A)$.

   The notation $P(B \mid A)$ is read "probability of B given A." This means you find the probability of B assuming that you know A happened. Show how to use the General Multiplication Rule to find:

   > $P(\textit{girl's name on first draw} \text{ and } \textit{boy's name on second draw})$.

6 Suppose you draw four names without replacement from the hat containing six boys' names and four girls' names on slips of paper.

   a. Find the probability that all names drawn are those of girls. That is, find $P(\textit{all four are girls' names})$.

   b. There are several methods for finding the probability in Part a. Compare your solution to those of other students until you find at least one other solution method. Explain your solutions to each other.

7 There are 50 people in a jury pool and 15 of them are Native Americans. You select two jurors at random from this pool.

   a. What is the probability that the two you select are a Native American and someone who is not a Native American?

   b. Find the probability of this event using a different method.

CHECK YOUR UNDERSTANDING

Consider the following probability and counting situations.

a. Revisit the names-in-a-hat situation one more time, to help you pull together what you have learned. You have the names of six boys and four girls on slips of paper in a hat. This time, you reach into the hat and pull out two names *at the same time*. What is the probability you get a boy's name and a girl's name?

b. In Iowa in 2011, a license plate had six characters. The first three characters were numbers and the last three were letters.

Suppose that no numbers or letters can be repeated. What is the probability that a randomly chosen Iowa license plate has three odd numbers?

Combinations, the Binomial Theorem, and Pascal's Triangle

In your previous work in algebra, you rewrote binomial expressions like $(x + y)^2$ and $(x + y)^3$ in equivalent expanded form. In this investigation, you will explore some of the properties of combinations and their applications in calculating binomial expansions.

As you complete the problems in this investigation, look for answers to these questions:

What is Pascal's triangle and what are some connections among combinations, Pascal's triangle, and expansions of binomial expressions of the form $(a + b)^n$?

How can you explain and prove some of these connections?

1 Think about expanding $(a + b)^n$. In particular, think about the coefficients of the terms in the expansion. For example, $(a + b)^2 = (a + b)(a + b) = a^2 + 2ab + b^2$. The coefficients are 1, 2, and 1. There is an important connection between combinations and the coefficients of the terms in the expansion of $(a + b)^n$. Investigate this connection by expanding $(a + b)^n$ for several values of n, as follows.

a. Without using technology, expand $(a + b)^n$ for $n = 0, 1, 2, 3$, and 4. See the computer algebra system (CAS) output below for the cases of $n = 5$ and $n = 6$.

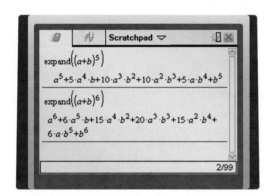

b. Examine and organize the coefficients of the terms of the expansions. Describe any patterns in the coefficients. Describe any connections you see to combinations.

c. What do you notice about the exponents on a and b for successive terms in the expansions?

Connections Between Pascal's Triangle and Expanding $(a + b)^n$ The coefficients of the terms in the expansion of $(a + b)^n$ have a close connection to an array of numbers called Pascal's triangle. In the next few problems, you will explore this connection.

2 You might organize your work from Problem 1 as follows.

coefficients of $(a + b)^0$				1				
coefficients of $(a + b)^1$			1		1			
coefficients of $(a + b)^2$		1		2		1		
coefficients of $(a + b)^3$	1		3		3		1	
coefficients of $(a + b)^4$	1	4		6		4		1

a. Continue this array of numbers using the coefficients of the terms in the CAS expansions of $(a + b)^5$ and $(a + b)^6$.

b. Describe how you could compute the numbers in a specific row of the array by using the numbers in the previous row.

c. Based on the pattern in the array, what do you think the coefficients are in the expansion of $(a + b)^7$? Check your conjecture.

The triangular array of numbers in Problem 2 is called *Pascal's triangle*. It is named for the French philosopher and mathematician Blaise Pascal (1623–1662). He explored many of its properties, particularly those related to the study of probability. Although the triangle is named for Pascal, other mathematicians knew about it much earlier. For example, the triangular pattern was known to Chu Shih-Chieh in China in 1303.

Blaise Pascal

Pascal's Triangle

row 0				1				
row 1			1		1			
row 2		1		2		1		
row 3	1		3		3		1	
row 4	1	4		6		4		1

The rules for constructing **Pascal's triangle** are as follows: The top row, which is the top vertex of the triangle, consists of the single number 1. Each succeeding row starts and ends with 1. The remaining entries are constructed by looking at the row above. Specifically, each number in a given row is found by computing this sum:

(*the number just above and to the left*) + (*the number just above and to the right*).

This is illustrated by the connector lines between 2, 1, and 3 in Pascal's triangle above.

3 You can add rows to Pascal's triangle indefinitely. Use the rules above to add rows 5 and 6 to the triangle. Compare to the rows of coefficients of the terms in the expansions of $(a + b)^5$ and $(a + b)^6$ that you determined in Problem 2. Resolve any differences.

Connections Between Pascal's Triangle and Combinations In Problems 2 and 3, you saw a remarkable connection. On the one hand, you have the coefficients of the terms in the expansion of $(a + b)^n$, which can be computed using algebraic multiplication. On the other hand, you have the numbers in Pascal's triangle, which are computed using the specific arithmetic rules given on the previous page. You have seen that these two very different procedures generate the same rows of numbers! Later in this lesson, you will see a reason for this connection. But first, consider a related connection between Pascal's triangle and combinations.

4 Notice that the rows of Pascal's triangle (shown on the previous page) are numbered starting with row 0. The entries in a given row can also be numbered beginning with 0. So, the initial entry in each row is labeled "entry 0," the next entry is labeled "entry 1," and so on.

   a. Compute $C(4, 2)$. Where is this number found in Pascal's triangle (which row and which entry)? What is the coefficient of a^2b^2 in the expansion of $(a + b)^4$?

   b. Compute $C(6, 4)$. Where is $C(6, 4)$ found in Pascal's triangle? What is the coefficient of a^2b^4 in the expansion of $(a + b)^6$?

   c. Now try to generalize your work in Parts a and b. Describe how to find $C(n, k)$ in Pascal's triangle. Describe where in Pascal's triangle you can find the coefficient of $a^{n-k}b^k$ in the expansion of $(a + b)^n$.

Connections Between Combinations and Expanding $(a + b)^n$ So far in this investigation, you have studied connections among three seemingly different mathematical topics: coefficients in the expansion of $(a + b)^n$, numbers in Pascal's triangle, and values of $C(n, k)$. One of the most important of these connections involves using combinations to expand $(a + b)^n$.

5 Study the following reasoning used by a group of students in Bertie STEM High School who were challenged to find the coefficient of $a^{54}b^{46}$ in the expansion of $(a + b)^{100}$. Discuss their reasoning with your classmates. Expand or clarify the reasoning as needed so that everyone understands.

> $(a + b)^{100} = (a + b)(a + b)(a + b) \cdots (a + b)$ (100 factors). To carry out this multiplication, you multiply each term in the first factor, that is, a and b, by each term in the second factor, then by each term in the third factor, and so on. You must multiply through all 100 factors. To get $a^{54}b^{46}$, you need to multiply by b in 46 of the factors. That is, you must choose 46 of the 100 factors to be those where you use b as the multiplier (and in the other factors, a will be the multiplier). So, the total number of ways to get $a^{54}b^{46}$ is the number of ways of choosing 46 factors from the 100 factors, which is $C(100, 46)$. So, the coefficient of $a^{54}b^{46}$ in the expansion of $(a + b)^{100}$ is $C(100, 46)$.

   a. Use similar reasoning to find the coefficient of $a^{29}b^{71}$ in the expansion of $(a + b)^{100}$.

   b. Based on this reasoning, use combinations to find the coefficients of the terms in the expansion of $(a + b)^5$. Confirm that your coefficients match those in the CAS display on page 577.

c. Use similar reasoning to find the coefficient of a^3b^5 in the expansion of $(a + b)^8$.

d. Now think about the general term in a binomial expansion. What is the coefficient of $a^{n-k}b^k$ in the expansion of $(a + b)^n$?

6 Your work in Problem 5 suggests the following general result, called the **Binomial Theorem**.

For any positive integer n,

$$(a + b)^n = C(n, 0)a^n + C(n, 1)a^{n-1}b + C(n, 2)a^{n-2}b^2 + \cdots + C(n, k)a^{n-k}b^k + \cdots + C(n, n-2)a^2b^{n-2} + C(n, n-1)ab^{n-1} + C(n, n)b^n.$$

a. Use the Binomial Theorem to expand $(a + b)^4$. Verify that you get the same answer as in Problem 1.

b. Use the Binomial Theorem to find the coefficient of a^3b^5 in the expansion of $(a + b)^8$. Compare to the answer you found using combinatorial reasoning in Part c of Problem 5.

c. Explain why the sum of the exponents of a and b in each term of the expansion of $(a + b)^n$ is n.

d. Explain why the coefficient of $a^{n-k}b^k$ is the same as the coefficient of a^kb^{n-k}.

e. Use the Binomial Theorem to expand $(2x - 3y)^5$.

Pascal's Triangle and Properties of Combinations Now that you have observed that the entries in Pascal's triangle are values of $C(n, k)$, you can make conjectures about properties of combinations by looking for patterns in Pascal's triangle.

7 Carefully write the first 10 rows of Pascal's triangle. Based on the symmetry and other patterns in Pascal's triangle, make at least two conjectures about properties of combinations. State your conjectures using $C(n, k)$ notation. Compare your conjectures to those of other classmates.

8 Consider the line symmetry in Pascal's triangle.

a. If you have not already done so in Problem 7, use the line symmetry to make a conjecture about the precise relationship between $C(n, k)$ and $C(n, n - k)$. You might find it helpful to examine a few examples using specific values of n and k.

b. State the relationship from Part a in the specific instance when $n = 8$ and $k = 3$. Prove this specific relationship in the following two ways:

  i. Using a factorial formula for combinations and algebraic reasoning.

  ii. Using combinatorial reasoning. That is, carefully explain how to choose and count combinations. In this case, you might find it helpful to think about how choosing 3 objects from 8 objects is the same as *not* choosing a particular number of objects.

c. Now prove the general property $C(n, k) = C(n, n - k)$, in two ways.

    i. Using factorial formulas and algebraic reasoning.

    ii. Using combinatorial reasoning by thinking about ways of choosing objects.

d. Which of the arguments in Part c was most convincing for you? Why?

SUMMARIZE THE MATHEMATICS

In this investigation, you explored several connections among combinations, the expansion of algebraic expressions of the form $(a + b)^n$, and Pascal's triangle.

a   Describe these connections.

b   Explain how you can reason with combinations to find the coefficient of a^2b^5 in $(a + b)^7$.

c   Describe how to use Pascal's triangle to find the coefficient of a^2b^5 in $(a + b)^7$.

d   Describe how to use Pascal's triangle to find $C(7, 5)$.

e   Use combinatorial reasoning to justify that $C(16, 4) = C(16, 12)$.

Be prepared to share your descriptions and reasoning with the class.

 CHECK YOUR UNDERSTANDING

Think about the relative advantages of algebraic, visual (Pascal's triangle), and combinatorial approaches to binomial expansions as you complete these tasks.

a. Expand $(x + 2)^3$ in the following three ways.

    i. Multiply by hand.

    ii. Use Pascal's triangle.

   iii. Use the Binomial Theorem.

b. Find the coefficient of a^4b^2 in the expansion of $(a + b)^6$ in the following three ways.

    i. Reason with combinations.

    ii. Use Pascal's triangle.

   iii. Use the Binomial Theorem.

APPLICATIONS

1 Monograms on jewelry, clothing, and other items consist of the initials of your name. Examine the three-initial monogram offer below.

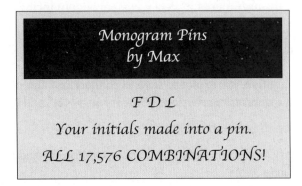

Monogram Pins
by Max

F D L
Your initials made into a pin.
ALL 17,576 COMBINATIONS!

   a. Is the number of different three-initial monograms given in the ad correct? Explain.

   b. Are these really "combinations" in the mathematical sense of the word?

   c. What is the probability that a randomly selected three-initial monogram has all three initials the same?

   d. Now consider the case where two of the initials are the same.

   i. How many three-initial monograms are possible if the first two initials are the same and the third is different?

   ii. How many three-initial monograms are possible if any two initials are the same and the other is different?

   iii. What is the probability that a randomly selected three-initial monogram will have two initials the same and the other initial different?

   e. What is the probability that a randomly selected three-initial monogram has all three initials different?

   f. What should be true about the probabilities in Parts c, d, and e?

2 Suppose you have 10 blue socks and 8 white socks in a drawer.

   a. You select two socks at random, in succession, and with replacement. Find the probability that both socks are blue.

   b. You reach in and pull out two socks at the same time, at random. Consider the following two possible solutions to finding the probability that both socks are blue.

   i. Mariam gives this correct solution: $\dfrac{C(10, 2)}{C(18, 2)}$. Describe reasoning that supports this solution.

   ii. John gives this correct solution: $\dfrac{10 \times 9}{18 \times 17}$. Describe reasoning that supports this solution.

   c. Explain why the two solution methods in Part b are equivalent.

3 The Chess Club at Asiniboyne High School consists of 6 seniors and 11 juniors. Presently, the club president and the club secretary are both seniors. (They must be different people.)

If the students were selected randomly for these offices, what is the probability both would be seniors?

a. Show how to find the answer to this question using the General Multiplication Rule for probability.

b. Show how to find the answer using the Multiplication Principle of Counting and the definition of probability.

4 Consider the experiment of flipping a fair coin four times and counting the number of heads.

a. Out of the possible sequences of heads and tails, how many sequences contain no heads? How many contain exactly one head? Exactly two heads? Exactly three heads? Four heads?

b. Show or describe where you can find the answers to these questions in Pascal's triangle.

c. Explain how you can reason about combinations to find the answers to the questions in Part a.

d. Find the probability of getting more than three heads.

5 There are many interesting patterns in Pascal's triangle. For example, consider the sum of each row of Pascal's triangle.

row 0				1		
row 1			1		1	
row 2		1		2		1
row 3	1		3		3	1
row 4	1	4		6	4	1
row 5	1	5	10	10	5	1

a. Compute the sum of each of the first five rows of Pascal's triangle. Describe any patterns you see. What kind of sequence is formed by the sums of the rows?

b. Make a conjecture about the sum of row n in Pascal's triangle.

c. Consider row 3 of Pascal's triangle. Express each entry as a combination; then express the sum of the entries as a sum of combinations. This should suggest a property of combinations. State that property.

6 Anaba is designing an agricultural experiment. The factors of interest to her are Fertilizer (F), Herbicide (H), and Pesticide (P), each of which can be either *present* or *absent* for the duration of the growing season. She is concerned that pairs of factors may lead to additional effects. For example, using both fertilizer and herbicide may lead to an effect that would not have been present with only fertilizer or with only herbicide.

a. If one of her plots must be a control (no factors present), how many plots of land must she obtain access to in order to account for all single factors (such as $\{P\}$) and all paired factors (such as $\{F, H\}$)?

b. Explain how Anaba could solve this problem by thinking in terms of four factors, one of which is an "empty factor," E.

c. How many plots will she need if she has n factors, instead of three?

CONNECTIONS

7 Re-examine the General Multiplication Rule: $P(A \text{ and } B) = P(A) \times P(B \mid A)$, where A and B are events.

a. You may recall that $P(B \mid A)$ is called a *conditional probability*. $P(B \mid A)$ is often found directly from information in the problem situation. Explain why $P(B \mid A)$ can also be found by calculating $\dfrac{P(A \text{ and } B)}{P(A)}$, provided $P(A) \neq 0$.

b. Use the result in Part a to show that if A and B are independent events, then $P(B \mid A) = P(B)$.

8 Prove that $C(2n, 2) = 2C(n, 2) + n^2$ for $n \geq 2$, using the two methods below.

a. The factorial formula for $C(n, k)$ and algebraic reasoning

b. Combinatorial reasoning

9 In Investigation 2, you saw that there are close connections among combinations, the coefficients of the expansion of $(a + b)^n$, and the entries in row n of Pascal's triangle. These connections allow you to use technology to quickly generate any row of Pascal's triangle.

a. Using combinations, express the entries in row 4 of Pascal's triangle.

b. How could you use the sequence command on your calculator to generate row 4 of Pascal's triangle?

c. Use the sequence command to generate row 10 of Pascal's triangle.

d. Use a CAS to expand $(a + b)^n$, for the appropriate value of n, to generate row 10 of Pascal's triangle.

10 In this lesson, you have primarily studied counting situations in probability and algebra. Combinatorial questions also often arise in geometry.

a. Given a set of n points, no three of which are collinear, how many distinct triangles can be formed with three of the n points as vertices? Does it make any difference if all the points are not in the same plane? Explain your reasoning.

b. Given a set of n points, no three of which are collinear and no four of which are coplanar (that is, no four of which lie in the same plane), how many distinct tetrahedra can be formed using four of the n points as vertices?

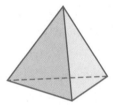

c. Suppose a map is formed by drawing n lines in a plane, no two of which are parallel and no three of which intersect in a common point. What is the fewest number of colors needed to color this map so that no two regions with a common boundary are the same color?

11 There are many counting problems related to vertex-edge graphs.

a. Recall that a **complete graph** is a graph in which there is exactly one edge between each pair of vertices. How many edges are there in a complete graph with 8 vertices? How many edges are there in a complete graph with n vertices?

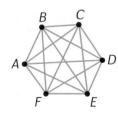

b. A **cycle graph** is a graph consisting of a single cycle, as in the graphs below.

In a previous course, you may have studied vertex coloring as a means to model situations involving potential conflicts. Now consider *edge coloring*. To **color the edges of a graph** means to assign colors to the edges in such a way that if two edges share a vertex, then they must be different colors. What is the fewest number of colors needed to color the edges of a cycle graph with n vertices ($n \geq 3$)?

c. What is the fewest number of colors needed to color the edges of a complete graph with n vertices ($n \geq 2$)?

REFLECTIONS

12 Describe connections between the General Multiplication Rule for probability and the ideas of permutations and combinations.

13 In their article, "The Evolution with Age of Probabilistic, Intuitively Based Misconceptions," in the *Journal for Research in Mathematics Education* (January 1997, pp. 96–105), Efraim Fischbein and Ditza Schnarch reported on a survey in which 100 high school students were asked the following question: "When choosing a committee composed of 2 members from among 10 candidates, is the number of possibilities smaller than, equal to, or greater than the number of possibilities when choosing a committee of 8 members from among 10 candidates?"

a. What is the correct answer? State this answer using combinations.

b. Eighty-five percent of the high school students surveyed in the study stated that the correct answer is "greater than." Why do you think that so many students believed that there are more two-member committees than eight-member committees?

14 At the beginning of this lesson, you were introduced to Paul Erdös, one of the most famous recent mathematicians in the area of combinatorics. Read a few chapters about Erdös in one of the following books. Write a short report about the man and his work.

- Paul Hoffman, *The Man Who Loved Only Numbers: Mathematical Truth.* New York: Hyperion, 1998.

- Bruce Schecter, *My Brain is Open: The Mathematical Journeys of Paul Erdös.* New York: Simon and Schuster, 1998.

15 A useful mathematical practice is to: *Look for and express regularity in repeated reasoning*. Look back at the work you did in Investigation 2: Combinations, the Binomial Theorem, and Pascal's Triangle. Identify at least one instance where you used this mathematical practice. Explain how you used the practice.

EXTENSIONS

16 Poll your class to find the number of students who have been to the movies in the last week. Suppose you select two students at random from your class using the following method: Write the name of each student on a slip of paper. Select a slip at random. Do not replace that slip. Select another slip at random.

a. What is the probability that both students you select have been to the movies in the past week? If you had replaced the first slip of paper, what would be the probability that each name you select is a student who has been to the movies in the past week?

b. In the remainder of this task, you will investigate whether the probability of selecting a student who has been to the movies in the last week is the same on the first draw as on the second draw if you do *not* replace the slip of paper.

   Make a conjecture about whether the probability of selecting a student who has been to the movies in the last week is the same on the first draw as on the second draw, when you do not replace the slip of paper. Then check your conjecture by completing the parts below.

   i. What is the probability of selecting a student on the first draw who has been to the movies in the last week?

   ii. What is the probability of selecting a student on the first draw who has not been to the movies in the last week and a student on the second draw who has been to the movies in the last week?

   iii. What is the probability of selecting a student on the first draw who has been to the movies in the last week and a student on the second draw who has been to the movies in the last week?

   iv. Use results from parts ii and iii to find the probability of selecting a student on the second draw who has been to the movies in the last week. Is this the same as, or different from, the first-draw probability you found in part i?

17 A small inland lake is stocked with 100 fish, 20 of which are tagged. Some time later, a fisherman catches five fish.

  a. Assuming that all 100 fish are still in the lake when he starts fishing and that this is the total population of the lake, how many ways are there for him to catch five fish of which two are tagged fish?

  b. Assuming that the fish are all equally likely to be caught, what is the probability of this event?

  c. Problems of this type can be challenging. How could you check your work?

18 Prove that $rC(n, r) = nC(n - 1, r - 1)$ for $n \geq r \geq 1$, using the following two methods.

  a. A factorial formula for $C(n, k)$ and algebraic reasoning

  b. Combinatorial reasoning (*Hint*: Think about choosing a committee of r people and a chairperson from a group of n people.)

19 Consider the following example of a property of combinations:

$$C(5, 3) = C(4, 2) + C(3, 2) + C(2, 2)$$

  a. Show that this statement is true by using a factorial formula for $C(n, k)$.

  b. Another example of this property is the following:

$$C(6, 2) = C(5, 1) + C(4, 1) + C(3, 1) + C(2, 1) + C(1, 1)$$

  Based on the above two examples, make a conjecture for a general statement of this property.

  c. Explain how this property appears as a pattern in Pascal's triangle.

  d. Prove this property using combinatorial reasoning.

  e. Justify the property by giving an argument in terms of coefficients in the expansion of $(a + b)^n$.

20 In this lesson, you have seen that there is a close correspondence between the entries of Pascal's triangle and values of $C(n, k)$, but you have only seen this correspondence as a pattern; you have not yet proven it. The reason for the close correspondence is because the construction rule for Pascal's triangle is the same as an important recursive property of combinations.

Rule for Pascal's triangle:	*(number in any row)*	=	*(number above and to the left)*	+	*(number above and to the right)*
Combination property:	$C(n, k)$	=	$C(n - 1, k - 1)$	+	$C(n - 1, k)$

  a. Think about this correspondence. Using the fact that entry k in row n of Pascal's triangle is $C(n, k)$, explain why $C(n - 1, k - 1)$ corresponds to the "number above and to the left" and $C(n - 1, k)$ corresponds to the "number above and to the right."

  b. Verify the combination property for some specific values of n and k.

  c. Prove the combination property. That is, prove
  $C(n, k) = C(n - 1, k - 1) + C(n - 1, k)$, where $k > 0$ and $k < n$.

21 Determine the value of each of the following without using technology.

a. $\ln e^3$

b. $\ln 1$

c. $\log 50 + \log 2$

d. $e^{\ln 5}$

e. $\dfrac{\log 25 - \log 250}{\log 100}$

22 Find all complex (including real) number solutions to each equation.

a. $2.6e^{x+3} - 12.6 = 86.2$

b. $4x^3 + 8x^2 - 52x = 24$

c. $\log (x + 2) - \log (x - 5) = 1$

d. $2 \sin^2 x + 3 \sin x - 2 = 0$

23 Consider the function $f(x) = \dfrac{\sin 2x + \sin x}{\cos x}$.

a. Identify the domain of $f(x)$. Explain your reasoning.

b. If $g(x) = 2 \sin x + \tan x$, show that $g(x)$ has the same domain as $f(x)$ and that $f(x) = g(x)$.

c. Determine all solutions to $f(x) = 0$.

24 Identify the conic section represented by each equation and write the equation in standard form. Then use properties of the conic to sketch its graph.

a. $x^2 - 4x + y^2 + 8y = 80$

b. $25x^2 - 4y^2 - 100 = 0$

c. $2x^2 + 16x + 5y^2 - 10y + 17 = 0$

25 Consider the function $f(x) = 2 \cos x - 1$.

a. Sketch a graph of the function for $-2\pi \le x \le 2\pi$.

b. Find all solutions to $f(x) = 0$.

c. Identify all values of x in the interval $[-2\pi, 2\pi]$ for which $f'(x) < 0$.

d. Estimate the value of $f'\left(\frac{\pi}{2}\right)$ using a difference quotient. Then check using technology.

e. Use four rectangles to estimate $\int_0^\pi f(x)\, dx$. Then check using technology.

26 The area of the Mular wetlands is decreasing by 3% per year. The current area of the Mular wetlands is 12 km².

a. Write a function rule of the form $f(t) = ab^t$ that will give the area of the Mular wetlands t years from now.

b. Write a function rule of the form $h(t) = ae^{kt}$ that is equivalent to the rule you wrote in Part a.

c. At the current rate of decrease, how long will it be before the area of the Mular wetlands is half of what it is now?

d. The area of the Crenshaw wetlands has decreased by 15% over the past five years. Assume that this rate continues for the next five years. Will the Mular wetlands or the Crenshaw wetlands experience a greater percent decrease in area over the next five years?

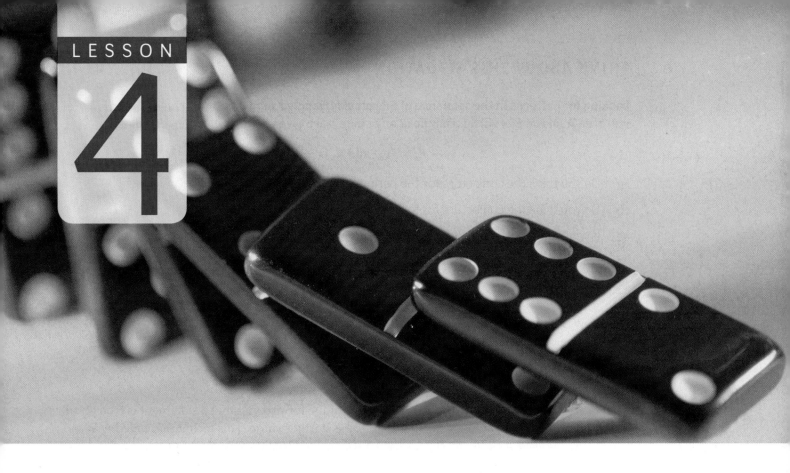

Proof by Mathematical Induction

In previous lessons and throughout *Core-Plus Mathematics*, you investigated and described general patterns that seem to hold for infinitely many integer values of *n*. For example, you made conjectures about the number of diagonals in a regular polygon with *n* sides, the number of two-scoop ice cream cones when *n* flavors are available, and the expansion of expressions of the form $(x + y)^n$.

It is not only important to detect patterns, but also to prove that a particular pattern *must* be true for *all* the stated values of *n*. You will see this proof method is analogous to tipping over an infinite sequence of dominos standing on end.

To begin, consider the Sierpinski triangle that you have investigated in previous courses. Starting with a solid-colored equilateral triangle with side lengths 1 unit, smaller and smaller equilateral triangles, formed by connecting the midpoints of sides, are cut out of the original. The first four stages in the construction are shown below.

Comstock Images/Jupiterimages

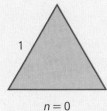
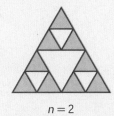
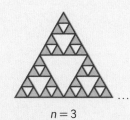

$n = 0$ $n = 1$ $n = 2$ $n = 3$

Looking for patterns in the sequence of Sierpinski triangles suggests that the perimeter P_n and area A_n of the Sierpinski triangle at any stage n are given by the following formulas:

$$P_n = \frac{3^{n+1}}{2^n} \text{ and } A_n = \frac{3^n\sqrt{3}}{4^{n+1}}$$

a Do you think the conjecture for the perimeter formula is correct? Why or why not?

b Do you think the conjecture for the area formula is correct? Explain.

c If you think that the conjectures are true, how might you convince a skeptic that each of the formulas is correct for *every* non-negative integer n?

d If some students think that one of these conjectures is false, how could they *prove* that it is false?

e How is the perimeter of the Sierpinski triangle at any stage $P_n (n \geq 1)$ related to the perimeter at the previous stage P_{n-1}? How is A_n related to A_{n-1}?

In this lesson, you will learn a technique for proving that a statement is true for all integer values of n larger than some starting value. This technique, called *proof by mathematical induction*, is closely connected to the work you have previously done with recursion.

INVESTIGATION 1

Infinity, Recursion, and Mathematical Induction

As you work on the problems in this investigation, look for answers to these questions:

How do you prove a statement by mathematical induction?

For what types of problems is mathematical induction often useful?

How is recursion used in and related to proof by mathematical induction?

You have to be very careful when thinking about a statement that is supposed to be true for infinitely many values of n. The statement may seem true, and in fact it may be true for many values of n, but then it can turn out to be false. As you know, a single counterexample will prove that a statement claimed for infinitely many integers n is not true.

1 **Conjectures, Counterexamples, and Proofs** Suppose the regions of a map are formed by drawing $n \geq 1$ lines in a plane. No two regions with a common boundary have the same color.

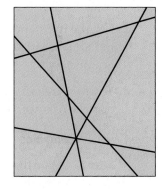

a. Copy and complete the table below. In each case, determine the minimum number of colors required to color the map.

Number of Lines	1	2	3	4	5
Number of Colors					

b. How many colors are needed to color the regions of a map formed by n lines? Compare your conjecture with your classmates.

c. How can you be sure that the conjecture holds for all cases?

2 Suppose you want to prove that a statement about infinitely many values of n is true. Using a computer, you might verify the statement for thousands of values of n. But that does not necessarily mean that the statement is always true. Maybe you have not yet found a counterexample; it might be the next value of n that does not work. You might generate a table and find a pattern as in Problem 1. But how do you know for sure that the pattern will always be valid?

  Consider a situation with exponential and factorial functions. As you have seen in your previous studies, these functions can generate numbers that are very large. Think about which type of function grows faster. In particular, compare the sizes of 5^n and $n!$ for positive integer values of n.

a. Consider the following conjecture: $5^n > n!$ for $n \geq 1$. Do you think this conjecture is true or false? Explain your reasoning.

b. Next, consider this conjecture: $n! > 5^n$ for $n \geq 1$. Is this conjecture true? Would it be true if the condition $n \geq 1$ was modified? If so, how would you modify it?

c. Now consider the conjecture: $n! > 5^n$ for $n \geq 12$. Is this conjecture true or false? How do you know for sure?

The Principle of Mathematical Induction In the next three problems, you will learn an ingenious technique for proving a statement about infinitely many integers. This powerful technique is based on a simple idea that is illustrated in the next problem.

3 Think about setting up a number of dominos on edge, and then making them all fall over, one after another. This is like proving a statement for all values of n, one after another.

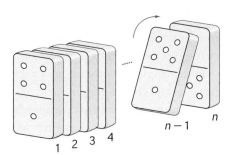

a. Set up five dominos or books and then make them all fall over, one after another, just by gently pushing the top of the first domino or book.

b. Suppose you set up a line of n dominos in such a way that for a given domino at any place in the line, if the previous domino falls then it will knock over the given domino. Now suppose you knock over the first domino. What will be the result?

c. List the two conditions that are required for the entire line of n dominos in Part b to fall down. Explain why neither one of the conditions alone is sufficient.

d. The idea in Parts b and c is the basic idea behind a *proof by mathematical induction*, as seen below.

Dominos	**Proof by Mathematical Induction**
Step 1. Set up the dominos in such a way that a given domino will fall over whenever the previous domino falls over.	Show that a statement is true for n whenever it is true for $n-1$.
Step 2. Knock over the first domino.	Show that the statement is true for some initial value of n.
Result	*Conclusion*
All the dominos fall over.	The statement is true for all values of n greater than or equal to the initial value.

Explain in your own words the analogy between a set of dominos falling over and proof by mathematical induction.

The validity of a proof by mathematical induction rests, in part, on the Principle of Mathematical Induction, stated below. This principle is just a formalization of the scheme in Problem 3 Part d above. The Principle of Mathematical Induction is a basic property of integers. It can be used to prove that a statement about integers, $S(n)$, is true for every integer $n \geq n_0$, where n_0 is the initial (starting) value.

Principle of Mathematical Induction

Suppose $S(n)$ is a statement about integers. If

(1) $S(n)$ is true whenever $S(n-1)$ is true, for each $n > n_0$, and

(2) $S(n_0)$ is true,

then $S(n)$ is true for all integers $n \geq n_0$.

From Inquiry to Proof In the next several problems, you will investigate how to carry out a proof by mathematical induction. The first problem involves vertex-edge graphs. Recall that a **complete graph** is a vertex-edge graph in which there is exactly one edge between every pair of vertices. The graph at the right is a complete graph on four vertices.

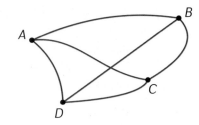

4 Consider this basic question in the study of graph theory.

How many edges does a complete graph with n vertices have?

Below is one possible way to proceed.

a. A common first step in discovering and proving the answer to a general question like the one above is to examine several special cases and look for patterns. Draw several complete graphs and count the number of edges in each. How many edges do you think there are in a complete graph with n vertices?

b. Let E_n be the number of edges in a complete graph with n vertices. As you have seen many times in *Core-Plus Mathematics*, there are two fundamental ways to think about E_n: You could find a function rule that describes E_n as a function of n, or you could find a recursive formula that describes E_n in terms of E_{n-1}. In the situation here, you need to do both. However, as you can see from condition (1) of the Principle of Mathematical Induction, an essential tool used in the proof is the recursive formula. It is the function rule that you will prove using the Principle of Mathematical Induction. (Note that a function rule is also commonly called an *explicit formula*.)

    i. Find a recursive formula that describes the relationship between E_n and E_{n-1}. Explain why this formula is valid.

    ii. Find a function rule that describes E_n as a function of n.

    iii. One function rule is $E_n = \dfrac{n(n-1)}{2}$. If necessary, transform your function rule from part ii into this form.

5 **Proof by Mathematical Induction** You can use the Principle of Mathematical Induction to *prove* that the function rule $E_n = \dfrac{n(n-1)}{2}$ is true for every $n \geq 1$. There are two steps in a proof by mathematical induction, corresponding to the two conditions of the Principle of Mathematical Induction. These steps may be done in either order.

I. Induction Step:
You must prove the statement is true for n whenever it is true for $n-1$. That is, you must show that *if* the statement is true for $n-1$, then it will be true for n. (In terms of dominos, you are showing that the dominos are set up in such a way that if the previous domino falls, then it will knock over the next domino.) Thus, in the induction step of a proof by mathematical induction, you may assume that the statement is true for $n-1$ while you try to prove it is true for n. (This assumption is often called the **induction hypothesis**.)

a. In this situation, this means that you may assume that the function rule for E_{n-1} is true. Write the function rule for E_{n-1}.

b. Your goal now is to prove that $E_n = \dfrac{n(n-1)}{2}$.

Start with E_n, use the recursive formula and the assumption of the function rule for E_{n-1}, and try to end with $\dfrac{n(n-1)}{2}$, as shown in the following outline of a proof.

Fill in the missing details.

$$E_n = E_{n-1} + (n-1)$$ The recursive formula from Problem 4 Part b

$$= \frac{(n-1)(n-2)}{2} + (n-1)$$ Why?

$$= \frac{n^2 - 3n + 2}{2} + (n-1)$$ Why?

$$= ?$$ Why?

$$= ?$$ Why?

$$\vdots$$ (You may use more or fewer steps.)

$$E_n = \frac{n(n-1)}{2}$$ Why?

The reasoning in Part b completes the induction step. Now complete the base step.

II. Base Step:
You need to show that the function rule is true for the initial value of n, in this case $n = 1$.

c. Prove that $E_n = \frac{n(n-1)}{2}$ is true for $n = 1$. That is, prove that a complete graph with $n = 1$ vertex has $\frac{n(n-1)}{2}$ edges.

d. Explain the base step in terms of the dominos analogy.

You have now completed the two steps of a proof by mathematical induction, and thus you have satisfied the two conditions of the Principle of Mathematical Induction. Therefore, you can conclude that $E_n = \frac{n(n-1)}{2}$ for every $n \geq 1$.

6 Here is another example of the process of experimenting, conjecturing, and then proving a conjecture by mathematical induction. Consider the following question.

How many regions are formed by n lines in the plane (n ≥ 1), all of which pass through a common point?

a. Experiment with this situation. Look for patterns and make a conjecture. Express your conjecture as a function of the number of lines n. Compare your conjecture to those of other students. Resolve any differences.

b. Find a recursive formula that describes the relationship between the number of regions formed by n lines and the number of regions formed by $n - 1$ lines. Compare your formula with that of your classmates and resolve any differences.

c. Use the Principle of Mathematical Induction to prove the function rule you conjectured in Part a.

CHECK YOUR UNDERSTANDING

Proof by mathematical induction is often used to prove statements about finite sums. For example, consider the sum of the first n odd positive integers:

$$S_n = 1 + 3 + 5 + 7 + \cdots + (2n - 1)$$

a. Compute S_n for a few values of n. Make a conjecture for a concise formula for S_n as a function of n.

b. Use mathematical induction to prove your conjecture.

INVESTIGATION 2

A Closer Look at Mathematical Induction

Now that you have some experience with proof by mathematical induction, you can consider some important details. First, you have seen that you must find a connection between the n case and the $n-1$ case in order to carry out a proof by mathematical induction. This connection is often expressed by a recursive formula. It is important to be able to explain (prove) that the recursive formula used in a proof by mathematical induction is valid.

Second, another important part of a proof by mathematical induction corresponds to knocking over the first domino. That is, you must prove that the statement is true for the first specified value of n. In the contexts used in Investigation 1, the initial value of n is $n = 1$. But, as you will see in this investigation, depending on the problem, the initial value of n could be any non-negative integer.

Finally, it is often valuable to consider different methods for proving the same conjecture. This will help you better understand the truth of the statement, the concepts involved, and various proof methods. In this investigation, you will consider the *Least Number Principle* and how to prove conjectures involving non-negative integers using this principle. Then you will compare this proof strategy to proofs by mathematical induction.

As you work on the problems in this investigation, look for answers to the following questions:

> *For proof by mathematical induction:*
>
> > *Why is it important to establish the first value for which a statement involving all non-negative integers is true?*
>
> > *Why is it important to be certain your recursive formula is true?*
>
> > *What is the Least Number Principle, how do you use it, and how do proofs using this principle compare to proof by induction?*

1 Think about the initial value for statements that are true for some subset of non-negative integers. Consider the power function $y = n^2$ and the exponential function $y = 2^n$.

a. Compare the growth of $y = n^2$ and $y = 2^n$. For large integer values of n, is $n^2 > 2^n$ or $2^n > n^2$? Make a conjecture.

b. What is the smallest integer m for which your conjecture in Part a is true for all $n \geq m$?

c. In Part b, you probably determined that the statement $2^n > n^2$ seems to hold for $n \geq 5$. Here is part of a proof by mathematical induction for this statement.

$$
\begin{aligned}
2^n &= (2)(2^{n-1}) && (1)\\
&> (2)(n-1)^2 && (2)\\
&> n^2 \text{ for } n \geq 4 \text{ (thus, also for } n > 5) && (3)
\end{aligned}
$$

i. Explain each step of this part of the proof. In particular, give an argument for why $2(n-1)^2 > n^2$ for $n \geq 4$.

ii. Complete the induction proof.

2 Now consider the importance of explaining (proving) why a recursive formula is true. For example, consider Problems 4 and 5 of Investigation 1 in which you were asked to find and prove a function rule for the number of edges in a complete graph. The recursive formula is $E_n = E_{n-1} + (n-1)$.

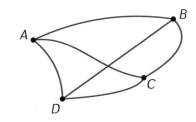

In this case, you might use the following reasoning to justify the formula.

The relation $E_n = E_{n-1} + (n-1)$ is true because you can construct a complete graph on n vertices from a complete graph on $n-1$ vertices by drawing a new vertex and then adding $n-1$ edges, one from the new vertex to each of the $n-1$ existing vertices.

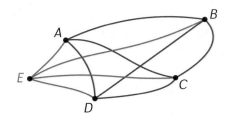

a. Look back at the induction proof you did in Problem 6 of Investigation 1. Identify the recursive formula used in the proof. Then explain why the recursive formula is true by reasoning about the context of the problem.

b. Now look back at the induction proof you did for the Check Your Understanding task on page 601. Identify the recursive formula used in the proof. Then explain why the recursive formula is true by reasoning about the context of the problem.

3 On the circle below, three points are marked and connected. Four non-overlapping regions are formed. Experiment with this situation to find the maximum number of non-overlapping regions formed by connecting $n \geq 2$ points on a circle.

a. Copy and complete the table below. In each case, draw a large circle. Mark points on the circle. Connect all pairs of points with segments. Count the number of non-overlapping regions.

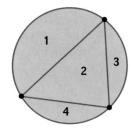

Number of Points	2	3	4	5
Number of Regions		4		

b. What patterns do you see in your completed table? Just consider the table up to 5 points on a circle. Compare your observations with those of your classmates.

   i. Based on the table, find a recursive formula that seems to describe the relationship between the n case and the $n - 1$ case. Also, make a conjecture for a rule that expresses the number of regions as a function of the number of points n.

   ii. You now have a recursive formula and you have made a conjecture for a function rule. Are you sure that the recursive formula is correct? If the recursive formula is correct, what would be your prediction for the maximum number of regions for 6 points? Make a large and carefully drawn diagram and count the number of regions for 6 points. What is your conclusion?

c. Since the case of $n = 6$ does not fit the pattern, you have discovered that this pattern is not valid for this circle problem.

   i. From part i above, you have a conjecture for a recursive formula ($R_n = 2R_{n-1}$) and for a function rule ($R_n = 2^{n-1}, n \geq 2$). Does the recursive formula fit with the function rule?

   ii. Explain why it is not enough to simply recognize patterns in a partially-completed table when using a recursive formula for doing a mathematical induction proof.

The Least Number Principle In Unit 4, *Trigonometric Functions and Equations*, you may have used a proof by contradiction involving the *Least Number Principle* to establish De Moivre's Theorem. The **Least Number Principle** is a simple intuitive idea: every nonempty set of non-negative integers has a least element. This statement is another basic property of integers.

In the following problem, you will use the Least Number Principle and the Principle of Mathematical Induction to prove the formulas for area and perimeter of a Sierpinski triangle introduced at the beginning of this lesson.

4 Reproduced below are the first four stages of the Sierpinski triangle you considered in the Think About This Situation on page 596.

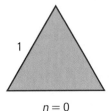 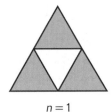 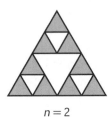 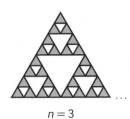

$n = 0$ $\qquad$ $n = 1$ $\qquad$ $n = 2$ $\qquad$ $n = 3$

a. The Least Number Principle can be used to provide an indirect proof (proof by contradiction) that the area of the figure at stage n is given by

$$A_n = \frac{3^n\sqrt{3}}{4^{n+1}} \text{ for } n \geq 0.$$

Give reasons for each of the following steps in the proof.

The formula $A_n = \frac{3^n\sqrt{3}}{4^{n+1}}$ is true for $n = 0$. $\qquad$ (1)

Assume the formula is not true for all $\qquad$ (2)
non-negative integers $n \geq 0$.

Let k be the smallest non-negative integer for $\qquad$ (3)
which the formula is false.

It follows that the formula is true for $k - 1$. $\qquad$ (4)
That is, $A_{k-1} = \frac{3^{k-1}\sqrt{3}}{4^k}$.

At each stage, the area is $\frac{3}{4}$ of the previous area. $\qquad$ (5)

Thus, $A_k = \left(\frac{3}{4}\right)\left(\frac{3^{k-1}\sqrt{3}}{4^k}\right) = \frac{3^k\sqrt{3}}{4^{k+1}}$. $\qquad$ (6)

But this contradicts the statement in Step 3. $\qquad$ (7)

Therefore, the assumption in Step 2 must be incorrect.
Hence, $A_n = \frac{3^n\sqrt{3}}{4^{n+1}}$ for $n \geq 0$. $\qquad$ (8)

b. Use the Least Number Principle to prove that the perimeter of the figure at stage n is given by $P_n = \frac{3^{n+1}}{2^n}$ for $n \geq 0$.

5 Now consider how mathematical induction can be used to prove the same results for the Sierpinski triangle.

a. Use mathematical induction to prove the area formula, $A_n = \frac{3^n\sqrt{3}}{4^{n+1}}$.

b. Use mathematical induction to prove the perimeter formula, $P_n = \frac{3^{n+1}}{2^n}$.

6 Discuss these two methods of proof with your classmates. How are they similar and how are they different? Which proof method makes the most sense to you? Why?

In this investigation, you examined proof by mathematical induction more closely and compared it to proof using the Least Number Principle.

a In a proof by mathematical induction, why is it important to establish the first value for which a statement involving all non-negative integers is true?

b In proof by mathematical induction, why is it important to be very sure that your recursive formula is true?

c Describe, in general, how you prove statements involving all non-negative integers by using the Least Number Principle.

Be prepared to share your strategies and reasoning with the class.

 CHECK YOUR UNDERSTANDING

The number of diagonals in a regular pentagon is 5 as shown in the diagram below.

a. Find an explicit formula (function rule) and a recursive formula that express the number of diagonals in a regular *n*-gon. Explain why your recursive formula is true.

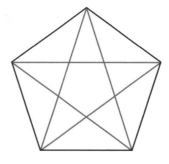

b. What is the smallest value of *n* for which your formula is true?

c. Prove by mathematical induction that your explicit formula is true for all *n* greater than or equal to the base value you found in Part b.

d. Use the Least Number Principle to prove your explicit formula in Part a.

APPLICATIONS

1 In Problem 6 of Investigation 1, you investigated the number of regions formed by n lines in the plane, all of which pass through a common point. Now remove the restriction that all the lines pass through a common point. What is the maximum number of regions formed by n lines in a plane if no two lines are parallel and no three intersect in a common point? Prove your conjecture using the Principle of Mathematical Induction.

2 The Towers of Hanoi is a mathematical game featured in an old story about when the world will end. As the story goes, a group of monks in a secluded temple is working on the game described below. When they have completed the game, the world will end!

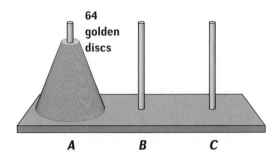

The game is played with three pegs mounted on a board and 64 golden discs of successively larger sizes with holes in the center. The game begins with all 64 discs stacked on one peg, graduated in size from largest to smallest. The goal is to move the discs one at a time from peg to peg until all the discs are stacked, largest to smallest, on another peg. The rules of the game are as follows.

- Only one disc may be moved at a time.

- You are not allowed to put a larger disc on top of a smaller disc.

- A disc may be moved more than once.

- During the course of the game, you may move a disc to any peg, including the one on which the discs were originally stacked.

a. What is the fewest number of moves needed to complete a game with two discs? How about a game with three discs? Four discs?

b. Consider the Towers of Hanoi game with n discs. Make a conjecture for the fewest number of moves needed to complete the game. Express your conjecture as a function of n.

c. Use the Principle of Mathematical Induction to prove your conjecture in Part b.

d. What is the fewest number of moves needed to finish a game with 64 discs, as in the original story? If the monks in the story move one disc every second and work nonstop, should you worry about the world ending soon? Explain.

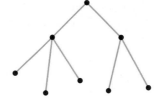

3 You may recall that a **tree** is a vertex-edge graph that is connected and has no circuits. Use mathematical induction to prove that a tree with n vertices has exactly $n - 1$ edges. You may find it useful to use the fact that every tree has at least one "dangling vertex," that is, a vertex of degree one.

4 Use mathematical induction to prove that the sum S_n of the first n positive integers is equal to $\dfrac{n(n + 1)}{2}$. That is, prove that

$$S_n = 1 + 2 + 3 + \cdots + (n - 1) + n = \frac{n(n + 1)}{2} \text{ for } n \geq 1.$$

5 In the Check Your Understanding on page 601, you proved that the sum of the first n odd positive integers is n^2. Now prove that statement using the Least Number Principle.

6 Prove that $n! > 2^n$ for $n \geq 4$.

CONNECTIONS

7 In Problem 4 Part b (page 599), you found a function rule for the number of edges in a complete graph on n vertices. Now, determine function rules for $C(n, 2)$ and for the sum of the first $n - 1$ positive integers. Explain the connections among the three function rules by reasoning about the three contexts.

8 In Course 3 Unit 8, *Inverse Functions*, you proved $\log x^m = m \log x$, where m is a non-negative integer. Use mathematical induction to prove this logarithm property.

9 In Course 3 Unit 7, *Recursion and Iteration*, you investigated geometric series of the form $a_0 + a_0 r + a_0 r^2 + \cdots + a_0 r^{n-1} + a_0 r^n$, where $r \neq 1$ and $n \geq 0$. Consider the sum S_n of the geometric series with $a_0 = 1$:

$$S_n = 1 + r + r^2 + \cdots + r^{n-1} + r^n$$

  a. Use the Principle of Mathematical Induction to prove that the following statement is true:

$$1 + r + r^2 + \cdots + r^{n-1} + r^n = \frac{1 - r^{n+1}}{1 - r} \text{ for } n \geq 0 \text{ and } r \neq 1.$$

  b. Use the Least Number Principle to prove the statement in Part a.

  c. There are often several ways to prove a mathematical statement. For the statement in Part a, explain why a reasonable alternate proof strategy would be to multiply both sides by $(1 - r)$. Use this strategy to carry out a proof.

10 Recall that a set A is a subset of a set B if every element of set A is an element of set B. For example, if $B = \{1, 4, 9\}$, then the subsets of B are:

$$\varnothing, \{1\}, \{4\}, \{9\}, \{1, 4\}, \{1, 9\}, \{4, 9\}, \{1, 4, 9\}$$

a. Examine some sets and their subsets and formulate a conjecture about the number of subsets of a set with n elements.

b. Prove your conjecture using the Principle of Mathematical Induction.

c. Prove your conjecture using the Least Number Principle.

11 In Unit 4, *Trigonometric Functions and Equations*, you may have used the Least Number Principle to prove De Moivre's Theorem about complex numbers:

If $z = r(\cos \theta + i \sin \theta)$ and n is any positive integer,
then $z^n = r^n(\cos n\theta + i \sin n\theta)$.

Now use mathematical induction to prove De Moivre's Theorem.

REFLECTIONS

12 Mathematical induction is a powerful method of proof, but it must be used correctly. What is wrong with the following "proof"?

Proposition: Every positive integer is equal to its successor. That is, $n = n + 1$.

Proof:
$n = (n - 1) + 1$ This is the recursive formula relating n to $n - 1$.
$= (n) + 1$ When proving by induction, you assume that the statement is true for the $n - 1$ case. Here, this means that $n - 1 = n$.
$= n + 1$ Done

13 How does proof by mathematical induction relate to the ideas of sequences and recursive and function formulas that you studied in Course 3 Unit 7, *Recursion and Iteration*?

14 The French mathematician Henri Poincaré (1854–1912) stated that the Principle of Mathematical Induction cannot be denied by the human mind. What do you think he meant? Do you agree?

15 Often there is more than one way to prove a statement. Look back at the mathematical induction proof from the Check Your Understanding task on page 605. Provide a counting argument that also proves this statement. (*Hint:* A diagonal can be determined by the two vertices that are its endpoints, so you can count the diagonals by counting pairs of appropriate vertices.) Which proof is more convincing to you? Which proof is easier for you? Why?

16 Look back at the Sierpinski triangle diagrams and formulas for the area and perimeter of the figures (pages 595 and 596). What happens to the area of the figures as n gets very large? What happens to the perimeter of the figures as n gets very large?

17 A useful mathematical practice is to: *Look for and make use of structure*. Look back at the work you did in this lesson. Find at least one instance where you found making use of mathematical structure to be helpful. Explain how you used the practice.

EXTENSIONS

18 In Lesson 3, you investigated the Binomial Theorem. Use the Principle of Mathematical Induction to prove this special case:

$$(x + 1)^n = C(n, 0)x^n + C(n, 1)x^{n-1} + C(n, 2)x^{n-2} + \cdots + C(n, n-1)x^1 + C(n, n)x^0$$

You may find it useful to use the fact from Extensions Task 20 in Lesson 3 (page 592) that $C(n, k) = C(n-1, k-1) + C(n-1, k)$.

19 In Course 3 Unit 7, *Recursion and Iteration*, you studied arithmetic sequences and series. By definition, for an arithmetic sequence with initial term a_1 and common difference d, the sum of the first n terms of the sequence is:
$$S_n = a_1 + (a_1 + d) + (a_1 + 2d) + \cdots + [a_1 + (n-1)d].$$
Consider the following conjecture: $S_n = \frac{1}{2}n[2a_1 + (n-1)d]$ for $n \geq 1$.

a. Verify that this conjecture is true for $n = 1, 2,$ and 3. That is, find the sum of the first 1, 2, and 3 terms using the definition. Show that you get the same answers if you substitute $n = 1, 2,$ and 3 into the conjectured formula.

b. Prove the conjecture using the Principle of Mathematical Induction.

c. Prove the conjecture without using the Principle of Mathematical Induction.

d. Prove that the formula $S_n = \dfrac{n(a_1 + a_n)}{2}$ is equivalent to the one given above in the conjecture.

20 Prove that $\displaystyle\sum_{k=0}^{n} C(n, k) = 2^n$, $n \geq 0$ using mathematical induction. Explain how this statement is related to counting subsets, as in Problem 7 on page 564 and Connections Task 10 on page 608. You may find it useful to use the fact from Extensions Task 20 on page 592 that $C(n, k) = C(n-1, k-1) + C(n-1, k)$.

21 Look back at your conjecture in Problem 1 (page 597) of Investigation 1. Prove your conjecture.

22 Look back at Problem 3 (page 603) of Investigation 2. It can be shown that for the cases of 7 and 8 points, the number of regions formed are 57 and 99, respectively.

a. What is this a reminder of when reasoning from patterns?

b. Find a polynomial function that relates the number of points on the circle to the maximum number of non-overlapping regions formed by connecting all the points.

23 Consider the Bernoulli inequality, named after the Swiss mathematician Jacob Bernoulli (1654–1705): $(1 + x)^n \geq 1 + nx$ for every integer $n \geq 0$ and every real number $x > -1$.

   a. Prove the Bernoulli inequality using mathematical induction.

   b. Use the Binomial Theorem to prove this inequality when $x > 0$.

   c. Prove Bernoulli's inequality using the Least Number Principle.

REVIEW

24 If $a > 0$, determine whether each of the following statements is always, sometimes, or never true. In each case, explain your reasoning.

   a. $\log 10a = 1 + \log a$

   b. $\log (3 + a) = \log 3 + \log a$

   c. $(a + 6)^2 = a^2 + 36$

   d. $\dfrac{a + 5}{5} = a + 1$

   e. $\sqrt{a}\sqrt{a} = a$

   f. $\log_3 27^a = 3a$

25 Without using technology, write an equation or function rule that matches each description.

   a. A line whose graph contains the points $(1, 7)$ and $(3, -5)$

   b. A line whose graph is parallel to the graph of $5x - 10y = 4$ and contains the point $(7, 2)$

   c. A quadratic function that contains the point $(4, 5)$ and has vertex at $(2, 9)$

   d. An exponential function representing a population with annual growth rate of 10% and a current population of 10,000

26 A group of friends were planning a beach reunion after having been at college for a year. They rented a large house and decided to share the cost of the rental. Each person would pay separately for personal expenses. The total cost of the rental is $800. Travis estimates that his personal expenses will be $225.

   a. Which rule will give Travis' total cost for the reunion as a function of the number of people n who attend the reunion?

   $$C(n) = \frac{225 + 800}{n} \qquad\qquad C(n) = 225 + \frac{800}{n}$$

   $$C(n) = 225 + 800(n) \qquad\qquad C(n) = 1{,}025n$$

   b. If five people attend the reunion, how much can Travis expect to spend?

c. Travis budgets a total of $325 for the reunion. How many people can attend in order for Travis to stay within his budget?

d. Find a function rule for $C^{-1}(x)$, where x is Travis's total cost, and describe how it could be used to answer the question in Part c.

27 If $\sin \theta = \frac{3}{5}$ and $\frac{\pi}{2} < \theta < \pi$, determine the value of each of the following.

a. $\cos \theta$
b. $\tan \theta$

c. $\sin^2 \theta + \cos^2 \theta$
d. $\sin (\theta + \pi)$

e. $\sin \left(\theta - \frac{\pi}{2}\right)$
f. $\sin 2\theta$

28 If $f(x) = 2x + 7$ and $g(x) = 1 - x^3$, determine each of the following.

a. $(f + g)(3)$
b. $g(f(-3))$

c. $g^{-1}(28)$
d. $f(x + 5)$

e. A rule for $f^{-1}(x)$
f. A rule for $f(g(x))$

29 Rewrite each expression in simplest form using only positive exponents.

a. $3x^{-2}$
b. $\dfrac{4x^2 y^{-3}}{12xy^4}$

c. $\left(\dfrac{6}{x^{-3}}\right)^2$
d. $(2x^{-2}y^3)(3x^5 y^{-1})$

30 The graph of $f(x)$ is shown below. Assume all important features of the graph are shown.

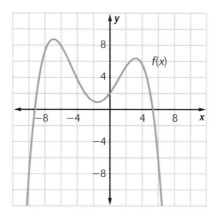

a. What is the global maximum of $f(x)$?

b. How many solutions does $f(x) = 2$ have?

c. Could $f(x)$ be a fifth-degree polynomial? Explain your reasoning.

d. Which of the following is a graph of $f(x - 3)$? Explain your reasoning.

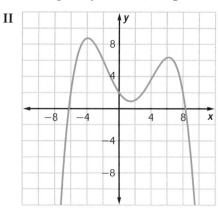

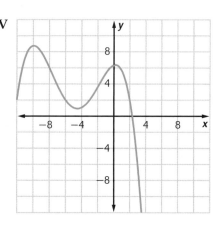

31 Without using technology, create a sketch of each function over the domain $-2\pi \le \theta \le 2\pi$. Then use technology to check your sketches.

a. $y = \sin \theta + 2$

b. $y = 2 \cos \theta$

c. $y = \sin 2\theta$

d. $y = \tan \theta$

e. $y = \cos \left(\theta + \dfrac{\pi}{4}\right)$

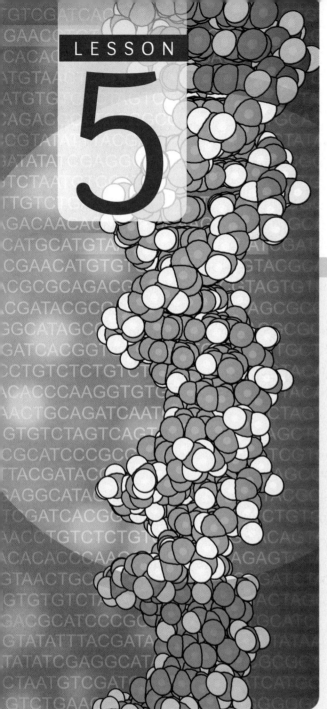

Chad Baker/Getty Images

LESSON 5

Looking Back

In this unit, you have learned useful counting strategies, including systematic lists, counting trees, the Multiplication Principle of Counting, the Addition Principle of Counting, and careful analysis of order and repetition when counting the number of choices from a collection. You have developed and applied formulas for counting permutations and combinations, and you have learned three new methods of proof—combinatorial reasoning, proof by mathematical induction, and proof using the Least Number Principle.

These strategies and skills enable you to solve counting problems both within and outside of mathematics. Outside of mathematics, you encountered applications of counting in many contexts. Within mathematics, your work led to several important results that have wide-ranging applications, including the Binomial Theorem, Pascal's triangle, and the General Multiplication Rule for probability.

In this final lesson, you will review and pull together all these ideas and apply them in new contexts.

1 DNA is part of every cell in every living organism. DNA in humans is arranged into 24 distinct molecules called chromosomes. Each chromosome is a pair of intertwined chains twisted into a spiral. Each spiral consists of a long sequence of pairs of bases. Each base is one of four chemicals: T (thymine), C (cytosine), A (adenine), and G (guanine). The order of the base chemicals determines the instructions, the genetic information, embedded in the DNA.

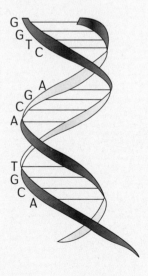

(**Source:** www.ornl.gov/sci/techresources/Human_Genome/project/info.shtml)

a. You have probably seen news reports or TV shows in which DNA is used to uniquely identify people. There are only four base chemicals that make up DNA, and yet there are so many sequences of these bases that no two people have exactly the same DNA. According to the Human Genome Project, there are about 3 billion DNA base pairs in the human genome, that is, in all of the 24 distinct chromosomes. Each chromosome has from about 50 million to 250 million base pair locations.

   i. A base pair is a pair of the base chemicals: T, C, A, and G. How many base pairs are possible? (The pair CG is different than the pair GC since C is on one chain in the twisted spiral in one case and on the other chain in the spiral in the other case.)

   ii. Suppose one chromosome has 120 million base-pair locations. So, there are 120 million locations for pairs of base chemicals to occur. How many different sequences of base pairs are possible for this chromosome?

b. Not all regions of a chromosome directly encode genetic information. Those regions that do are the functional units of heredity, called genes.

   i. Genes comprise only about 2% of the human genome, that is, about 2% of the roughly 3 billion base-pair locations in the human genome. How many base-pair locations is this?

   ii. It is estimated that there are about 22,000 different genes. Suppose a given gene has about the average number of base-pair locations. How many base-pair locations are in this gene?

   iii. How many different sequences of base pairs are possible for this gene? Compare this number to an estimate of the number of atoms in the universe.

2 Examine the following information from a state lottery ticket where each play costs $1 including sales tax.

How to Play
To play, choose two numbers from 1 to 21 from each section (Red, White, and Blue).

How to Win
Every day, two numbers will be drawn from each of three colored ball sets (Red, White, and Blue). Each set has 21 balls numbered from 1 to 21. To win a prize for the following matches, you *must* match both number *and* color.

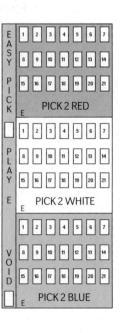

Match	Prize	Number of Winning Choices
6	$1,000,000	1 (out of 9,261,000)
5	$5,000	114 (out of 9,261,000)
4	$100	4,845 (out of 9,261,000)
3	$5	93,860 (out of 9,261,000)
2	$2	828,495 (out of 9,261,000)

a. Verify at least two of the entries in the column entitled "Number of Winning Choices." Explain your reasoning and show your calculations.

b. Determine the probability of each of the different types of matches. State the probability definition that you are using and explain why it is appropriate to use in this situation.

c. Do you think a state lottery like this is an effective way for the state to make money? Do you think it is an effective way for players to make money? Explain.

3 An international relations committee consists of five Democrats, six Republicans, and four Independents. Some of the committee members will be selected to attend an economic conference in Africa.

a. Suppose three members will be chosen to attend the conference. How many different groups of three people can be chosen?

b. How many groups of three people can be chosen if the group must contain exactly one Democrat?

c. Suppose three people will be chosen for the conference and each will take on a role that could be filled by anyone—one will attend all meetings at the conference related to banking, one will attend all meetings related to energy, and one will attend all meetings related to water. How many different role-specific groups of three people can be selected?

d. Suppose that due to budget cuts, only two members will go to the conference. The two members will be chosen at random. What is the probability that both members chosen are Republicans?

e. Think of another counting problem that might arise in this situation. Pose the problem and give the solution. Compare your problem and solution with those of some other students.

4 The following identity exhibits several connections among combinatorial ideas.

$$C(n, 0) + C(n, 1) + \cdots + C(n, n) = 2^n (n \geq 0)$$

To discover some of these connections, complete at least two of the following parts.

a. Explain how this statement appears as a pattern in Pascal's triangle.

b. Use the Binomial Theorem to prove this statement.

c. Use combinatorial reasoning about the number of subsets of a set with *n* elements to prove this statement.

©Corbis

5 Prove that $C(n, 3) = C(n, n - 3)$ for $n \geq 3$, using the following two methods.

a. Use the factorial formula for $C(n, k)$.

b. Use combinatorial reasoning (by reasoning about ways to choose combinations).

6 *Spirals* are a common form in nature and in the man-made world. Pine cones, spider webs, seashells, drill bits, DNA models, screws, and many works of art are examples of spiral designs.

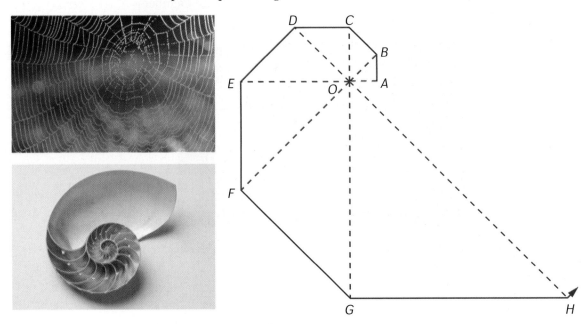

Examine the spiral-like design at the right above. This design is created by starting with the isosceles right triangle OAB in which $OA = AB = 1$ unit. Each of the remaining triangles outlined in the figure also is an isosceles right triangle.

a. Assume that the design begins with $\overline{AB}$. Let L_n be the length of the nth segment. So, $L_0 = 1$. Find rules that describe the patterns relating L_n and L_{n-1} and relating n and L_n. For what values of n do you think your conjectures are true?

b. Use the Principle of Mathematical Induction to prove your conjectured function rule relating n and L_n.

c. Use the Least Number Principle to prove the rule relating n and L_n.

d. Consider the total length of a growing spiral like the one shown in the figure.

  i. What is the total length of such a spiral with five segments, beginning with $\overline{AB}$?

  ii. Explain why the sequence of segment lengths in this spiral design is a geometric sequence.

iii. In Connections Task 9 (page 607) of Lesson 4, you were asked to prove the following formula for the sum of a geometric series with initial term 1 and constant multiplier r.

$$1 + r + r^2 + \cdots + r^n = \frac{1 - r^{n+1}}{1 - r} \text{ for } n \geq 0 \ (r \neq 1)$$

Use this formula to determine the total length of a spiral like the one in the figure, but with 25 segments.

SUMMARIZE THE MATHEMATICS

In this unit, you have investigated many situations where systematic counting methods and combinatorial reasoning are useful.

a Below is a list of some of the combinatorial ideas you have studied. Add other important ideas and topics to the list. Then give a brief explanation and describe an application for each topic on the list.

- Multiplication Principle of Counting
- Permutations
- Combinations
- Combinatorial Reasoning
- Binomial Theorem
- General Multiplication Rule for Probability

b Why are order and repetition important when deciding how to count the number of possible choices from a collection? How are these ideas related to permutations and combinations?

c Describe at least one counting application in each of the following areas of mathematics: algebra, probability, geometry, and graph theory.

d How are Pascal's triangle and the Binomial Theorem related to each other and to the ideas of counting?

e When is mathematical induction a useful method of proof? How is the Principle of Mathematical Induction used in a proof?

f How do you prove a statement is true for all positive integers using the Least Number Principle?

Be prepared to share your ideas and reasoning with the class.

 CHECK YOUR UNDERSTANDING

Write, in outline form, a summary of the important mathematical concepts and methods developed in this unit. Organize your summary so that it can be used as a quick reference in future courses.

English	Español

A

Absolute value of a complex number (pp. 215, 315) A complex number $a + bi$ can be expressed as a position vector. The length r of the position vector is called the **absolute value** or **modulus** of $a + bi$ and is denoted $|a + bi|$. $|a + bi| = \sqrt{a^2 + b^2}$.

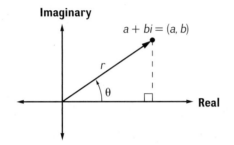

El valor absoluto de un número complejo (pág. 215, 315) Un número complejo $a + bi$ puede ser expresado como un vector posicional. La longitud r del vector posicional se llama *valor absoluto* o *modulo* de $a + bi$ y se representa $|a + bi|$. $|a + bi| = \sqrt{a^2 + b^2}$.

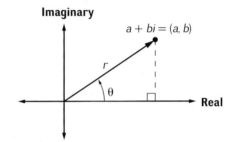

Absolute value function (p. 28)

$$f(x) = |x| = \begin{cases} x, & \text{for } x \geq 0 \\ -x, & \text{for } x < 0 \end{cases}$$

La función del valor absoluto (pág. 28)

$$f(x) = |x| = \begin{cases} x, & \text{por } x \geq 0 \\ -x, & \text{por } x < 0 \end{cases}$$

Addition Principle of Counting (p. 542) The total number of outcomes from two tasks is the sum of the number of outcomes from each task minus the number of outcomes that are common to both tasks, if there are any.

El Principio de Adición de Contar (pág. 542) El número total de resultados de dos tareas es la suma del número de los resultados de dicha tarea menos el número de resultados que son comunes en cada tarea.

Addition of vectors Geometric: (p. 109) When vectors $\vec{u}$ and $\vec{v}$ are represented as arrows, then their sum (or resultant) $\vec{w} = \vec{u} + \vec{v}$ can be found joining the head of $\vec{u}$ to the tail of $\vec{v}$. The sum $\vec{u} + \vec{v}$ is given by the arrow drawn from the tail of $\vec{u}$ to the head of $\vec{v}$.

(See also **Parallelogram Law**.)

La sumación de vectores Geométrica: (pág. 109) Cuando los vectores $\vec{u}$ y $\vec{v}$ están representados como flechas, su suma (o resultante) $\vec{w} = \vec{u} + \vec{v}$ puede ser encontrada conectando el principio de $\vec{u}$ y el fin de $\vec{v}$. La suma $\vec{u} + \vec{v}$ está dada por la flecha dibujada al final de $\vec{u}$ al principio de $\vec{v}$.

(Ver también la **Ley de Paralelogramos**.)

Coordinate: (p. 134) When position vectors $\vec{u}$ and $\vec{v}$ are represented by coordinates $\vec{u} = (u_1, u_2)$ and $\vec{v} = (v_1, v_2)$, then $\vec{u} + \vec{v} = (u_1 + v_1, u_2 + v_2)$.

Coordenada: (pág. 134) Cuando los vectores posicionales $\vec{u}$ y $\vec{v}$ están representados por coordenadas $\vec{u} = (u_1, u_2)$ y $\vec{v} = (v_1, v_2)$, entonces $\vec{u} + \vec{v} = (u_1 + v_1, u_2 + v_2)$.

Arithmetic sequence (p. 20) A sequence of numbers in which the difference between any two consecutive terms is a fixed nonzero constant d. Symbolically, $a_n = a_{n-1} + d$.

Sequencia aritmética (pág. 20) Una sequencia de números dónde la diferencia entre dos términos consecutivos es una constante (desigual a cero, o nocero) d. Simbolicamente, $a_n = a_{n-1} + d$.

English	Español

Associative Property of Addition For any numbers a, b, and c: $a + (b + c) = (a + b) + c$.

La Propiedad Asociativa de Sumatoria Para cualquier número a, b, y c: $a + (b + c) = (a + b) + c$.

Associative Property of Multiplication For any numbers a, b, and c: $a(bc) = (ab)c$.

La Propiedad Asociativa de Multiplicación Para cualquier número a, b, and c: $a(bc) = (ab)c$.

B

Binomial Theorem (p. 584) For any real numbers a and b and positive integer n, $(a + b)^n = C(n, 0)a^n + C(n, 1)a^{n-1}b + C(n, 2)a^{n-2}b^2 + \cdots + C(n, k)a^{n-k}b^k + \cdots + C(n, n-2)a^2b^{n-2} + C(n, n-1)ab^{n-1} + C(n, n)b^n$.

Teorema de binomial (pág. 584) Para cualesquiera números reales a y b y entero positivo n, $(a + b)^n = C(n, 0)a^n + C(n, 1)a^{n-1}b + C(n, 2)a^{n-2}b^2 + \cdots + C(n, k)a^{n-k}b^k + \cdots + C(n, n-2)a^2b^{n-2} + C(n, n-1)ab^{n-1} + C(n, n)b^n$.

C

Circle (p. 407) The set of points in a plane that are a fixed distance r, called the *radius*, from a given point O, called the *center*. A circle with center at the origin and radius r can be represented by an equation of the form $x^2 + y^2 = r^2$, where r is a nonzero real number.

Círculo (pág. 407) El conjunto de puntos en un plano que es de una distancia fija r, llamada *radio*, de un punto dado O, llamada *centro*. Un círculo con centro en el origen y radio r puede ser representada por una ecuación de la forma $x^2 + y^2 = r^2$, donde r es un número real no nulo.

Circular functions (p. 12) Functions (sine and cosine) pairing points on the real number line with coordinates of points on a unit circle by "mapping" the real number line around the unit circle— non-negative real axis is wrapped *counterclockwise* starting at $(1, 0)$; negative real axis is wrapped in *clockwise* direction. For any real number r, if the wrapping procedure pairs r with (x, y), then $\cos r = x$ and $\sin r = y$.

Funciones circulares (pág. 12) Puntos de apareamiento de las funciones (seno y coseno) en una recta numérica real con coordenadas en el círculo unitario mediante "aplicación"—el eje real no negativo se envuelve en *dirección contraria a las manecillas del reloj* comenzando en $(1, 0)$; el eje real negativo se envuelve en *dirección de las manecillas del reloj*. Para cualquier número real r, si el procedimiento de envolver empareja r con (x, y), entonces $\cos r = x$ y $\sin r = y$.

Closure Property of Addition for complex numbers (p. 223) The sum of any two complex numbers is another complex number.

La Propiedad Clausurativa de Adición para números complejos (pág. 223) La suma de cualquier número complejo es otro número complejo.

Closure Property of Multiplication for complex numbers (p. 223) The product of any two complex numbers is a complex number.

La Propiedad Clausurativa de Multiplicación para números complejos (pág. 223) El producto de cualquier número complejo es un número complejo.

Combination (p. 558) A selection of distinct objects or numbers in which order does not matter; that is, different orderings are not counted as different selections (repetitions are not allowed)

Combinación (pág. 558) Una selección de diferentes objetos o números en donde el orden no tiene significado; esto es, ordenes diferentes no están tomadas en cuenta como selecciones diferentes (repeticiones no son permitidas)

Commutative Property of Addition For any numbers a and b: $a + b = b + a$.

La Propiedad Conmutativa de Suma Para cualquier número a y b: $a + b = b + a$.

Commutative Property of Multiplication For any numbers a and b: $ab = ba$.

La Propiedad Conmutativa de Multiplicación Para cualquier número a y b: $ab = ba$.

English

Complex conjugates (p. 212) Pairs of complex numbers in the form $a + bi$ and $a - bi$. The conjugate of complex number z is denoted $\bar{z}$.

Complex number (p. 211) A number that can be expressed in the form $a + bi$ where a and b are real numbers and $i = \sqrt{-1}$. The notation $a + bi$ is called the standard form of a complex number.

Complex number, graphical representation (p. 214) In a coordinate plane (called the complex number plane), the complex number $a + bi$ can be represented as a point or a position vector with coordinates (a, b). See the examples on the grid below where scales on the axes are 1 unit.

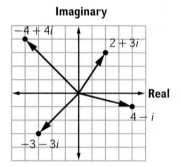

Congruent figures Two figures are congruent if and only if one figure is the image of the other under a rigid transformation (line reflection, translation, rotation, glide reflection) or a composite of rigid transformations. Congruent figures have the same shape and size, regardless of position or orientation.

Conic sections (p. 406) The two-dimensional figures that can be obtained from the intersection of a double cone and a plane: circle, ellipse, parabola, hyperbola. They can also be defined as loci satisfying distance constraints.

Contour diagram (p. 394) By analogy with a map, a diagram that illustrates how the value of one quantity, such as elevation, varies with changes in the values of two variables, such as those representing position.

Español

Las conjugaciones complejas (pág. 212) Pares de números complejos en la forma $a + bi$ y $a - bi$. La conjugación de un número complejo z se representa $\bar{z}$.

Número complejo (pág. 211) Un número que puede ser expresado en la forma $a + bi$ donde a y b son números reales y $i = \sqrt{-1}$. La notación $a + bi$ se llama la forma estándar de un número complejo.

Número complejo, representación gráfica (pág. 214) En un plano coordenado (llamado el plano complejo de números), el número complejo $a + bi$ puede ser representado como un punto o una posición vectoreal con coordenadas (a, b) Observa el ejemplo en la cuadricula de arriba donde las escalas en los ejes son una unidad.

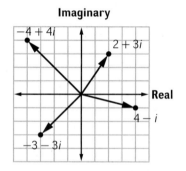

Figuras congruentes Dos figuras son congruentes si y sólo si una figura es la imagen de la otra bajo una transformación rígida (línea de reflexión, traslación, rotación, deslizamiento reflexión) o un compuesto de transformaciones rígidas. Figuras congruentes tienen la misma forma y tamaño, a pesar de la posición u orientación.

Secciones cónicas (pág. 406) Las figuras de dos dimensiones que pueden ser obtenidas de la intersección de un cono doble y un plano: círculo, elipse, parábola, hipérbola. También se definan como lugares completando las restricciones de distancia.

Diagrama de curvas de nivel (pág. 394) Por analogía con un mapa, un diagrama que muestra cómo el valor de una cantidad, como la altitud, varía con los cambios en los valores de dos variables, como las que representan la posición.

English	Español

Converse of an if-then statement Reverses the order of the hypothesis and conclusion of the if-then statement. Given the original statement $p \Rightarrow q$, its converse is $q \Rightarrow p$.

El recíproco de una frase "si-entonces" Invierte el ordén de la hipótesis y la conclusion de una frase "si-entonces". Dada la frase original $p \Rightarrow q$, su recíproco es $q \Rightarrow p$.

Cosecant function (p. 281) If $P(x, y)$ is any point (other than the origin) on the terminal side of an angle θ in standard position and $r = \sqrt{x^2 + y^2}$, then:

$$\text{cosecant of } \theta = \csc \theta = \frac{r}{y}, y \neq 0.$$

Función cosecante (pág. 281) Si $P(x, y)$ es cualquier punto (cualquiera menos el de origen) en el lado terminal de un ángulo θ en la posición estandar y $r = \sqrt{x^2 + y^2}$, entonces:

$$\text{cosecante de } \theta = \csc \theta = \frac{r}{y}, y \neq 0.$$

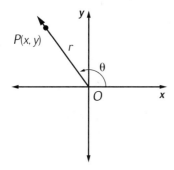

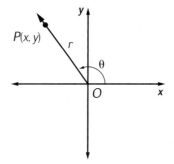

The cosecant function is the reciprocal of the sine function and is not defined for $\theta = k\pi$, for all integers k.

La función cosecante es el recíproco de la funcíon seno y no es definida para $\theta = k\pi$, para todos los números enteros k.

Cotangent function (p. 281) If $P(x, y)$ is any point (other than the origin) on the terminal side of an angle θ in standard position and $r = \sqrt{x^2 + y^2}$, then:

$$\text{cotangent of } \theta = \cot \theta = \frac{x}{y}, y \neq 0.$$

La función cotangente (pág. 281) Si $P(x, y)$ es cualquier punto (cualquiera menos el de origen) en el lado terminal de un ángulo θ en la posición estandar y $r = \sqrt{x^2 + y^2}$, entonces:

$$\text{cotangente de } \theta = \cot \theta = \frac{x}{y}, y \neq 0.$$

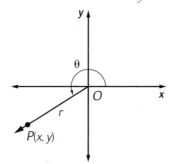

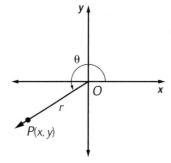

The cotangent function is the reciprocal of the tangent function and is not defined for $\theta = k\pi$, for all integers k.

La función cotangente es el recíproco de la funcíon tangente y no es definida para $\theta = k\pi$, para todos los números enteros k.

Cylindrical surface (p. 448) A surface generated by moving a line (the generator) along the path of a plane curve. As it moves, the generator must stay parallel to a fixed line (or axis) not in the plane of the curve.

Superficie cilíndrica (pág. 448) Una superficie generada por mover una recta (el generador) junto con el camino de una curva plana. Mientras se mueve, el generador tiene que permanecerse a una línea fijada (o eje) que no está en el plano de la curva.

English	Español

D

Definite integral of $f(x)$ from a to b (p. 512) For a continuous function $f(x)$ defined over an interval $[a, b]$, the area of regions bounded by the x-axis and the graph of the function above the x-axis minus the area of regions bounded by the x-axis and the graph of the function below the x-axis.

Integral definida de $f(x)$ desde a a b (pág. 512) Para una función continua $f(x)$ definida sobre un intervalo $[a, b]$, el área de regiones atadas por el eje-x y la gráfica de la función encima del eje-x menos el area de regiones atadas por el eje-x y la gráfica de la función debajo del eje-x.

Degree of a polynomial (p. 191) The highest power of the variable in the polynomial.

Grado de un polinomial (pág. 191) La potencia más alta de un variable en el polinomial.

Derivative of $f(x)$ at a (p. 479) The instantaneous rate at which a function $f(x)$ is changing at some particular point $x = a$ in the domain of the function.

Derivada de $f(x)$ en a (pág. 479) La tasa instantánea de la cual una función $f(x)$ está cambiando en algún punto particular $x = a$ en el dominio de la función.

Direction of a position vector (p. 130) The direction of a position vector is the measure of the positive directed angle from the positive x-axis to the vector.

Dirección de un vector posicional (pág. 130) La dirección de un vector posicional es la medida del ángulo positivo del eje positivo-x al vector.

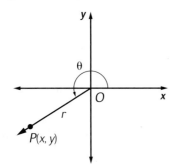

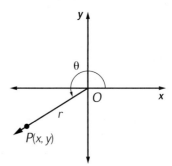

Distance formula in three dimensions (p. 433) Formula for calculating the distance between two points (x_1, y_1, z_1) and (x_2, y_2, z_2)

Formula de distancia en tres dimensiones (pág. 433) Formula para calcular la distancia entre dos puntos (x_1, y_1, z_1) y (x_2, y_2, z_2)

Distributive Property of Multiplication over Addition For any numbers a, b, and c: $a(b + c) = ab + ac$ and $ac + bc = (a + b)c$.

La Propiedad Distributiva de Multiplicación sobre Sumación Para cualquier número a, b, y c: $a(b + c) = ab + ac$ y $ac + bc = (a + b)c$.

Dot product (inner product) (p. 136) The dot product (or inner product) of two position vectors $\vec{a} = (x_1, y_1)$ and $\vec{b} = (x_2, y_2)$ is written $\vec{a} \cdot \vec{b}$ and defined as: $(x_1, y_1) \cdot (x_2, y_2) = x_1x_2 + y_1y_2$.

Producto de punto (producto interior) (pág. 136) El producto de punto (o producto interior) de dos vectores posicionales $\vec{a} = (x_1, y_1)$ y $\vec{b} = (x_2, y_2)$ está escrito $\vec{a} \cdot \vec{b}$ y se representa: $(x_1, y_1) \cdot (x_2, y_2) = x_1x_2 + y_1y_2$.

E

e (pp. 337–338) e is an irrational number whose value is approximately 2.71828. It is the limit of $\left(1 + \frac{1}{n}\right)^n$ as n becomes very large. Formally, $\lim_{n \to \infty} \left(1 + \frac{1}{n}\right)^n = e$.

e (pág. 337–338) e es un número irracional cuyo valor es de aproximadamente 2,71828. Es el limite de $\left(1 + \frac{1}{n}\right)^n$ mientras n crezca a ser muy grande. Formalmente, $\lim_{n \to \infty} \left(1 + \frac{1}{n}\right)^n = e$.

GLOSSARY/GLOSARIO

English

Ellipse (p. 409) The set of points in a plane for which the sum of the distances from two fixed points, called the *foci*, is a constant. An ellipse with center at the origin can be represented by an equation of the form $\frac{x^2}{a^2} + \frac{y^2}{b^2} = 1$, where a and b are nonzero real numbers.

Ellipsoid (p. 441) The three-dimensional analogue of an ellipse. An ellipsoid with center at the origin can be represented by an equation of the form $\frac{x^2}{a^2} + \frac{y^2}{b^2} + \frac{z^2}{c^2} = 1$, where a, b, and c are nonzero real numbers.

End behavior of a function (p. 191) The behavior of the graph of a function as x approaches positive infinity $(+\infty)$ or negative infinity $(-\infty)$.

Equal vectors (p. 106) Vectors with the same magnitude and direction.

Even function (p. 202) An even function is a function f for which $f(-x) = f(x)$ for all x in the domain of f.

Function A relationship between two variables in which each value of the independent variable x corresponds to exactly one value of the dependent variable y. The notation $y = f(x)$ is often used to denote that y is a function of x.

Function composition (p. 80) For any given functions f and g where the domain of f is a subset of the range of g, values of the composite function $f \circ g$ are calculated by evaluating $f(g(x))$. For example, if $f(x) = 3.5x$ and $g(x) = 5 - x$, $f \circ g = f(g(x)) = 3.5(5 - x)$.

Geometric sequence (p. 20) A sequence of numbers in which the ratio of any two consecutive terms is a fixed nonzero constant r. Symbolically, $a_n = r \cdot a_{n-1}$.

Español

Elipse (pág. 409) El conjunto de puntos en un plano cuya suma de distancias a dos puntos fijos, llamados *focos*, es constante. Una elipse con centro en el origen puede ser representada por una ecuación de la forma $\frac{x^2}{a^2} + \frac{y^2}{b^2} = 1$, donde a y b son números reales no nulos.

Elipisoide (pág. 441) La elipse tridimensional análoga. Una elipisoide con centro en el origen puede ser representada por una ecuación de la forma $\frac{x^2}{a^2} + \frac{y^2}{b^2} + \frac{z^2}{c^2} = 1$, donde a, b, y c son números reales no nulos.

Comportamiento final de una función (pág. 191) El comportamiento del gráfico de una función cuando x se acerca a infinito positivo $(+\infty)$ o a infinito negativo $(-\infty)$.

Vectores iguales (pág. 106) Vectores con la misma magnitud y dirección.

Función equitativa (pág. 202) Una función equitativa es una función f para que $f(-x) = f(x)$ para toda x en el dominio f.

Función Una relación entre dos variables donde cada valor de la variable independiente x corresponde a exactamente un valor de la variable dependiente y. La notación $y = f(x)$ es usada mucho para expresar que y es una función de x.

Función de composición (pág. 80) Para cualquiera función f y g donde el dominio de f es un subconjunto del rango de g, los valores de la función compuesta $f \circ g$ son calculados evaluando $f(g(x))$. Por ejemplo, si $f(x) = 3.5x$ y $g(x) = 5 - x$, $f \circ g = f(g(x)) = 3.5(5 - x)$.

Secuencia geométrica (pág. 20) Una secuencia de números donde la ración de cualquiera de los dos términos consecutivos es una constante desigual a cero r. Simbolicamente, $a_n = r \cdot a_{n-1}$.

English

Español

H

Half-turn symmetry (p. 442) In three dimensions, a figure has half-turn symmetry if there is a line (called the *axis of symmetry*) about which the figure can be turned 180° in such a way that the rotated figure appears to be in exactly the same position as the original figure.

Media vuelta simetría (pág. 442) En tres dimensiones, una figura tiene media vuelta simetría si hay una línea (llamada el *eje de simetría*) acerca del que la figura puede girar 180° de tal manera que la girado figura parece estar en la misma posición que el figura original.

Horizontal asymptote (p. 11) A function f has a horizontal asymptote at $y = b$ if and only if $f(x)$ approaches b as $|x|$ becomes very large.

Asíntota horizontal (pág. 11) Una función f tiene una asíntota horizontal en $y = b$ si y sólo si $f(x)$ b enfoques como $|x|$ se hace muy grande.

Horizontal stretch or compression (p. 59) Let f be a function and k a constant. The graph of $y = f(kx)$ can be obtained from the graph of f by a horizontal stretch if $|k| < 1$ or a horizontal compression if $|k| > 1$. If $k < 0$, then the graph of f must also be reflected across the y-axis to obtain the graph of $y = f(kx)$.

Una compresión o extensión horizontal (pág. 59) Deje que f sea una función y k una constante. El gráfico $y = f(kx)$ puede ser obtenido del gráfico f por una extensión horizontal si $|k| < 1$ o una compresión horizontal si $|k| > 1$. Si $k < 0$, entonces el gráfico de f tiene que ser reflejado sobre el eje de y también para obtener el gráfico de $y = f(kx)$.

Horizontal and vertical vector components (p. 111) Any vector $\vec{a}$ can be written as the sum of a horizontal vector $\vec{h}$ and vertical vector $\vec{v}$ (components of the vector).

Componentes vectoriales horizontal y vertical (pág. 111) Cualquier vector $\vec{a}$ puede ser escrito como la suma de un vector horizontal $\vec{h}$ y un vector vertical $\vec{v}$ (componentes del vector).

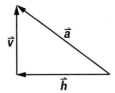

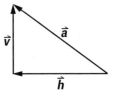

If $\vec{a} = (a_1, a_2)$ is a position vector, then $\vec{h} = (a_1, 0)$, and $\vec{v} = (0, a_2)$.

Si $\vec{a} = (a_1, a_2)$ es un vector posicional, entonces $\vec{h} = (a_1, 0)$, y $\vec{v} = (0, a_2)$.

Hyperbola (p. 412) The set of points in a plane for which the absolute value of the differences of the distances from two fixed points, called the *foci*, is a constant. A hyperbola can be represented by an equation of the form $\frac{x^2}{a^2} - \frac{y^2}{b^2} = 1$ or $\frac{y^2}{a^2} - \frac{x^2}{b^2} = 1$, where a and b are nonzero real numbers.

Hipérbola (pág. 412) El conjunto de puntos en un plano cuyo valor absoluto de las diferencias de las distancias a los puntos fijos, llamados *focos*, es constante. Una hipérbola puede ser representada por una ecuación de la forma $\frac{x^2}{a^2} - \frac{y^2}{b^2} = 1$ o $\frac{y^2}{a^2} - \frac{x^2}{b^2} = 1$, donde están a y b son números reales no nulos.

Hyperboloid (p. 458) A hyperboloid (of one sheet) can be represented by an equation of the form $\frac{x^2}{a^2} + \frac{y^2}{b^2} - \frac{z^2}{c^2} = 1$, $\frac{x^2}{a^2} + \frac{z^2}{c^2} - \frac{y^2}{b^2} = 1$, or $\frac{y^2}{b^2} + \frac{z^2}{c^2} - \frac{x^2}{b^2} = 1$, where a, b, and c are nonzero real numbers. When two of the terms in the equation are negative, the figure is a hyperboloid of two sheets.

Hiperboloide (pág. 458) Un hiperboloide (de una sola hoja) puede ser representada por una ecuación de la forma $\frac{x^2}{a^2} + \frac{y^2}{b^2} - \frac{z^2}{c^2} = 1$, $\frac{x^2}{a^2} + \frac{z^2}{c^2} - \frac{y^2}{b^2} = 1$, o $\frac{y^2}{b^2} + \frac{z^2}{c^2} - \frac{x^2}{b^2} = 1$, donde están a, b, y c son números reales no nulos. Cuando dos de los términos de la ecuación son negativos, la figura es de un hiperboloide de dos hojas.

English	Español

I

Imaginary number (p. 211) Any imaginary number can be expressed in the form bi, where b is a real number and $i = \sqrt{-1}$.

Número imaginario (pág. 211) Cualquier número imaginario puede ser expresado en la forma bi donde b es un número real y $i = \sqrt{-1}$.

Inner product of two vectors See **dot product**.

El producto interior de dos vectores Ver **el producto de punto**.

Inverse functions (p. 88) Functions f and g are inverses of each other if and only if $f(g(x)) = x$ for every x in the domain of g and $g(f(x)) = x$ for every x in the domain of f. The inverse of $f(x)$ is denoted $f^{-1}(x)$.

Funciones inversas (pág. 88) Funciones f y g son inversas entre sí, si y sólo si $f(g(x)) = x$ para todo x en el dominio de g y $g(f(x)) = x$ para todo x en el dominio de f. El inverso de $f(x)$ se denota $f^{-1}(x)$.

L

Least Number Principle (pp. 326 and 603) Every nonempty set of positive integers has a least element.

El Principio del Número Mínimo (pág. 326 y 603) Cada conjunto de enteros positivos (non vacío) tiene un elemento mínimo.

Line reflection A transformation that maps each point P of the plane onto an image point P' as follows: If point P is not on line ℓ, then ℓ is the perpendicular bisector of $\overline{PP'}$. If point P is on ℓ, then $P = P'$.

Línea de refracción Una transformación que mapea cada punto P del plano en un punto de imagen P' así: Si el punto P no está en la línea ℓ, entonces ℓ es el bisector perpendicular de $\overline{PP'}$. Si el punto P está en ℓ, entonces $P = P'$.

Local maximum (p. 29) A function f has a local maximum at b if and only if $f(x)$ changes from increasing values to decreasing values at $x = b$.

Máximo local (pág. 29) Una función f tiene un máximo local en b si y sólo si $f(x)$ cambia de valores crecientes a valores decrecientes en $x = b$.

Local minimum (p. 29) A function f has a local minimum at a if and only if $f(x)$ changes from decreasing values to increasing values at $x = a$.

Mínimo local (pág. 29) Una función f tiene un mínimo local en a si y sólo si $f(x)$ cambia de valores decrecientes al aumentar los valores en $x = a$.

Log transformation (p. 367) For bivariate data, taking the log of each value of the dependent variable. If this linearizes the data, the original points follow an exponential pattern.

Transformación logarítmica (pág. 367) Para los datos de dos variables, de tomar el logaritmo de cada valor de la variable dependiente. Si este lineariza los datos, los puntos originales siguen un patrón exponencial.

Log-log transformation (p. 370) For bivariate data, taking the log of each value of both variables. If this linearizes the data, the original points follow a pattern of the form $y = ax^b$.

Transformación loglog (pág. 370) Para los datos de dos variables, de tomar el logaritmo de cada valor de ambas variables. Si este lineariza los datos, los puntos originales siguen un patrón de la forma $y = ax^b$.

Logarithm, common If $a = 10^b$, then b is the common or base-10 logarithm of a. This relationship is often indicated by the notation $\log_{10} a = b$ or $\log a = b$.

Logaritmo, común Si $a = 10^b$, entonces b es el logaritmo común o base-10 de a. Esta relación es indicada por la notación $\log_{10} a = b$ o $\log a = b$.

Logarithm, natural (p. 341) If $a = e^b$, then b is the natural or base-e logarithm of a. This relationship is often indicated by the notation $\log_e a = b$ or $\ln a = b$.

Logaritmo, natural (pág. 341) Si $a = e^b$, entonces b es el logaritmo natural o base-e de a. Esta relación es indicada frecuentemente con la notación $\log_e a = b$ o $\ln a = b$.

GLOSSARY/GLOSARIO

English	**Español**

— M —

Magnitude of a vector (pp. 105, 130) The length of an arrow that represents the vector. For a position vector $\vec{u} = (x, y)$, the magnitude is denoted $|\vec{u}|$ and defined as $|\vec{u}| = \sqrt{x^2 + y^2}$.

Magnitud de un vector (pág. 105, 130) La longitud de una flecha que representa un vector. Para un vector posicional $\vec{u} = (x, y)$, la magnitud se expresa $|\vec{u}|$ y se defina como $|\vec{u}| = \sqrt{x^2 + y^2}$.

Midpoint formula in three dimensions (p. 435) Formula for calculating the midpoint of a segment with endpoints (x_1, y_1, z_1) and (x_2, y_2, z_2)

Fórmula del punto medio en tres dimensiones (pág. 435) Fórmula para calcular el punto medio de un segmento con los extremos (x_1, y_1, z_1) y (x_2, y_2, z_2)

Modulus of a complex number (p. 315) See **absolute value of a complex number**.

Módulo de un número complejo (pág. 315) Ver **el valor absoluto de un número complejo**.

Multiplication Principle of Counting (p. 540) If you want to count all the combined outcomes from a sequence of tasks in which all these combined outcomes are distinct, and if the first task has n_1 outcomes and for each of these the second task has n_2 outcomes and for each of these the third task has n_3 outcomes and so on, then the number of combined outcomes from the whole sequence of tasks is $n_1 \times n_2 \times n_3 \times \cdots$.

El Principio de Multiplicación de Contra (pág. 540) Si desea contar todos los resultados combinados de una secuencia de tareas en las que todos estos resultados combinados son distintas, y si la primera tarea tiene n_1 resultados y para cada una de estas tareas el segundo tiene n_2 resultados y para cada una de estas tareas la tercera tiene n_3 resultados y así sucesivamente, entonces el número de resultados combinados de toda la secuencia de tareas es $n_1 \times n_2 \times n_3 \times \cdots$.

Multiplication Rule for Probability (p. 579) If A and B are two events, $P(A \text{ and } B) = P(A)P(B \mid A)$.

Regla de Multiplicación Para Probabilidad (pág. 579) Si A y B son dos eventos, $P(A \text{ y } B) = P(A)P(B \mid A)$.

— N —

Nested multiplication form of a polynomial (p. 188) A special kind of expression for a polynomial function that does not use exponents. For example, $(((x - 1)x + 3)x - 2)x + 4 = x^4 - x^3 + 3x^2 - 2x + 4$

Multiplicación andidada de un polinomio (pág. 188) Una clase de expression especial para una función polynomial que no utiliza los exponentes. Por ejemplo, $(((x - 1)x + 3)x - 2)x + 4 = x^4 - x^3 + 3x^2 - 2x + 4$

— O —

Odd function (p. 202) An odd function is a function f for which $f(-x) = -f(x)$ for all x in the domain of f.

Función impar (pág. 202) Una función impar es una función f en la que $f(-x) = -f(x)$ para toda x en el dominio de f.

One-to-one function (p. 19) A function f is one-to-one if for a and b in its domain, $f(a) = f(b)$ implies that $a = b$.

Función una-a-una (pág. 19) Una función f is una-a-una si para a y b en su dominio, $f(a) = f(b)$ implica que $a = b$.

English	**Español**

Opposite vectors (p. 107) The **opposite of vector** $\vec{a}$, denoted $-\vec{a}$, is the vector having the same magnitude as $\vec{a}$, but opposite direction.

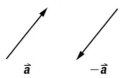

For a position vector, $\vec{a} = (a_1, a_2)$, the opposite vector is $-\vec{a} = (-a_1, -a_2)$.

Vectores opuestos (pág. 107) El **opuesto de vector** $\vec{a}$, expresado $-\vec{a}$, es el vector con la misma magnitud que $\vec{a}$, pero en la dirección opuesta.

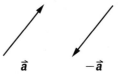

Para un vector posicional, $\vec{a} = (a_1, a_2)$, el vector opuesto es $-\vec{a} = (-a_1, -a_2)$.

P

Parabola (pp. 248, 407) The set of points in a plane equidistant from a fixed line, called the *directrix*, and a fixed point, not on the line, called the *focus*. A parabola can be represented by an equation of the form $y = ax^2$ or $x = ay^2$, where a is a nonzero real number.

Parábola (pág. 248, 407) El conjunto de puntos en un plano de distancia igual de una recta fija, llamada *directrix*, y un punto fijo, no en la recta, llamada *foco*. Una parábola puede ser representada por una ecuación de la forma $y = ax^2$ o $x = ay^2$, donde a es un número real distinto de cero.

Paraboloid (pp. 442, 458) The three-dimensional analogue of a parabola. A paraboloid can be represented by an equation of the form $\frac{x^2}{a^2} + \frac{y^2}{b^2} - z = 0$ where the z-axis is the axis of symmetry, $\frac{x^2}{a^2} + \frac{z^2}{b^2} - y = 0$ where the y-axis is the axis of symmetry, and $\frac{y^2}{a^2} + \frac{z^2}{b^2} - x = 0$ where the x-axis is the axis of symmetry, and a and b are nonzero real numbers.

Parabaloide (pág. 442, 458) La parábola tridimensional análoga. Una parabaloide puede ser representada por la ecuación de la forma $\frac{x^2}{a^2} + \frac{y^2}{b^2} - z = 0$ donde el eje-z es el eje de simetría, $\frac{x^2}{a^2} + \frac{z^2}{b^2} - y = 0$ donde el eje-y es el eje de simetría, y $\frac{y^2}{a^2} + \frac{z^2}{b^2} - x = 0$ donde el eje-x es el eje de simetría, y a y b son números reales no nulos.

Parallelogram law (p. 121) The sum of two vectors $\vec{u}$ and $\vec{v}$ can be represented as a diagonal of the parallelogram determined by placing $\vec{u}$ and $\vec{v}$ tail-to-tail. The resultant vector $\vec{u} + \vec{v}$ has the same initial point as $\vec{u}$ and $\vec{v}$. See diagram below.

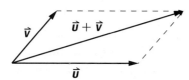

La ley del paralelogramo (pág. 121) La suma de dos vectores $\vec{u}$ y $\vec{v}$ puede ser representada como un diagonal del paralelogramo determinado por poniendo $\vec{u}$ y $\vec{v}$ punta a punta. El vector resultante $\vec{u} + \vec{v}$ tiene el mismo punto inicial que $\vec{u}$ y $\vec{y}$. Ver el diagrama a continuación.

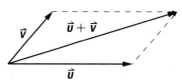

Parametric equations (pp. 139–140) A pair of functions $x = f(t)$ and $y = g(t)$ that describe the coordinates $(x(t), y(t))$ of the graph of some curve (or line) in the plane in terms of a parameter t.

Equaciones paramétricas (pág. 139–140) Un par de funciones $x = f(t)$ y $y = g(t)$ que describan las coordenadas $(x(t), y(t))$ del gráfico de alguna curva (o línea) en el plano en termino de un parámetro t.

English	**Español**
Period of a trigonometric function The length of a smallest interval (in the domain) that corresponds to a portion of the graph from one point to the point at which the graph starts repeating itself.	**Punto de una función trigonométrica** La longitud del intervalo más pequeña (en el dominio) que corresponde a una porción del gráfico desde un punto a un punto en donde el gráfico empieza a repetir.
Permutation (p. 557) An arrangement of distinct objects or numbers in which order matters	**Permutación** (pág. 557) Un conjunto de distintos objetos o números donde el orden tiene significancia
Piecewise-defined function (p. 19) A function that is written using two or more expressions.	**Función definida por tramos** (pág. 19) Una función que está escirta usando dos o más expresiones.
Plane, equation of (p. 438) The three-dimensional analogue of a line. The standard form equation of a plane is $Ax + By + Cz = D$, where A, B, C, and D are real numbers, not all zero.	**Un plano, ecuación de** (pág. 438) El tridimensional anológico de una línea. La ecuación de la forma estándar de un avión es $Ax + By + Cz = D$, donde A, B, C, y D son números reales, no todos de cero.
Polar form of a vector (p. 105) The form [*magnitude, direction*] or [r, θ] that represents a position vector. See diagram below.	**La forma polar de un vector** (pág. 105) La forma [*magnitud, dirección*] o [r, θ] que representa el vector posicional. Ver el diagrama a continuación.

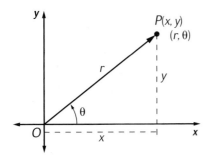

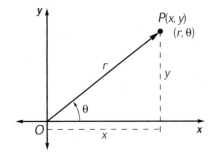

Position vector (p. 131) A vector with its initial point at the origin of a coordinate system is said to be in **standard position** and is called a **position vector**. Since the terminal point of every position vector has unique rectangular coordinates (x, y), the ordered pair is often identified with the vector. (See the diagram for **polar form of a vector**.)	**Vector posicional** (pág. 131) Un vector con su punto inicial en el origin de un sistema coordenadas, está en su **posición estander** y se llama un **vector posicional**. Hasta que el punto terminal de cada vector posicional tiene coordenadas rectangulares únicas (x, y), el par ordenado es identificado con el vector. (Ver el diagrama para **forma polar de un vector**.)
Principle of Mathematical Induction (p. 598) A property of integers that states that if a statement about integers is true for any integer n whenever it is true for $n - 1$ and if the statement is true for an initial integer n_0, then the statement is true for all integers greater than or equal to n.	**Prueba Matemática por Inducción** (pág. 598) Una propiedad de enteros dice que si la declaración de enteros es cierto para cualquier entero n en cualquier momento cuando es cierto para $n - 1$ y si la declaración es cierto para un entero inicial n_0, entonces la declaración es cierto para todos los enteros mayores o iguales a n.

English	**Español**

Q

Quadratic formula The formula
$x = \dfrac{-b \pm \sqrt{b^2 - 4ac}}{2a}$ that gives the solutions of any quadratic equation in the form $ax^2 + bx + c = 0$, where a, b, and c are constants and $a \neq 0$.

Formula quadrática La formula
$x = \dfrac{-b \pm \sqrt{b^2 - 4ac}}{2a}$ que da las soluciones de cualquier equación quadrática en la forma $ax^2 + bx + c = 0$, donde a, b, y c son constantes y $a \neq 0$.

R

Radian The measure of a central angle of a circle that intercepts an arc equal in length to the radius of the circle. One radian equals $\dfrac{180}{\pi}$ degrees, which is approximately $57.2958°$.

Radianes La medida de un ángulo central de un círculo que intercepta un arco igual en longitud al radio de un círculo. Un radian es igual a $\dfrac{180}{\pi}$ grados, que es aproximadamente $57.2958°$.

Reflection symmetry in three dimensions (p. 441) In three dimensions, a figure has reflection symmetry if there is a plane (called the *symmetry plane*) that divides the figure into mirror-image halves. Also called *plane symmetry*.

Simetría axial tridimensional (pág. 441) En tres dimensiones, una figura tiene simetría axial si hay un plano (llamada el *plano de simetría*) que divida la figura en dos partes que se reflejan. También llamada *simetría del plano*.

Roots of unity (p. 319) For every positive integer $n \geq 2$, there are n complex numbers that are solutions to the equation $z^n = 1$. These are the nth roots of unity. In the complex number plane, the points that represent these roots are equispaced on the unit circle, with one root at $(1, 0)$.

Raíces de unidad (pág. 319) Para cada número entero positivo $n \geq 2$, hay n números complejos que son soluciones a la equación $z^n = 1$. Estos son las raíces de unidad "nth". En el plano del número complejo, los puntos que representan estas raíces están puestos en posicion con la distancia igual entre uno al otro en el círculo de unidad, con una raíz en $(1, 0)$.

Rotation A transformation that "turns" all points in a plane through a directed angle α about a fixed point C called the rotation center. That is, if points P' and Q' are the images of points P and Q under this rotation about point C, then $CP = CP'$, $CQ = CQ'$, and $\alpha = \text{m}\angle PCP' = \text{m}\angle QCQ'$.

Rotación Una transformación que "gira" todos los puntos en un plano a través de un ángulo dirigido α sobre un punto fijo C que se llama el centro rotacional. Es decir, si los puntos P' y Q' son los imágenes de los puntos P y Q expuestos a esta rotación sobre el puento C, entonces $CP = CP'$, $CQ = CQ'$, y $\alpha = \text{m}\angle PCP' = \text{m}\angle QCQ'$.

Rotational symmetry in three dimensions (p. 442) In three dimensions, a figure has rotational symmetry if there is a line (called the *axis of symmetry*) about which the figure can be turned less than $360°$ in such a way that the rotated figure appears in exactly the same position as the original figure.

Simetría de rotación en tres dimensiones (pág. 442) En tres dimensiones, una figura tiene simetría rotacional, si hay una línea (llamada el *eje de simetría*) acerca del que la figura puede girar menos de $360°$ de tal manera que la girada figura aparece en la misma posición como la figura original.

English	**Español**

Scalar multiple of a vector (p. 107) When a vector $\vec{a}$ is multiplied by a real number k, the number is called a **scalar** and the product, $k\vec{a}$, is a **scalar multiple** of $\vec{a}$. $k\vec{a}$ is the vector whose length is $|k|$ times the length of $\vec{a}$. The direction of $k\vec{a}$ is the same as the direction of $k\vec{a}$ when $k > 0$ and opposite to that of $\vec{a}$ when $k < 0$.

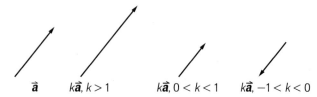

Secant function (p. 281) If $P(x, y)$ is any point (other than the origin) on the terminal side of an angle θ in standard position and $r = \sqrt{x^2 + y^2}$, then:

$$\text{secant of } \theta = \sec \theta = \frac{r}{x}, x \neq 0.$$

The secant function is the reciprocal of the cosine function and is not defined for $\theta = (2k + 1)\frac{\pi}{2}$, for all integers k.

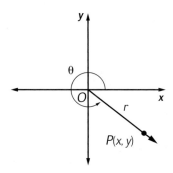

Size transformation A transformation with center C and magnitude $k > 0$ that maps each point P of the plane onto an image point P' as follows: point C is its own image, and for $P \neq C$, the image point P' is on $\overrightarrow{CP}$ and $CP' = k \cdot CP$.

Sphere (p. 434) The three-dimensional analogue of a circle. A sphere with center at the origin and radius r can be represented by an equation of the form $x^2 + y^2 + z^2 = r^2$, where r is a nonzero real number.

La escala multiple de un vector (pág. 107) Cuando un vector $\vec{a}$ es multiplacado por un número real k, el número se llama **escalar** y el producto, $k\vec{a}$, es un **múltiplo escalar** de $\vec{a}$. $k\vec{a}$ es el vector cuya longitud es $|k|$ más la longitud de $\vec{a}$. La dirección de $k\vec{a}$ es igual a la dirección de $k\vec{a}$ cuando $k > 0$ y el opuesto de $\vec{a}$ cuando $k < 0$.

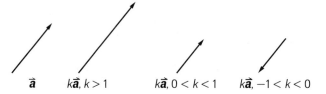

Función secante (pág. 281) Si $P(x, y)$ es cualquier punto (menos el origen) en el lado terminal de un ángulo θ en la posición estandar y $r = \sqrt{x^2 + y^2}$, entonces:

$$\text{secante de } \theta = \sec \theta = \frac{r}{x}, x \neq 0.$$

La función secante es el recíproco de la función coseno y no es definida para $\theta = (2k + 1)\frac{\pi}{2}$, para todos los números enteros k.

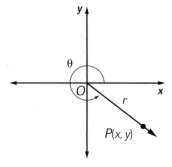

Transformación de tamaño Una transformación con el centro C y la magnitud $k > 0$ que mapea cada punto P del plano sobre un punto de imagen P' así: el punto C es su propia imagen, y para $P \neq C$, el punto de imagen P' está sobre $\overrightarrow{CP}$ y $CP' = k \cdot CP$.

Esfera (pág. 434) El círculo tridimensional análogo. Una esfera con el centro en el origin y el radio r puede ser representada por una equación de la forma $x^2 + y^2 + z^2 = r^2$, donde r es un número real no nulo.

English	Español

Surface of revolution (p. 446) A surface formed by rotating (revolving) a curve about a line.

Superficie de una revolución (pág. 446) Una superficie formada por girar (revolver) una curva sobre una linea.

T

Topographic profile (p. 402) A vertical cross-section view along a line drawn across a portion of a map.

Perfil topográfico (pág. 402) Una vista vertical de una sección transversal a lo largo de una recta dibujada por una porción de un mapa.

Traces (p. 437) Cross sections of a surface formed by the intersection of the coordinate planes with the surface.

Trazas (pág. 437) Las secciones transversales de una superficie formada por la intersección del plano coordenado con la superficie.

Transformation A one-to-one correspondence (function) between all points of a plane and themselves.

Transformación Una correspondencia de uno-a-uno (función) entre todos los puntos de un plano y los mismos.

Translation A transformation that "slides" all points in the plane the same distance (magnitude) and same direction. If points P' and Q' are the images of points P and Q under a translation, then $PP' = QQ'$ and $\overline{PP'} \parallel \overline{QQ'}$. For any point P, the vector $\overrightarrow{PP'}$ is called the *translation vector.*

Traslación Un movimiento que desliza todos los puntos en un plano con la misma distancia (magnitud) y misma dirección. Si los puntos P' y Q' son las imágenes de los puntos P y Q bajo traslación, por lo tanto $PP' = QQ'$ y $\overline{PP'} \parallel \overline{QQ'}$. Para cualquier otro punto P el vector $\overrightarrow{PP'}$ es llamado vector de traslación.

Trigonometric form of a complex number (p. 315) A complex number $a + bi$ can be represented as $r(\cos \theta + i \sin \theta)$, where $r = \sqrt{a^2 + b^2}$ and θ is the directed angle from the positive real axis to the vector representing the complex number.

La forma trigonométrica de un número complejo (pág. 315) Un número complejo $a + bi$ puede ser representado $r(\cos \theta + i \sin \theta)$, donde $r = \sqrt{a^2 + b^2}$ y θ es el ángulo dirigido del eje real positivo al vector representando el número complejo.

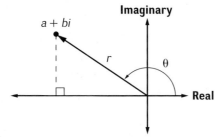

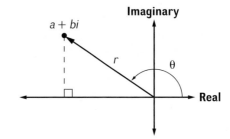

Trigonometric identity (p. 271) A statement involving trigonometric functions which is true for all values of the variable(s) for which the terms are defined.

Identidad trigonométrica (pág. 271) Una declaración que involucra funciones trigonométricas que es cierto para todos los valores de la variable(s) para los que los términos están definidos.

English	Español
V	

Vector (p. 105) A quantity with magnitude and direction, represented geometrically by an arrow.

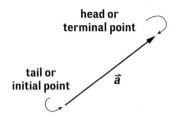

(See also **position vector**.)

Vertical asymptote (p. 11) A function f has a vertical asymptote at $x = a$ if and only if $|f(x)|$ increases without bound as x approaches a.

Vertical stretch or compression (p. 53) If f is any function and k is a positive constant, then the graph of $y = kf(x)$ is a vertically stretched version of the graph of $y = f(x)$ when $k > 1$, or a vertically compressed version of the graph of $y = f(x)$ when $0 < k < 1$.

Vector (pág. 105) Una cantidad con magnitud y dirección, representado geométricamente por una flecha.

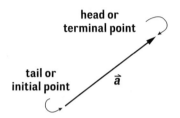

(Ver también **vector posicional**.)

Asíntota vertical (pág. 11) Una función f tiene una asíntota vertical en $x = a$ si y sólo si $|f(x)|$ crece sin límite cuando x tiende a a.

Una compresión o extensión vertical (pág. 53) Si f es una función y k una constante positiva, entonces el gráfico $y = kf(x)$ es una versión estirada verticalmente del gráfico $y = f(x)$ cuando $k > 1$, o una versión comprimida verticalmente del gráfico $y = f(x)$ cuando $0 < k < 1$.

Z	

Zero of multiplicity k (p. 192) A zero that is repeated k times. For the function $f(x) = (x - 1)^2(x + 5)^4$, 1 is a zero of multiplicity 2 and -5 is a zero of multiplicity 4.

Zero vector (p. 134) The vector $\vec{0}$ has magnitude 0 and no direction. For any vector $\vec{a}$, $\vec{a} + \vec{0} = \vec{0} + \vec{a}$. As a coordinate vector, $\vec{0} = (0, 0)$.

Multiplicación por cero k (pág. 192) Un cero que está repitido k veces. Para la función $f(x) = (x - 1)^2(x + 5)^4$, 1 es un cero de multiplicidad 2 y -5 es un cero de multiplicidad 4.

Vector cero (pág. 134) El vector $\vec{0}$ tiene un magnitud de 0 y no tiene dirección. Para cualquier vector $\vec{a}$, $\vec{a} + \vec{0} = \vec{0} + \vec{a}$. Como un vector coordinado, $\vec{0} = (0, 0)$.

Math Online A mathematics multilingual glossary is available at www.glencoe.com/apps/eGlossary612/grade.php. The Glossary includes the following languages:

Arabic	English	Korean	Tagalog
Bengali	Haitian Creole	Russian	Urdu
Cantonese	Hmong	Spanish	Vietnamese

INDEX OF MATHEMATICAL TOPICS

Alignment of *Core-Plus Mathematics*: Course 4 with the CCSS

In the following charts, CCSS indicators in bold font are those "focused on" in the Investigation. Indicators not in bold are "connected to" in the Investigation. The Mathematical Practices permeate every lesson. A separate *CCSS Guide to Core-Plus Mathematics* provides, for each high school mathematics standard, corresponding page references in Courses 1–4.

Course 4: Preparation for Calculus	CCSS Content Standards
Unit 1 – Families of Functions	
Lesson 1 **Function Models Revisited**	
Investigation 1 Modeling Atmospheric Change	**F-IF.1, F-IF.7, F-IF.8, F-LE.5,** A-CED.1, A-CED.2, F-IF.2, F-LE.4, G-GMD.3, G-MG.2
Investigation 2 It's All in the Family	**F-IF.3, F-IF.4, F-IF.5, F-IF.7, F-IF.9, F-BF.2,** F-BF.1, F-TF.5
Lesson 2 **Customizing Models by Translation and Reflection**	
Investigation 1 Vertical Translation	**A-SSE.1, A-SSE.3, F-IF.4, F-IF.7, F-BF.1, F-BF.3, F-LE.5,** A-REI.10, F-IF.2, S-ID.2, G-CO.2, S-ID.6
Investigation 2 Reflection Across the *x*-Axis	**A-SSE.1, A-SSE.3, F-IF.4, F-IF.7, F-BF.1, F-BF.3, F-LE.5,** A-REI.10, F-IF.2, G-CO.2, S-ID.6
Investigation 3 Horizontal Translation	**A-SSE.1, A-SSE.3, F-IF.4, F-IF.7, F-BF.1, F-BF.3, F-LE.5,** A-REI.3, A-REI.10, F-IF.2, G-CO.2, G-GPE.1, S-ID.2, S-ID.6
Lesson 3 **Customizing Models by Stretching and Compressing**	
Investigation 1 Vertical Stretching and Compressing	**A-SSE.1, A-SSE.3, F-IF.7, F-BF.1, F-BF.3, F-LE.5, T-FT.5,** A-REI.10, F-IF.2, G-CO.2, S-ID.2
Investigation 2 Horizontal Stretching and Compressing	**A-SSE.1, A-SSE.3, F-IF.7, F-BF.1, F-BF.3, F-LE.5, F-TF.5,** N-VM.12, A-REI.10, F-IF.2, G-CO.2
Lesson 4 **Combining Functions**	
Investigation 1 Arithmetic with Functions	**A-SSE.1, A-SSE.3, A-APR.1, A-APR.6, F-IF.7,** N-VM.8, N-VM.12, A-APR.1, F-LE.5
Investigation 2 Composition of Functions	**A-SSE.1, A-SSE.3, F-IF.7, F-IF.8, F-BF.1,** F-BF.2, F-BF.3, F-BF.4, F-LE.5
Unit 2 – Vectors and Motion	
Lesson 1 **Modeling Linear Motion**	
Investigation 1 Navigation: What Direction and How Far?	**N-VM.1, N-VM.3, N-VM.5,** A-SSE.1, N-VM.2, G-CO.5, G-CO.10, G-SRT.1
Investigation 2 Changing Course	**N-VM.2, N-VM.4, N-VM.5,** A-SSE.1, N-VM.1, N-VM.3, G-SRT.8
Investigation 3 Go with the Flow	**N-VM.1, N-VM.3, N-VM.4, G-SRT.10, G-SRT.11,** A-SSE.1, G-SRT.8

Course 4	CCSS Content Standards
Lesson 2 Vectors and Parametric Equations	
Investigation 1 Coordinates and Vectors	**N-VM.1, N-VM.2, N-VM.3, N-VM.4,** G-SRT.8, G-GPE.4, G-MG.1, G-MG.3
Investigation 2 Vector Algebra with Coordinates	**N-VM.1, N-VM.3, N-VM.4, N-VM.5,** N-VM.7, N-VM.11, G-CO.5, G-CO.10, G-SRT.8, G-SRT.10, G-SRT.11, G-GPE.4
Investigation 3 Follow That Dot	**N-VM.1, N-VM.3, A-SSE.1, G-MG.1, G-MG.3,** A-SSE.3, A-CED.2
Lesson 3 Modeling Nonlinear Motion	
Investigation 1 What Goes Up, Must Come Down	**N-VM.1, N-VM.3, A-SSE.1, A-CED.2, G-MG.1, G-MG.3,** G-GPE.1
Investigation 2 Representing Circles and Arcs Parametrically	**N-VM.1, N-VM.3, A-CED.2, F-TF.1, G-MG.1, G-MG.3**
Investigation 3 Simulating Orbits	**N-VM.1, N-VM.3, A-SSE.1, A-CED.2, G-MG.1, G-MG.3,** F-TF.5, G-GPE.1, G-GPE.3
Unit 3 – Algebraic Functions and Equations	
Lesson 1 Polynomial Function Models and Operations	
Investigation 1 Constructing Polynomial Function Models	**A-CED.2, A-CED.3, A-REI.6, A-REI.8, A-REI.9, F-IF.2, F-IF.4, F-IF.5,** N-VN.10, G-MG.1, G-MG.3
Investigation 2 Zeroes and Factors of Polynomials	**A-SSE.1, A-SSE.2, A-SSE.3, A-APR.1, A-APR.3, A-APR.4, A-REI.3, F-IF.7,** A-REI.4
Investigation 3 Division of Polynomials	**A-SSE.1, A-SSE.2, A-SSE.3, A-APR.1, A-APR.2, A-APR.4, A-APR.6,** A-SSE.4, A-REI.1, A-REI.4, F-IF.4
Lesson 2 The Complex Number System	
Investigation 1 A Complex Solution	**N-CN.1, N-CN.2, N-CN.3, N-CN.7, A-SSE.3, A-REI.4,** A-SSE.1, A-SSE.2
Investigation 2 Properties of Complex Numbers	**N-CN.1, N-CN.2, N-CN.4, N-CN.5, N-CN.6, N-CN.8, A-SSE.3,** A-SSE.1, A-SSE.2
Lesson 3 Rational Function Models and Operations	
Investigation 1 Rational Function Models	**A-SSE.1, A-SSE.2, A-APR.6, A-APR.7, A-CED.1, F-IF.1, F-IF.2, F-IF.4, F-IF.5, F-IF.7, F-IF.8, F-BF.1,** A-CED.10, G-MG.1, G-MG.3, S-ID.6
Investigation 2 Properties of Rational Functions	**A-SSE.1, A-SSE.3, A-APR.6, A-APR.7, A-CED.1, F-IF.1, F-IF.2, F-IF.4, F-IF.7, F-IF.8,** A-CED.10, A-CED.12, A-REI.1, G-MG.1, G-MG.3, S-ID.6
Lesson 4 Algebraic Strategy	
Investigation 1 Dealing with Radicals	**A-SSE.1, A-CED.1, A-REI.2, F-IF.7, F-IF.8,** N-VM.8, A-CED.4, G-GPE.4
Investigation 2 Seeing the Big Picture	**A-SSE.1, A-SSE.3, A-APR.3, A-APR.6, A-REI.6, F-IF.1, F-IF.4, F-IF.7,** A-CED.4, A-REI.1, F-IF.2

Course 4	CCSS Content Standards
Unit 4 – Trigonometric Functions and Equations	
Lesson 1 **Reasoning with Trigonometric Functions**	
Investigation 1 Proving Trigonometric Identities	**A-SSE.1, A-SSE.2, A-SSE.3, F-TF.3, F-TF.4, F-TF.5, F-TF.8, F-TF.9,** A-REI.1
Investigation 2 Sum and Difference Identities	**A-SSE.1, A-SSE.2, A-SSE.3, F-TF.3, F-TF.5, F-TF.9, G-SRT.8,** A-REI.1, G-SRT.10, G-SRT.11
Investigation 3 Extending the Family of Trigonometric Functions	**A-SSE.1, A-SSE.2, A-SSE.3, F-IF.4, F-IF.7, F-TF.5,** A-REI.1
Lesson 2 **Solving Trigonometric Equations**	
Investigation 1 Solving Linear Trigonometric Equations	**A-SSE.1, A-SSE.2, A-SSE.3, F-TF.5, F-TF.6, F-TF.7,** A-CED.2, A-REI.1, A-REI.10, A-REI.11, F-IF.4, F-BF.1, F-BF.3
Investigation 2 Using Identities to Solve Trigonometric Equations	**A-SSE.1, A-SSE.2, A-SSE.3, A-REI.1, F-IF.8, F-TF.5, F-TF.7,** A-CED.2. A-REI.1, A-REI.4, A-REI.10, F-IF.4, F-BF.1, F-BF.3
Lesson 3 **The Geometry of Complex Numbers**	
Investigation 1 Trigonometric Form of Complex Numbers	**N-CN.3, N-CN.4, N-CN.5, A-SSE.1, A-SSE.2, G-CO.2,** N-VM.4, N-VM.5, A-REI.4
Investigation 2 De Moivres's Theorem	**N-CN.2, N-CN.5, N-CN.9, A-SSE.1, A-SSE.2,** N-VM.4, N-VM.5, A-SSE.3, A-REI.4
Unit 5 – Exponential Functions, Logarithms, and Data Modeling	
Lesson 1 **Exponents and Natural Logarithms**	
Investigation 1 What is e^x?	**A-SSE.1, A-SSE.2, F-IF.4, F-IF.7,** F-IF.3
Investigation 2 Applications of e^x and ln x	**A-SSE.1, A-SSE.2, F-IF.4, F-IF.7, F-IF.8, F-BF.5, F-LE.5,** A-REI.1
Investigation 3 Properties of e^x and ln x	**A-SSE.1, A-SSE.2, F-IF.4, F-IF.7, F-IF.8, F-BF.5, F-LE.5,** A-REI.1
Lesson 2 **Linearization and Data Modeling**	
Investigation 1 Assessing the Fit of a Linear Model	**F-IF.4, S-ID.6**
Investigation 2 Log Transformations	**F-IF.4, F-IF.8, F-BF.5, F-LE.4, F-LE.5, S-ID.6**
Investigation 3 Log-Log Transformations	**F-IF.4,** G-MG.2

Course 4	CCSS Content Standards
Unit 6 – Surfaces and Cross Sections	
Lesson 1 Three-Dimensional Representations	
Investigation 1 Using Data to Determine Surfaces	**G-GMD.4, G-MG.1,** S-ID.1
Investigation 2 Visualizing and Reasoning with Cross Sections	**G-GMD.4, G-MG.1, G-MG.3,** N-VM.4
Investigation 3 Conic Sections	**A-SSE.3, A-REI.4, F-IF.8, G-GPE.1, G-GPE.2, G-GPE.3,** N-VM.4, A-SSE.1
Lesson 2 Equations for Surfaces	
Investigation 1 Relations Among Points in Three-Dimensional Space	**A-SSE.3, A-CED.2,** A-CED.1, G-GPE.4, G-GMD.4
Investigation 2 The Graph of $Ax + By + Cz = D$	**A-CED.2, G-GMD.4,** A-CED.3
Investigation 3 Surfaces Defined by Nonlinear Equations	**A-SSE.1, A-CED.2, F-IF.7, F-IF.8, G-GMD.4,** A-CED.1, F-BF.3
Investigation 4 Surfaces of Revolution and Cylindrical Surfaces	**A-SSE.1, G-GMD.2, G-GMD.4, G-MG.1, G-MG.2,** F-IF.7, G-CO.5, G-CO.6
Unit 7 – Concepts of Calculus	
Lesson 1 Introduction to the Derivative	
Investigation 1 Instantaneous Rates of Change I	**F-IF.1, F-IF.2, F-IF.4, F-IF.6, F-IF.7, F-IF.9,** S-ID.1
Investigation 2 Instantaneous Rates of Change II	**F-IF.1, F-IF.2, F-IF.4, F-IF.6, F-IF.7, F-IF.9, F-TF.5,** F-IF.8
Investigation 3 The Derivative	**A-SSE.1, F-IF.1, F-IF.2, F-IF.4, F-IF.6, F-IF.9,** A-REI.1, F-IF.3, F-BF.3, F-TF.5
Investigation 4 From Function Graph to Derivative Graph	**F-IF.1, F-IF.2, F-IF.4, F-IF.6, F-IF.9,** A-SSE.1, F-IF.7, F-TF.5
Lesson 2 Introduction to the Definite Integral	
Investigation 1 What is the Total?	**F-IF.2, G-GMD.1,** A-CED.1, F-IF.1, F-IF.7, F-IF.8
Investigation 2 Velocity and Net Change	**G-GMD.1,** F-IF.8, G-CO.2
Investigation 3 The Definite Integral	**F-IF.2, G-GMD.1,** A-CED.1, F-IF.1, F-IF.7, F-IF.8

Course 4	CCSS Content Standards
Unit 8 – Counting Methods and Induction	
Lesson 1 **Systematic Counting**	
Investigation 1 Methods of Counting	S-CP.1
Investigation 2 Principles of Counting	S-CP.1, S-CP.2, S-CP.4
Lesson 2 **Order and Repetition**	
Investigation 1 Permutations and Combinations	**A-SSE.1, A-SSE.2**
Investigation 2 Collections, Sequences, This or That	S-CP.1
Lesson 3 **Counting Throughout Mathematics**	
Investigation 1 Counting and Probability	**S-CP.3, S-CP.6, S-CP.7, S-CP.9**
Investigation 2 Combinations, the Binomial Theorem, and Pascal's Triangle	**A-APR.5, S-CP.3, S-CP.6, S-CP.8, S-CP.9**
Lesson 4 **Proof by Mathematical Induction**	
Investigation 1 Infinity, Recursion, and Mathematical Induction	**F-BF.1, F-BF.2**, A-SSE.1, A-REI.1
Investigation 2 A Closer Look at Mathematical Induction	**F-BF.1, F-BF.2**, A-SSE.1, A-REI.1